THE GLOBAL AIRLINE INDUSTRY

THE GLOBAL AIRLINE INDUSTRY

Edited By

Peter Belobaba
Department of Aeronautics & Astronautics, MIT, USA
Amedeo Odoni
Departments of Aeronautics & Astronautics and Civil & Environmental Engineering, MIT, USA
Cynthia Barnhart
Department of Civil & Environmental Engineering, MIT, USA

A John Wiley and Sons, Ltd., Publication

This edition first published 2009
© 2009, John Wiley & Sons, Ltd

Registered office
John Wiley & Sons Ltd, The Atrium, Southern Gate, Chichester, West Sussex, PO19 8SQ, United Kingdom

For details of our global editorial offices, for customer services and for information about how to apply for permission to reuse the copyright material in this book please see our website at www.wiley.com.

The right of the author to be identified as the author of this work has been asserted in accordance with the Copyright, Designs and Patents Act 1988.

Reprinted with corrections November 2009
Reprinted April 2010 and November 2010

Wiley also publishes its books in a variety of electronic formats. Some content that appears in print may not be available in electronic books.

Library of Congress Cataloging-in-Publication Data
Record on file

A catalogue record for this book is available from the British Library.

ISBN: 978-0-470-74077-4 (H/B)

Typeset in 10/12 Times by Laserwords Private Limited, Chennai, India
Printed and bound in Great Britain by CPI Antony Rowe, Chippenham, Wiltshire

Contents

List of Contributors

EDITORS

Peter P. Belobaba
MIT Department of Aeronautics and
  Astronautics
Room 33–217
77 Massachusetts Avenue
Cambridge, MA 02139
USA
belobaba@mit.edu

Amedeo R. Odoni
MIT Department of Aeronautics and
  Astronautics
Room 33–219
77 Massachusetts Avenue
Cambridge, MA 02139
USA
arodoni@mit.edu

Cynthia Barnhart
MIT School of Engineering
Associate Dean of Engineering
Room 1–235
77 Massachusetts Avenue
Cambridge, MA 02139
USA
cbarnhart@mit.edu

CONTRIBUTORS

Greg J. Bamber
Department of Management
Monash University
PO Box 197, Caulfield East
Melbourne
Victoria 3145
Australia
gregbamber@gmail.com

Arnold Barnett
MIT Sloan School of Management
Room E53–39
77 Massachusetts Avenue
Cambridge, MA 02139
USA
abarnett@mit.edu

Jody Hoffer Gittell
The Heller School for Social Policy and
  Management
Brandeis University
Waltham, MA 02454
jgittell@brandeis.edu

R. John Hansman
MIT Department of Aeronautics and
   Astronautics
Room 33–303
77 Massachusetts Avenue
Cambridge, MA 02139
USA
rjhans@mit.edu

Thomas A. Kochan
MIT Sloan School of Management
E52–583
77 Massachusetts Avenue
Cambridge, MA 02139
USA
tkochan@mit.edu

Karen Marais
School of Aeronautics and Astronautics
Purdue University
ARMS 3219
701 West Stadium Avenue
West Lafayette, IN 47907
USA
kmarais@purdue.edu

Robert McKersie
MIT Sloan School of Management
Room E52–503
77 Massachusetts Avenue
Cambridge, MA 02139
USA
rmckersi@mit.edu

Alan H. Midkiff
MIT Department of Earth, Atmospheric
   and Planetary Sciences
54–410
77 Massachusetts Avenue
Cambridge, MA 02139
USA
ahmidkiff@aol.com

Tom G. Reynolds
Institute for Aviation and the Environment
   (AIM Group)
University of Cambridge
1–5 Scroope Terrace
Cambridge
CB2 1PX
UK
tgr25@cam.ac.uk

William S. Swelbar
MIT Department of Aeronautics and
   Astronautics
Room 33–217
77 Massachusetts Avenue
Cambridge, MA 02139
USA
swelbar@mit.edu

Andrew von Nordenflycht
Faculty of Business Administration
Simon Fraser University
500 Granville Street
Vancouver, BC
Canada
V6C 1 W6
vonetc@sfu.ca

Ian A. Waitz
MIT Department of Aeronautics and
   Astronautics
Room 33–207
77 Massachusetts Avenue
Cambridge, MA 02139
USA
iaw@mit.edu

Series Preface

The Aerospace Series of books by John Wiley and Sons provides a practical source of aerospace system topics aimed at engineering and business professionals, industry operators and users. The series, which draws from experts from within the aerospace industry as well as from universities and learning institutions from around the world, also provides a valuable insight into wide range of aerospace systems technologies and best practices for the benefit of prospective engineers, graduates and educators from the academic community.

The Global Airline Industry is an important topic that fits well within the scope and intent of the series. Here the authors, Peter Belobaba, Amedeo Odoni and Cynthia Barnhart, and their team of contributors provide a thorough and complete coverage of the evolution of the airline industry from its inception to the present day from a worldwide perspective. Material content includes a detailed treatment of the economic, operational and infrastructure issues to achieve an excellent reference source for anyone directly or indirectly involved in the airline industry.

Roy Langton, Allan Seabridge and Ian Moir

Notes on Contributors

EDITORS

Cynthia Barnhart is Associate Dean for Academic Affairs for the MIT School of Engineering, Professor of Civil and Environmental Engineering and Engineering Systems, and Co-Director of the Operations Research Center at the Massachusetts Institute of Technology. She has developed and teaches courses including carrier systems, optimization of large-scale transportation systems, airline schedule planning and the airline industry. Her research activities have focused on the development of optimization models and methods for designing, planning and operating transportation systems. She currently serves or has served as Area Editor (Transportation) for *Operations Research*, as Associate Editor for *Operations Research* and for *Transportation Science*, as President of the INFORMS Women in Operations Research/Management Science Forum, as President of the INFORMS Transportation and Logistics Section, as President of INFORMS, and as Co-Director of MIT's Center for Transportation and Logistics. Professor Barnhart has been awarded the Franz Edelman 2nd Prize for Achievement in Operations and the Management Sciences, the Junior Faculty Career Award from the General Electric Foundation, the Presidential Young Investigator Award from the National Science Foundation, the First Prize Award for Best Paper in *Transportation Science & Logistics*, and the INFORMS award for the Advancement of Women in Operations Research and Management Science.

Peter P. Belobaba is Principal Research Scientist at MIT's International Center for Air Transportation, in the Department of Aeronautics and Astronautics. He holds a Master of Science in Transportation and a Ph.D. in Flight Transportation Systems from the Massachusetts Institute of Technology. He currently teaches graduate-level courses on the airline industry and airline management at MIT and is the Program Manager of the MIT Global Airline Industry Program. He is also Adjunct Professor of Aviation Management in the International Aviation MBA Program at Concordia University's John Molson School of Business in Montreal, where he teaches courses on airline economics, airline management, and air carrier planning and operations. He directs the Passenger Origin–Destination Simulator (PODS) MIT Research Consortium funded by seven international airlines to explore issues of demand forecasting, seat inventory optimization and competitive impacts of revenue management. Dr. Belobaba has also worked as a consultant to over 40 airlines and other companies worldwide. He has published articles dealing with airline pricing, revenue management, competition, operating costs and productivity analysis in *Airline Business, Operations Research, Transportation Science, Decision Sciences, Journal of Revenue and Pricing Management, Transportation Research* and the *Journal of Air Transport Management*.

Amedeo R. Odoni is Professor of Aeronautics and Astronautics and of Civil and Environmental Engineering at MIT and one of the Co-Directors of the Airline Industry Program. He has also served as Co-Director of the FAA's National Center of Excellence in Aviation Education, Co-Director of MIT's Operations Research Center, Editor-in-Chief of *Transportation Science*, and consultant to numerous international airport and aviation-related organizations and projects. The author, co-author or co-editor of eight books and more than 100 other technical publications, he is a Fellow of the Institute for Operations Research and Management Science (INFORMS) and has received several distinctions, among them the INFORMS Lifetime Achievement Award for Contributions to Transportation Science, the T. Wilson Endowed Chair at MIT, the US Federal Aviation Administration's (FAA) National Award for Excellence in Aviation Education, an Honorary Ph.D. from the Athens University of Economics and Business, and four MIT awards for excellence in teaching, mentoring and advising. His students have also received many prizes for their research and dissertations.

CONTRIBUTORS

Greg J. Bamber is a Professor in the Department of Management at Monash University, in Melbourne, Australia. His (joint) publications include: *Up in the Air* (Cornell University Press); *International and Comparative Employment Relations* (Sage): *Employment Relations in the Asia Pacific* (Thomson); *Managing Managers* (Blackwell); and *Organisational Change Strategies* (Longman). He has published many articles and is on the editorial boards of international refereed journals. He is conducting research on high-performance human resource management in hospitals. He researches and consults with international agencies, governments, companies and other organizations. He has served as an arbitrator for the British Advisory, Conciliation and Arbitration Service and as President of the Australian & New Zealand Academy of Management and of the International Federation of Scholarly Associations of Management. He was educated at Manchester University, the London School of Economics and Heriot-Watt University, Edinburgh. He is Adjunct Professor, Griffith University, Australia, and Visiting Professor, Newcastle University, England. He has been a visitor at MIT and a range of universities in other countries.

Arnold I. Barnett is George Eastman Professor of Management Science at MIT's Sloan School of Management. He holds a B.A. in Physics from Columbia College and a Ph.D. in Mathematics from MIT. His research specialty is applied mathematical modeling on issues of policy importance; aviation safety is one of his primary areas of emphasis. Professor Barnett has authored or co-authored nearly 100 published papers. His research articles about aviation safety have been extensively summarized in, among others, *The New York Times, The Wall Street Journal, Scientific American, The Economist* and *Newsweek*. He has served many times as consultant to the FAA and its contractors, and to five airports and fourteen airlines. Professor Barnett has studied passenger mortality risk in commercial aviation, public perceptions of and reactions to the risks of flying, and such specific safety issues as weather hazards, runway collision risk, adoption of free flight routings, and the dangers of terrorism. He was chair over 1996–1998 of the FAA's Technical Team about Positive Passenger Bag Match, and was recently hired by the Transportation Security Administration. In 2002, he received the President's Citation from the Flight Safety Foundation for "truly outstanding contributions on behalf of safety."

Jody Hoffer Gittell is Associate Professor of Management and MBA Program Director at Brandeis University's Heller School for Social Policy and Management. Her research explores how coordination by front-line workers contributes to quality and efficiency outcomes in service settings, with a particular focus on the airline and healthcare industries. Dr. Gittell is the author of dozens of articles and chapters, a book entitled *The Southwest Airlines Way: Using the Power of Relationships to Achieve High Performance*, and a forthcoming book called *Up In the Air: How the Airlines Can Improve Performance by Engaging Their Employees*. She is currently writing *High Performance Healthcare: How Relationships Drive Results*. Dr. Gittell has won the Best Book Award from the Alfred P. Sloan Foundation, Best Paper Award from the Human Resource Division of the Academy of Management, and an Honorable Mention for the Douglas McGregor Award for Best Paper on Leadership. Before joining the faculty at Brandeis University, Dr. Gittell received her Ph.D. from the Massachusetts Institute of Technology and taught at the Harvard Business School.

R. John Hansman, Jr. is the T. Wilson Professor of Aeronautics and Astronautics at MIT, and Director of the MIT International Center for Air Transportation. He conducts research on information technologies applied to air transportation in several areas related to flight vehicle operations, air traffic control and safety. Professor Hansman holds six patents and has authored over 250 technical publications. He has over 5300 hours of pilot-in-command time in aircraft, helicopters and sail planes, including meteorological, production and engineering flight test experience. Professor Hansman chairs the FAA Research and Development Advisory Committee. He is a Fellow of the American Institute of Aeronautics and Astronautics. He received the 1996 FAA Excellence in Aviation Award, the 2004 Dryden Award for Aeronautics Research and the 1994 Losey Atmospheric Sciences Award from the American Institute of Aeronautics and Astronautics, and the 2006 Kriske Award for Career Contributions from the Air Traffic Control Association.

Thomas A. Kochan is the George M. Bunker Professor of Management at MIT's Sloan School of Management. He has done research on a variety of topics related to industrial relations and human resource management in the public and private sector. His recent books include: *Restoring the American Dream: A Working Families' Agenda; Up in the Air: How Airlines Can Improve Performance by Engaging their Workforce* (2009); and *Healing Together: The Kaiser Permanente Labor Management Partnership* (2009). His 1986 book *The Transformation of American Industrial Relations* received the annual award from the Academy of Management for the best scholarly book on management. Professor Kochan is a Past President of both the International Industrial Relations Association (IIRA) and the Industrial Relations Research Association (IRRA). In 1996, he received the Heneman Career Achievement Award from the Human Resources Division of the Academy of Management. He was named the Centennial Visiting Professor from the London School of Economics in 1995. From 1993 to 1995 he served as a member of the Clinton administration's Commission on the Future of Worker/Management Relations.

Karen Marais is an Assistant Professor in the School of Aeronautics and Astronautics at Purdue University. Her research interests include safety analysis and risk assessment of complex socio-technical systems in general and aerospace systems in particular. In addition, she conducts research on the environmental impact of aviation within the FAA PARTNER Center of Excellence, and is currently growing her research platform to include maintenance and operations of engineering systems. Dr. Marais received her Ph.D. from the Department of Aeronautics and Astronautics at MIT in 2005. She also holds a Master's degree in Space-Based Radar from MIT. Prior to graduate school, she worked in South Africa as an electronic engineer. She holds a B.Eng. in Electrical and Electronic Engineering from the University of Stellenbosch and a B.Sc. in Mathematics from the University of South Africa. She is the author or co-author of some twenty technical publications including six journal publications and one book chapter. She is currently working on an introductory textbook on reliability and risk analysis.

Robert B. McKersie is Professor (Emeritus) at MIT's Sloan School of Management. He has been at MIT since 1980. Prior to that he served as Dean of the New York State School of Industrial and Labor Relations at Cornell University, and prior to that he was on the faculty of the Graduate School of Business. His research interests have been in labor–management relations with particular focus on bargaining activity. With Richard Walton, he co-authored *A Behavioral Theory of Labor Negotiations* in 1965. Subsequently, he focused his attention on the subject of productivity (authoring a book with Lawrence Hunter entitled *Pay, Productivity and Collective Bargaining*) and participated in a multi-year project at the Sloan School that resulted in the award-winning book by Thomas Kochan, Harry Katz and Robert McKersie entitled *The Transformation of American Industrial Relations*. More recently, he has returned to the subject of the bargaining process and co-authored with Richard Walton and Joel Cutcher-Gershenfeld a book entitled *Strategic Negotiations*. He continues to do research on strategies being pursued in different industries to bring about more effective organizational arrangements.

Alan H. Midkiff is currently a Boeing 767 captain for a major US airline, where he has been employed for the past 22 years. He began flying at age 14 and his civilian background includes flight experience ranging from light single-engine piston trainers, to "steam gauge" three-crew turbojets and, more recently, modern glass cockpit wide-body aircraft on both domestic and international routes. After receiving an S.M. in Aeronautics and Astronautics from MIT in 1992, he worked for 12 years at the MIT International Center for Air Transportation as a research engineer. His undergraduate education includes a B.S. in Electrical Engineering from Lehigh University and an A.S. in Aviation Administration from Hawthorne College. In addition to his airline flying, he is also employed as the site manager of the MIT George R. Wallace Jr. Astrophysical Observatory.

Tom G. Reynolds has expertise in air transportation systems engineering, with a particular focus on air traffic control and its role in mitigating environmental impacts of aviation. He is a senior researcher at the University of Cambridge Institute for Aviation and the Environment, where he is project manager of the Aviation Integrated Modelling Project which is developing a policy assessment tool for aviation, environment and economic interactions. Previously he was with the Cambridge–MIT Institute Silent Aircraft Initiative, where he developed novel aircraft approach procedures for the ultralow-noise concept aircraft design. In addition, he coordinated the design, flight trials and operational analysis of a set of advanced approach procedures for current aircraft types at a regional UK airport which have delivered major environmental benefits. He has a Ph.D. in Aerospace Systems from the MIT International Center for Air Transportation.

William S. Swelbar is a Research Engineer in MIT's International Center for Air Transportation, where he is affiliated with the Global Airline Industry Program and Airline Industry Consortium. Prior to accepting his research position at MIT, he was President and a Founding Partner of the former Eclat Consulting, Inc. He also serves as a member of the Board of Directors of Hawaiian (Airlines) Holdings, Inc. He has a long track record in identifying industry trends in their formative stages and forecasting the consequences. Over the past 25 years, he has represented airlines, airports, investors, manufacturers and labor groups in a consulting role. His work has included competitive assessments, cost–benefit analyses, and other economic and financial advisory services in support of strategic planning exercises, corporate communications efforts, asset valuation projects and labor negotiations activities. He is a much sought-after speaker and has provided expert witness testimony before various tribunals and before the US Congress regarding the economics of commercial air transport. He holds a B.S. degree with honors from Eastern Michigan University and an MBA from The George Washington University.

Andrew von Nordenflycht is Assistant Professor of Strategy at Simon Fraser University in Vancouver, Canada, and a Research Fellow of the CIBC Center for Corporate Governance and Risk Management. He received a B.A. in History from Stanford University and his Ph.D. in Management from the MIT Sloan School of Management. He researches the structure and management of human-capital-intensive firms, with a focus on professional services and airlines. He is the co-author of *Up In the Air: How the Airlines Can Improve Performance by Engaging Their Employees* (2009) and has published articles in *Industrial and Labor Relations Review, Academy of Management Journal* and *Monthly Labor Review*.

Ian A. Waitz, the Jerome C. Hunsaker Professor of Aeronautics and Astronautics at MIT, is Head of the Department of Aeronautics and Astronautics and Director of PARTNER (the Partnership for AiR Transportation Noise and Emissions Reduction), an FAA/NASA/Transport Canada-sponsored Center of Excellence. His principal areas of interest are the modeling and evaluation of climate, local air quality, and noise impacts of aviation, including the assessment of technological, operational and policy options for mitigating these impacts. Professor Waitz has written approximately 70 technical publications including a report to the US Congress on aviation and the environment, holds three patents and has consulted for many organizations. He has also served as an Associate Editor of the *AIAA Journal of Propulsion and Power*. In 2003 Professor Waitz received a NASA "Turning Goals Into Reality" Award for Noise Reduction. He was awarded the FAA 2007 Excellence in Aviation Research Award. He is a Fellow of the AIAA, and an ASME and ASEE member. He teaches graduate and undergraduate courses in the fields of thermodynamics and energy conversion, propulsion and experimental projects.

Acknowledgements

This book is based on materials developed for an introductory-level interdepartmental graduate course called "The Airline Industry" which we started teaching at MIT in September 2001. The impetus for this course was provided by the establishment at MIT of the Global Airline Industry Program, an inter-disciplinary program of research and education that brings together faculty, researchers and students from the Departments of Aeronautics and Astronautics, Civil and Environmental Engineering, Economics and the Sloan School of Management. The Program also includes affiliated faculty and researchers from other universities, several of whom have contributed to this book.

The Global Airline Industry Program was made possible through a generous grant from the Alfred P. Sloan Foundation to MIT for the purpose of developing a center devoted to airlines and the air transportation sector as part of the Foundation's Industry Studies Program. The fruitful and innovative collaborations that resulted from the Program, including the ones that led to the preparation of this volume, would not have been possible without this support. We are especially grateful to the recently-retired President of the Foundation, Dr. Ralph Gomory, and to its Executive Directors, Dr. Hirsh Cohen and his successor, Dr. Gail Pesyna, for their guidance and wise counsel over the years.

During the decade of its existence, the Global Airline Industry Program has grown appreciably. It now brings together more than a dozen faculty, research staff and faculty affiliates, as well as many graduate students working on airline industry and air transportation research projects. It has also become a financially self-sustaining program – very much in line with what the Foundation and MIT were hoping for. This is largely due to gifts to the Program by major aviation-related companies, as well as the establishment of an Airline Industry Consortium at MIT, which has attracted a large number of industry and government members. We acknowledge, in this respect, the financial support to the Program from: Air Canada; the Air Transport Association of America; Alitalia; Amadeus, S.A.; American Airlines; American Express; the Boeing Commercial Airplane Company; the U.S. Federal Aviation Administration; Jeppesen Systems; JetBlue Airways; Lufthansa German Airlines; the Massachusetts Port Authority (Massport); the Metropolitan Washington Airports Authority; Scandinavian Airlines System (SAS); SITA; and United Airlines.

Finally, the authors and editors of this book are grateful to the roughly 300 very bright MIT students, mostly at the graduate level, who have taken our course over the past eight years and contributed materially to this book through their comments and suggestions concerning the course notes and other documents we have been distributing and progressively revising. Despite the economically difficult times that the global airline industry has been experiencing in recent years, it remains a vital and dynamic industry that continues to present intellectually fascinating issues for students, researchers and practitioners worldwide.

1

Introduction and Overview

Peter P. Belobaba and Amedeo Odoni

This chapter presents an introduction to the global airline industry, its evolution and current status. The major forces shaping the industry are described, including deregulation and liberalization worldwide, along with some important recent industry challenges, such as the severe financial problems that the industry has faced since 2000, which were followed by restructuring of some of the industry's largest airlines. Looking ahead, the industry still faces major challenges, including unprecedented fuel price volatility, a global financial crisis and weakening demand for air travel as economic growth slows. Infrastructure capacity poses a major constraint worldwide and threatens continued evolution and long-term profitability. The final section provides a brief overview of each of the other chapters of this book.

1.1 Introduction: The Global Airline Industry

The global airline industry provides a service to virtually every country in the world, and has played an integral role in the creation of a global economy. The airline industry itself is a major economic force, in terms of both its own operations and its impacts on related industries such as aircraft manufacturing and tourism, to name but two. Few other industries generate the amount and intensity of attention given to airlines, not only by those directly engaged in its operations, but also by government policy makers, the news media, as well as its billions of users, who, almost to a person, have an anecdote to relate about an unusual, good or bad, air travel experience.

During much of the development of the global airline industry, its growth was enabled by major technological innovations such as the introduction of jet aircraft for commercial use in the 1950s, followed by the development of wide-body "jumbo jets" in the 1970s. At the same time, airlines were heavily regulated throughout the world, creating an environment in which technological advances and government policy took precedence over profitability and competition. It has only been in the period since the economic deregulation of airlines, beginning with the USA in 1978, that cost efficiency, operating

The Global Airline Industry P. Belobaba, C. Barnhart and A. Odoni
© 2009 John Wiley & Sons, Ltd

profitability and competitive behavior have become the dominant issues facing airline management. Airline deregulation or, at least, "liberalization" has now spread far beyond the USA to most of the industrialized world, affecting both domestic air travel within each country and, perhaps more importantly, the continuing evolution of a highly competitive international airline industry.

Today, the global airline industry consists of over 2000 airlines operating more than 23 000 commercial aircraft, providing service to over 3700 airports (ATAG, 2008). In 2007, the world's airlines flew more than 29 million scheduled flights and transported over 2.2 billion passengers (IATA, 2008). The growth of world air travel has averaged approximately 5% per year over the past 30 years, with substantial yearly variations due both to changing economic conditions and to differences in economic growth in different regions of the world. Historically, the annual growth in air travel has been about twice the annual growth in GDP. Even under relatively conservative assumptions concerning economic growth over the next 10–15 years, a continued 4–5% annual growth in global air travel will lead to a near-doubling of total air travel during this period.

The annual growth rates in passenger air traffic, measured in revenue passenger kilometers (RPK) (see Chapter 3 for definitions) are shown in greater detail in Figure 1.1, for the period 1987–2007. The principal driver of air travel demand is economic growth: over the period shown in Figure 1.1, the 5–6% average annual growth in air travel has been fed by an average 2–3% annual growth in GDP worldwide. However, there has been substantial variability in different years, as well as differences between US and non-US

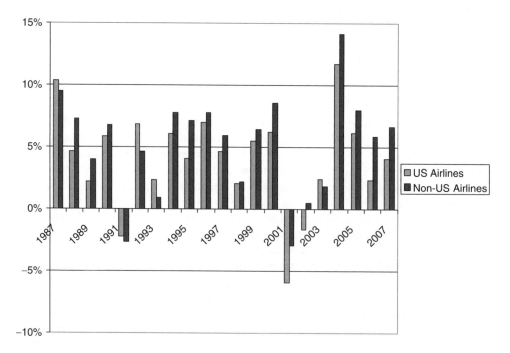

Figure 1.1 Annual RPK growth rates, 1987–2007
Data sources: Air Transport Association, ICAO

airlines. World passenger air traffic growth has been positive in all years shown, with only two exceptions. Traffic declined in 1991 due to the first Gulf War, the subsequent fuel crisis and economic recession, and then again in 2001 due to the effects of the 9/11 terror attacks in the USA. Figure 1.1 also shows that the annual growth rates experienced by non-US airlines have consistently outpaced those of US carriers. As a result, the proportion of world passenger traffic carried by US airlines has declined, from over 40% in 1987 to less than 32% in 2007.

The growth of passenger air traffic by region of the world is illustrated in Figure 1.2. North America continues to be the leading region in terms of air traffic, followed by Europe and Asia–Pacific. North American air travel was clearly the most affected by the after-effects of 9/11, as evidenced by the sharp decrease in RPK from 2001 through 2003. Growth rates in the Asia–Pacific region have been substantially higher than those of North America and Europe, with the result that total passenger air traffic in the Asia–Pacific region reached almost the same level as in Europe in 2007. With continued high growth rates expected, the Asia–Pacific region will soon become the second-largest world region for air traffic.

Figure 1.3 provides a similar plot of the growth of air freight by world region since 1971, measured in freight tonne kilometers (FTK) – defined in Chapter 3. The relative size and growth of air freight in each world region differs from that of passenger air traffic: Europe generated slightly more air freight than North America in the first part of the period shown in Figure 1.3, followed by Asia–Pacific. However, the growth of air freight in the Asia–Pacific region has surged over the past three decades and the region has led

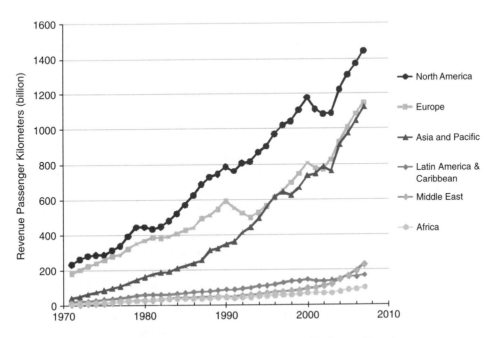

Figure 1.2 Growth of airline passenger traffic by world region
Data sources: ICAO (1971–2005), IATA (2006–2007)

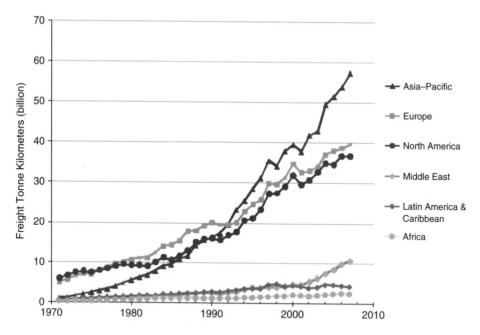

Figure 1.3 Growth of air freight volumes by world region
Data sources: ICAO (1971–2005), IATA (2006–2007)

the world in terms of total air freight volume since 1992. In both Figures 1.2 and 1.3, it is also worth noting the recent growth of the Middle East region. Despite ranking well behind North America, Europe and Asia–Pacific in both passenger and cargo air traffic, the Middle East has shown a surge in both categories since 2000.

In the US airline industry, approximately 100 certificated passenger airlines operate over 11 million flight departures per year, and carry approximately one-third of the world's total air passengers – US airlines enplaned 769 million passengers in 2007. US airlines (both cargo and passenger) reported over $170 billion in total operating revenues, with approximately 561 000 employees and over 8000 aircraft operating 31 000 flights per day (ATAG, 2008). The economic impacts of the airline industry range from its direct effects on airline employment, company profitability and net worth to the less direct but very important effects on the aircraft manufacturing industry, airports and tourism industries, not to mention the economic impact on virtually every other industry that the ability to travel by air generates.

Commercial aviation contributes 8% of US GDP, according to recent estimates (ATAA, 2004). Worldwide, the global economic impact of aviation has been estimated to be about 7.5% of world GDP, or more than $3.5 trillion per year (ATAG, 2008).

The economic importance of the airline industry and, in turn, its repercussions on so many other major industries make the volatility of airline profits and their dependence on good economic conditions a serious national and international concern. This has been particularly true since the start of airline deregulation, as stable profits and/or government assistance were the rule rather than the exception for most international airlines prior to the 1980s. As shown in Figure 1.4, the total net profits of world airlines have been cyclical

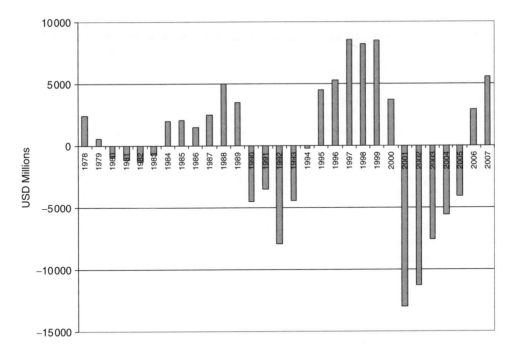

Figure 1.4   World airline net profits, 1978–2007
Data sources: Air Transport Association, ICAO

and extremely variable over the past 30 years. After the world airline industry posted four consecutive years of losses totaling over $22 billion from 1990 to 1993, as a result of the Gulf War and subsequent economic recession, it returned to record profitability in the late 1990s, with total net profits in excess of $25 billion being reported by world airlines from 1995 to 1999. Even more dramatic was the industry's plunge into record operating losses and a financial crisis between 2000 and 2005.

1.1.1   Deregulation and Liberalization Worldwide

Since the deregulation of US airlines in 1978, the pressure on governments to reduce their involvement in the economics of airline competition has spread to most of the rest of the world. US airline deregulation is perceived as a success by most other countries, as the overall benefits to the vast majority of air travelers have been clearly demonstrated. While US domestic air travel grew at rates significantly greater than prior to deregulation, average real fares have declined significantly since deregulation and today remain at less than half of 1978 levels (ATA, 2008). Successful new entrant and low-fare airlines had a great impact both on airline pricing practices and on the public's expectations of low-priced air travel. And, despite worries at the time of deregulation that competitive cost pressures might lead to reduced maintenance standards, there is no statistical evidence that airline safety has deteriorated (Chapter 11).

At the same time, deregulation in the USA has had some negative impacts. The pressure to cut costs, combined with increased profit volatility, mergers and bankruptcies of several large airlines, led to periodic job losses and reduced wages. Furthermore, the benefits of deregulation have not been enjoyed equally by all travelers. Residents of some small US cities saw changes in the pattern of air service to their communities, as smaller regional airlines replaced previously subsidized jet services. And, despite a substantial decrease in the average real fare paid for air travel in US domestic markets, the disparity between the lowest and highest fares offered by airlines increased, irritating business travelers forced to pay the higher fares. The development of large connecting hubs by virtually all US major airlines also raised concerns about the pricing power of dominant airlines at their hub cities (GAO, 1993).

The management strategies and practices of airlines were also fundamentally changed by increased competition within the industry. Cost management and productivity improvements became central goals of US airlines with the shift to market competition. Non-US airlines have more recently been forced by competitive realities to face up to this challenge as well. A by-product of the quest for lower costs and increased productivity has been the pursuit of economies of scale by both US and non-US airlines. In the past, internal growth and/or mergers were the primary ways in which airlines hoped to take advantage of scale economies. With growing government concerns about industry consolidation, further mergers began to receive greater regulatory scrutiny. And, with regulations limiting foreign ownership of airlines still enforced by many countries (Chapter 2), complete mergers between airlines of different countries continue to face legal barriers. The response of airlines has been to expand their networks and to achieve at least some economies of scale through partnerships and "global alliances" designed to offer a standardized set of products and to project a unified marketing image to consumers.

1.1.2 Recent Industry Evolution

On a global scale and especially in the USA, the airline industry has been in a financial crisis for much of the twenty-first century. The problems that began with the economic downturn at the beginning of 2001 reached almost catastrophic proportions after the terror attacks of September 11, 2001. In the USA alone, the industry posted cumulative net losses of over $40 billion from 2001 to 2005, and only in 2006 was it able to return to profitability with a total net profit of just over $3 billion (ATA, 2008). Many of the same forces affected non-US airlines, which as a group recorded losses in the years 2001–2003, but posted modest net profits in 2004 and 2005. Airlines outside the USA were in particular affected by international military and political events, as well as the SARS-related health crisis in 2003.

The events of 9/11 resulted in immediate layoffs and cutbacks by US airlines of almost 20% in total system capacity, in anticipation of the inevitable decline in passenger traffic due to concerns about the safety of air travel. However, the airlines were in serious trouble even before that date, as the start of an economic downturn had already affected negatively the volume of business travel and average fares. At the same time, airline labor costs and fuel prices had been increasing faster than the general rate of inflation for several years. To make matters worse, airlines faced deteriorating labor/management relations, aviation infrastructure constraints that led to increasing congestion and flight delays, and dissatisfied customers due to perceptions of poor service.

Thus, the recent poor performance of the US airline industry cannot be attributed solely to the impacts of 9/11. In fact, the events of 9/11 provided a temporary reprieve from some of the industry's fundamental problems. Reductions in flight schedules alleviated some of the pressure on the aviation infrastructure, resulting in fewer flight delays; faced with massive layoffs and tremendous uncertainty about the financial future of the airlines, labor unions moved toward a more conciliatory position, and passengers became more willing to lower their service expectations in exchange for improved security. In the period after 9/11, passenger traffic made a slow recovery and returned to pre-9/11 levels by mid-2004. With total US domestic airline capacity substantially lower than before 9/11, average load factors soared to historically record levels.

As shown in Figure 1.5, the proportion of available seats filled with revenue passengers (i.e., the load factor) has increased steadily since the mid-1990s for both US and world airlines, with particularly dramatic increases since 2001. By 2007, average load factors for US airlines reached almost 80%, or 10 percentage points higher than in 2000. At the same time, world airlines as a group saw their load factors increase to 77%, an unprecedented level that might imply significant financial success. Yet, despite operating flights that were becoming increasingly full, many traditional or "legacy" airlines around the world still struggled to make an operating profit given that much of the increase in the proportion of seats sold stemmed from fare discounting in the face of increasing competition.

The ability of the legacy airlines to generate adequate revenues to cover their operating costs was severely impacted by major shifts in passenger choice behavior, particularly on the part of business travelers. The volume of business air travel demand decreased in early 2001 due to an economic downturn. Business air travel was then further affected by the increased "hassle factor" and greater uncertainty in passenger processing times caused

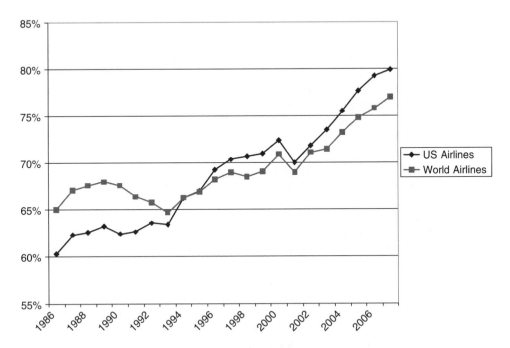

Figure 1.5 US and world airline passenger load factors, 1986–2007
Data sources: Air Transport Association, ICAO

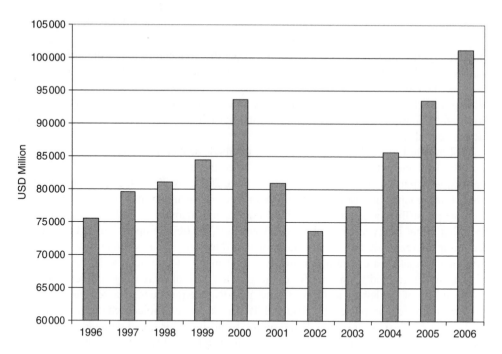

Figure 1.6 US airline industry passenger revenues, 1996–2006
Data source: ATA

by increased security requirements after 9/11. The combination of reduced business travel budgets and substantial cutbacks in airline passenger service quality led more business travelers to look for alternatives to paying premium airfares – teleconferencing and other travel substitutes, alternative travel modes and, especially, low-fare airlines. As a result, in the US airline industry alone, passenger revenues dropped by over 20% between 2000 and 2002, and did not recover to 2000 levels until 2006, as shown in Figure 1.6.

The recent growth of low-fare air travel options combined with a reduced willingness on the part of business travelers to pay the higher airfares charged by legacy carriers contributed in a major way to the poor financial performance of traditional airlines, both in the USA and in many other countries. In the USA, low-fare airlines (also known as low-cost carriers or "LCCs") exhibited slow but steady growth in the first decade after deregulation, but still accounted for less than 7% of US domestic air passengers in 1991. As shown in Figure 1.7, LCCs have grown far more rapidly in the USA since the mid-1990s, to the point where they carried 25% of US domestic passengers in 2005 (US DOT, 2007).

The LCC concept, born in the USA several decades ago, has now taken root in many other parts of the world where LCCs have grown even more rapidly. With the liberalization of European airline markets in the mid-1990s, LCCs proliferated to the point that the number of LCCs in Europe (62) was more than triple that in the USA (19) in 2006, as shown in Figure 1.8. The growth of LCCs in Asia is even more recent, and by 2006 there were already 40 such carriers operating in that region. The LCCs have changed the competitive landscape in most regions with liberalized airline markets, affecting pricing structures with their substantially lower fares and, in turn, the revenues

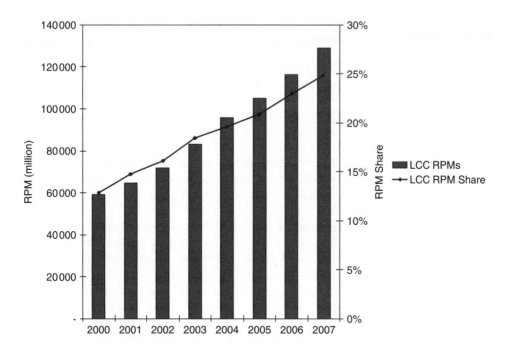

Figure 1.7   US domestic traffic carried by low-cost carriers, 2000–2007
Data source: US DOT, Form 41

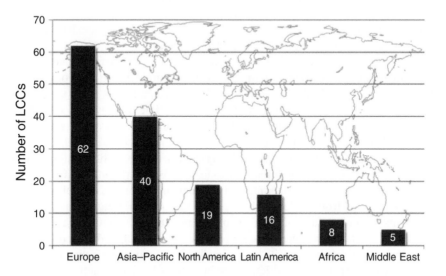

Figure 1.8   Low-cost carriers around the world, 2006
Data sources: Airline websites and Wikipedia Low-Cost Carrier list

of traditional legacy airlines which are effectively forced to match these lower prices to remain competitive.

While it is true that the legacy airlines around the world encountered a revenue problem with the emergence of LCCs as a formidable competitive force, it also became clear that they had fundamental operating cost and productivity disadvantages compared to their low-cost challengers. The differences in the cost structures between legacy airlines and LCCs reflected substantial differences in the productivity of both aircraft and employees. LCCs tend to operate in a "point-to-point" manner in which they can minimize aircraft ground times, in contrast to the hub-and-spoke networks of the largest legacy airlines. Shorter ground times translate directly into higher aircraft utilization rates. As a result, LCCs have been able to achieve aircraft utilization rates more than 45% higher than what legacy carriers achieve for the same aircraft type, which contribute to 35% lower unit aircraft operating costs (US DOT, 2007).

Perhaps the most critical element of the successful low-cost airline business model is significantly higher labor productivity than traditional legacy carriers. The differences can be attributed to labor productivity per se, not simply to employee unionization or even to wage rates. The oldest LCC in the USA, Southwest Airlines, is the most heavily unionized US airline and its salary rates are considered to be at or above average compared to the industry. The LCC labor advantage lies in much more flexible work rules that allow cross-utilization of virtually all employees (except where disallowed by licensing and safety standards). Such cross-utilization and a long-standing culture of cooperation among labor groups translate into lower unit labor costs. These strategies, used for several decades by Southwest Airlines to increase both labor and aircraft productivity, have now been replicated in some respects by virtually every new entrant airline in every world region.

The challenges described above led four out of the six US legacy carriers (US Airways, United, Delta and Northwest) into Chapter 11 bankruptcy between 2001 and 2005. Under bankruptcy protection, these carriers were able to focus on downsizing, cutting operating costs and improving productivity as part of their restructuring efforts. The other two legacy carriers, American and Continental, used the threat of bankruptcy to do the same. Much of the cost-cutting strategy focused on labor: legacy airline employment was reduced by 30% in just five years, representing over 100 000 jobs lost, while average wage rates were also cut by 7% (US DOT, 2007).

Non-US airlines were not immune to bankruptcies and even liquidations during this same period. Financial problems forced Swissair to cease operations and restructure itself under new ownership as Swiss International Airlines. In the process, employees lost jobs, aircraft were removed from its fleet and routes were eliminated. Belgian flag carrier Sabena shut down its operations after it was unable to restructure. Air Canada, after merging with Canada's other flag carrier, Canadian Airlines International, declared bankruptcy in 2003, but has since restructured its costs and business model to return to profitability. Ansett Airlines of Australia was liquidated in 2002 while, in South America, notable legacy airlines such as Varig and Aerolineas Argentinas have experienced bankruptcies and continue to deal with enormous financial instability.

In response to these challenges and in an effort to restructure, legacy airlines in many regions of the world have been able to achieve additional productivity gains by introducing new technologies (e.g., Internet ticket distribution, web check-in) and by reallocating capacity (e.g., moving capacity from domestic to international routes in an effort to improve aircraft utilization with increased stage lengths). Some of these carriers have also attempted to imitate strategies of the LCCs, e.g., by eliminating meals on shorter flights to reduce costs and by reducing aircraft turnaround times to improve aircraft productivity.

The LCC sector of the global airline industry in many respects took advantage of the weaknesses of the legacy carriers during their financial crises and restructuring efforts. Many LCCs were able to expand their networks rapidly and captured significant market share, not only in the USA but in Europe, Canada and South America. They expanded into new markets with new aircraft, more flights and, of course, lower fares. However, during this same period some of the more established LCCs also began to face increasing operating costs, driven by aging fleets and personnel with increasing seniority. Furthermore, the LCCs could not escape the impacts of increasing fuel costs – even the successful fuel hedging strategy of Southwest provided only a temporary reprieve. In fact, the concerted cost-cutting efforts of both legacy and LCC airlines were not sufficient to offset the increased fuel prices, which tripled between 2002 and 2007 (see Figure 1.9).

In summary, the cost and productivity improvements made by legacy airlines between 2001 and 2006 changed the competitive environment of the global airline industry yet again. And there is increasing evidence of cost and productivity convergence between the legacy and LCC airlines, particularly in the USA. The financial results for 2006 were positive for most of the largest legacy airlines around the world, while some of the LCCs struggled financially. The world airline industry as a whole posted an aggregate net profit of $3 billion in 2006, followed by almost $6 billion net profit in 2007. The painstaking

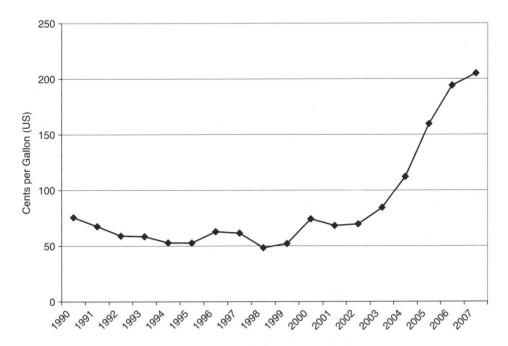

Figure 1.9   US airline jet fuel prices, 1990–2007
Data source: Air Transport Association

restructuring efforts and cost reductions of the legacy airlines appeared to be paying off, and the industry also benefited from an improving revenue environment and from little or no growth in available capacity.

1.1.3 Looking Ahead: Industry Challenges

The airline industry continues a dramatic restructuring that involves even more fundamental changes than those experienced following its deregulation in 1978. Yet, three decades after the initial deregulation of US airline markets and the subsequent liberalization of many other markets around the world, the industry remains fragile. Multiple cycles of financial successes and failures have left airlines struggling to find a business model that can ensure sustained profitability. Competitive pressure from LCCs, the loss of consumer confidence in the air transportation system's reliability and operating performance, and the transparency of pricing facilitated by the Internet and online travel distribution channels have all contributed to a precipitous decline in average fares and have had a significant impact on airline revenues.

In 2008, after appearing to recover from its latest cycle of financial struggles with positive financial results in two consecutive years, the global airline industry entered another turbulent period. With the surge in oil prices since 2006, fuel emerged as the single largest industry expense, surpassing labor costs for the first time (US DOT, 2007). Even as fuel prices dropped rapidly in late 2008, the very real threat of a deep economic recession driven by unstable financial markets began to weaken the demand for air travel worldwide. Given higher fuel costs and lower demand, the International Air Transport Association, the worldwide airline industry trade association, predicted net losses for the world airline industry of over $5 billion in 2008 and $4 billion in 2009 (IATA, 2008).

While there has been much progress on issues of aviation safety and security since 9/11, with increases in the number of airport security screeners, improved screener training, and movement toward explosives screening for all checked baggage, the questions "are we doing enough?" and "are we doing the right things?" remain unanswered. Demand for air travel, particularly in short-haul markets, has been reduced by passenger perception of the "hassle factor" of increased security and the uncertainty of passenger processing times at airports. For the airlines, the new security procedures have increased operating costs and induced more security-related flight disruptions and delays. The Director-General of the IATA has stated that "our passengers have been hassled for 6 years ... that's far too much" (Bisignani, 2006). Some experts, however, have expressed concern that cutbacks in existing security measures might increase the risk of future terrorist acts that could devastate the industry.

The temporary reprieve from congestion and flight delays experienced immediately after 9/11 has effectively ended at the world's busiest airports. The number of delayed flights once again reached record levels in 2007, and media reports of chronic and excessive airline passenger delays have again become commonplace. Several factors, including the lack of coordination of airline flight schedules at some of the most congested airports, outdated air traffic control systems, finely tuned airline flight schedules with little slack to dampen delay propagation, and record-high load factors preventing timely reaccommodation of passengers who misconnect or whose flights are canceled, all combine to create passenger disruptions and lengthy passenger delays that exceed even the record-high levels of flight delays. Solutions to the problem will require a mix of improved management of airspace and airport demand, and an increase in airport capacity brought about primarily by improved management and utilization of existing capacity.

These important challenges – sustaining airline profitability, ensuring safety and security, and developing adequate air transportation infrastructure – are not confined to any single country. Airlines around the world are encountering a growing wave of liberalization, if not outright deregulation, and as a result are facing competitive pressures, both from new entrant low-cost airlines and restructured legacy carriers. The rapid growth of the global airline industry and the continued threat of terrorist attacks make safety and security issues critical to every airline and every airline passenger. And the need for expanded aviation infrastructure, both airports and air traffic control, is of particular importance to emerging economies of the world such as India, China, Africa and the Middle East, where much greater rates of demand growth are forecast for both passenger and cargo air transportation.

1.2   Overview of Chapters

This book provides a comprehensive introduction to the global airline industry for air transportation professionals worldwide, for students in transportation programs and, more generally, for anyone with a serious interest in the subject. Following the brief description in this chapter of the evolution, recent developments and future challenges facing the industry, Chapters 2 through 15 of this book offer more detailed coverage of airline economics, management, scheduling and operations, as well as reviews of the topics of aviation safety and security, airports, air traffic control, environmental impacts, and the international regulatory environment in which the industry operates. The perspective is international, drawing on the authors' extensive experience with airline and air transport issues around the world.

The chapters were written by a group of faculty and research staff affiliated with the Massachusetts Institute of Technology (MIT), recognized for their expertise on the multiple facets of the air transport sector. All are associated with MIT's Global Airline Industry Program (web.mit.edu/airlines). Much of the content of this book is based on materials developed for a first-year graduate subject at MIT, "The Airline Industry," that the co-authors have been teaching together since 2001.

To provide the reader with adequate background for several subsequent chapters, Chapter 2 describes the regulatory and institutional context within which air transportation must operate. It first presents a brief account of the accomplishments of the Chicago Convention and explains the "Freedoms of the Air" for international air transportation services, in layman's terms. It then reviews the emergence of economic deregulation, privatization of airline companies and liberalization of international aviation agreements. Similarly, recent trends toward airport privatization and the semi-privatization of air traffic management (ATM) services are discussed, with emphasis on the emergence of infrastructure capacity constraints that can act as a global restraint to competition.

The fundamentals of airline economics, markets and demand are introduced in Chapter 3. This chapter introduces basic airline terminology and definitions, along with the concepts of air transportation markets and the demand for air travel. The joint supply of service to multiple origin–destination (O-D) markets by any single flight in a network is explained, along with the importance of service quality and total trip time, in addition to price, in determining air travel demand volumes. A brief overview of air travel demand models is then provided, focusing on the effects of price and time elasticity, as well as of the cross-elasticity of demand. Finally, airline competition in terms of flight frequency is introduced and the classical S-shaped market share/frequency model explained.

Chapter 4 continues the discussion of airline economics with an introduction to recent developments in airline pricing and revenue management, based on the demand fundamentals introduced in Chapter 3. The theoretical discussion examines price discrimination and market segmentation techniques used by airlines to maximize revenue through "differential pricing" of air transportation services in the same O-D market. The discussion of actual practices explains fare product restrictions with examples of traditional airline fare structures (pre-2000), followed by recent trends in airline pricing aimed at simplifying fares in response to LCC competition. The basics of revenue management (RM), used by airlines to control the number of seats sold at different fares, are then presented. Different revenue management techniques are described, along with the evolution of computerized RM systems to implement these techniques. The changes in RM practices required under the simplified fare structures offered more recently by LCCs are also discussed.

The review of airline economics continues in Chapter 5, which is devoted to airline operating costs and productivity measures. This chapter first describes the challenges in categorizing airline operating costs and presents alternative categorization schemes. Based on recent data reported by airlines, total airline operating costs are broken down by category, and comparisons are made both across airlines (legacy vs. LCC) and across aircraft types. The most commonly used measures of aircraft and labor productivity are then defined, and recent trends for legacy and LCC airlines are used to illustrate these measures and their impacts on unit operating costs.

The next several chapters are devoted to the processes of planning, optimization and operation of a schedule of services for an airline. Chapter 6 first provides an overview of the airline planning process, as it describes major airline planning decisions, from the longest-range strategic decisions involving aircraft acquisition to medium-term decisions related to route planning and scheduling. Approaches to fleet planning are described, and the most important aircraft characteristics considered in fleet selection are discussed. Methods for evaluating route profitability are then presented, focusing on the importance of hub economics in the network structure of most large airlines. The advantages and incremental costs of hub-and-spoke operations are also discussed. Finally, the schedule development process is described qualitatively, to introduce the major steps of frequency planning, timetable development, fleet assignment and aircraft rotation planning.

Building on the qualitative introduction to the schedule development process in Chapter 6, Chapter 7 focuses on the models used by airlines to find optimal solutions for several scheduling problems. The fleet assignment model (allocation of the aircraft fleet to scheduled flight legs) is presented first, with both a detailed description of the problem and an advanced section describing network optimization formulations, extensions and solution methods. The application of these methods to schedule design and retiming of flights is then discussed, and the basics of several related scheduling problems are presented, including the crew scheduling problem (how to minimize the crew costs associated with covering all flight legs in a schedule through "crew pairing" and "crew rostering") and the aircraft maintenance routing problem (ensuring that aircraft rotate through maintenance bases at the required times).

The smooth execution of the flight schedules developed during the planning process, as described in Chapters 6 and 7, is the responsibility of the flight operations departments of airlines. Chapter 8 presents an overview of airline operations and of the sequence of actions needed to coordinate each airline's complex combination of aircraft, cockpit and cabin crews, maintenance facilities and ground service personnel. The work regulations and scheduling of flight crews are described first, providing important background for the remainder of the chapter, in which the activities taking place during the different phases of a typical flight are described in more detail. The steps in preparing a flight for departure, followed by push-back, takeoff, cruise, descent and landing, culminating with arrival at the destination airport, are all discussed with a focus on the role of regulatory requirements and the activities of airline personnel in each step.

Despite an airline's best efforts to optimize and execute a schedule plan, unforeseen disruptions to this plan occur inevitably on every single day. These result in "irregular operations." Chapter 9 is devoted to describing the problems associated with irregular operations and with "schedule recovery" by the airlines. The chapter identifies the various reasons for such disruptions, and the resulting delays and economic impacts. It then focuses on optimization-based techniques that have been developed for schedule recovery – replanning aircraft, crew and passenger routings. An alternative approach that aims at preparing "robust" airline schedules is also described. Robust planning approaches are explicitly aimed at reducing the complexity and/or need for schedule recovery by generating, during the planning process, schedules that are resilient to disruptions. An overview of schedule recovery and robust planning approaches concludes the chapter, along with a summary of recent successes and remaining challenges.

Chapters 10 through 15 provide a review of several of the most critical issues facing airlines in the planning and delivery of safe and profitable air transportation services – labor, safety and security, airport infrastructure, air traffic control, and the impacts of information technology. Chapter 10 surveys different approaches to labor relations and human resource management used by airlines around the world. The chapter begins by presenting a general framework for understanding the key aspects of employment relationships in airlines, and then discusses the traditional management–labor paradigm in the airline industry and the unique regulatory setting that governs labor relations in it. Given this historical context, the discussion turns to alternative models for airline human resource management that have proven to be effective in various contexts.

Aviation safety and security is essential to the economic viability of airlines, a fact made painfully clear by the events of 9/11. Chapter 11 first focuses on aviation accidents not attributable to terrorist attacks, and presents a set of statistics that address in a global context the critical question of "how safe is it to fly?" Measures for quantifying the mortality risk of passenger air travel are developed and then used to make safety-related comparisons between First and Third World airlines and between domestic and international services, over a nearly 50-year period that begins in 1960. In its second half, the chapter turns to aviation security, with an overview of approaches used internationally to protect against terrorist acts. The chapter concludes with several examples that illustrate the complexity of making decisions about how much security to provide at airports and on flights.

Chapter 12 presents an overview of the characteristics, operations and finances of large commercial airports. The objective is to familiarize the reader with a broad range of issues and terminology. Fundamental characteristics of airports in different regions of the world are described first, followed by a discussion of airside and landside facilities, including international design standards and specifications. The critical topic of airport capacity is then addressed, both airside and landside, including a brief review of airport demand management techniques. Finally, the controversial subject of user charges for airport facilities and services is discussed in the context of airport economics and the financing of capital projects.

Along similar lines, Chapter 13 provides an introduction to the characteristics and operation of air traffic management (ATM) systems – better known as air traffic control (ATC). The main components of ATC infrastructure are described, including its communications, navigation and surveillance subsystems. In view of the role of runway capacity as a major constraint on airline operations, the principal factors that affect the capacity of runway systems at major airports are also reviewed in some detail. Similarly, the operations of the air traffic flow management (ATFM) subsystem are described. ATFM attempts to control the flow of air traffic to best match demand with available capacity on a day-to-day basis. Operations in the various types of airspace, such as terminal areas and high-altitude en route sectors are also discussed with emphasis on identifying some of the challenges that the next generation of ATM systems must address.

Aviation affects the environment primarily through noise and emissions, leading to a dramatic increase of public and political pressure on the industry to better manage and mitigate such impacts. Chapter 14 provides a review of civil aviation's impacts on community noise, local air quality, water quality and climate change, along with a summary of the main ways in which these impacts can be mitigated. It begins by describing briefly the local, national and international bodies that address the environmental impacts of aviation. The different types of environmental impacts are then discussed in some detail, along with an overview of recent trends and of possible mitigation strategies for each type of impact. Emerging issues concerning the different types of impacts are also explored, with particular reference to recent research and its implications for both regulatory policy and technological development.

Airlines are leaders in the use of information technology, with the development of databases and decision support systems introduced in several of the early chapters of the book. Chapter 15 first provides a review of the role of computerized systems in airline planning and operations, and then focuses on the evolution of information technology applications for airline distribution and passenger processing. The development of computer reservations systems is described, as background for understanding the dramatic changes in airline distribution that have occurred over the past decade – a shift away from travel agencies to websites for booking and ticketing. The differences between traditional and emerging airline distribution channels are explained, and the cost reductions achieved in this area by airlines are presented. In addition, electronic ticketing and recent innovations in passenger processing, including airport kiosks and web check-in, upgrade notification and baggage tracking, are discussed, with a focus on the implications for airline economics and passenger satisfaction.

Chapter 16 concludes this book with a discussion of the critical issues and prospects for the global airline industry, drawing together the insights provided by the preceding chapters concerning the many constraints, decision processes and stakeholders that contribute to industry evolution.

References

ATAG – Air Transport Action Group (2008) "The Economic and Social Benefits of Air Transport 2008," www.iata.org.

ATAA – Air Transport Association of America (2004) "Statement on the State of the Airline Industry, Statement for the Record of the Sub-committee on Aviation, Transportation and Infrastructure Committee," US House of Representatives, June.

ATAA – Air Transport Association of America (2008) Economic Report 2008, www.airlines.org.

Bisignani, G. (2006) "State of the Air Transport Industry," Address to the Annual General Meeting, International Air Transport Association, Vancouver, June, www.iata.org.

GAO – General Accounting Office (1993) "Airline Competition: Higher Fares and Less Competition Continue at Concentrated Airports," Report to Committee on Commerce, Science and Transportation, US Senate, Report GAO/RCED-93-171, Washington, DC, July.

IATA – International Air Transport Association (2007) Fact Sheet: World Industry Statistics, www.iata.org.

IATA – International Air Transport Association (2008) "Airlines to Lose \$5.2 billion in 2008 – Slowing Demand and High Oil to Blame," Press Release, September 3.

ICAO – International Civil Aviation Organization (1971 through 2005) *Statistical Yearbook and Annual Reviews of Civil Aviation*.

US DOT – United States Department of Transportation (2007) "Air Carrier Financial Reports (Form 41 Financial Data)," Bureau of Transportation Statistics, Washington, DC, http://www.transtats.bts.gov/.

2

The International Institutional and Regulatory Environment

Amedeo Odoni

2.1 Introduction

Very few global industries are as deeply affected by changes in the international and domestic regulatory environment as the airline industry. The world's airlines have experienced dramatic regulatory changes over the past half-century and today are subject to a wide variety of rules and regulations in different parts of the world. To complicate things further, a large number of organizations, agencies and associations – international and national, governmental and non-governmental – play important regulatory, oversight or advocacy roles on critical issues affecting air transport, such as safety, economics, security and even national defense. In this light, the objective of this chapter is to present a summary of the regulatory and institutional context within which air transport operates – and thus to provide adequate background for several chapters in this book that require an understanding of this context. The presentation is in layman's terms and omits many of the (often important) distinctions and details that pervade the highly specialized area of air transport regulation and law.[1]

Section 2.2 provides a brief account of the Chicago Convention, the agreement that laid the foundation for today's global air transport system. It also describes the nine "Freedoms of the Air," which are used in international bilateral and multilateral agreements to specify the rights of access to international and domestic air travel markets. Section 2.3 is concerned with the regulation of airline markets. It describes briefly the movement toward airline privatization, which has become dominant in most parts of the world, as well as the deregulation of the US domestic market after 1978. It then reviews the evolution of

---

[1] The volume of scholarly writings on the subject over at least 70 years is enormous, with several journals, e.g., the *Journal of Air Law and Commerce*, dedicated solely to developments in this area.

The Global Airline Industry   P. Belobaba, C. Barnhart and A. Odoni
© 2009 John Wiley & Sons, Ltd

international aviation agreements with respect to access to markets, airline designation, capacity offered, and the setting of airfares. The fundamental provisions contained in the three main categories of existing bilateral and multilateral agreements are outlined and several examples of the continuing movement toward a "liberalized" international environment are provided.

Section 2.4 focuses on airports and discusses two important relevant developments: the trend toward airport privatization and issues that arise as a result; and the emergence of airport capacity constraints and attendant allocation procedures as a significant restraint on competition in international air transportation markets. Section 2.5 summarizes the more limited movement toward the "corporatization" or "commercialization" of the provision of air traffic management (or "air navigation") services. Section 2.6 presents brief descriptions of several important international and national organizations and agencies in the air transport sector, outlining the role and some of the characteristics of each. Finally, Section 2.7 contains some general conclusions.

2.2   Background on the International Regulatory Environment

This section provides background for the remainder of this chapter. To understand why the regulations that govern international air transport exhibit such major differences from country to country, it is necessary to have a historical perspective on the framework within which these regulations have been established. The "Chicago Convention," whose main contributions are summarized in Section 2.2.1, developed the main elements of this framework. The resulting regulations and international agreements make constant reference to the "Freedoms of the Air," which are described in Section 2.2.2.

2.2.1   The Chicago Convention

In 1944, before the end of World War II, representatives of 54 States[2] attended the International Convention on Civil Aviation, a conference on the future of international air transport that took place in Chicago. This conference – and the international treaty that was signed as a result – became known as the "Chicago Convention." The treaty marks a critical milestone in the history of aviation, as it laid the foundation for today's global air transportation system. The Convention made several fundamental contributions to the conduct of domestic and, especially, international civil aviation and underpinned the industry's enormous growth over time.

First, at a general level, the Convention confirmed the emerging realization that civilian air transportation was an activity of potentially enormous global importance, deserving to be nurtured and promoted through a set of internationally accepted rules for the rights of access to markets. In trying to define what these rules would be, the Convention, for the first time in the history of aviation, had to confront seriously the choice between a "liberal" and a "protectionist" regulatory environment for international services. This fundamental choice has been at the center of much controversy ever since. The USA advocated at the

---

[2] The term "State" is used in this chapter interchangeably with the less formal "country" or "nation."

time a liberal framework based on a regime that would place few restrictions on access to the world's air transportation markets and permit open competition among airlines, including the right to set market-driven flight frequencies and airfares. However, most other States, led by the UK, advocated a much more restrictive system for reasons of national security and of preserving airspace sovereignty, as well as of facilitating the growth of the then-nascent airline industry. An apparent underlying concern of this group was that a liberal competitive environment might lead to dominance by US airlines, which were believed to be much stronger, financially and in many other respects, than the airlines of other nations at the time. Using the threat of refusing landing rights as their bargaining tool, the protectionist group largely prevailed.

Rather than establishing a universal set of rules, the Convention thus decided to simply create a *framework* within which such rules could be established for regulating air transport services on a *bilateral* basis, i.e., between *pairs of countries*. As a result, bilateral *air service agreements* (ASAs) between States emerged as the instrument for initiating or modifying international transportation services and for regulating these services. The principal initial "model" for such agreements was developed only two years later, through the 1946 "Bermuda I" Agreement, which established the ground rules for US–UK air services. Bilateral ASAs continue to be prevalent today, but multilateral ASAs have also become increasingly common and important in recent years. The key provisions of the principal types of such agreements will be discussed in Section 2.3.

In another fundamental contribution, the Convention recognized the critical need for international commonality in airport and air traffic control facilities, equipment and procedures to ensure the safety and operability of aircraft across national boundaries. It recommended the establishment of a permanent international body charged with coordinating the rules guiding air transport operations around the world, developing international standards for aviation facilities and equipment, and overseeing adherence to these rules. This body became a reality in 1947 in the form of the International Civil Aviation Organization (ICAO), headquartered in Montreal (Section 2.6).

2.2.2 *"Freedoms of the Air"*

The Chicago Convention also originated the concept of the so-called "Freedoms of the Air." The Convention specified only the first five of the nine freedoms described below. The remaining four were subsequently defined informally over time, in response to the development of additional types of international aviation services. The freedoms refer to the rights that an airline of any State may enjoy with respect to another State or States (e.g., "the bilateral agreement between States A and B includes Fifth Freedom rights for carriers of State A").

Consider then a commercial carrier X registered in a State A:

- The First Freedom refers to the right of carrier X to fly over another State B without landing.
- The Second Freedom refers to the right of carrier X to land in another State B for technical (e.g., maintenance or refueling) or other reasons, without picking up or setting down any revenue traffic.

First and Second Freedom rights are granted essentially automatically in all but exceptional cases: they are exchanged among States under the so-called International Air Service Transit Agreement. Any exceptions are typically due to reasons of national security and are applied in a selective way by individual States – for example, a particular State may bar the airlines of one or more specified States from flying in its airspace.

The remaining freedoms refer to commercial rights and are best understood in the context of an ASA between two States A and B:

- The Third Freedom (Figure 2.1a) refers to the right of carrier X to enplane revenue traffic from its home State A for transport to an airport of State B.
- The Fourth Freedom (Figure 2.1b) refers to the right of carrier X to enplane revenue traffic at an airport of the agreement partner, State B, for transport to its home State A.

(a) Third Freedom flight   (b) Fourth Freedom flight

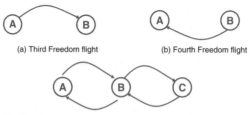

(c) Fifth Freedom: the flight in each direction is a continuing one (A to B to C and C to B to A); in picking up passengers at B to carry on to C (and vice versa), the carrier of State A exercises Fifth Freedom rights

(d) Sixth Freedom: the carrier of State A uses different flights to carry traffic from B to A and then from A to C (and vice versa)

(e) Seventh Freedom: the flight of the carrier of State A carries traffic between two other States B and C; the flights do not originate from or terminate at A

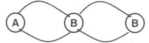

(f) Eighth Freedom: on the continuing leg of the flight from an airport of State A (own State) to an airport of State B, domestic traffic of State B is carried between two airports of State B and vice versa on the return flight

(g) Ninth Freedom: a carrier of State A transports purely domestic traffic within State B without restrictions

Figure 2.1   The Nine Freedoms of the Air; in all cases, the "home" State of the carrier involved is State A

Third and Fourth Freedom rights are fundamental to any bilateral ASA. The Fifth and Sixth Freedoms involve third States, in addition to A and B:

- The Fifth Freedom (Figure 2.1c) refers to the right of carrier X to enplane revenue traffic at an airport of the agreement partner, State B, for transport to a third State C, and vice versa, *as part of the continuation of a service (flight) originating or terminating in its home State A.*

As an example, consider a flight by US carrier Delta Airlines from New York to Paris, continuing to Mumbai. Fifth Freedom rights would make it possible for Delta to transport passengers from Paris to Mumbai on this flight and vice versa. Note that, although Fifth Freedom rights are agreed to between pairs of States (A and B in Figure 2.1c), it is necessary to also obtain the agreement of a third State (C) in order to implement such service.

- The Sixth Freedom (Figure 2.1d) refers to the right of carrier X (with home State A) to use separate sets of Third and Fourth Freedom rights with two other States (B and C in Figure 2.1d) in order to transport revenue traffic between these two other States (B and C) *using an airport in A as a connection point.*

Sixth Freedom rights are a natural consequence of Third and Fourth Freedom rights and provide the basis for many international "hubbing" operations. Lufthansa, for example, can pick up passengers in Milan, Italy, fly them to Frankfurt, Germany, and transfer them there to a Lufthansa flight to Boston, USA. Note that, once Third and Fourth Freedom rights are in place between States A and B and between States A and C, it becomes very difficult, in practice, for States B and/or C to curtail "Sixth Freedom traffic," should they ever wish to do so. In the example, Lufthansa could easily circumvent any restrictions that Italy and/or the USA might try to impose on flying passengers between their two countries via Frankfurt, e.g., by issuing separate sets of tickets for the two segments of the trip. Sixth Freedom rights are therefore typically taken for granted and are rarely mentioned or specified explicitly in an ASA.

Finally, the Seventh, Eighth and Ninth Freedoms are more "advanced" in that they greatly diminish the importance of the concept of the "home State" of an airline:

- The Seventh Freedom (Figure 2.1e) refers to the right of carrier X of State A to transport revenue traffic between a pair of airports in States B and C on a service that operates entirely outside State A.

Note that the home State A does not figure at all in the service provided. For example, Air Canada could serve the Los Angeles–Mexico City market, without the flights having to originate or terminate in Canada.

The Eighth and Ninth Freedoms involve "cabotage," i.e., the transport of domestic revenue traffic within a State other than a carrier's home State.

- The Eighth Freedom (Figure 2.1f) refers to the right of carrier X of State A to carry *domestic* revenue traffic between two points within State B *on a service originating or terminating in State A.*

- The Ninth Freedom (Figure 2.1g) is the same as the Eighth, but with no requirement that the flight of carrier X originate or terminate in X's home State A.

Eighth Freedom rights would allow a Toronto–Chicago–Los Angles flight of Air Canada to pick up Los Angeles-bound passengers in Chicago, while the Ninth Freedom would allow Air Canada to serve the Chicago–Los Angeles domestic market with flights originating from Chicago and from Los Angeles, or from any other point in the USA, for that matter.

Thus, the Ninth Freedom removes all restrictions on cabotage. The Eighth Freedom is sometimes referred to as "consecutive" or "fill-up" cabotage, and the Ninth Freedom as "full" or "pure" cabotage.[3]

2.3 Airline Privatization and International Economic Regulation

This section is concerned with the ownership of airlines and with their rights of access to international markets. It first discusses briefly (Section 2.3.1) the relatively recent international movement toward airline privatization. It then identifies the three principal types of ASAs currently in existence and describes their main characteristics with respect to four principal attributes: access to markets; airline designation; capacity offered; and the setting of airfares (Sections 2.3.2–2.3.4).

2.3.1 Airline Privatization

As recently as the early 1980s the great majority of the world's major airlines were government owned – with the notable exception of those in the USA. Perhaps the most important of the many reasons for this state of affairs was the perception of the airline industry as a nascent one, too fragile to be exposed to the rigors of competition.[4] Government-owned, and often government-subsidized, "flag carriers" were deemed to be the best way to sustain the growth of the industry to economic maturity. The flag carriers operated mostly as monopolists in domestic markets and in a protectionist environment on their international routes (Section 2.3.3). Even in the USA, where government-owned airlines did not exist, the airline industry was tightly regulated until 1978, to prevent "excessive" competition (Section 2.3.2). In addition, and even in some of the wealthiest countries in the world, the private sector was generally reluctant to assume the financial cost and economic risk associated with starting and operating an airline. Other contributing factors were the popularity with the general public of the notion of the "national flag" carrier, the potential role of such carriers at times of national emergency, and the use of coveted jobs at many of these national carriers as an instrument of political patronage.

All this changed dramatically when the industry began reaching economic maturity, first in the USA and then in much of the rest of the world. In economically developed

---

[3] The Eighth and Ninth Freedoms are often referred to simply as "cabotage rights" and not identified as separate "Freedoms." Alternatively, they are also merged into a single Freedom – the Eighth – with "consecutive cabotage" being just a special case of "full cabotage."

[4] This is still the case in many parts of the world.

countries, very few airlines currently remain fully in government hands[5] and more and more private airlines are emerging in the less developed parts of the world. There has also been a veritable explosion in the number of private, low-cost carriers (LCCs) – "no-frill" airlines that emphasize low fares to attract passengers. However, the remaining (primarily or fully) government-owned flag carriers continue to compete with privatized carriers in many markets and regions. In such cases, the burden of "leveling the playing field" and ensuring a fair competitive environment for all market participants necessarily falls on regulatory authorities. The European Commission, for example, has had to address several contentious cases since the mid-1990s involving such flag carriers as Sabena (Belgium), Air France, Alitalia and Olympic Airways (Greece), all of which have been accused of receiving overt or covert subsidies from their national governments or of being treated preferentially by these governments.

From a long-term perspective, the complete or partial privatization of many government-owned airlines since the mid-1980s has been one of the most important transformations that the industry has undergone in its history. In light of the steadily increasing "liberalization" of the international economic regulatory environment (Sections 2.3.2 and 2.3.3), the global trend toward airline privatization would seem to be irreversible for the foreseeable future.

2.3.2   Types and Critical Aspects of Air Service Agreements (ASAs)

As noted earlier, the Chicago Convention established a framework for regulating international air transportation services on the basis of ASAs. All bilateral and multilateral ASAs make reference to four critical aspects of the services to be provided:

(a) Market access: the *potential* city pairs to be served between the States involved in the ASA, as well as any Freedoms beyond the Third and Fourth, which may be granted under the ASA.
(b) Airline designation: the number of airlines[6] from each of the States that have the right to provide service in each city pair included in the agreement.
(c) Capacity: the frequency of flights and the number of seats that can be offered on each city pair.
(d) Airfares ("tariffs"): the manner in which passenger fares and/or cargo rates to be charged are determined and any steps necessary for government approval of these fares.

Depending on how it deals with each one of the above four aspects, an ASA can be classified into one of the following three categories (Doganis, 2001):

1. Traditional: The earliest example is the 1946 "Bermuda I" Agreement between the USA and the UK; more "liberal" examples of traditional ASAs emerged

---

[5] A few of the most notable airline privatizations include those of Air Canada, Air France, Air New Zealand, British Airways, Iberia, Japan Airlines, Lufthansa, Qantas and SAS. In some of these cases, the State still retains partial ownership.
[6] Note that, by specifying the number of airlines, but not necessarily their identity, the ASA allows a State to change designated airlines over time, if needed.

subsequently starting with the 1973 "Bermuda II" Agreement between the USA and the UK.

2. Open market: The earliest examples are the ASAs between the USA and The Netherlands, the USA and Singapore, and the UK and The Netherlands, all signed in 1978–1979.

3. Open skies: The earliest examples are the ASAs between the USA and The Netherlands, the USA and Singapore, and New Zealand and Chile in 1992.

Agreements of *all three types* are still being signed today. Many countries, especially those outside North America and Western Europe, continue to enter into bilateral ASAs of the traditional type. In fact, the majority of ASAs currently in force around the world are of the traditional type – although these are no longer the most important in terms of impact, as measured by the number of passengers affected and the economic significance of the markets involved.

In addition to the ASAs mentioned above, there are several other landmarks in the regulatory history of the airline industry. By far the most important before the 1990s is the deregulation of the US domestic market. Under the Airline Deregulation Act that went into effect in 1978, US airlines (*only*) have complete freedom to enter or exit any US domestic market. Each airline can also determine for itself the frequency of flights and the number of seats it offers in any market, the airfares it charges and the number of seats it will allocate to each airfare class on each flight. The role of the federal government – the US Department of Justice (DOJ), the US Department of Transportation (DOT) and the Federal Aviation Administration (FAA) – is limited to monitoring and policing the air transportation system in such matters as: potential predatory or monopolistic pricing (DOJ and DOT); potential collusion in setting airfares (DOJ); compliance with fair competitive practices in advertising or in listing flights on computer reservation systems (DOT); and compliance with safety and security regulations (FAA and, since 2001, the Transportation Security Administration, TSA).

Other landmarks in international airline regulation, to be described in Section 2.3.4, include the three-stage "liberalization" of the intra-European Union market (1988–1993), the Asia–Pacific Economic Community (APEC) multilateral ASA (2001) and the Open Skies Agreement between the EU, as a whole, and the USA (2007).

2.3.3 Typical Content of Bilateral and Multilateral ASAs

The typical content of each of the three ASA categories (traditional, open market and open skies) can now be summarized. The reader is warned, however, that significant variations may exist among agreements within each of the three classes. The discussion is structured with reference to each of the four main characteristics of ASA (market access, airline designation, capacity and airfares) identified in Section 3.2.

Market Access

Under *traditional* agreements, which are always bilateral, access is permitted to only a limited number of specified city-pair markets. Moreover, the number of airlines that may operate in these markets are specified in the ASA. In practically every case, no Fifth Freedom rights are granted. Rights for charter flights between airports in the two States are granted only on a flight-by-flight basis – there is no general provision for charter flights.

Under *open market* agreements, access is generally open to all potential city pairs in the two signatory countries. However, some bilateral open market agreements signed by the USA have restricted access by foreign airlines to only a limited number of US airports, specified in the agreement, while US airlines have been granted unlimited access to all (qualified) airports in the other signatory country.[7] Open market agreements usually provide unlimited rights for charter flights from each side, as well as offer Fifth Freedom rights to/from a specified set of third-party countries.

Under *open skies* agreements, access to city pairs is unlimited at both ends. So are Fifth Freedom rights (subject to concurrence from third-party countries) and the right to organize and operate charter flights. It is possible that some Seventh, Eighth or Ninth Freedom rights will soon become part of some multilateral open skies agreements, as suggested by the APEC Agreement of 2001 and the US-EU Open Skies Agreement (Section 2.3.4).[8]

Airline Designation

Under *traditional* agreements, each country may typically designate only one of its airlines for the right to operate flights between any specific pair of cities. Double designation may be allowed for major markets with high volumes of demand, e.g., New York–London, under more "liberal" ASAs such as Bermuda II. Under *open market* and *open skies* agreements, multiple designations are permitted – and indeed are the rule, when feasible.

Implicit in all this is the association of each airline with a "nationality" ("home State A") – a practice which has remained unchanged throughout the modern history of air transport. "Substantial ownership and effective control by the State's nationals" is still the basis for qualifying an airline for designation by its home State in an ASA, as well as for allowing an airline to operate domestic flights. Every State, with very few exceptions, specifies limits on "foreign ownership," i.e., on the percentage of shares of its airlines that can be owned by foreign nationals. Table 2.1 summarizes these restrictions for a number of countries. As can be seen, the USA has one of the most restrictive sets of requirements: not only is foreign ownership limited to 25% of the shares, but also at least two-thirds of the members of the board of directors, including its chairman/woman, must be US nationals. Various proposals over the years to relax these requirements have been rejected through the political process in Washington.[9]

The importance of "nationality" in the global airline industry is something of an international anomaly. In most other major global industries, the nationality of the shareholders and of the directors of companies operating across national borders has become largely immaterial over the years. The exceptional treatment of airlines in this respect is mostly due to two (real or perceived) concerns: their potential role at times

---

[7] The rationale for this obvious asymmetry has been that the USA offers a very large number of attractive origin and destination airports, whereas the other signatory country often offers only a single one (e.g., as in the cases of Singapore or The Netherlands) or, at most, very few. It is therefore reasonable, according to this argument, to limit access by foreign airlines to a small number of mutually agreed airports in the USA.

[8] The term "open area" is now increasingly used to refer to multilateral agreements that include Seventh Freedom rights, in addition to Fifth Freedom. This distinguishes such agreements from open skies agreements.

[9] See, however, Section 3.4 regarding the 2007 US–EU Open Skies Agreement.

Table 2.1 Restrictions regarding foreign ownership of airlines in selected countries

Country	Limits on foreign ownership
Australia	49% for airlines engaged in international operations (25% on any single shareholder); 100% for solely domestic airlines
Canada	25%
China	35%
Chile	100% as long as airline's principal place of business is in Chile
European Union	49%, applies to non-EU citizens
India	49%, but foreign airlines cannot hold shares in Indian airlines
Japan	33.33%
Korea	49%
Malaysia	45%
New Zealand	49% for airlines engaged in international operations; 100% for solely domestic airlines
Singapore	27.51%
Taiwan	33.33%
Thailand	30%
USA	25%; one-third of the board of directors; chairman/woman must be US national

Source: Based on Chang and Williams (2001) with updates/additions.

of national emergency or military mobilization; and the potential loss of highly prized airline jobs to foreign workers. Another factor may be the lingering fondness in many countries for the notion of "flag carriers."

In light of the movement toward a liberalized regulatory environment for air transportation, the realities of globalization, and the ease with which capital can flow into ventures around the world, there is growing international pressure to ease restrictions on foreign ownership of airlines, which many regard as an anachronism. For example, Chile and, for airlines engaged in purely domestic operations, Australia and New Zealand have removed all restrictions on the percentage of foreign-owned shares (Table 2.1). Additionally, several airlines, such as SAS and Gulf Air, have ownership and principal loci of business that are distributed among groups of nations.[10] And, most important in terms of impact, the concept of the "community carrier" (Section 3.4) allows any group of EU nationals to own and operate an airline registered in any one of the (currently) 27 EU Member States.

Capacity Constraints

Under *traditional* agreements, strict control is exercised on the frequency of flights and the number of seats offered in each one of the served markets. Typically, in the least "liberalized" agreements, the capacity provided is allocated on a 50–50 basis to each

---

[10] Another interesting case is Aerolineas Argentinas, which has been owned at various times since the early 1990s by an often-changing combination of foreign investors, banks and airlines. Yet this airline has been consistently treated as an Argentine carrier for ASA purposes.

side's designated airlines. Even more remarkably, revenues from each market are, in some cases, pooled and shared in proportion to the capacity offered, irrespective of how many passengers actually fly on each of the designated airlines! In more liberal versions of traditional agreements, more flexible capacity controls (or even no capacity controls) are applied, but provisions are typically made for possible government reviews aimed at protecting airlines that may find themselves "at a disadvantage" due to more capacity being offered by the other airline(s) designated to serve the same markets. Revenue sharing is not an element of these more liberal versions.

Under *open market* agreements, no restrictions are imposed on the capacity an airline may offer in a market. An interesting feature of some of these agreements is that they permit a change in aircraft type on Fifth Freedom flights. For example, under an open market regime, a US airline might operate a New York–Frankfurt–Athens flight, with Fifth Freedom rights on the Frankfurt–Athens leg, and use a 400-seat Boeing 747 on the New York–Frankfurt leg and a 130-seat Boeing 737 on the Frankfurt–Athens leg, with both legs having the same flight number. The change in aircraft "gauge" might be critical in making the Frankfurt–Athens leg (and possibly the entire New York–Frankfurt–Athens service) economically viable.

Under *open skies* agreements, no capacity restrictions are likewise imposed. A change of aircraft gauge on Fifth Freedom flights is also typically permitted. To add further flexibility, code sharing is also often permitted, so that consecutive legs of the flight may actually be flown by two different alliance partners.

Airfares/Tariffs

Under *traditional* agreements, all airfares are subject to approval by both governments involved ("double approval"). Airfares are often determined through an agreed-upon "cost plus profit" formula. In such cases, use of IATA's tariff-setting procedures (Section 2.6) is sometimes encouraged or required in the text of the ASA.[11]

Under *open market* agreements, a proposed airfare cannot be rejected unless both governments disapprove ("double disapproval"). This allows for price cutting by airlines as long as at least one of the governments does not object. In some cases, double disapproval is replaced by "country of origin rules," i.e., airfares are subject to approval rules in force in the country whence the trip originates.

Under *open skies* agreements, airfares are not subject to government approval with the exception of the threat of potential government intervention in cases of alleged predatory or monopolistic practices.

In conclusion, the transition from *traditional* to *open market* and then to *open skies* agreements clearly marks a movement toward increased liberalization with respect to each of the four principal aspects of bilateral and multilateral ASAs. The USA has played a central role in this process, initially by negotiating the first "liberal" version of traditional agreements (Bermuda II) and subsequently the first "open market" and "open skies" agreements. Existing policies that severely restrict foreign ownership of US airlines are at odds with this otherwise strongly "pro-market" record.

---

[11] The USA prohibits use of these procedures, which it considers as anticompetitive and illegal.

2.3.4   The Unified European Union Market and Other Major Developments

We turn next to three further important examples of increasingly relaxed regulation of international air transport markets. First and foremost, the *liberalization and, essentially, unification, of the air transport market within the European Union* marks a regulatory breakthrough that goes well beyond the content of any existing multilateral ASA. The global significance of the EU agreement is comparable to that of the 1978 deregulation of the domestic US market.

In contrast to the overnight change that took place on January 1, 1978 in the USA, liberalization in the EU took place in several steps over a period of more than 10 years as a result of three "packages" of measures that were approved by Member States of the (then) European Economic Community in 1987, 1990 and 1992. The last of these ("Third Package") was implemented gradually, beginning on January 1, 1993, when its most important provisions went into effect, and ending in 1997 with activation of Ninth Freedom rights for EU airlines. When the EU was expanded from 15 to 25 Member States in 2004 and then to 27 in 2007, the newly admitted countries automatically became participants in the unified intra-EU market that now encompasses close to 500 million people.

The principal regulatory provisions that currently apply to the intra-EU market are based in large part on the notion of the *community air carrier* or simply "community carrier," defined as *any carrier registered in an EU Member State*, thus necessarily satisfying the airline ownership requirements of the EU. These provisions are remarkably simple and designed to transcend nationality distinctions: any group of nationals of one or more EU Member States may apply to establish an airline within any EU Member State, as long as the group holds 51% or more of the voting stock. Such applications must be approved by that State, unless the proposed airline does not meet the requisite technical qualifications.

The main provisions currently governing the EU unified market can be summarized as follows:

1. Market access and airline designation: From the very beginning (January 1, 1993) of the applicability of the Third Package *all* community carriers were granted full Seventh Freedom rights (and, obviously, Fifth and Sixth, as well) for flights *within the EU*. Thus, Air France may provide service between, for example, Milan and Barcelona, without the flight having to originate from or terminate at an airport in France. Eighth Freedom rights ("consecutive cabotage") were also granted at the time, thus permitting the transport of domestic passengers on continuations of intra-EU international flights, but for only up to 50% of the aircraft's capacity.[12] However, all cabotage restrictions were gradually removed and EU air carriers today enjoy full Ninth Freedom rights. For example, easyJet, registered in the UK, has become an important competitor in the large market between Paris and Nice in France.
2. Capacity: There are no restrictions on the capacity offered by an airline on any market.
3. International airfares: Essentially no restrictions exist. Proposed airfares can be challenged only if they are judged by one of the States involved to constitute a case of

---

[12] For example, a Lufthansa flight from Munich to Milan, Italy, and then to Naples, Italy, could fill up to 50% of the aircraft's seats with Milan-to-Naples passengers on the Milan-to-Naples leg of the flight.

monopolistic or of predatory pricing. A process of appeal to the EU Commission is available in such cases. The Commission is the ultimate arbiter of airfare disputes.

4. Other provisions: A number of other provisions are in place, primarily for the purpose of preventing anticompetitive practices. They mandate "strict enforcement" against: collusion between carriers, e.g., leading to fare fixing or capacity fixing; anticompetitive joint ventures by airlines; discrimination in computer reservations systems (CRSs); anticompetitive mergers or takeovers; and subsidies to airlines by national governments.[13]

A second, important, recent development in international liberalization has been the multilateral open skies Asia–Pacific Economic Community (APEC) Agreement, signed in May 2001 (Findlay, 2003). The parties to the agreement were the USA and five APEC States (Singapore, Brunei Darussalam, Chile, New Zealand and Peru). Bilateral agreements between most pairs of these countries, mostly with liberal provisions, already existed before 2001, but were all merged into and superseded by the multilateral agreement. The APEC Agreement provides for First through Sixth Freedom rights for all signatory States. There are no approval requirements for airfares. Code sharing among airline alliance partners is permitted on all flights within the scope of the agreement. All-cargo flights also enjoy Seventh Freedom rights within this group of nations. There is no provision, however, for Eighth and Ninth Freedom rights.

The APEC ASA deals with the issue of airline designation in an interesting way. It stipulates that each State may designate any set of airlines that have their principal locus of business in that State; however, each State may impose more restrictive requirements of its own regarding eligibility for designation. In this way, very liberal foreign ownership requirements (e.g., those of Chile – see Table 2.1) can be reconciled with more restrictive ones (e.g., of the USA).

Each party to the APEC ASA is also free to negotiate independent bilateral or multilateral agreements with third parties. Finally, the Agreement stipulates that other States can be invited in, thus allowing for the possibility that participation in one of the most liberal existing multilateral ASAs will grow in the future.

A third international regulatory breakthrough was prompted by a landmark decision on November 5, 2002 by the European Court of Justice. The Court found that the nationality clauses guiding airline designation that were contained in then-existing bilateral agreements between individual EU Member States and the USA were in violation of EU legislation (i.e., of the Third Package, see above) which mandates that all community carriers should receive equal treatment in all Member States.[14] At the time, 11 of the (then) 15 Member States of the EU had open skies agreements with the USA. All existing agreements were thus declared invalid[15] and the Court authorized the European Commission to negotiate a new multilateral agreement with the USA on behalf of all EU Member States.

---

[13] The Commission typically reviews instances of such alleged violations on a case-by-case basis. Experience has shown that, in practice, adjudication is time consuming. The internal consistency of some of the Commission's decisions regarding government subsidies has also been questioned widely.

[14] The argument is that, for example, a US–France ASA, which allows only US and French airlines to fly between New York and Paris, discriminates against other EU community carriers in that it deprives these other carriers of rights that French airlines enjoy.

[15] The then-existing agreements remained temporarily in effect, until March 2008 – see below.

Negotiations toward this goal began in November 2003 and concluded in April 2007 with a *provisional* open skies agreement between the EU *as a whole* and the USA that took effect in March 2008.[16] The negotiations were extremely difficult and complex, as the Commission had to take into account the often diverging interests of each of the EU's 27 different Member States, while the US negotiators had to cope with a shifting and often skeptical political environment in Washington. Moreover, issues of general principle had to be resolved in parallel with such specific points of controversy as the question of London Heathrow slots (Section 2.4) that had been a long-standing cause of conflict between the USA and the UK.

The provisional agreement contained several landmark clauses. Most important is the fundamental provision that any US airline or community air carrier can serve any city pair between the USA and the EU. Fifth Freedom rights were also granted to both sides, for continuation of flights between the EU and the USA to third countries. Importantly, the continuation of a flight can be performed by an alliance partner of the airline that flew the US–EU part. Community air carriers also obtained Seventh Freedom rights for services between cities in the USA and certain non-EU European States. Seventh Freedom rights were also granted to all-cargo flights performed by Community air carriers between the USA and third countries and, similarly, for certain all-cargo flights performed by US carriers between the EU and third countries.[17] At the same time, the 2007 provisional ASA did not resolve some of the most contentious issues in the negotiations.[18] These will presumably be resolved in a future permanent agreement.[19]

Despite some probably excessive claims regarding the potentially enormous economic benefits it will generate and its job-creation potential, there is little question that the US–EU Open Skies Agreement will indeed stimulate passenger and cargo traffic growth in the North Atlantic market through more competition, lower fares and a significant increase in the number of city pairs served. Equally important, the US–EU Agreement offers a model for other similar future agreements, possibly with even more liberal provisions. For instance, negotiations between the EU and Canada were initiated for this purpose.

2.4 Airports

The global system of airports constitutes one of the two principal components of the world's aviation infrastructure, the second being the world's air traffic management systems. As described in Section 2.3, regulatory restrictions on airline competition in

---

[16] Because the agreement treats the EU as a single entity, it is generally referred to as a *bilateral*, not multilateral, ASA.

[17] The agreement also contains numerous provisions of a more technical nature. Most important among them is a granting of broad antitrust immunity in a number of matters to airline alliance partners on the two sides of the Atlantic.

[18] Chief among these is the issue of relaxing the restrictions that the USA imposes on foreign ownership and control of US carriers (Table 2.1). The Bush administration made a number of attempts to move in this direction, but then backtracked in the face of strong opposition in the US Congress, which was, in turn, prompted by the objections of labor unions and of some airlines. The US Department of Transportation has suggested that the 25% limit should apply only to "effective control" by foreigners and not to share ownership – for which the limit should be raised to 49% as requested by the Europeans.

[19] Negotiations toward a permanent agreement began in 2008, but had not been fruitful as of this writing. However, the provisional agreement of 2007 would not seem to be in jeopardy for the foreseeable future.

international and domestic markets around the world are gradually receding and have largely been eliminated in certain important geographical regions, contributing to air traffic growth. Concurrently, major airports everywhere have become increasingly congested, as traffic has grown rapidly over the years. To prevent further congestion and delays, restrictions have been imposed on access to many of these airports. Because of the capacity allocation procedures used in this connection, airport access is now threatening to turn into a new form of market regulation and an impediment to the functioning of a competitive marketplace. This is the subject of Section 2.4.1. Section 2.4.2 addresses a second major development that has been taking place at major airports since the late 1980s: the movement toward privatization of airport ownership and management. This is a trend that has generated widespread institutional and organizational change, as well as international interest in relevant regulatory measures.

2.4.1   Restrictions on Airport Access

Airport capacity is widely believed to be one of the most important long-term constraints on the growth of air traffic. While any element of an airport (taxiway system, apron stands, passenger buildings, etc.) can be a capacity "bottleneck," it is usually the system of runways that acts as the ultimate constraint,[20] especially at the world's busiest and most important airports.

Airport capacity constraints are reflected in the long-term upward "creep" of the average airport-related delay experienced by flights in the parts of the world where air transport is most developed. Just as important, the high variability of runway system capacity with weather (cf. Chapter 13) creates an unstable operating environment at those airports where traffic demand is close to capacity even under good weather conditions: delays become long and schedule reliability deteriorates on days when weather conditions are less than ideal.

One obvious way to deal with these problems, at least in the short and medium run, is through *demand management*. This term refers to any set of administrative and/or economic measures and regulations aimed at *restraining the demand* for access to an airport during certain parts of the day and/or of the year, when congestion might otherwise be experienced. Demand management is a "proactive" approach, in the sense that the associated measures and regulations are set months in advance and for extended periods (e.g., for six months at a time). This is different from the more dynamic (or reactive) *air traffic flow management* measures, such as delaying aircraft on the ground before takeoff or rerouting them when airborne, which are taken on a daily basis by ATM service providers (e.g., the FAA in the USA, EUROCONTROL in Europe) to address the specific demand/capacity relationships that prevail on any given day[21] and to alleviate congestion on that particular day.

Airport demand management is practiced very extensively outside the USA. The fundamental mechanism used for this purpose is the *Schedule Coordination* (SC) process that has been developed by the International Air Transport Association (IATA) and is applied

---

[20] See Chapters 12 and 13 for a more detailed discussion of airport capacity and congestion.

[21] Demand management and air traffic flow management, far from being mutually exclusive, can work in a complementary way, with demand management restricting demand to a priori reasonable levels (compared to the typical capacity of an airport) and air traffic flow management "smoothing out" everyday problems.

almost universally with some local or regional variations. The description of this process provided below omits many important details that can be found in IATA (2005) or in de Neufville and Odoni (2003).

The SC process classifies certain airports as *fully coordinated*. A fully coordinated airport is one "where demand exceeds capacity during the relevant period[22] and it is impossible to resolve the problem through voluntary co-operation between airlines" (IATA, 2005). At these airports, the task of resolving scheduling conflicts and allocating available "slots" among the airlines that have requested them is assigned to a *schedule coordinator*, who is typically supported by a small staff and, in many cases, a committee of experts and stakeholders. A *slot* is an interval of time reserved for the arrival or the departure of a flight and is assigned to an airline or other aircraft operator for a specified set of dates, e.g., "Monday through Friday of every week of the upcoming season." Two slots are required for an airline to operate a flight into and out of an airport. Once a slot is allocated, it typically becomes associated with a specific flight operation, e.g., "arrival of Flight 1234 of Airline X." As of 2006, approximately 150 airports worldwide were designated as fully coordinated, including the great majority of the busiest airports of Europe, Asia and the Pacific Rim.

To participate in the SC process, each fully coordinated airport must specify a *declared capacity*, which indicates the number of slots available per hour (or other specified unit of time), i.e., the number of aircraft movements that the airport can accommodate in each such interval. Responsibility for determining the declared capacity of each airport rests with local and national authorities. The capacity that an airport declares does not have to be dictated solely by the capacity of the runway system. Constraints due to the availability of aircraft stands, passenger processing capacity and even aircraft ramp servicing capacity can be taken into consideration.

The SC process is repeated every six months during conferences that take place in November for the allocation of slots for the following summer season and in June for the following winter season. Airlines must submit, in advance of each conference, requests for every slot they are seeking at each coordinated airport. The schedule coordinator then allocates the declared capacity according to a set of criteria. Among these, the principal and overriding criterion is *historical precedent*: if an airline was assigned a slot in a particular season ("summer" or "winter") and utilized that slot for at least 80% of the reserved times during that season, then that airline is automatically entitled to continued use of that "historic" (or "grandfathered") slot during the respective next season. The recipient of a slot awarded on the basis of historical precedent may, in fact, use it in the new season to serve a different destination from the one served in the previous season. Second priority is assigned to requests for changing the time of historic slots. Slot exchanges between airlines, on a one-for-one, non-monetary basis, are also allowed, as are short-term leases to code-sharing partners.

Slots that were not used at least 80% of the time during the previous season or are abandoned by their holders are returned to a "slot pool" for reallocation. Any new slots made available through increases in airport capacity are also placed in the slot pool. All requests for new slots are then served from this slot pool. To encourage the development

---

[22] This means a period deemed sufficiently long (e.g., several hours during a typical day of the upcoming season – see below) to warrant taking measures to prevent unacceptable congestion.

of additional competition at airports, at least 50% of the slots in this pool are designated for assignment to airlines qualifying as *new entrants*. However, the definition of a "new entrant" is very restrictive: an airline qualifies for this designation as long as it does not hold more than four slots in a day, *after* receiving any new slots from the slot pool (IATA, 2005). Thus, at best, a new entrant can be awarded two flights (or four runway operations) per day, hardly sufficient to establish a significant foothold at a major airport.[23]

In summary, SC, as practiced under the IATA rules outlined above, is based entirely on an administrative procedure and is almost completely removed from economic considerations: prospective users who may attach a high economic value to slots at any given airport(s) have no assurance that they will be able to obtain them at any price. The SC approach may then work reasonably well in instances where demand exceeds airport capacity by a relatively small margin during a few non-contiguous hours in a day. Under such conditions, a limited number of airline schedule adjustments, effected through judicious slot assignments, may lead to a smoother daily demand profile and thus to more reliable and less delay-prone flight operations at the airport. However, the SC approach may distort seriously the workings of the marketplace whenever a significant excess of demand over capacity exists. In such cases, the current, purely administrative slot allocation rules may act, in effect, as a form of market regulation by default. This may well be the case already at some of the most important (and seriously congested) European[24] and Asian airports. Through its heavy reliance on historical precedent, the SC process largely protects the status quo by preserving the dominance of older airlines – in most instances the former "flag carriers" – at their principal national and regional airports.

In the USA there are essentially no airport- or government-mandated restrictions on access to commercial airports.[25] A partial exception were four "high density rule" (HDR) airports – New York's LaGuardia and Kennedy, Chicago's O'Hare and Washington's Reagan – that, for historical reasons, had been operating since 1968 under strict limits regarding the number of movements that can be scheduled per hour.[26] In April 2000, the US Congress mandated the phasing out of the HDR by January 1, 2007.[27]

---

[23] In 2004, the EU relaxed considerably the IATA definition of "new entrant," as it applies to EU airports: a new entrant can hold up to 5% of the slots at an individual airport and must satisfy a number of conditions beyond those required by IATA (see EU Commission, 2004).

[24] A recent survey of European airports (Madas, 2007) indicated that more than 75% of all available slots during the *winter season* at the main airports of Amsterdam, Frankfurt, Helsinki, London (Heathrow and Gatwick), Manchester, Milan (Linate and Malpensa), Munich, Paris (de Gaulle and Orly) and Stockholm were "grandfathered" and retained from year to year by the same airlines. Note that the 75% refers to the share of *all* the slots available during an entire week. When it comes to the most desirable times for scheduling flights, close to 100% of all slots at many of these airports are grandfathered.

[25] Partly because of the absence of any slot restrictions, air traffic delays at major US airports may be significantly higher, on average, than those at their counterparts elsewhere. The principal constraint that airlines face when wishing to schedule additional flights at a US airport concerns the availability of terminal gates, aircraft stands and other terminal facilities.

[26] A form of slot coordination was practiced at these four airports until 1985. At that time, the approach was abandoned in favor of a "buy-and-sell" regime, which made it possible for incumbent slot holders to sell slots, individually or in groups, to new owners.

[27] Congress mandated that the HDR will continue to be applied at the Washington Reagan Airport beyond 2007; in addition, due to severe congestion problems, the FAA has maintained caps on flight scheduling at Chicago, LaGuardia and Kennedy Airports even after January 1, 2007.

In response to the fundamental deficiencies of the SC approach – and to complaints from airlines which have been partly excluded, in effect, from some of the most attractive markets in the world – the European Commission and a number of national governments have been considering alternatives that would (partially or fully) replace or supplement the "historical precedent" criterion through use of so-called *market-based mechanisms*.[28] Prominent examples of such mechanisms include *congestion pricing*,[29] the creation of *slot exchange markets* ("buying and selling" of slots) and the *auctioning* of airport slots. In the USA, the Department of Transportation, the FAA and a few individual airport operators have also given serious consideration to some of these alternatives. But, despite numerous studies[30] and several specific implementation proposals, practical experience with market-based mechanisms is scarce to date. Some forms of congestion pricing have been applied in a number of airports over the years. Somewhat more extensive experience has also been obtained with slot trading, primarily at the HDR airports in the USA since 1986. Little or no practical experience exists with slot auctions, outside of some simulation games. Nonetheless, pressure has been mounting worldwide in the directions of developing alternatives to IATA's SC approach and of injecting economic criteria into the process of allocating access to the busiest commercial airports – for a comprehensive review see Czerny *et al.* (2008).

Finally, it should be noted that, to a lesser extent than SC, certain other airport-related practices currently act as impediments to market access and thus to liberalization and competition. A list would include: various government-imposed legal and administrative requirements for use of airports; some types of customs and currency restrictions; excessive user charges and fees; lack of adequate aircraft maintenance and repair facilities and of other technical support; and access to, quality and cost of airport ground handling services.

2.4.2 Airport Ownership and Management

Until the mid-1980s, practically all airports with scheduled airline service were owned by government[31] and operated either by government organizations or, in a small number of cases, by independent companies called "airport authorities." This is not surprising as airports perform a public service by providing infrastructure for air travel and were thus considered a government responsibility. Moreover, until the 1980s, even the busiest airports in the world had to rely heavily on direct or indirect government subsidies, especially when it came to capital expenditures.[32]

However, beginning with the privatization of the British Airports Authority in 1986, there has been growing private participation in airport ownership. Equally important, private-sector management practices are rapidly replacing traditional government-style management of airports. These changes have been brought about in response to a

---

[28] See de Neufville and Odoni (2003) for a detailed description and review.

[29] For classical papers on congestion pricing, see Vickrey (1969) and Carlin and Park (1970).

[30] For recent ones, see, for example, DotEcon (2001), NERA (2004) and Ball *et al.* (2007).

[31] The owners were either the national government or a regional or a local governmental entity (single municipality or a group of municipalities in a metropolitan area) or some combination thereof.

[32] This is still the case in most less developed economies.

confluence of factors, each of which applies to varying degrees to different groups of airports. These include the following:

1. The realization that, with continuing traffic growth, airports have become a "mature" industry, able in many cases to achieve economic self-sufficiency,[33] including paying for their capital costs.
2. The potential of ever-expanding non-aeronautical and commercial activities at airports to generate large revenues and profits for airport operators.
3. Pressure from the airlines and other airport users, including passengers, for operational and economic efficiencies.
4. Fast-changing conditions in the air transport sector that require rapid response mechanisms and management flexibility on the part of airports.
5. The inability of many governments to raise the large amounts of capital often required to finance the development of major airport infrastructure.
6. Competition for traffic among some types of airports, especially major "hubs" handling large numbers of connecting passengers.

After a slow beginning, the airport privatization movement accelerated on an international scale, reaching its apex during the period 1997–2000. A number of private or semi-private investors – companies or individuals – acquired majority or minority interests at many individual airports or groups of airports. The trend slowed down in the few years immediately after September 2001, but regained full speed by 2005. The investors come from a variety of fields and include: operators of some large international airports (mostly European); construction companies; providers of transportation services and of airport handling services; and banks and other financial institutions. A typical example was the 1998 agreement through which a consortium of companies took control of some 30 major and secondary airports in Argentina – practically all the significant airports in the country – for a period of 25 years in exchange for a sizable annual payment to the Argentine government and a commitment to make certain infrastructure investments in these airports. The consortium consisted of the Milan (Italy) Airport Authority, an Argentine construction company and a US company specializing, among other areas, in airport handling services.[34]

Quite often, privatization takes the form of a public–private partnership and involves a "build, operate and transfer" (BOT) contract. For example, in 1996, the government of Greece signed a 30-year BOT agreement with a consortium of German companies for the construction and operation of the new Athens International Airport, which opened in March 2001. The government acquired 55% of the shares of the company that was created for this purpose and the German consortium 45%.

Beginning with the 1986 privatization of the BAA, shares of many airport operating companies have also been offered to small investors in the world's stock exchanges, with

---

[33] Traffic volume is a key parameter in this respect. Current conventional wisdom has it that airports with 3 million or more passengers per year can largely pay their own way, if managed properly. In truth, the traffic level at which self-sufficiency can be reached depends strongly on local circumstances.

[34] Each of the three partners held one-third of the shares initially; as is often the case with airport privatization ventures, the composition of the consortium has changed since, as has the allocation of shares.

the fraction of shares floated ranging from a few percent to 100%. Overall, more than 150 airports worldwide,[35] including many of the most important ones outside the USA, are currently privatized to some extent.

Perhaps surprisingly, the USA has not been a part of the airport privatization movement, as large commercial US airports are, by law, owned by state or local government.[36] On the other hand, it is true that US airports are among the most "privatized" in the world, in the sense that they outsource most of their financing, planning and operating activities to private companies. The emphasis on outsourcing is reflected in the relatively small number of persons employed directly by the operators of major airports in the USA.[37] The operators largely confine themselves to managing airport assets and projects and to overseeing and monitoring the day-to-day operations of the airlines, concessionaires, etc., on the airport's premises.[38] Based on the successful "US model," a worldwide trend toward outsourcing airport services has developed since the 1990s.

The privatization movement has led to growing awareness of the need for regulatory safeguards to protect the public interest from potential conflicts with the goals of private airport investors. The possibility of such conflicts is becoming increasingly clear, as experience with privatized airports accumulates. For example, private investors may be less willing than public entities to invest in capacity expansion projects that may entail large capital expenditures. They may also be more likely to increase charges at an airport from year to year, if given the latitude to do so.

One major goal of airport economic regulation is to prevent monopolistic practices[39] on the part of airport operators in pricing access to aeronautical facilities and services. The principal means of such regulation to date has been the setting of (a) limits on the rates of return on assets that airport operators may earn, and (b) various types of caps on year-to-year changes in the unit prices charged for aeronautical facilities and services – see Chapter 12 for more details.

Finally, airport privatization has given added impetus to two other important trends in airport management and operations: (a) the increased emphasis on the provision of commercial and other non-traditional services, resulting in fast-growing non-aeronautical revenues; and (b) ventures and investments away from the airport operator's home base, which sometimes include activities, such as real-estate development, that are largely unrelated to air transport. As a consequence, the organizational structure and "business models" of many airport authorities have been changing rapidly in ways designed to accommodate these new operating environments and activities.

---

[35] In case this number sounds small, note that there are only about 350 airports in the world with more than 2 million passengers annually.

[36] Airport privatization is permitted, but requires a lengthy and difficult approval procedure, including paying back to the government all federal airport development grants that the airport may have received in the past.

[37] The numbers are particularly striking when compared to those of some overseas airport operators who also engage in aircraft and passenger handling (Chapter 12). US operators do not engage in such activities but outsource these services to the airlines themselves ("self-handling") or to specialized companies.

[38] A small number of US airports have subcontracted their airport management functions to specialized companies, which, in several cases, are based outside the USA.

[39] With a few exceptions, airports are natural monopolies when it comes to origin/destination traffic from/to any particular city or metropolitan area.

2.5 Air Traffic Management

The institutional and regulatory environment in which providers of air traffic management (ATM) services operate has also been undergoing important changes. Traditionally, ATM services[40] have been provided by national civil aviation authorities, i.e., agencies of national governments that are subject to the usual budgetary processes, civil service rules and other constraints. In addition to providing ATM services, civil aviation authorities have historically been charged with such diverse tasks as regulating user charges for ATM services and airport services, overseeing and monitoring aviation safety, investigating accidents, collecting user charges from airspace users, and even operating airports.[41] The appropriateness of assigning to a single agency responsibility for all these tasks – which often generate internal conflicts of interest – has been questioned extensively.

The provision of ATM and related services has become an increasingly complex and demanding task for several reasons:

- The provider must make major investments in human resources and capital assets (facilities and equipment).
- ATM systems must keep pace with rapid technological change and advances.
- ATM personnel, especially air traffic controllers, are generally thought of as deserving compensation and other benefits commensurate with the specialized nature of the tasks they perform: civil service regulations and pay scales are usually not sufficiently flexible to accommodate such special treatment.

As a result, a growing number of developed countries have resorted to "corporatizing" (or "commercializing") the provision of air navigation services, i.e., have entrusted what was previously a government service to *autonomous corporate entities that largely operate according to private-sector principles*. Beginning with Switzerland in 1991 and Germany in 1992, a total of 38 countries, as of 2005, have set up national corporations for this purpose. These countries include most EU Member States, as well as Australia, Canada, New Zealand and South Africa, among others. The new corporate entities, often referred to as Air Navigation Service[42] Providers (ANSPs), are thought to be better able than regular government agencies to cope with the special demands mentioned above. Because of the worldwide growth of air traffic, many ANSPs are also in a position to be self-sufficient economically: they can impose user charges sufficient to cover costs, including a reasonable return on net assets, and can adjust these charges over time in response to traffic developments and needs.

The corporatized ANSPs are generally *wholly government owned*. This, however, has not been the case in two important recent instances (GAO, 2005). NAV CANADA, established in 1996, is a privately owned corporation that purchased Canada's ATM system from the national government for approximately $1.5 billion. Its board of directors

---

[40] Historically, "ATM services" were called "air traffic control (ATC) services" (see Chapter 13).

[41] The exact set of responsibilities of civil aviation authorities varies widely from State to State.

[42] "Air navigation services" is a more formal term for "air traffic management"; the term covers "air traffic control" and all related services aimed at ensuring the safe and efficient movement of air traffic though airspace and on the ground at airports.

includes three government representatives among its 15 members. Airlines, general and business aviation and NAV CANADA's employees (air traffic controllers and engineers) also hold board seats. In the UK, National Air Traffic Services Ltd. (NATS), established in 2001, is a public–private partnership in which the government holds 49%, airlines 42%, NATS staff 5% and the BAA 4%. Germany has also announced plans to sell about 75% of DFS, its ANSP, to private investors.

As one might expect, corporatized ANSPs are subject to strict economic regulation, because they are natural monopolies. For example, in the UK, the Civil Aviation Authority's Economic Regulation Group sets caps for five-year periods on the fees that NATS can charge for its services, while in Germany the Transport Ministry must review and approve any changes in user fees (GAO, 2005). Other duties of ANSP regulators, according to ICAO, should include ensuring that: user fees are non-discriminatory; the services offered are of high quality; investments necessary to meet future demand are made; and users are regularly consulted regarding the quality and cost of services.

In another important recent development, the EU approved in March 2004 the Single European Sky (SES) initiative whose overall aim is to restructure European airspace, create additional capacity, and increase the overall efficiency of the ATM system.[43] As part of this initiative, the EU mandated the separation of the regulatory functions of civil aviation authorities from the provision of air navigation services.[44] The EU requires the creation of distinct bodies, called national supervisory authorities (NSAs), to perform the regulatory supervision of ANSPs:[45] "The national supervisory authorities shall be independent of air navigation service providers ... . Member States shall ensure that national supervisory authorities exercise their powers impartially and transparently." The ANSPs[46] themselves "should be organized *under market conditions* whilst taking into account the special features of such services and maintaining a high level of safety." Each ANSP must prepare a business plan, update it regularly, and consult with users and other stakeholders regarding performance targets and user charges.

Finally, a truly path-breaking feature of the SES initiative is that it offers Member States the possibility of contracting with any EU ANSP for the provision of air navigation services in that State's airspace. For instance, Poland may contract with DFS, the German ANSP, to provide air navigation services in Polish airspace; or, a joint German–Polish corporation might be created to provide air navigation services in German and Polish airspace and possibly in the airspace of other Member States as well. This feature is intended to encourage efficiency in the provision of air navigation services, as well as some consolidation of the ANSPs in Europe.[47]

2.6   Key Organizations and Their Roles

A large number of public or private institutions and organizations play a role in shaping policies regarding economic, regulatory and technical matters concerning the air transport

---

[43] The EU's four SES Regulations, 549/2004, 550/2004, 551/2004 and 552/2004, are available at http://www.eurocontrol.int/ses/public/standard_page/sk_regulations.html

[44] SES Regulations 549/2004 and 550/2004.

[45] Article 4, SES Regulation 549/2004.

[46] Preamble of SES Regulation 550/2004.

[47] In 2008, Ireland and the UK signed the first such agreement for the joint provision of air navigation services.

sector. This section presents brief descriptions of a selected (and far from comprehensive) group of these institutions and organizations, which are particularly influential internationally (Section 2.6.1) and in the USA (Section 2.6.2). For more details, the reader is directed to the respective websites, some of which are very informative.

2.6.1 International Organizations

On the international side, this section provides thumbnail sketches of the International Civil Aviation Organization and the International Air Transport Association, as well as brief mentions of a number of other airport-related and airline-related organizations. It also summarizes the current roles of a number of influential EU-level agencies.

International Civil Aviation Organization – ICAO

As noted in Section 2, ICAO was established in 1947 in response to the recommendations of the Chicago Convention. In addition to its world headquarters, located in Montreal, Canada, ICAO maintains seven regional headquarters around the globe. ICAO can be likened to a "United Nations of Civil Aviation" – and, in fact, its official status is that of a specialized agency of the UN. Its many important functions include the development, approval and updating of international technical standards and recommended practices for airports and air traffic control (ATC), as well as the preparation and publication of broad regulatory guidelines and of economic and environmental policy statements regarding international air transport. Such statements are often expressed in quite broad and unspecific terms.

As of 2007, ICAO had 190 Member States, i.e., it included practically every nation in the world engaging in aviation activities of any significant level. All Member States participate in the ICAO Assembly which meets every three years. The Assembly, among its other responsibilities, elects a Council which consists of 36 Member States and has a three-year term. The Council meets on a regular basis and exercises most of the policy-making and overseeing prerogatives of ICAO. Like the Security Council of the United Nations, the Council has some permanent members and others which change every three years. Permanent members are countries which have large air transport sectors and/or have traditionally been influential in the sector. Day-to-day tasks and responsibilities are carried out by a permanent Secretariat that has a large staff.

ICAO has concentrated primarily on technical regulation over the years, developing a set of 18 "Annexes" that contain international standards and recommended practices on such matters as aviation safety and security, airport design standards and air traffic management requirements, facilities and equipment. It also organizes occasional conferences on economic issues and regulation and publishes "statements" that contain general guidelines in these respects.

All bilateral and multilateral ASAs (Sections 2 and 3) must be registered with ICAO. About 4000 such ASAs are currently on file.

International Air Transport Association – IATA

IATA is the trade association of most of the international airlines in the world, with headquarters in Geneva, Switzerland, and Montreal, Canada. In 2008, membership stood at about 230 airlines from 140 different countries that accounted for about 93% of total *international* scheduled passenger traffic.

IATA was founded in 1945, the year after the Chicago Convention, primarily in response to the perceived need to coordinate international airfares, a topic not addressed by the Convention. Indeed IATA assumed responsibility for this task and coordinated the setting of international airfares during annual "traffic conferences." By extension, the organization played a critical role in the development of international air transportation over more than three decades. However, in 1978, the US government concluded that coordinated airline price setting was anticompetitive[48] and asked IATA to "show cause" why "the antitrust immunity that the IATA tariff-setting machinery had enjoyed for the previous 33 years" should not be terminated (Shane, 2005).

In response, and after considerable international skirmishing, IATA adopted a two-tier organizational structure, which is still in existence today. At one tier, it operates as a trade association offering various technical, legal and financial services. Examples include: defining "conditions of carriage", i.e., the legal responsibilities of carriers vis-à-vis passengers and cargo; advising airlines regarding such issues as the transportation of dangerous goods or the availability, condition and costs of airport and ATC facilities; and organizing twice-a-year airport schedule coordination conferences (Section 4.1). All member airlines of IATA avail themselves of these services.

At the second tier, in which only about one-third of the airline members participate, IATA still operates as a "tariff coordination" organization, assisting in the setting of passenger airfares and cargo rates, commissions for travel agents, etc. Tariff coordination is performed during "traffic conferences" and is based on a "cost plus" formula. The guiding principle, according to IATA, is that "fares and rates should not involve cut-throat competition, but should be as low as possible." The activities of this second tier still cover a significant number of international markets, but few major ones.[49]

With the spread of liberalization and deregulation in international air transportation (Section 3) the influence of IATA has been steadily diminishing over the past three decades. However, the organization is still treated in many countries as a semi-official international body, rather than a trade association.

Other International Associations

A large number of other international associations are influential in shaping economic and other regulatory policies concerning specific segments of the air transportation sector. A few prominent examples include: the **Airports Council International – ACI**, an increasingly active and important trade association of airport operators (567 members in 2006) with regional branches in Africa, Asia–Pacific, Europe, Latin America and the Caribbean, and North America plus an ICAO Bureau; the **International Federation of Airline Pilots Associations – IFALPA** that, in addition to representing the interests of unionized pilots internationally, plays a significant role in matters involving global air navigation services and ATC; and the **International Council of Aircraft Owners and Pilot Associations – ICAOPA**, which is one of many organizations representing the general aviation community internationally.

---

[48] In 1946 the US government had granted antitrust immunity to IATA to carry out airfare coordination.

[49] For example, no markets involving the US or intra-EU international travel markets are included.

Numerous airline associations are also active at the regional level. For example, the **Association of European Airlines – AEA** (31 airline members in 2006), the **European Regions Airline Association – ERAA** (78 airline members) and the **European Low-Fares Airline Association – ELFAA** (11 airline members) all represent different segments of the European airline industry.

European nations as a group have been steadily increasing their involvement in air transportation issues in general. The **Directorate-General – Transportation and Energy (DG-Tren)** of the European Commission in Brussels is constantly expanding its role in international airline and airport regulation and policy, e.g., as the EU negotiating party on the subject of the US–EU open skies agreement, and in preparing a series of regulations regarding airport slot allocation. It has also adopted an aggressive stance recently on the subjects of the rights of air transportation consumers and, most importantly, of environmental protection and mitigation (Chapter 14).

The **European Aviation Safety Agency – EASA** is an agency of the EU which was set up in 2002 to take over some of the functions of the European **Joint Aviation Authorities – JAA**, as well as many additional ones in the field of civil aviation safety and environmental protection. Its main tasks include, among others: inspections, training and standardization programs to ensure uniform implementation of EU aviation safety legislation in all Member States; safety and environmental certification of aircraft, engines and parts; and the safety assessment of foreign aircraft using EU airports.

Finally, **EUROCONTROL** is an increasingly important European organization (not limited to EU Member States) whose principal mission is to "harmonize and integrate air navigation services in Europe" and create a uniform ATM system for civil and military users. EUROCONTROL has 38 Member States – importantly, not including Russia – and acts, in many respects, as the coordinator and central (short-term and long-term) planner of ATM services in Europe, leaving to the individual Member States the task of operating the national ATC systems on a daily basis.

2.6.2 Organizations in the USA

In the USA, responsibility for domestic and international economic policy and regulation rests with the **US Department of Transportation – US DOT**. For example, the US DOT monitors the domestic airline industry for indications of anticompetitive practices, such as price gouging to take advantage of market dominance at certain hub airports or predatory pricing. Its broad range of tasks includes such diverse issues as preventing deceptive advertising of airfares, ensuring fairness in the listing of flights on travel agent computer reservation systems, and the public reporting of on-time performance statistics for airline flights. The US DOT also oversaw the program of assistance to US carriers – in the form of $5 billion in direct compensation payments and $10 billion in government loan guarantees – which was implemented following the September 11, 2001 events. Most important for the purposes of this review, the Office of the Assistant Secretary for Aviation and International Affairs that reports to the Under Secretary for Policy formulates US policies concerning international aviation and coordinates and leads negotiations on bilateral and multilateral ASAs on behalf of the government. The Under Secretary for Policy is also charged with monitoring compliance of foreign governments and aviation authorities with existing agreements. Support in the area of international aviation policy and regulation is also provided by the **US Department of State**, while the **US Department of Justice** is responsible for issues involving antitrust legislation, including the prosecution of alleged violators.

When it comes to the actual operation of the air transportation system, the **Federal Aviation Administration – FAA** plays a dominant role. The FAA was established in 1958, as an independent government agency, but has been part of the US DOT since 1968. In 2007, it had 44 000 employees of whom roughly 14 000 were air traffic controllers. The FAA maintains many regional offices, as well as a European office in Brussels. The FAA's budget for fiscal year 2007 was approximately $14 billion, of which about $8 billion were for operations and about $3.5 billion were designated for the Airports Improvement Program (AIP), i.e., was distributed among US airports to support capital expenditures. The FAA is organized along six lines of activity: regulation and certification; research and acquisitions; airports; administration; commercial space transportation; and air traffic services. By far the largest part of the organization, in terms of staff and resources, is dedicated to the provision of air traffic services.

Two other government agencies, the **National Transportation Safety Board – NTSB** and the **Transportation Security Administration – TSA**, have major roles in the areas of aviation safety and security, respectively. The former is an independent government agency charged with investigating every civil aviation accident in the USA, as well as all other major transportation accidents, and with making recommendations on improving transportation safety. The TSA, now part of the Homeland Security Department, was created in late 2001.

In addition to these government organizations, several industry associations and labor unions, all with headquarters in Washington, play a significant role in various aspects of policy making regarding air transport. Prominent examples include the following:

- **Air Transport Association – ATA**, representing the major and national carriers, with 19 member airlines, both legacy and low cost, in 2006.
- **Regional Airlines Association – RAA**, representing regional carriers engaged in short- and medium-haul scheduled services, often with regional jets, and with about 50 member airlines.
- **Airports Council International, North America – ACI-NA**, the North American regional branch of ACI, the global association of airport operators (Section 6.1).
- **American Association of Airport Executives – AAAE**, also representing airport interests.
- **Aircraft Owners and Pilots Association – AOPA**, the main association representing general aviation interests, with very strong political ties in Washington.
- **National Business Aircraft Association – NBAA**, representing owners and operators of general aviation business aircraft, with interests and activities which are often aligned with those of the AOPA.
- **Airline Pilots Association – ALPA**, the union representing many airline pilots with about 60 000 members in 2006.
- **International Association of Machinists and Aerospace Workers – IAMAW**.

- **Association of Flight Attendants – AFA**, representing 55 000 members in 2006.
- **National Air Traffic Controllers Association – NATCA**, representing 14 500 members in 2006.

2.7 Summary and Conclusions

The principal points addressed in this chapter can now be summarized:

1. The global airline industry operates in an international regulatory environment that ranges from strict regulation and protectionism in some countries or regions to almost complete deregulation in others. The long-term trend would seem to be in the direction of further deregulation and liberalization, with the unified market created by the EU and the proliferation of bilateral and multilateral "open skies" agreements marking major advances in this direction since the early 1990s. National ownership requirements, even in the economically developed regions of the world, persist as an important barrier to a full "globalization" of the industry. There is also a strong trend toward privatizing the many government-owned national carriers ("flag carriers") that had long dominated air travel outside the United States.

2. Airport capacity constraints are becoming increasingly severe in many regions of the world, resulting in serious problems of flight delays and cancellations and of low reliability of flight schedules. Using IATA's schedule coordination system, many major international airports largely rely on historical precedent as the primary criterion for allocating scarce airport "slots" among airlines that request them. This approach to airport demand management is restraining competition and is increasingly viewed as an indirect form of industry regulation. Market-based mechanisms for effecting slot allocation are currently being considered as possible alternatives to the IATA approach. Private participation in airport ownership and private-sector-style airport management is another important trend that emerged in the late 1980s and the 1990s outside the USA. US airports are owned by state or local governments, but rely heavily on outsourcing most of their financing, planning and operating activities to private companies. This approach to airport management is now being widely imitated by both privatized and government-owned airports around the world. Airport privatization has also stimulated interest in the economic regulation of airports – particularly as regards the rates they charge for aeronautical services.

3. In response to the growing complexity and cost of air navigation services (or "air traffic management"), a number of economically developed countries have "corporatized" (or "commercialized") the provision of these services, through establishment of autonomous corporate entities that largely operate according to private-sector principles. These entities, mostly 100% government owned, have assumed responsibility for what was previously a government service. As is the case with airports, economic regulatory measures have been put in place in response to these developments.

4. A large number of international and national institutions and organizations, public and private, play a central role in policy making regarding economic, regulatory and

technical matters that profoundly affect the air transport sector. Some of the most important among them are described briefly in Section 2.6.
5. The study of any aspect of the global airline industry must be cognizant of the complex regulatory, legal and institutional contexts within which the industry operates.

References

Ball, M.O., Ausubel, L.M., Berardino, F., Cramton, P., Donohue, G., Hansen, M., and Hoffman, K. (2007) "Market-Based Alternatives for Managing Congestion at New York's LaGuardia Airport," Proceedings of AirNeth Annual Conference, Amsterdam.

Carlin, A. and Park, R. (1970) "Marginal Cost Pricing of Airport Runway Capacity," *American Economic Review*, 60, 310–318.

Chang, Y.-C. and Williams, G. (2001) "Changing the Rules – Amending the Nationality Clauses in Air Services Agreements," *Journal of Air Transport Management*, Vol. 7, pp. 207–216.

Czerny, A.I., Forsyth, P., Gillen, D., and Niemeier, H.-M. (eds.) (2008) *Airport Slots: International Experiences and Options for Reform*, Ashgate, Aldershot, Hants.

de Neufville, R. and Odoni, A. (2003) *Airport Systems: Planning, Design and Management*, McGraw-Hill, New York.

Doganis, R. (2001) *The Airline Business in the 21st Century*, Routledge, London and New York.

DotEcon Ltd (2001) *Auctioning Airport Slots*, Report for HM Treasury and the Department of the Environment, Transport and the Regions, United Kingdom, http://www.dotecon.com.

EU Commission – Council of the European Communities (2004) *Regulation (EC) No. 793/2004 of the European Parliament and of the Council*, Official Journal of the European Union, 21 April, http://europa.eu.int/eur-lex/pri/en/oj/dat/2004/l_138/l_13820040430en00500060.pdf.

Findlay, C. (2003) "Plurilateral Agreements on Trade in Air Transport Services," *Journal of Air Transport Management*, Vol. 9, pp. 211–220.

GAO – US Government Accountability Office (2005) *Air Traffic Control: Characteristics and Performance of Selected International Air Navigation Service Providers and Lessons Learned from Their Commercialization*, Report GAO-05-769, Washington, DC.

IATA – International Air Transport Association (2005) *Worldwide Scheduling Guidelines*, 11th edition, Montreal, Canada.

Madas, M. (2007) "A Critical Assessment of Airport Demand Management: Strategies, Implications, and Potential for Implementation," Ph.D. thesis, Athens University of Economics and Business.

NERA – National Economics Research Associates (2004) *Study to Assess the Effects of Different Slot Allocation Schemes*, Final Report for the European Commission, DG-Tren, London, http://europa.eu.int/comm/transport/air/rules/doc/2004_01_24_nera_slot_study.pdf.

Shane, J.N. (2005) *Air Transport Liberalization: Ideal and Ordeal*, Second Annual Assad Kotaite Lecture, Royal Aeronautical Society (Montreal Branch), US Dept. of Transportation, Office of Public Affairs, Washington, DC, December 9, www.dot.gov/affairs/briefing.htm.

Vickrey, W. (1969) "Congestion Theory and Transport Investment," *American Economic Review*, Vol. 59, pp. 251–260.

3

Overview of Airline Economics, Markets and Demand

Peter P. Belobaba

The provision of air transportation service is driven primarily by the demand for air travel, as well as the demand for the shipment of goods by air. Virtually all of the interrelated decisions of the many stakeholders in the airline industry stem from the need to accommodate the historically growing demand for air transportation. And many of the activities of governments, airlines, airports and aircraft manufacturers are determined by the interaction of supply and demand in a variety of different markets associated with the airline industry. This chapter provides a foundation for the discussion of the many facets of the airline industry addressed in the remainder of this book by introducing some basic airline terminology and definitions, along with the concepts of air transportation markets and the demand for air travel.

3.1 Airline Terminology and Definitions

In the airline industry, there exist standard measures of passenger traffic and airline output, which are also combined to generate several common measures of airline performance. As we shall see later in this section, some of these performance measures are not particularly useful on their own, and in fact are often misinterpreted. At this point, we introduce the measures and their definitions.

Airline Traffic and Revenue

Measures of "airline traffic" quantify the amount of airline output that is actually consumed or sold. Traffic carried by airlines consists of both passengers and cargo, which can include air freight, mail and passenger baggage. All-cargo airlines transport primarily air freight, whereas passenger or "combination" airlines transport a mix of traffic that can include all four types of traffic mentioned. Combination carriers can operate a mix of all-cargo

The Global Airline Industry   P. Belobaba, C. Barnhart and A. Odoni
© 2009 John Wiley & Sons, Ltd

("freighter") and passenger aircraft, but even the passenger aircraft can carry one or more types of cargo in the belly compartments. In the following paragraphs, the definitions and examples focus on passenger airlines, although there is a parallel terminology for cargo airlines.

For passenger airlines, "traffic" refers to enplaned passengers, as opposed to "demand," which includes both those consumers that boarded the flight(s) and those who had a desire to travel on the flight but could not be accommodated due to insufficient capacity. Thus, at a given price level (or set of prices), there exists a total potential demand for air transportation between cities. Given a limited total capacity (available seats) offered by airlines, this potential total demand includes both passengers carried (traffic) and passengers unable to find seats, also known as "rejected demand" or "spill."

Passenger airline traffic can be measured in terms of the number of passengers transported, but the most common measure of airline traffic is a revenue passenger kilometer (RPK) or, alternatively, a revenue passenger mile (RPM). In the following examples, we use kilometers. Thus, 1 RPK is defined as one paying passenger transported 1 kilometer. For example, a flight carrying 140 passengers over a distance of 1000 km generates 140 000 RPK of airline traffic.

The fare paid by passengers to travel by air varies by distance, by season, and according to the conditions and characteristics of the fare product purchased (e.g., business class or advance purchase excursion fares). Yield is a measure of the average fare paid by all passengers per kilometer (or mile) flown, in a market, on a set of routes, or a region of operation for an airline. Yield is calculated by dividing the total passenger revenues collected by the number of RPK carried. In our example, if the flight that carried 140 000 RPK generates $16 000 of total passenger revenue, its yield would be $0.114 per RPK (i.e., $16 000/140 000).

Airline Output and Operating Expense

As we shall see later in this book, the output of a passenger airline can be represented in a variety of ways, including the number of flight departures operated and number of seats flown. Similar to RPK, the most common measure of airline output is an available seat kilometer, ASK (or available seat mile, ASM). Thus 1 ASK is defined as one available seat flown 1 kilometer. If our example flight operates over a distance of 1000 km with a 200-seat aircraft, it generates 200 000 ASK of airline output.

In generating its output, the airline incurs a variety of operating expenses, as will be detailed in Chapter 5. The average operating expense per unit of output, ASK, is the unit cost of the airline, an important measure of cost efficiency, both over time and across airlines. Unit cost is defined as the total operating expense divided by the ASK produced by an airline, for a route, region or total network. If the airline incurs $15 000 of expense to operate our example flight, the unit cost for this flight would be $0.075 per ASK (i.e., $15 000/200 000).

Load Factor

Load factor refers to the ratio of traffic to airline output, representing the proportion of airline output that is sold or consumed. For a single flight leg (i.e., a non-stop operation), its load factor can simply be defined as the number of passengers divided by the number of seats on the flight. For our example flight, the load factor can be calculated as passengers carried divided by available seats, or 70% (140/200).

Because most airlines operate many flights, each with different distances flown, the proportion of output consumed is better represented by an average load factor, defined as the ratio of RPK to ASK (or RPM/ASM). Our example flight's load factor can thus also be calculated as 140 000 RPK divided by 200 000 ASK, or 70%.

Now, assume that the same airline operates another flight leg using the same 200-seat aircraft over a distance of 2000 km, and this second flight carries 170 passengers. The load factor of this second flight leg is:

$$LF = Passengers/Capacity = 170/200 = 85\%$$

What is the total average load factor of this small "network" of two flights? There are two different (and correct) answers.

- The *average leg load factor* is the simple mean of the load factors of the two flight legs:

$$ALLF = (70\% + 85\%)/2 = 77.5\%$$

- The *average network or system load factor* is the ratio of total RPK to ASK, as defined earlier:
$$ALF = (140\,000 + 340\,000)/(200\,000 + 400\,000) = 80.0\%$$

Both measures of load factor are correct, but are used in different ways. The ALLF is more appropriate for analysis of demand to capacity or passenger service levels on a series of flight leg departures (on a particular route over a month, for example). The ALF is the more common measure, and is used in most financial and traffic reports of system-wide airline performance.

These five measures – traffic (RPK), yield, capacity (ASK), unit cost and load factor – are the most common measures of passenger airline performance and will be referred to throughout the rest of this book. Of course, there are numerous other measures of cost efficiency, productivity and financial performance, which will be introduced in the relevant chapters. And although we have defined these measures for a passenger airline, there are parallel and very similar measures that apply to cargo air transportation. For example, cargo traffic carried is measured in units of a revenue tonne kilometer (RTK) (or revenue ton mile, RTM), meaning 1 tonne of cargo transported 1 kilometer. Cargo airlines provide output in units of an available tonne kilometer (ATK) and make use of both yield and unit cost measures in analyzing their performance.

3.1.1 Basic Airline Profit Equation

As for any industry, operating profit for an airline is defined as total revenue minus total operating expense. For passenger airlines, the revenue and expense terms can be broken down into the measures of output and sales defined above, as follows:

$$Operating\ profit = RPK \times Yield - ASK \times Unit\ cost \qquad (3.1)$$

This basic airline profit equation illustrates how the use of any of the individual terms defined above to measure airline performance can be misleading. For example, high yield is often (incorrectly) used as an indicator of airline success and even profitability. A high yield is clearly not desirable if only a few passengers pay a very high fare and leave a large proportion of seats unused, resulting in a low average load factor (ALF) and total revenues that do not cover total operating expenses. As a general rule, yield is a poor indicator of airline profitability by itself.

Low unit costs are also often mentioned as a measure of airline success. Although low costs of production provide a competitive advantage in any industry, low unit costs alone are of little value to an airline if yields are low and/or load factors are low, with total revenues falling short of covering total operating expenses. Even ALF on its own tells us little about profitability, as high ALF could be the result of selling a large proportion of seats at extremely low fares (yields). A high ALF does not guarantee operating profit, as many high-cost airlines have realized in the recent past.

Given this basic profit equation, the obvious airline profit-maximizing strategy is to increase revenues and/or decrease costs. However, there exist important interactions among the terms in the equation, so that no single term can be varied without affecting other terms and, in turn, overall operating profit. For example, a strategy designed to increase revenues requires the airline to increase its traffic carried (RPK) and/or increase its average fares charged (yield). Either tactic can have unintended and potentially negative impacts on other terms in the airline profit equation, as explained below.

In order to increase traffic, an airline may decide to cut fares (average yields) to stimulate demand, but the revenue impact of such a price cut depends on the elasticity of demand for air travel. For revenues to increase, the price cut must generate a disproportionate increase in total demand (i.e., "elastic demand"). Alternatively, the volume of traffic carried (RPK) can be increased if the airline increases its frequency of flights or improves its passenger service quality to attract passengers, but both of these actions will also increase operating costs. Increases in flight frequency, all else being equal, will increase total ASK and, in turn, total operating expenses. Improvements in passenger service quality will increase unit costs.

Increasing total revenues by simply increasing fares (yields) is another option. However, economic theory tells us that any price increase will inevitably lead to a traffic decrease. A price increase can still be revenue positive if demand is "inelastic" (i.e., the percentage decrease in passengers is less than the percentage increase in price).

Airline efforts to improve profitability by reducing operating expenses include tactics that reduce unit costs (cost per ASK) and those that reduce the airline's output (ASK). Both strategies can lead to lower total operating expenses but, once again, there can be negative impacts on other terms in the airline profit equation. A common airline tactic is to reduce unit costs by cutting back on passenger service quality, e.g., eliminating meals, pillows and extra flight attendants. However, excessive cuts of this type can affect consumers' perceptions of the airline's product, leading to a reduced market share and, in turn, RPK. A tactic for reducing unit costs indirectly is to actually increase ASK by flying more flights and/or larger aircraft, which can lower unit costs by spreading fixed costs over a larger volume of output. But such an approach will still lead to higher *total* operating costs and potentially lower load factors and reduced profitability.

Finally, an airline might decide to reduce its total operating expenses by decreasing its level of output (ASK). Cutting back on the number of flights operated will clearly reduce total operating costs, but lower frequencies might lead to market share losses (lower RPK and lower revenues). At the same time, reduced frequencies and/or use of smaller aircraft can lower ASK and total operating expenses, but can also lead to higher unit costs, as the airline's fixed costs are now spread over fewer ASK.

The basic airline profit equation introduced here incorporates the five most common measures of passenger airline performance introduced above, and illustrates the interdependence among these measures in airline management decisions. Perhaps more important, it provides preliminary insights into the difficulties of finding strategies to improve and sustain airline profitability.

3.2 Air Transportation Markets

The second section of this chapter is devoted to the description of markets in air transportation. The discussion begins with a description of a typical air passenger trip, followed by alternative definitions of markets for air travel, focusing on scheduled air services for passengers. The objective is to establish geographic or spatial definitions of air transport markets, taking into account the characteristics of a typical trip by passengers that use scheduled air transportation services.

3.2.1 Typical Air Passenger Trip

The spatial definitions of air transportation markets involving consumers and air carriers (or "airlines"), as well as much of the economic modeling of demand and supply in these markets, depend on the characterization of a typical trip by an air passenger. This characterization was originally proposed by Simpson (1995), and provides the basis for our definitions here.

As shown in Figure 3.1, a typical air passenger trip starts not from an airport, but from an origin point such as a residence or place of business. The *ground access* portion of the trip from the passenger's origin point to the originating airport can involve travel by

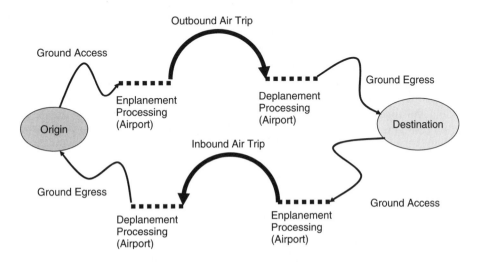

Figure 3.1 Representation of a typical air passenger trip

private car, taxi or public transport. The *origination airport region* containing the origin points of all travelers departing from an airport can have a radius ranging from a few kilometers to several hundred kilometers. Travel times for ground egress can therefore range from minutes to several hours.

Enplanement processing consists of purchasing tickets (if this has not been done in advance), obtaining boarding passes, checking baggage, undergoing security inspection and boarding the aircraft. In short-haul domestic markets, this portion of the total trip time can be as short as 15–20 minutes at some small airports. However, heightened security requirements in the recent past have increased enplanement processing times at most airports around the world, and especially in the USA, to 1 hour or more in many cases. For long-haul international services, enplanement processing can take even longer, as airlines can require minimum check-in times of 2–3 hours before flight departure.

The aircraft portion of the *outbound air trip* lasts for 1 hour or more and covers distances of 200 to 14 000 kilometers (125 to 9000 miles) or more. In 2008, Singapore Airlines operated the longest non-stop flight offered in scheduled passenger service between New York (Newark) and Singapore, covering 15 283 kilometers (9524 miles) in a scheduled time (gate-to-gate or "block time") of almost 19 hours (Official Airline Guides, 2008). The average length of a typical airline passenger trip worldwide is approximately 1824 kilometers (1140 miles) (*Airline Business*, 2005). In the USA, about one-fifth (21.5%) of all domestic air trips are shorter than 500 miles in length (*Aviation Daily*, 2004). This proportion decreased significantly in the aftermath of 9/11, as short-distance travelers have shifted to other modes of travel to avoid security and other airport delays.

After the flight arrives at the destination airport, there is *deplanement processing*, which can take from just a few minutes for the passenger to exit the airport terminal to over an hour when baggage retrieval, immigration and customs inspection are required. The trip then concludes with a *ground egress* portion involving travel from the airport to a destination point in the *destination airport region*.

Each airline trip has a *duration of stay* at the destination point that can range from a few hours to several months or more, before the passenger returns to the originating airport or region and makes a final ground egress trip to a location in the traveler's origination airport region (e.g., home or place of business).

This description of a typical air passenger trip raises several points important to the definition of scheduled passenger air transportation markets and demand. First, the purpose of each air trip is to move from the "true" origin to the "true" destination of the passenger, not simply to travel from one airport to another. The characteristics of the total trip, including the time required for each of its components in addition to the actual times spent on board the aircraft, will affect the total demand for air transportation between two airports.

Second, there is typically an outbound and inbound portion of passenger air trips, such that consumers in an air transport market start their trip in the origination airport region and return there after a trip of varying duration. As a result, every air travel market has an *opposite market* consisting of passengers who originate their trips from the destination airport region of the market described above. This opposite market is serviced by the same airline flights as the original market (Simpson, 1995). That is, the outbound flights for the original market are at the same time the inbound flights in the opposite market. As we shall see later in the discussion, the supply of air service is typically shared by demand from many markets, as passengers use various multi-stop or connecting itineraries in any given market.

3.2.2 Spatial Definitions of Airline Markets

Another way of defining air transportation markets is through their spatial boundaries, as shown in Figure 3.2. The origination region around airport A contains all the origin points of travelers, also referred to as that airport's *catchment area*. An airport's catchment area can extend for hundreds of kilometers and can vary with the destination and trip purpose of the traveler. For example, a traveler flying a short distance for a business trip with a short duration of stay is more likely to minimize the travel distance and time of the ground egress portion to an airport. On the other hand, a vacation traveler flying a much longer distance and staying at the destination for several weeks is more likely to be willing to travel much further to an originating airport, perhaps to take advantage of lower fares.

Similarly, airport B has a destination region that contains the destination points for passengers originating in region A. As was the case with the origination airport region, the size of destination airport region B can vary with trip purpose. For example, for London Heathrow Airport, the majority of destinations for non-resident business travelers are in a very small business area, while the destination region for pleasure travelers, visitors and residents of the surrounding area is much larger.

In Figure 3.2, the market for air services from A to C and back is *distinct and separate* from the market ABA (Simpson, 1995). Improvements in the quality of airline service or changes in the fares charged in the market ACA should not affect the demand for air travel in the ABA market. These are clearly two different markets, although the potential passengers in both markets are residents of originating airport region A. There are also two "opposite" markets shown in Figure 3.2. Market BAB has origination region B for consumers wishing to travel to points in a destination region A, and who use the same air services as market ABA.

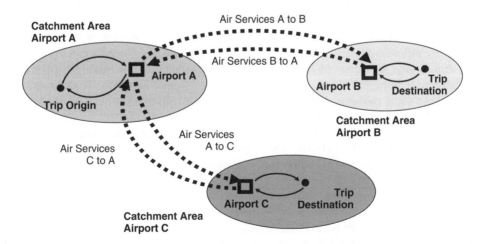

Figure 3.2 Distinct and separate origin–destination markets (adapted from Simpson, 1995)

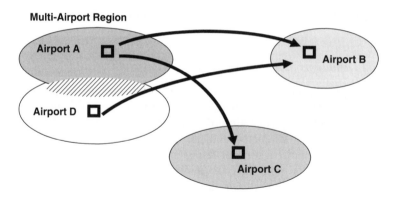

Figure 3.3 Parallel markets and overlapping airport regions (adapted from Simpson, 1995)

Opposite markets can have different characteristics. The volume of demand of opposite markets can be different, but since nearly all air trips are eventually round trips, the flow of traffic in each direction will be approximately equal over an extended time period. It might also be the case that the prices are not equal in each opposite market even though both markets use the same airline flights. This can occur when the origins and destinations are in different countries and fares are determined in different currencies, for example.

Airport regions can overlap when two or more airports provide alternative fight options for travelers in origin or destination regions. Figure 3.3 illustrates a scenario in which there exist flight options from airports A and D to airport B, while the only flights to airport C are provided from airport A. In this example, the airport catchment area for market ACA is the total shaded area around A and D, while the airport catchment areas for markets ABA and DBD overlap. Passengers making trips originating in the overlap area must choose which airport they access in order to travel by air to B.

Markets ABA and DBD are called *parallel markets*, and the flight options serving each parallel market are to some extent substitutes for each other within the larger region. For example, if the availability or speed of ground access/egress to and from airport A improves, the catchment area of airport A will expand. The pricing of air services in parallel markets will also affect the volume of demand using each market. With competitive pricing by airlines, passengers have the option to increase the ground egress portion of their trip to take advantage of lower fares in a parallel market. For example, if the fares available to a European destination are much lower from Montreal's Trudeau Airport, then at least some passengers from the Ottawa region (160 km away) can be expected to drive to Montreal rather than flying out of Ottawa Airport.

As shown in Figure 3.4, a traveler in market AB can connect between the flights being provided in markets AC and CB. The flight from A to B is thus providing a shared supply to both the AB and AC markets at the same time (as well as many other markets, depending on the extent of the airline's network). It is possible that the fares for travel from A to B via C are lower than the non-stop AB fare. If the fare for travel via C is cheaper, it will affect non-stop demand in market ABA. It is also possible to find examples where the fares from A to B via C are actually lower than the published AC fares. This is an outcome of the fact that AB and AC are economically distinct and separate markets, in which prices for air travel are determined by the demand and competitive characteristics

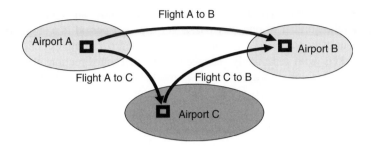

Figure 3.4   Non-stop vs. connecting service from A to B (adapted from Simpson, 1995)

of the different markets, not necessarily the distances traveled. These and other airline pricing implications of the nature of air transportation markets are discussed in more detail in Chapter 4.

The spatial definitions of air travel markets developed here suggest that there can be multiple, even overlapping, geographical delineations of origination and destination regions for air travel. The most common representation of origin–destination demand is with reference to a *city-pair* market – for example, the potential number of passengers per day wishing to travel between Boston and Chicago. However, because Chicago is served by two airports, the city-pair demand can be disaggregated to two (parallel) *airport-pair* markets – between Boston's Logan Airport and Chicago's O'Hare and Midway Airports, respectively. On the other hand, there is a broader market for air travel between the larger Boston metropolitan region and Chicago metropolitan region, which can include additional airport pairs such as Providence (Rhode Island)–Midway, Providence–O'Hare, Manchester (New Hampshire)–Midway and Manchester–O'Hare. This broader *region-pair* market thus includes six airport-pair markets, all of which are parallel and interrelated.

In summary, the spatial definitions of origin–destination markets presented here are based on consideration of the total trip characteristics for a typical airline passenger. Demand for air transportation is generated for a particular origin–destination (or O-D) market. However, with the existence of overlapping airport regions, parallel markets, and the sharing of scheduled airline supply on connecting flights, even "distinct and separate" origin–destination markets are interrelated.

3.3   Origin–Destination Market Demand

Air travel demand is defined for an origin–destination (O-D) *market*, not for a flight leg in an airline network. Air travel demand is typically measured in terms of (potential) flow per time period in one or both directions of an O-D market. For example, the number of persons wishing to travel from origin A to destination B during a given time period (e.g., per day) is the total O-D market demand in a single direction.

Note that the number of persons traveling from A to B during a time period will include both those travelers starting their trip at A (and belonging to the A–B–A round-trip market) as well as travelers returning home and finishing their trip (belonging to the B–A–B round-trip market). Although it is possible to separate these opposite round-trip markets, most data sources typically aggregate these demands. The total O-D demand can be measured as the number of one-way passenger trips per period in a single direction or, alternatively, one-way passenger trips per period summed over both directions of the O-D market. In markets with reasonably balanced flows in each direction, the latter will be twice the former. A common measure of O-D market demand used in the airline industry is the number of *passenger trips per day each way*, or "PDEW."

3.3.1 Dichotomy of Airline Demand and Supply

The supply side of air transportation is characterized to a large degree by joint production – many markets are simultaneously served by a single flight in an airline network. Each leg operated as part of an airline network will have passengers from many markets on board, not just the passengers from the local O-D market. The total number of passengers on board most flight legs is usually much larger than the "local" traffic belonging to the O-D market being served on a non-stop basis. The additional passengers belong to other O-D markets that are effectively distinct in economic terms, but are making use of the same flight leg as part of a multi-stop or connecting travel itinerary.

When the notion of O-D market demand is combined with the realities of how airlines provide air transportation services on a network of flights, several complications result. First, not all A–B O-D passengers will fly on non-stop flights from A to B, as some will choose one-stop or connecting flight itineraries, or *paths*. At the same time, a single non-stop flight leg A–B not only carries passengers belonging to the AB O-D market (i.e., A–B–A and B–A–B round-trip markets), but also serves many other O-D markets, as part of connecting or multi-stop paths.

A vivid illustration of the shared supply provided by a single non-stop flight to multiple O-D markets is provided by the data shown in Figure 3.5. Over the course of several months of operation, this non-stop flight from Boston to an airline's connecting hub carried passengers belonging to at least 64 different O-D markets, the largest 20 of which are shown in Figure 3.5. That is, on average, this Boston–Hub flight carried only 35 passengers per day in the "local" Boston–Hub O-D market, out of a total load of approximately 120 passengers. The remaining 70% of the passengers on this flight belonged to many

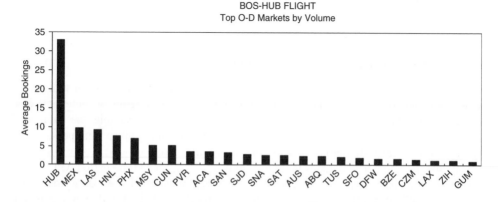

Figure 3.5 Joint supply of service to multiple O-D markets

other O-D markets served by this flight as part of a multi-stop or connecting itinerary chosen by the passenger. For example, in Figure 3.5, passengers traveling to destinations beyond the hub to Mexico City, Las Vegas and Honolulu comprised the next three-largest components of the total passenger load on this Boston–Hub flight.

There exists a *dichotomy of demand and supply* in airline economic analysis, meaning that there is an inherent inability to directly compare demand and supply in an individual O-D market. Demand is generated at the level of a passenger's O-D trip, whereas supply is inherently provided in the form of flight leg departures on a network of scheduled flight operations. One flight leg provides a *joint supply* of seats to many O-D markets simultaneously, so it does not make sense to consider the total number of seats on a flight leg as representing the "supply" of air transportation to a single O-D market, nor is it practical to determine accurately the actual number of seats supplied to each O-D market (Simpson and Belobaba, 1992b).

A single O-D market is typically served by many competing airline paths (non-stop, one-stop and connecting). As a result, the volume of O-D market demand cannot be determined simply by counting the number of passengers on non-stop flights operating between the origin and destination under consideration. Instead, detailed ticket samples must be used to make sure that all passengers originating at A and destined for B are included, regardless of the path they chose to fly.

This dichotomy of demand and supply makes it very difficult to answer some seemingly simple economic questions related to the supply and demand in an O-D market. For example, if it is not possible to identify the number of seats available to a single O-D market in a network of connecting flights, then it is impossible to determine whether the market is in "equilibrium," whether the prices charged for that O-D market are too high or too low, and even whether an airline's service is "profitable" in an individual O-D market. These questions can only be answered at the level of an entire network of flights, in theory at least.

3.3.2   Factors Affecting Volume of O-D Demand

The demand for air travel in an O-D market is affected by many different variables, although economic models of air travel demand typically include only those variables that have the greatest impact on demand and those that can be measured. In this section, the major factors affecting the volume of demand in an O-D market are briefly reviewed.

Socioeconomic and demographic variables reflect the characteristics of the potential passengers in an O-D market. The populations of cities A and B will clearly have an impact on air travel demand – the larger the populations, the greater the potential demand for air travel. The amount and type of economic interaction between cities A and B will also affect air travel demand – two cities with common industries will generate more demand for air travel (e.g., much of the Detroit–Seoul business travel is related to the auto manufacturing industry). Characteristics such as disposable income, levels of education and age of the populations in each city will also have an impact on the volume of air travel demand. Cities with greater disposable income will generate greater demand for air transportation. ·

Trip purpose characteristics associated with travel between A and B will also determine the volume of demand. An O-D market in which business travel dominates will have a lower overall volume of demand than a market in which passengers travel for business, vacation and personal "VFR" (Visiting Friends and Relatives) reasons, all else being equal.

Prices of travel options will clearly have an important impact on total air travel demand in each O-D market. The monetary (out-of-pocket) prices of airline fares, along with the implied disutility costs of fare restrictions (such as Saturday night minimum stay requirements and non-refundability), are perhaps the most critical determinants of the volume of O-D market demand. In addition, the prices of competing modes (train, bus, car) relative to the price of air travel can have a large impact on air travel demand, particularly in short-haul markets where competing modes provide a viable alternative to air travel.

Finally, a variety of "quality of service" considerations will contribute to the overall attractiveness of air transportation, both in absolute terms and relative to other modes. As will be described in more detail below, the frequency of flight departures and the time spent flying determine "total trip time," which includes schedule displacement or "wait times." The lower total travel time for a trip made by air, relative to other modes, is a primary reason for air travel in the first place. Other quality of service variables include the comfort, safety and ease of travel by air relative to other modes.

3.3.3   Quality of Service Factors

A passenger will consider the quality of service offered by various modes in deciding which mode to use for a particular trip. In fact, the quality of air transportation services offered in an O-D market will have a significant effect on the total volume of demand in the market. Airline "quality of service" factors include on-board service amenities, trip reliability, convenience of ticketing, and total trip time (Simpson and Belobaba, 1992a). Among these factors, total trip time is perhaps the most important in determining the volume of O-D market demand for air travel.

As described in Section 3.2, the typical passenger trip involves more than the actual flight time spent in the aircraft. *Total trip time* is measured from the "true" origin (e.g., home in city A) to the "true" destination (e.g., hotel in city B). As described in more detail below, the concept of total trip time is designed to capture elements of the frequency of airline departures, as well as the directness of the air services provided (e.g., non-stop versus connecting flights). For example, the high frequency of non-stop air services between London and Paris makes it more convenient to travel by air. In turn, the number of air trips taken is higher than would be the case if there were only three flights per day in this market (even if the aircraft used were large enough to accommodate all demand), or if there were only connecting flights available.

Although often taken for granted, reliability measures such as the likelihood of schedule completion, on-time performance, and safety will affect the total demand for air travel. For example, if every second flight between London and Paris were routinely cancelled, and many of the flights that departed had an accident, the demand for air travel between these two cities would certainly be reduced substantially!

The quality of on-board services, seating space, meals, and even entertainment can affect the volume of air travel demand in a market to some degree, whether considered in absolute terms or relative to competing modes. The ability of a passenger to easily make a reservation, obtain a ticket and meet any conditions of travel by air will also have an impact on total demand. For example, demand for air travel between London and Paris would undoubtedly be lower if airlines required all passengers to line up at a central ticket office to make reservations and purchase tickets.

3.3.4   Total Trip Time and Frequency

As introduced above, next to the price of air travel, total trip time is the most important factor affecting demand for airline services. Based on the description of a typical passenger trip in Section 3.2, total trip time includes:

- Access and egress times to/from airports at origin and destination.
- Pre-departure and post-arrival processing times at each airport.
- Actual flight times plus connecting times between flights.

Not included in the previous description of a typical passenger trip is one final component of total trip time – the "schedule displacement" or "wait time" associated with the fact that there is not likely to be a flight departure *exactly* at the time that each passenger is ready to depart. If a passenger's preferred flight departure time is 15:30 but the scheduled flight departures in the market are at 13:00 and 17:00, then the "schedule displacement" time for this passenger is either 2.5 hours or 1.5 hours, depending on whether the passenger ultimately chooses to take the earlier or later flight.

In an aggregate analysis of schedule displacement in an O-D market, it is assumed that a large number of passengers have desired departure times spread over the day. In the above example, if the desired departure times of all passengers wishing to fly between 13:00 and 17:00 are assumed to be uniformly distributed, then we can estimate that the average "schedule displacement" time or "wait time" associated with this schedule for this subset of passengers would be 1 hour: the longest displacement time (assuming passengers are indifferent between the earlier and later flights) is 2 hours, for the passenger wishing to depart at 15:00; the shortest displacement time is zero for those wishing to depart exactly at 13:00 or 17:00; and the average of all passengers' displacement times, assuming a uniform distribution of desired departure times, would thus be 1 hour.

Therefore, the concept of *total trip time* captures not only the effects of path quality (non-stop versus connecting flights), but also the effects of flight frequency. Increased frequency of departures and the existence of non-stop flights both reduce total trip time in an O-D market. Intuitively, a reduction in total trip time (represented as an average value over all potential passengers in an O-D market) should lead to an increase in total air travel demand.

Total Trip Time Model (Simpson and Belobaba, 1992a)

Total trip time for a passenger air trip is denoted as T, where:

$$T = t\,(\text{fixed}) + t\,(\text{flight}) + t\,(\text{schedule displacement}) \qquad (3.2)$$

The fixed time elements include access and egress, as well as airport processing times, and are not likely to change in the short term. Flight time includes aircraft *block times* plus connecting times at intermediate stops. Note that "block time" is defined as the difference between scheduled gate departure and arrival times for a flight, and thus includes time spent pushing the aircraft away from the gate, taxiing to the runway, actual air time, and taxiing from the runway to the gate at the arrival airport.

The schedule displacement component of total trip time can be expressed approximately as having a simple inverse relationship with the frequency of service. For example:

$$t(\text{schedule displacement}) = K/\text{Frequency} \qquad (3.3)$$

where K is a constant expressed in hours and frequency is the number of flight departures (per day) in the market. Schedule displacement time (or "wait time") thus decreases with increases in the frequency of departures.

As a very simple example, consider a non-stop market in which the desired departure times of passengers are distributed uniformly between 06:00 and 22:00, a 16-hour period of "reasonable" departure times. In this case, if only one flight per day were offered at 14:00, the mean schedule displacement time for a typical passenger would be 4.0 hours (the possible "wait times" range from 0 for those passengers wishing to depart at 14:00, to a maximum of 8 hours for passengers with departure time preferences of 06:00 or 22:00). As the airline increases its frequency of service to two daily flights (assuming the departure times are chosen to minimize mean wait times, with flights at 10:00 and 18:00), the overall average schedule displacement time drops to 2.0 hours. Further increases in frequency to three and four flights (under the same assumption of mean wait time minimization) would reduce average schedule displacement time to 1.33 and 1.0 hours, respectively. Thus, under these simplifying assumptions, K would be equal to 4 in expression (3.3).

This total trip time model is useful in explaining why non-stop flights are preferred by passengers over connecting itineraries, due to lower flight times and lower total T. In addition, the model supports the notion that more frequent departures can increase total travel demand in an O-D market, due to reduced schedule displacement times and a lower total T. It also demonstrates why frequency of departures is more important in short-haul markets than in long-haul markets. Schedule displacement time is a much larger proportion of total T in short-haul markets, whereas in long-haul markets, even extremely frequent departures will not reduce the relatively large flight time components of T.

Finally, in the context of airline hub networks, the total trip time model suggests that many connecting departures might in fact provide a better service quality to passengers than one non-stop flight per day. Despite a longer total flight time for connecting departures, the reduced schedule displacement time resulting from frequent connecting departures can lead to a lower total T for the average passenger.

3.4 Air Travel Demand Models

Demand models are mathematical representations of the relationship between demand and selected explanatory variables. By definition, models do not and cannot replicate all of the factors that affect demand for air travel in a market. Therefore, the objective in demand modeling is to "explain" as much of the variation in demand as possible, either across markets or over time within the same market, by identifying the explanatory variables that have the greatest and most direct impact on the volume of O-D market demand.

All demand models are based on assumptions of what affects air travel demand. Given the previous discussion, the major factors affecting the volume of air travel in an O-D market are the price of air travel, total trip time (T) and demographic variables related to the market itself. The specification of a mathematical model of demand reflects *a priori* expectations of how demand responds to changes in the explanatory variable(s). In the simplest example, air travel demand can be modeled as being affected only by the prices being charged. An inverse relationship is expected, based on economic theory (e.g., when prices rise, demand should decrease), and the mathematical formulation could be additive (linear) or multiplicative (nonlinear).

An additive or linear price–demand model takes the form $D = a - bP$, where D is the market demand, P is the average market price, and a and b are parameters that represent the intercept and slope of the demand function with respect to price. In this linear model, each \$1 increase in market price is assumed to lead to a demand decrease equal to b.

Most demand models of airline markets assume a multiplicative relationship among the independent (explanatory) variables (Doganis, 1991). A multiplicative price–demand model is of the form $D = aP^b$, where $b < 0$. This function is still downward sloping, but the relationship between demand and price is constant in percentage terms, rather than absolute terms. That is, a 1% increase in market price is assumed to lead to a demand decrease of $b\%$.

A properly estimated demand model allows airlines to analyze more accurately and even forecast demand in an O-D market, perhaps as a function of changes in average fares, given recent or planned changes to frequency of service, or to account for changes in market demographics or economic conditions.

3.4.1 Elasticity of Air Travel Demand

The price elasticity of demand is the percentage change in total market demand that occurs with a 1% increase in average price charged. Price elasticity is *negative* for "normal" (as opposed to "luxury") goods and services. A 10% price increase will cause an $X\%$ demand *decrease*, all else being equal (e.g., with no change to frequency or other market variables). Historically, it has been assumed that business air travel demand tends to be slightly "inelastic" ($0 > E_p > -1.0$), meaning that the volume of demand does not change by as much as a change in price, in percentage terms. On the other hand, leisure demand for air travel is assumed to be much more "elastic" ($E_p < -1.0$), meaning the percentage response of total demand is greater than the percentage change in price.

Over the years, many different empirical studies of air travel demand have found a typical range of airline O-D market price elasticities from -0.8 to -2.0 (Gillen *et al.*, 2003). That is, air travel demand tends to be elastic, or relatively responsive to changes in the average prices of air travel. Of course, the elasticity of demand in specific O-D markets will depend on the mix of business and leisure travel in each market. Predominantly business markets will appear to be less elastic than markets in which leisure travel represents the larger proportion of total demand.

The price elasticity of demand in each O-D market has important implications for airline pricing. Inelastic (-0.8) business demand for air travel means less sensitivity to price changes. A 10% price increase leads to only an 8% demand reduction. As a result, total airline revenues will increase, despite the price increase. On the other hand, elastic (-1.6) leisure demand for air travel means a greater sensitivity to price changes. A 10% price increase causes a 16% demand decrease. For the airline, total revenues *decrease* given a price increase, and vice versa.

A large part of the development of airline pricing practices over the past two decades can be explained by differences in price elasticities between business and leisure travelers. An airline will try to increase fares for inelastic business travelers in order to increase total revenues. At the same time, the airline should decrease fares for elastic leisure travelers if it wants to increase revenues from this demand segment.

Similar to price elasticity, the time elasticity of demand is the percentage change in total O-D demand that occurs with a 1% increase in total trip time (T), as defined previously. Time elasticity is also negative, as a 10% increase in total trip time will cause an $X\%$ demand *decrease*, all else being equal (e.g., no change in prices). Business air travel demand is assumed to be more time elastic $(E_t < -1.0)$, because more business passengers will travel as total trip time is reduced (due to increased frequency, or the introduction of non-stop flights to replace connecting alternatives). Leisure demand is assumed to be time inelastic $(0 > E_t > -1.0)$, because price-sensitive vacationers are more willing to endure longer total trip times (less frequent service, connecting flights) if necessary to secure a lower fare.

Airline market time elasticities are affected by the proportion of business demand in the market, as well as the level of existing flight frequency. That is, in markets with a high level of flight frequency, the incremental demand resulting from further improvements to frequency will be smaller than in a market with relatively little frequency to begin with.

There exists a "saturation frequency" in each market, defined as the point at which additional frequency does not increase demand, even for business travel (Simpson and Belobaba, 1992a). For example, in the short-haul Boston–New York shuttle market, two competitors currently each offer non-stop flights every hour (one at the top of the hour and the other at 30 minutes past each hour), such that flights in this market depart every half hour. It would be difficult to argue that increasing the frequency of departures to a flight every 15 minutes would have any measurable positive impact on the total volume of demand in this market.

The concept of time elasticity of demand for airline travel in an O-D market has implications for airline scheduling decisions. Business demand responds more than leisure demand to reductions in total travel time in a market. Increased frequency of departures is the most important way for an airline to reduce schedule displacement time and, in turn, total travel time for business travelers in the short run. Reduced flight times can also have an impact. Flight times can be reduced by offering non-stop flights in markets previously served only by connecting flights. To a lesser extent, differences in the airborne speeds of different aircraft also have an impact on business travel demand (e.g., jet vs. turboprop aircraft). Because leisure demand is not nearly as time sensitive, frequency of departures and non-stop services are not as important as price. In predominantly leisure markets, there is thus less incentive for the airline to provide frequent non-stop services.

3.4.2 Air Travel Demand Segments

The business versus leisure classification of travelers has been the traditional basis for identifying different air travel demand segments, used widely in both research and industry practice. As mentioned above, airlines have made use of the distinctions between business and leisure passengers in terms of price and time sensitivity in their pricing and scheduling strategies. This simple business versus leisure model of demand segmentation overlooks the reality that not all business travelers or leisure travelers have the same travel characteristics and, in turn, air travel preferences. For example, leisure (vacation) travel is but one component of the non-business or personal travel segment, which also includes travel to visit friends or relatives (VFR travel) and personal emergency travel. Each of these subsets of non-business travel can in fact have very different characteristics and preferences.

The concepts of price and time elasticity, or sensitivity, can be used to identify different segments of the total demand for air travel. Travelers very sensitive to price will base air travel decisions almost exclusively on a lowest-fare criterion and will be willing to contend with the restrictions, reduced service amenities, and perhaps less convenient flight times or routings in order to minimize the out-of-pocket costs of air travel. Conversely, passengers with severe time constraints and service-sensitive travelers will value level of service factors such as schedule convenience, travel flexibility and in-flight amenities, to the point that price might not even be a factor in selecting an air travel alternative. Of course, there exists a continuum between these two extremes, along which the majority of air travelers are likely to fall.

Given this continuum between extreme price sensitivity and extreme service-level sensitivity, it is difficult to divide the total demand in an airline market into well-defined segments. Belobaba (1987) proposed a demand segmentation model that incorporates explicitly the notions of both the price–service trade-off and the value of a trip as determined by the consumer's time sensitivity with respect to a given trip. By separating time sensitivity from price sensitivity, this segmentation model characterizes consumer groups without reference to trip purpose.

The availability of different flight and schedule options at various price levels requires consumers to make a trade-off between the higher-priced options associated with high levels of service amenities and fewer restrictions, and the lower-priced options with greater restrictions and, in some cases, reduced service amenities. This choice is related to the price sensitivity of the consumer for a given trip, and is not independent of the time sensitivity of the trip being considered. The time sensitivity of a trip is determined by the length of the "time window" over which a trip may be taken and still provide the consumer with a certain value of being at the desired location. A totally non-discretionary trip is one that must be taken at a specified time, meaning the acceptable window for the trip is very short. On the other hand, a totally discretionary trip is one for which the acceptable time window for travel is extremely long.

A consumer's location along both the price-sensitivity and time-sensitivity scales can differ from one planned trip to the next. An individual can be insensitive to price for an extremely time-sensitive trip, be it for business or personal purposes. The same individual can be extremely price sensitive for a trip that has no time sensitivity associated with it, again regardless of trip purpose. We can define four generalized demand segments by dividing each scale into two sections, as shown in Figure 3.6.

The characteristics of most of the travelers falling into each segment can be described as follows:

- *Type 1*: *Time sensitive and insensitive to price.* This demand segment represents the traditional characterization of business travelers who prefer to travel on flights that meet their schedule requirements, and are willing to pay a higher fare to do so. They might even be willing to pay a premium price for the extra amenities of a business

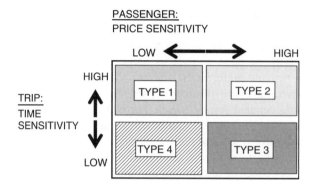

Figure 3.6 Air travel demand segments
Source: Belobaba (1987)

or first-class service. Travel flexibility and last-minute seat availability are extremely important to this segment.

- *Type 2: Time sensitive but price sensitive*. This segment was not explicitly recognized in the traditional business–leisure classification, yet a large proportion of business travelers probably belong to this demand segment. These consumers must make a trip but are willing to be somewhat flexible in order to secure a reduced fare. They cannot book far enough in advance to obtain the lowest fares, although they might be willing to rearrange a trip, accept less convenient flight times and/or routings if the savings are great enough.
- *Type 3: Price sensitive and insensitive to time constraints*. This segment contains the classical leisure or vacation travelers, willing to change their time and day of travel, and even destination airports, to find a seat at the lowest possible fare. This group is willing to stop or make connections and can meet virtually any travel or ticketing conditions associated with a low fare.
- *Type 4: Insensitive to both price and time constraints*. This segment includes the relatively few consumers that have little or no time constraints for travel, yet are willing to pay for high levels of service. Type 4 consumers can be combined with the Type 1 group, since both are willing to purchase a high-priced fare product to secure a high level of service and/or travel flexibility, regardless of their trip purpose.

The number of demand segments need not be limited to the four categories described here. In fact, many airlines, through their marketing and pricing strategies, have attempted to identify many additional segments of air travel demand, in order to tailor their product offerings (and fares) more specifically to each segment. Identifying clear and distinct segments of demand can be difficult, however, because the characteristics of any air travel demand segment are defined largely by the two dimensions of price and time sensitivity.

The segmentation model illustrated here could be expanded to include more demand segments by assuming more than two segments along each scale, but the specific divisions become more difficult to quantify. Nonetheless, the use of air travel demand segmentation for pricing purposes has progressed beyond the original two-segment model due to a realization that the characteristics of consumer demand for air travel are more complex than a simple distinction between business and leisure travelers.

3.4.3   O-D Market Demand Functions

The concepts of price and time elasticity of demand for air travel can be incorporated into a relatively simple O-D market demand function. Consider the following multiplicative model of demand for travel *(D)* in a given O-D market per period (e.g., per day):

$$D = M \times P^a \times T^b \tag{3.4}$$

where: $M =$ market sizing parameter (constant) that represents populations and economic interaction between the cities

$\quad\quad P =$ average price of air travel

$\quad\quad T =$ total trip time, reflecting changes in frequency

$\quad a, b =$ price and time elasticities of demand

We can estimate values of M, a and b from a historical data sample of D, P and T for the same market, or from a sample of similar markets over a period of time. The required historical data sample consists of previous observations of demand levels (D) under different combinations of price (P) and total trip time (T). Statistical estimation techniques like ordinary least squares (OLS) regression applied to the historical data provide us with the "best fit" curve, based on estimates of parameters M, a and b. Note that a and b are also the "constant elasticity" estimates with respect to changes in P and T, respectively.

This simple air travel demand model can be extended to more accurately represent the different components of air travel demand in a market and the fact that there can be more than one air travel "product" offered to travelers (Simpson and Belobaba, 1992a). Airlines typically offer different classes of service with different price levels and service qualities. The total demand for air travel can be categorized by trip purpose, which in turn reflects different reactions to variations in price and/or quality of service.

In this "multiple product" air travel demand model originally proposed by Simpson, it is assumed that airlines offer three classes of service, with different quality levels and/or conditions of use – first class (f), coach class (c) and discount class (d). Each class of service has a different price level for travel in the same O-D market – P_f, P_c and P_d, respectively.

At the same time, the total demand for air travel in a market can be separated according to trip purpose. The most common distinction made in air transportation economics is between business (b) and leisure or "personal" travelers (p), as introduced earlier. With three different products available for air travel and two different categories of air travel demand, the total demand for air travel in a market can be split among the following components:

D_{fb} demand for first-class service by business travelers

D_{cb} demand for coach-class service by business travelers

D_{cp} demand for coach-class service by personal travelers

D_{dp} demand for discount-class service by personal travelers

In the interest of simplicity, we exclude two additional components that are assumed to be small enough to ignore in this disaggregate multiple product demand model, namely

D_{db} and D_{fp}. Two pairs of demand functions represent the four components of air travel demand, based on the above categorization. For business travel:

$$D_{fb} = M_b Q_f P_f^{a1} T_f^{b1} P_c^{c1} \tag{3.5}$$

$$D_{cb} = M_b Q_c P_c^{a1} T_c^{b1} P_f^{c1} \tag{3.6}$$

where: M_b = the market sizing parameter for business travel demand (constant)
Q_f, Q_c = quality of service factors for first- and coach-class services
P_f, P_c = prices of first- and coach-class services
T_f, T_c = total travel times for first- and coach-class services
$a1$ = price elasticity of demand for business travelers
$b1$ = time elasticity of demand for business travelers
$c1$ = cross-elasticity of business travel demand for first-class service with respect to the price of coach-class service, and vice versa

Similarly, for personal travel:

$$D_{cp} = M_p Q_c P_c^{a2} T_c^{b2} P_d^{c2} \tag{3.7}$$

$$D_{dp} = M_p Q_d P_d^{a2} T_d^{b2} P_c^{c2} \tag{3.8}$$

where: M_p = the market sizing parameter for personal travel demand (constant)
Q_c, Q_d = quality of service factors for coach- and discount-class services
P_c, P_d = prices of coach- and discount-class services
T_c, T_d = total travel times for coach- and discount-class services
$a2$ = price elasticity of demand for personal travelers
$b2$ = time elasticity of demand for personal travelers
$c2$ = cross-elasticity of personal travel demand for coach class service with respect to the price of discount-class service, and vice versa.

The specification of the above two pairs of demand equations creates a substantially more complex (but also more realistic) mathematical representation of air travel demand. This set of disaggregate demand models reflects the assumptions that business travelers will only choose between first- and coach-class services while personal travelers will only choose between coach and discount services.

The relative prices and qualities of the services offered will determine the proportion of business and personal travelers that choose each class of service. If discount-class prices were to be lowered, the demand for the discount class would be expected to grow, not only by attracting new travelers with the lower price, but also by diverting some of the previous coach-class passengers to the lower-priced discount class. Similarly, if the quality of coach-class service were to be improved while prices did not change, some business passengers would be expected to divert to coach class (away from first class) and some discount passengers would divert to coach class (away from discount class).

This disaggregate model of air travel demand includes six separate and different elasticities of demand, three variables representing the differences in quality of service between the three product offerings, and two market sizing parameters which reflect different volumes of business and personal air travel demand in each O-D market. There are two different price elasticities of demand, one for each segment of demand. As mentioned earlier, the price elasticity for business travelers is likely to be more inelastic while the price elasticity for personal travelers will be more elastic, or even highly elastic. On the other hand, the time elasticity for business travelers is expected to be more elastic but relatively inelastic for personal travelers.

The two "cross-elasticity" terms in the equations are expected to be positive, given that first class is a substitute for coach class in the choice being made by business travelers, and coach class is a substitute for discount class in the choice facing personal travelers. As with all substitutes, an increase in the price of one of the alternatives should lead to an increase in the demand for the other (substitute) alternative. Conversely, a decrease in the price of one alternative should lead to a decrease in the volume of demand for its substitute. For example, if the parameter $c1$ in Equations (3.5) and (3.6) were estimated from a sample of historical data to have a value of $+0.6$, then a 10% increase in the coach-class fare would be expected to result in a 6% increase in demand for first-class travel, among business travelers. As a rule, the greater the magnitude of cross-elasticity parameters, the greater the extent to which they are perceived to be substitutes by consumers.

This more complex and disaggregate model of air travel demand in a market incorporates some of the complexity of air travel demand in the real world. At the same time, it is still an imperfect model that is based on simplifications of real-world behavior and several assumptions that might not always be true. This is but one example of many different demand models that could be formulated based on our perceptions of the most important factors affecting air travel demand. These models could also be estimated, given the appropriate historical data and statistical estimation techniques.

To summarize, the important characteristics of air travel demand in a single O-D market that such models attempt to capture in the form of mathematical equations include:

- The capacity offered by a set of flights in an air travel market can be used by the airline to provide different classes or qualities of service, at different price levels.
- The total demand for air travel in a market can be segmented into two or more components based on trip purpose or other characteristics that result in different price and time elasticities for each component.
- The different components of air travel demand in a market are not independent of each other, as the same travelers choose between two or more alternatives offered by the airline based on relative prices and quality of service. This interdependence is represented in demand models as cross-elasticities of demand.

3.5 Airline Competition and Market Share

Airlines compete for passengers and market share based on the following factors:

- Frequency of flights and departure schedule on each route served.
- Price charged, relative to other airlines, to the extent that regulation allows for price competition.

• Quality of service and products offered, including airport and in-flight service amenities and/or restrictions on discount fare products.

Passengers choose the combination of flight schedules, prices and product quality that minimizes their total disutility. Each passenger would like to have the best service on a flight that departs at the most convenient time, for the lowest price. However, passengers are seldom able to find the perfect itinerary and the highest service quality for the lowest possible price, so they must trade off these factors to minimize their disutility, subject to budget constraints.

3.5.1   Market Share/Frequency Share Model

To this point, we have focused on the total demand per period in an O-D market, the factors that affect it, and mathematical model representations of it. We now introduce a simple model of airline competition for market share in an O-D market. The "market share" of airline A is defined as the proportion of total market demand that is captured by airline A. For this discussion, market share is expressed in terms of the share of passengers carried by an airline, although it could also be expressed in terms of the airline's share of market RPK or revenues.

With all else being equal, airline market shares will approximately equal their frequency shares, expressed in terms of competing non-stop flight departures in an O-D market. This observation is based on analysis of historical schedule data and airline traffic data, and can be used as a relatively simple "rule of thumb." The assumption of "all else being equal" requires that the price and service quality differences (apart from frequency of service) among competing carriers are negligible. In the real world, it could be argued that these differences are in fact close to negligible most of the time, as air transportation has come to be viewed as a "commodity" with no real differences between major airlines competing in developed markets.

However, there is widespread acceptance in the airline industry of an "S-curve" relationship between airline market share and frequency share, as shown in Figure 3.7. As Button and Drexler (2005) note, it is difficult to document the origins and evolution of this model from the published literature. Early theoretical development and empirical evidence that higher-frequency shares are associated with disproportionately higher market shares was provided in the 1970s, before deregulation (Simpson, 1970; Taneja, 1976). After deregulation, there exist references to the "familiar" S-curve (Kahn, 1993) and to the "well-known" S-curve (Baseler, 2002) by, respectively, one of the architects of US deregulation and a senior Boeing executive.

The S-curve relationship between frequency and market share helps to explain the use by airlines of flight frequency as an important competitive weapon. For example, in a two-airline competitive market, if one airline offers 60% of the non-stop flights it is likely to capture more than 60% of the market share. Conversely, the other airline (with 40% frequency share) will see less than 40% market share. The extent of this disproportionate response of market share to frequency share will depend on the degree to which the S-curve bends away from the market share = frequency share diagonal line. The postulated S-curve makes immediate intuitive sense at three points on Figure 3.7: (1) when an airline offers zero frequency, it will receive zero market share; (2) at 100% frequency share, it must receive 100% market share; and (3) when both carriers offer 50%

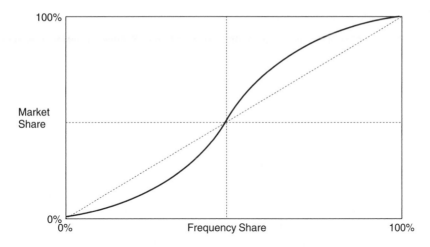

Figure 3.7   Market share vs. frequency share S-curve model

of the frequency, they should expect 50% market share, again assuming no significant differences in price or other service factors.

The S-curve shape of the model at other points on Figure 3.7 can also be explained as follows. Consider a distribution of passengers' desired departure times over the course of a day, and assume that passengers will choose those flights closest to their desired departure time. An airline with greater frequency of departures will capture all passengers wishing to fly during periods when only it offers a flight, while it will compete for and share the demand wishing to depart at times when both airlines offer flights.

For a simple example, consider a long-haul international market in which airline A offers daily non-stop service (seven flights per week), while airline B offers only three flights per week (Simpson and Belobaba, 1992a). If we assume that the demand is spread equally over the days of the week (not strictly true), one-seventh of the demand would like to depart each day. The market shares of the two airline are therefore:

$$\text{MS(A)} = 4 \text{ days} \times 1/7 + 3 \text{ days} \times 1/7 \times 50\% = 79\%$$

$$\text{MS(B)} = 3 \text{ days} \times 1/7 \times 50\% = 21\%$$

In this example, the weekly frequency share of the two carriers is 70% for airline A and 30% for airline B. Yet, airline A is expected to capture 79% of the market share, while airline B must settle for only 21%. Although this example is for weekly frequency in a long-haul international market, the same logic applies to an analysis of daily frequency in a shorter-haul market, if we assume each portion of the day to represent an equal block of demand with desired departure times.

From the S-curve and the example above, it is easy to see why there exists a tendency for competing airlines to *match* flight frequencies in many non-stop markets, to retain market share. As long as both airlines (in a two-carrier market) offer the same number of departures, they will both capture approximately half of the market demand. In a three-carrier market, the tendency would be for all three airlines to offer 33% of the frequency share, to retain approximately 33% of the market share.

3.5.2   S-curve Model Formulation

The mathematical model formulation that gives us the S-curve relationship between non-stop flight frequency share and airline market share can be expressed as:

$$MS(A) = \frac{FS(A)^a}{FS(A)^a + FS(B)^a + FS(C)^a + \ldots} \qquad (3.9)$$

where: $MS(i)$ = market share of airline i
$\quad\ FS(i)$ = non-stop frequency share of airline i
$\quad\quad\ a$ = exponent greater than 1.0, and generally between 1.3 and 1.7

In this model the value of the exponent a determines how far the S-curve bends away from the linear diagonal model. With $a = 1.0$, the denominator on the right-hand side is equal to 1 and we revert to the linear model where $MS(A) = FS(A)$. The higher the value of a, the more the S-curve will bend away from the linear diagonal, and the greater the market share return for an airline that offers more frequencies than its competitor. As mentioned, empirical studies have shown that a generally falls in the range of 1.3 to 1.7.

Note that the value of a will depend on the importance of frequency in a particular market. Based on the previous discussion, higher values of a are more likely to be associated with short-haul business markets in which frequency of departures is important, and lower values with longer-haul and/or leisure-dominated markets.

The simple model presented here accounts for non-stop frequency competition only. In the real world, where most O-D markets are served not just by non-stop flights but by one-stop and connecting flights as well, we would have to modify the model to recognize that one-stop and connecting flight services also contribute to an airline's market share.

One approach to doing this would require additional "weighting" parameters to be included and estimated in the model. That is, if a non-stop flight has a weight of 1.0, then a one-stop through flight departure would be assigned a lesser weight, while a connecting flight departure would be assigned an even lower weight. The weighted combination of all non-stop, one-stop and connecting flight departures (per day or per week) would then represent each airline's adjusted frequency share in the above model.

This approach has been tried in the past, with estimates of the relative weights of one-stop flights falling in the range of 0.2 to 0.3, and as low as 0.05 for connecting flight departures. This type of modified market share model is also very similar to "quality of service index" or QSI models used by airlines in determining the expected market shares for a planned schedule of departures in each market.

3.6   Chapter Summary

This chapter has introduced basic airline terminology and definitions, as well as several concepts related to air transportation markets and the demand for air travel. The most common measures of airline performance – RPK, yield, ASK, unit cost and load factor – were defined. They were then incorporated into a basic airline profit equation that illustrates the interdependence among these measures in airline management decisions.

Our discussion of markets in air transportation began with a description of a typical passenger airline trip, followed by several spatial definitions of origin–destination (O-D) markets. Different O-D markets are considered to be distinct and separate, but can be interrelated through "parallel markets." Multiple O-D markets also can share the joint supply of seats on a single flight leg, given the existence of multiple leg and connecting passenger itineraries. Because demand for air travel is generated at the level of a passenger's O-D trip, while a joint supply is provided to multiple O-D markets by a set of flight leg departures, there is an inherent inability to directly compare demand and supply in an individual O-D market.

The demand for air travel in an O-D market is affected by many different variables, including various socioeconomic characteristics of the origin and destination regions, prices of air services (and competing modes) and quality of service factors. The concept of total trip time was discussed as being the most important quality of service factor affecting the volume of O-D market demand for air travel.

The concepts of price and time elasticity were defined, and related to the segmentation of total demand by airlines, which have traditionally separated business and leisure demand for pricing and scheduling purposes. These elasticity concepts were then incorporated into several examples of O-D market demand functions, including a model that reflected the different segments of air travel demand in a market and the fact that there can be more than one air travel "product" offered to travelers.

Finally, the nature of competition for market share on the basis of airline frequency of service was described. The "S-curve" relationship between airline market share and frequency share reflects that higher airline frequency shares are associated with disproportionately higher market shares in competitive markets. This model explains why competing airlines use frequency as a competitive weapon, perhaps to a greater extent than other quality of service elements, to retain market share.

References

Airline Business (2005) *The Airline Industry Guide 2005/06*, September.

Aviation Daily (2004) "Change in Domestic Origin and Destination Traffic by Mile Group," October 11, p. 7.

Baseler, R. (2002) "Airline Fleet Revenue Management – Design and Implementation," *Handbook of Airline Economics*, 2nd Edition, Aviation Week, Washington, DC.

Belobaba, P.P. (1987) "Air Travel Demand and Airline Seat Inventory Management," Ph.D. thesis, Massachusetts Institute of Technology.

Button, K. and Drexler, J. (2005) "Recovering Costs by Increasing Market Share: An Empirical Critique of the S-Curve," *Journal of Transport Economics and Policy*, Vol. 39, Part 3, pp. 391–404.

Doganis, R. (1991) *Flying Off Course: The Economics of International Airlines*, Harper Collins, London, p. 246.

Gillen, D., Morrison, W., and Stewart, C. (2003) "Air Travel Demand Elasticities: Concepts, Issues and Measurement," Final Report, Department of Finance, Ottawa, Canada, www.fin.gc.ca/consultresp/Airtravel/airtravStdy$_e$.html.

Kahn, A. (1993) "Change, Challenge and Competition: A Review of the Airline Commission Report," *Regulation*, Vol. 3, pp. 1–10.

Official Airline Guides (2008) *OAG Pocket Flight Guide North American Edition*, June.

Simpson, R.W. (1970) "A Market Share Model for US Domestic Airline Markets," Flight Transportation Laboratory Memorandum M70-5, MIT Department of Aeronautics and Astronautics.

Simpson, R.W. (1995) "Markets in Air Transportation," Unpublished Notes for Air Transportation Economics Course 16.74, Massachusetts Institute of Technology.

Simpson, R.W. and Belobaba, P.P. (1992a) "The Demand for Air Transportation Services," Unpublished Notes for Air Transportation Economics Course 16.74, Massachusetts Institute of Technology.

Simpson, R.W. and Belobaba, P.P. (1992b) "The Supply of Scheduled Air Transportation Services," Unpublished Notes for Air Transportation Economics Course 16.74, Massachusetts Institute of Technology.

Taneja, N.K. (1976) *The Commercial Airline Industry*, D.C. Heath, Lexington, MA.

4

Fundamentals of Pricing and Revenue Management

Peter P. Belobaba

There are few facets of the airline business that generate as much discussion and confusion, among industry observers and consumers alike, as airline pricing and revenue management practices. "Pricing" refers to the process of determining the fare levels, combined with various service amenities and restrictions, for a set of fare products in an origin–destination market. "Revenue management" is the subsequent process of determining how many seats to make available at each fare level. Together, airline pricing and revenue management interact to create what can be a bewildering array of fare quotes for a consumer who simply wants to know how much it will cost to travel by air from one point to another.

In this chapter, we build on the concepts of origin–destination (O-D) demand, elasticities and demand segmentation, introduced in Chapter 3, to provide an overview of the economic rationale behind airline pricing and revenue management practices. The chapter begins with a brief discussion of airline pricing concepts and definitions of relevant terms, including product differentiation and price discrimination. The theory and practice of airline differential pricing is then explored, using several examples of airline fare structures. Recent trends in airline pricing, including the move to "simplified" and totally unrestricted fares by some low-cost carriers (LCCs), are then discussed.

The last section of this chapter is devoted to airline revenue management (RM), its underlying objectives, and the systems and models used by airlines to maximize revenues through the control of seat inventory availability. The evolution of airline RM models and systems, from traditional leg-based systems to recent advances toward network RM, is described. The impacts of recent changes in pricing strategies, including fare simplification, on the RM process are also discussed.

The Global Airline Industry   P. Belobaba, C. Barnhart and A. Odoni
© 2009 John Wiley & Sons, Ltd

4.1 Airline Prices and O-D Markets

As was the case with air travel demand, airline fares are defined for an O-D market, not for an airline flight leg. That is, airline prices are established for travel between origination point A and destination point C, where AC (or CA) is the relevant market. Given the "dichotomy of supply and demand" described in Chapter 3, it is possible for travelers in the AC market to choose from many itinerary (or path) options that can involve non-stop, one-stop or connecting flights. At the same time, a single flight leg serves many different O-D markets, each with its own set of prices.

The fact that airline prices are defined only for an O-D market gives rise to the following additional observations about airline pricing. Airline prices for travel A–C depend primarily on the volume and characteristics of the O-D market demand for travel between A and C (e.g., trip purpose and price elasticity of demand), as well as the nature of airline supply between A and C (frequency and path quality of flights) and the competitive characteristics in that market (number and type of airline competitors).

There is therefore no inherent theoretical reason for prices in market AC to be related to prices in another distinct and separate market AD, based strictly on distance traveled (even though this was more likely to be the case under previously regulated airline pricing regimes). The distance to be traveled is certainly an important element in determining the cost of providing airline service, and is thus reflected in price differences between markets in many cases. However, because the other market characteristics mentioned (demand elasticity, airline supply, nature of competition) all affect airline prices, it could well be the case that prices for travel A–C are actually lower than prices for A–D, even though A–C involves a greater travel distance. As defined in Chapter 3, these are *distinct and separate markets* with different demand characteristics, which might just happen to share the joint supply of seats on a flight leg.

4.1.1 Regulated vs. Liberalized Pricing

Under historical conditions of airline regulation, prices were subject to controls by a government agency. In the USA, the Civil Aeronautics Board (CAB) used a mileage-based formula to ensure equal prices for equal distances. A passenger wishing to fly on a non-stop flight from Boston to Seattle (approximately 4000 km or 2500 miles) would pay the same price as a passenger traveling on a double-connection service from Boise, Idaho, to Miami, Florida, covering about the same distance. Airlines were required to charge the same price for either passenger, despite the fact that the Boise–Miami O-D market is substantially smaller, and the costs to the airline of providing double-connection service on smaller aircraft are substantially higher on a per passenger basis. In terms of different price levels, airlines were allowed to offer only first-class and unrestricted economy fare (coach- or "tourist-"class) products, both of which were tied to the mileage-based fare formula.

With deregulated or liberalized airline pricing in the USA and in many countries throughout the world, this strict relationship between airline fares and distance traveled has become less apparent. Different O-D markets can have prices not related to distance traveled, or even the airline's operating costs, as airlines match low-fare competitors to maintain market presence and share of traffic. It is also possible that low-volume O-D

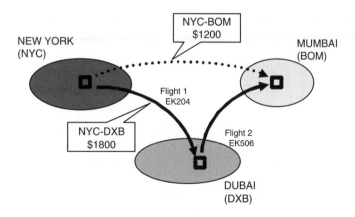

Figure 4.1 Example of O-D market price differences

markets that are more costly to serve on a per passenger basis will see higher prices than high-density O-D markets, even if similar distances are involved.

The relationship between O-D markets and airline prices is illustrated by the example shown in Figure 4.1. There are two distinct and separate O-D markets shown, New York (NYC) to Dubai (DXB) and NYC to Mumbai (BOM). As described in Chapter 3, distinct and separate markets have different demand volumes, different travelers with different price and time elasticities, and perhaps even different trip purposes and currency valuations. The one-way unrestricted economy-class prices shown on Figure 4.1 were in effect in November 2007 for travel in each O-D market: $1800 one way for travel on Emirates Airlines from NYC to DXB, and $1200 one way for travel on the same airline from NYC to BOM (with a connection in DXB). The comparable fare for NYC to BOM was substantially lower than for NYC to DXB, despite the substantially greater distance between NYC and BOM.

This type of apparent inconsistency in airline prices occurs because the two markets are distinct and separate, with different demand characteristics, as mentioned. Moreover, competition can also explain many such inconsistencies – if a non-stop competitor offers a $1200 fare NYC–BOM, then Emirates is likely to match that fare to retain its market share of the NYC–BOM demand, even if the fare is lower than what the airline charges for the shorter-distance market NYC–DXB. In economic terms, such pricing is entirely reasonable – different markets with different demand characteristics and competitive environments are priced differently. For passengers, however, it can be perplexing, given that the NYC–BOM passenger makes use of the same NYC–DXB flight and can sit next to a passenger who paid much more in the shorter O-D market. This is another vivid illustration of the dichotomy of demand and supply in air transportation networks.

4.1.2 Theoretical Pricing Strategies

The different theoretical bases that an airline might use for establishing the prices that it wishes to charge for air transportation services are introduced in this section. In theoretical terms, for determining the prices to charge in an O-D market, airlines can utilize one of following economic principles (Simpson and Belobaba, 1992):

- Cost-based pricing
- Demand-based pricing
- Service-based pricing.

In practice, most airline pricing strategies reflect a mix of these theoretical principles. As mentioned, prices are also highly affected by the nature of competition in each O-D market. The presence of a low-fare airline in an O-D market is perhaps one of the most important determinants of average fare levels. In recent years, severe competition from low-fare airlines in some markets has led airlines to follow what might even be called "price-based costing," in which airlines try to reduce costs to be able to match low-fare competitors and passengers' expectations of low prices.

Cost-Based Airline Pricing

Microeconomics textbooks make reference to the practice of "marginal cost pricing," in which the producer sets its prices equal to the marginal cost of producing an incremental unit of output. This practice is one of the theoretically optimal conditions of "perfectly competitive" markets, which do not exist in the real world.

In the short run, the costs to an airline of operating a schedule of flights are effectively fixed. The commitment to operate a scheduled service irrespective of the number of passengers on board means that not only aircraft ownership costs, but crew costs and even fuel costs, can be considered as fixed for a planned set of flights. The marginal costs of carrying an incremental passenger are therefore very low – essentially the cost of an additional meal and a very small amount of incremental fuel. Therefore, airlines could not possibly cover their total operating costs under a strict marginal pricing scheme in which marginal costs are attributed to an incremental passenger carried on a flight.

An alternative approach to cost-based pricing is that of average-cost pricing. Under this pricing principle, an airline would set its prices in all O-D markets based on system-wide operating cost averages per flight or per available seat kilometer (ASK). This is in fact the pricing principle used under regulated airline regimes, as described above, so its use is feasible in airline markets, but its shortcomings are what led to the deregulation of airline pricing in the first place.

Average-cost pricing ignores airline cost differences in providing services to different O-D markets. It allows smaller markets to benefit (with artificially low prices) at the expense of higher-density markets that airlines can serve more efficiently (e.g., with larger aircraft). Some have argued that average-cost pricing ensures "equity," if we believe that air travel at equal prices per kilometer traveled represents a form of air transportation equity.

Demand-Based Pricing

The principle of "demand-based" pricing is based on consumers' "willingness to pay," as defined by the price–demand curve in each O-D market. The underlying assumption is that there are some consumers who are "willing" to pay a very high price for the convenience of air travel, while others will only fly at substantially lower prices. Under this approach, airlines charge different prices to different consumers with different price sensitivity. The price elasticities of different demand segments and different O-D markets reflect their sensitivity to the prices of air travel, and the airline sets different prices for each segment in an attempt to maximize its total revenues.

Demand-based pricing results in different prices for different O-D markets, even for different demand segments within the same market. These price differences are not related to cost differences experienced by the airline in providing services to the different demand segments, only to the differences in price sensitivity, demand elasticity and "willingness to pay." This practice is referred to as strict "price discrimination" by economists.

Service-Based Pricing

The third theoretical pricing principle uses differences in the quality of services (and, in turn, in the cost of providing these services) as a basis for pricing. Even under US regulation of airline prices, some service distinctions were allowed in airline pricing structures (i.e., first class vs. economy class) due to the different costs to the airlines of providing them. In theory (and in practice), the notion of *fare product differentiation* can be extended beyond this simple first- vs. economy-class distinction.

Unlike demand-based pricing, service-based pricing has a differential cost basis for the airline. Because higher-quality services generally cost the airline more to produce, this approach cannot be considered "price discrimination." Even if the on-board product (i.e., economy seat and meal service) is the same, lower fares with advance purchase requirements actually represent an opportunity cost savings to the airline, as the airline is better able to reduce uncertainty about loads on future departures and reduce the risk of lost revenue potential from empty seats.

4.1.3 Price Discrimination vs. Product Differentiation

In the preceding discussion of theoretical airline pricing principles, references were made to both "price discrimination" and "product differentiation." It is important to recognize the difference between these terms, as the discussion moves toward understanding how airlines apply these principles in practice.

Price discrimination is the practice of charging different prices for the same (or very similar) products that have the same costs of production, based solely on different consumers' "willingness to pay" (Tirole, 1988). On the other hand, *product differentiation* involves charging different prices for products with different quality of service characteristics and therefore different costs of production (Botimer and Belobaba, 1999).

Most airline fare structures reflect both of these strategies. Product differentiation is clearly evident in the variety of *fare products* offered by airlines in the same O-D market. Fare product differentiation by airlines involves not only differences in tangible quality of services (e.g., first vs. economy class), but differences in the purchase and travel conditions associated with different fare products, most notably those with the lowest price levels.

At the same time, the large differences in price levels charged for the different fare products offered by many airlines within the economy class of service cannot be explained by product differentiation principles alone. The substantially higher prices that airlines charge for unrestricted fare products targeted at business travelers are also based on their greater willingness to pay, suggesting that price discrimination is a component of these pricing strategies. The term "differential pricing" will be used throughout the remainder of this discussion to refer to current airline pricing practices, which reflect both product differentiation and price discrimination principles.

4.2   Airline Differential Pricing

The use of differential pricing by airlines in an O-D market is designed to present a range of fare product options to consumers, who must make a trade-off between the inconvenience of fare restrictions associated with lower fares and the higher prices of unrestricted fares. In microeconomic terms, airline fare structures allow each consumer to maximize his or her utility (or minimize disutility) subject to a budget constraint.

Business travelers are assumed to be willing to pay higher fares in return for more convenience and fewer restrictions on the purchase and use of tickets, meaning price is less important to them than the disutility of these restrictions. Leisure travelers are less willing to pay higher prices, but accept the disutility "costs" of restrictions on low-fare products, longer travel times associated with connecting flights, and a lower quality of on-board service.

The economic concept of "willingness to pay" (WTP) is defined by the theoretical price–demand curve. The price–demand curve can be interpreted as the maximum price that any given number of consumers will all pay for a specified product or service. The use of differential pricing principles by airlines is an attempt to make those with higher WTP purchase the less restricted, higher-priced fare product options.

In Figure 4.2, a price–demand curve for an O-D market is illustrated. If the airline offers an unrestricted fare P1 to those consumers with higher WTP, we would expect that Q1 consumers will purchase this fare because they have a WTP equal to P1 or greater. If the airline also offers a lower or "discount" fare P2 to those consumers with a lower WTP, then Q2 – Q1 additional consumers would be expected to purchase this lower fare, as they have a WTP greater than P2 but less than P1. This simple model assumes that the

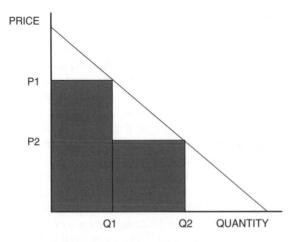

Figure 4.2   Differential pricing model

airline has a perfect ability to segment demand according to WTP, such that all consumers with high WTP purchase the higher fare P1.

Accepting for the moment the assumption that the airline does have the perfect ability to segment its market demand by WTP as shown in Figure 4.2, the advantages of differential pricing for both the airline and consumers can be identified. For the airline, offering two different fares instead of a single fare for all passengers (which could be between P1 and P2) allows it to increase total flight revenues with little impact on total operating costs. Incremental revenue will clearly be generated by discount fare passengers who otherwise would not fly at a single fare between P1 and P2. Incremental revenue will also be generated from the high-fare passengers willing to pay P1, which is more than what the airline would charge under a single price strategy. At a single fare level, many "legacy" airlines with high costs would be unable to attract enough passengers (and revenue) to cover the total operating costs of their flights.

Consumers can also benefit from the airlines' use of differential pricing. Obviously, the discount passengers paying P2 who otherwise would not fly at the single fare level benefit from the practice. While it is true that some high-fare passengers paying P1 are paying more than they would if the airline offered a single price level between P1 and P2, it is also conceivable that these high-fare passengers actually end up paying less and/or enjoy more frequency of flights given the presence of low-fare passengers. This argument is based on the premise that, without low-fare passengers to contribute incremental revenue to the operating costs of the airline, high-fare passengers would have to pay even higher fares and/or have a reduced set of flight departure options.

The above discussion is the basis of much disagreement among consumers and even some government regulators, as the common perception is that the airlines' practices of charging business travelers substantially more than leisure travelers are blatantly unfair to business travelers. It is important, however, to recognize that economic theory supports and helps to explain these pricing practices.

4.2.1 Market Segmentation

The successful use of differential pricing principles depends on the airline's ability to identify different demand groups or *segments*. In theory, total revenue in an O-D market (or even on a single flight) is maximized when *each* customer pays a different price equal to his or her WTP. In the context of the price–demand curve shown in Figure 4.2, the entire triangular area under the curve represents the total potential revenue available in a market. If the airline could charge a different price for each consumer based on his or her maximum WTP, its revenues would be close to this theoretical maximum total potential revenue. In practice, such a theoretical segmentation is clearly impossible to achieve as airlines cannot determine each individual's WTP for a given trip, nor can they publish different fares available only to specific individuals.

Instead, airlines identify segments of the total market demand with similar characteristics, in terms of trip purpose, price sensitivity and time sensitivity. Business and leisure travelers are the two traditional segments targeted by airlines in their differential pricing efforts. Even with recent shifts in demand patterns and the decreasing proportion of business versus leisure passengers, this is still the most important distinction made between air travel demand segments for pricing purposes. It is possible for the airline to further increase revenues with more prices and products targeted at additional demand segments, but it becomes more difficult to identify differences in purchase and travel behaviors between these additional segments.

To achieve the required demand segmentation in practice, airlines can physically differentiate their fare products by offering clearly identifiable different products with different quality of service, such as first class and business class in addition to economy (or coach) class. On the other hand, restrictions on the advance purchase, use and refundability of lower-priced fare products within the economy-class cabin, although not physical product differentiators, are designed to reduce the attractiveness (increase the disutility) of these fare products, particularly to business travelers.

The combination of greater service amenities and lack of restrictions on the so-called "full coach fares" makes these fare products more attractive to business travelers, relative to the more restricted discount fare products. For example, even though travelers on full coach fares typically receive the same quality of on-board service as those paying reduced fares, some airlines provide priority seat assignment and special check-in services for full-fare travelers, increasing the attractiveness of the unrestricted full-fare product. At the same time, the complete absence of restrictions on the full-fare product is in itself a differentiating factor (compared to the lower, restricted fares) that is highly valued by some business travelers.

The overall goals for an airline establishing a differentiated fare structure in any O-D market can be summarized as follows. A wide enough range of fare product options at different price levels should be offered to capture as much of the revenue potential from the market price–demand curve as possible, while ensuring that each fare product can be targeted at specific demand segments with different levels of WTP. At the high end of the fare structure, the airline should offer enhanced service amenities that improve the attractiveness of the fare products to travelers who are not price sensitive and willing to pay for these amenities. And, at the low end of the fare structure, prices low enough to *stimulate* new demand for low-fare travel should be offered to fill empty seats that would otherwise remain empty.

The most difficult goal is to find mechanisms to prevent the *diversion* of consumers with higher WTP (who were expected to buy the higher-fare products) to the lower-fare products, given that they were planning to fly anyway and that they could well be aware of the lower-priced options.

4.2.2 Fare Product Restrictions

As introduced earlier, the application of progressively more severe restrictions on low-fare products is the primary mechanism used by airlines to prevent diversion, as differences in service amenities are generally not enough to prevent many high WTP travelers from buying lower fares. The types of restrictions applied to the lower fares in most O-D markets are familiar to most air travelers. The lowest fares usually have advance purchase and minimum stay requirements, as well as cancellation and change fees. These restrictions increase the inconvenience or "disutility cost" of low fares to travelers with high WTP, causing them to choose higher fares when they minimize their own disutility of air travel. Studies have shown that the "Saturday night minimum stay" condition is among the most effective in keeping business travelers from purchasing low fares (Boeing, 1988). Longer minimum stay conditions (e.g., 7 days) are more common in longer-haul international markets, as the Saturday night minimum stay restriction is not sufficient to prevent diversion of business passengers to lower fares on longer-haul trips.

Even with the use of a variety of fare restrictions, it is impossible to achieve the perfect segmentation of demand implied by the model in Figure 4.2. Some proportion of travelers with high WTP will be able to meet even the most severe restrictions, or alternatively will replan their trips to allow them to meet these restrictions. Airline data shows that some business travelers have long been able to purchase restricted fares by rearranging their travel plans. This practice became even more common as the price differences between the highest unrestricted full coach fares and the lowest available restricted fares increased, and as business travelers became less willing to pay the highest fare levels.

A major challenge for airlines that practice differential pricing is the establishment of a set of fare product restrictions that can reduce the potential for diversion, especially as passengers' valuations of restriction disutilities change. Almost any set of fare restrictions will be imperfect in terms of demand segmentation. In the simple differential pricing model shown in Figure 4.3, some percentage of those in the higher demand segment will buy the lower fare at P2, which is lower than their WTP. If, for example, 50% of those with a high WTP that the airline had expected to buy the higher fare at P1 actually divert to P2, revenue is lost (lighter tinted area in Figure 4.3). This example illustrates the critical importance of effective restrictions to support fare product differentiation. At the extreme, inadequate restrictions on lower-fare products, or "fare fences," can result in so much diversion that the revenue benefits of differential pricing disappear.

On the other hand, if they are unable to purchase a low fare, some passengers will "sell up" to a higher fare if they are able to meet the conditions of the lower fare, but cannot find low-fare seat availability. This behavior is also known in the airline industry as "trade-up" and "buy-up." Sell-up is really the opposite of diversion. Only those with a higher WTP will even consider a higher fare if they are unable to obtain a booking at the lower fare. Those with a higher WTP were only interested in the lower fare in the first place because of inadequate restrictions on it.

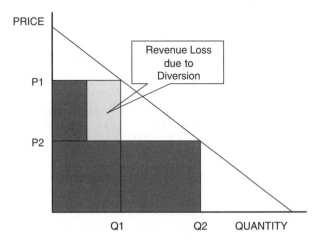

Figure 4.3 Revenue impact of diversion

4.2.3   Traditional Airline Fare Structures

As introduced above, multiple "fare products" with different characteristics are offered to different demand segments assumed to have varying levels of price and time sensitivity, as well as WTP. Given the ability to set their own fare structures in deregulated markets, airlines moved toward differentiated fare structures based on both service amenities and fare restrictions. The service amenities designed to make higher-priced fare products more attractive to consumers with a higher WTP include extra seating space and premium meals to physically differentiate the first- and business-class products, for example. Many airlines also offer special check-in lines, preassigned seats and/or lounge access for full economy fare passengers. As mentioned earlier, the complete flexibility of purchase and travel associated with unrestricted fares has effectively become an "amenity" relative to the more restricted lower-priced fares.

Fare product restrictions are designed to make low fares less attractive to those with a higher WTP, while still offering those with lower WTP a viable travel option. As a result, a Saturday night (or longer) minimum stay has historically been associated with most discount fares. Advance booking and ticket purchase requirements for discount fares range from 7 to 21 days in most O-D markets. In addition, the lower-priced fare products carry non-refundability conditions and/or change and cancellation fees.

Figure 4.4 shows a simplified example of a traditional fare structure that relies on fare restrictions to differentiate the fare products offered within the economy class of service (i.e., not including first or business classes). In this hypothetical example, each of the fares lower than the full economy ("Y") fare has restrictions requiring advance purchase, a Saturday night minimum stay for the traveler, as well as various change fee and/or non-refundability conditions. The restrictions become more severe as the level of discount from the full economy fare increases.

This type of fare structure reflects substantial differentiation of fare products based on the use of restrictions rather than service amenities. It is effective in preventing business travelers with high WTP from making use of the lower fares. In fact, any business traveler who is not able to or does not wish to stay over Saturday night on his or her business trip has little choice but to purchase the highest "Y" fare. Even if a business traveler is willing to stay over Saturday night, the lower fares are not an option if the trip cannot be booked more than 7 days in advance or if the traveler wants to retain the flexibility to make changes and/or obtain a refund should the trip have to be cancelled.

Fare Code	Price (USD)	Advance Purchase	Saturday Night Min. Stay	Change Fee	Non-Refundable
Y	$800	--	--	--	--
B	$400	7 day	Yes	Yes	--
M	$300	14 day	Yes	Yes	Yes
Q	$200	21 day	Yes	Yes	Yes

Figure 4.4   Traditional fare structure with restrictions

Round-trip Fare ($)	Cls	Advance Purchase	Minimum Stay	Change Fee?	Comment
458	N	21 days	Sat. Night	Yes	Tue/Wed/Sat
707	M	21 days	Sat. Night	Yes	Tue/Wed
760	M	21 days	Sat. Night	Yes	Thu-Mon
927	H	14 days	Sat. Night	Yes	Tue/Wed
1001	H	14 days	Sat. Night	Yes	Thu-Mon
2083	B	3 days	None	No	2 X OW Fare
2262	Y	None	None	No	2 X OW Fare
2783	F	None	None	No	First Class

Figure 4.5 Boston–Seattle fares – American Airlines, October 1, 2001
Data source: Travelocity.com

Such a differentiated fare structure has come to be known as a "traditional" or "restricted" fare structure, as compared to the less restricted fare structures to be discussed below. A real-world example of such a fare structure is provided in Figure 4.5, which summarizes the fare products offered in 2001 by American Airlines for travel in the Boston–Seattle market. This fare structure is reasonably representative of the type of fare product differentiation that was very common in US domestic markets throughout most of the 1990s, especially in markets dominated by traditional "network" carriers in which no low-fare competitor was present.

The highest unrestricted economy fare (Y) is almost five times that of the lowest discount fare with restrictions, although this ratio can be as great as eight times the lowest fare in some similar markets. The first-class fares are even higher, relative to the "full" economy fare. Most noteworthy is the fact that all fares with any meaningful discount from the unrestricted fare require both an advance purchase *and* a Saturday night minimum stay. Furthermore, the lowest fares are limited to specific flights, days of the week and/or have a limited "seat sale" duration for booking and ticketing, as outlined in the "Comment" column of Figure 4.5.

Contrary to popular belief, the use of differential pricing practices is not limited to traditional network airlines, in the USA or elsewhere. Virtually every "low-fare" airline in the world offers multiple fare products at different fare levels for travel in the same O-D market and on the same flight(s). For example, Figure 4.6 shows the published fare structure of Southwest Airlines for travel between Providence (Rhode Island) and Seattle, for the same day as the American Airlines example above. Southwest's fare structure offers six different fare products, with increasing advance purchase requirements and a minimum stay restriction for the lower fare levels. Admittedly, the range of fare values is lower than that of the Boston–Seattle example (with a ratio of highest to lowest fare of about 3.5 to 1) and the severity of the restrictions is somewhat reduced. Nonetheless, it is important to realize that even low-fare airlines make use of the differential pricing model to increase their total revenues.

It is also important to note that the same approaches to differential pricing and use of traditional restricted fare structures became widespread in international markets as well. Figure 4.7 shows an example of a fare structure in the Los Angeles–Auckland market

Round-trip Fare ($)	Cls	Advance Purchase	Minimum Stay	Non-Refund?	Comment
178	M	3 days	1 Night	Yes	Special Sale
402	H	7 days	1 Night	Yes	
438	Q	14 days	None	Yes	2 X OW Fare
592	B	7 days	1 Night	Yes	Off-peak
592	B	7 days	1 Night	Yes	Peak
634	Y	None	None	No	Unrestricted

Figure 4.6   Providence-Seattle fares – Southwest Airlines, October 1, 2001
Data source: Southwest.com

Round-trip Fare (US$)	Cls	Advance Purchase	Min Stay	Max Stay	Comments
1148/1218	Q	21 days	7	30	Off-peak/Peak
1548/1618	K	14 days	7	90	Off-peak/Peak
1698/1768	H	14 days	7	180	Off-peak/Peak
1958	M	14 days	None	None	2 X OW Fare
2258	B	None	7	365	
3481	Y	None	None	No	2 X OW Fare
5336	C	None	None	No	Business Class
8844	F	None	None	No	First Class

Figure 4.7   Los Angeles–Auckland fares, October 1999
Data source: Travelocity.com

from 1999, in which advance purchase restrictions are combined with both minimum and maximum stay requirements to make the lowest fares most attractive to leisure travelers able to plan their trips in advance and stay at least 7 days at their destination. Note that the minimum stay requirement is 7 days in this long-haul market, given that even business travelers are likely to stay at the destination for 3–4 days and are more likely to include a Saturday night in their trip.

4.2.4   Recent Trends in Airline Pricing

The use of the differential pricing principles described in previous sections improved the ability of airlines to generate revenues and cover what in retrospect were relatively high-cost operations. The combination of multiple price levels, fare products with different restrictions and control of the number of seats made available at lower fares through RM controls contributed significantly to the record airline industry profits of the late 1990s. These practices led to higher load factors and increased unit revenues (passenger revenue/ASK), as airlines embraced the notion of pricing based on their perception of passengers' WTP.

In recent years, however, the ability of legacy airlines to extract such revenue premiums has changed dramatically. Beginning in early 2001, the revenue model of the legacy airlines started to unravel dramatically. Business passengers became much less willing to pay 5, 8 or even 10 times the lowest available fare in a market for travel on the same airline in the same economy-class seat, irrespective of the differences in fare product restrictions. The rapid growth of Internet distribution channels, both airline websites and electronic travel agencies such as Expedia, Orbitz in the USA and Opodo in Europe gave more consumers more information about alternative fare and airline options than they ever had before. And the emergence of low-fare airlines increased the number of price, product and itinerary options available to consumers.

Beginning in early 2001, it started to become apparent to airline pricing departments that an unprecedented shift in consumer demand and booking patterns might be under way. Business air travel demand dropped by 30% in the first half of 2001, due in part to the dot.com bust, stock market slide and resulting economic downturn. The decline in business travel was exacerbated by the 9/11 terror attacks and subsequent security hassles that affected frequent travelers more than vacationers. The remaining business travelers, faced with reduced travel budgets, became more willing to consider restricted lower-price fare products in order to avoid paying the highest unrestricted fares. At the same time, cost cutting by legacy carriers after 9/11 led to service quality cutbacks that all but eliminated any remaining perceptions of value difference between legacy carriers and low-fare airlines. Combined with the rapid growth of both Internet distribution channels and low-fare airlines, these forces presented the airline industry with an unprecedented revenue challenge.

The heavily restricted fare structures described in the previous section allowed legacy airlines to effectively segment their demand and to force passengers to pay fares closer to their "WTP" – that is, to extract additional revenues from consumer surplus. In the absence of the revenue challenges described above, there would be little incentive for the high-cost legacy airlines to abandon what proved to be a very profitable approach to pricing. However, with these significant shifts, especially with the growth of low-cost competition that does not rely on this traditional pricing model to make a profit, a pricing trend toward "simplified" fare structures began to spread to air travel markets, not only in the USA but around the world.

"Fare simplification" refers to airline pricing strategies that involve fewer fare levels, less onerous restrictions and more compressed ratios of the highest to lowest fares offered in any given O-D market. Typically, the restriction that tends to be removed in the interest of fare simplification is that of the "Saturday night minimum stay." Figure 4.8 shows a

Fare Code	Price (USD)	Advance Purchase	Saturday Night Min. Stay	Change Fee	Non-Refundable
Y	$490	--	--	--	--
B	$325	7 day	--	Yes	--
M	$250	14 day	--	Yes	Yes
Q	$175	21 day	--	Yes	Yes

Figure 4.8 Simplified fare structure

"simplified" version of the hypothetical restricted fare structure introduced in Figure 4.4. The Saturday night stay restrictions have been removed, the prices for each product have been lowered, and the range of prices has been compressed. The advance purchase restrictions, however, still remain as a demand segmentation tool. Also, change fees and non-refundability restrictions have been retained for the intermediate and lowest fares.

While consumers, and business travelers in particular, welcomed the elimination of the minimum stay restriction on lower fares, its removal also took away from the airline its most effective way of segmenting business and leisure demand, as described earlier. Simulation studies have shown that removal of this powerful segmentation restriction, with all else equal, can lead to airline revenue losses of 10–15% as more business travelers are more frequently able to meet the reduced restrictions of the lower fares (Dar, 2006).

A real-world example of a simplified fare structure for the Boston–Seattle O-D market discussed previously is shown in Figure 4.9. This fare structure was introduced by Alaska Airlines, which implemented a similar simplified fare structure virtually system-wide in the USA in 2004. Compared to the more restricted fare structure shown for this market in Figure 4.4, two major changes are noteworthy. First, none of the fare products have a Saturday night minimum stay restriction, although the lowest fares still have a one-day minimum stay restriction that effectively forces consumers to purchase a round-trip ticket to make use of those fares. Second, the ratio of the highest to lowest fares in the economy cabin has dropped from the previous 5:1 to no more than 3.5:1. The unrestricted "full coach fare" has decreased by 43% to $948 round-trip. On the other hand, the number of fare products offered has not decreased to a great extent – from seven different fare levels/product types in 2001 to six in the new structure (including first class).

The move to such "simplified" fare structures, however, does not reflect an abandonment of the differential pricing and demand segmentation principles described earlier. In Figure 4.9, the advance purchase restrictions are still in place, and are in effect segmenting consumers by their ability to plan and commit to their trips in advance. This segmentation by booking time prior to departure, although not as effective in segmenting business from leisure travelers, is nonetheless a proxy for such segmentation. The fare structure still assumes that business travelers tend to book closer to departure and/or are unwilling to risk the purchase of a ticket that is non-refundable or difficult to change far in advance.

Round-trip Fare ($)	Cls	Advance Purchase	Minimum Stay	Change Fee?	Comment
374	V	21 days	1 day	Yes	Non-refundable
456	L	14 days	1 day	Yes	Non-refundable
559	Q	14 days	1 day	Yes	Non-refundable
683	H	7 days	1 day	Yes	Non-refundable
827	B	3 days	None	No	2 X OW Fare
929	Y	None	None	No	2 X OW Fare
1135	F	None	None	No	First Class

Figure 4.9   Boston–Seattle fares – Alaska Airlines, May 2004
Data source: SABRE GDS

At the same time, the advance purchase segmentation scheme attempts to capture the increased WTP of consumers who book closer to departure date, irrespective of their trip purpose.

Impacts on Differential Pricing Model

The differential pricing model introduced in this section was the basis for the evolution of pricing practices at many airlines over the past 30 years. Yet, despite the major changes in demand patterns, consumer choice behavior and competitive fare structures that have occurred since 2000, the fundamental theory of differential pricing remains valid and is still reflected in the simplified fare structures described above. However "simplified," fare structures like the one shown in Figure 4.9 are still designed to extract higher fares from consumers with a higher WTP closer to departure.

Simpler fare structures have tended to be associated with low-fare carriers, and the greatest motivation of legacy carriers to simplify their own fare structures has clearly been the desire to stop the loss of market share to low-fare competitors. It is important to note, however, that virtually all low-fare carriers do indeed offer differentiated price levels for seats on the same flight in the same economy cabin. At the extreme, low-fare airlines such as easyJet and Ryanair offer perhaps the simplest fare structures in terms of restrictions. While some might call their fare structures "unrestricted," a more accurate characterization would be to call them "undifferentiated."

For example, easyJet offers as many as 13 different price levels on a single flight departure, but each fare has the same restrictions (non-refundable with a change fee). And, although this low-fare airline does not associate explicit advance purchase requirements with its lowest fares, most travelers familiar with easyJet recognize that the quoted price for travel increases closer to departure, meaning that there are effectively advance purchase requirements being imposed by the airline's RM system, which closes availability of lower fare classes closer to departure.

The perception that low-fare airlines all have substantially simpler fare structures than their network airline counterparts has been an effective marketing tool for them. However, not only do the vast majority of low-fare airlines offer multiple price levels for travel on a single flight in an O-D market, but many also impose advance purchase and/or cancellation or change restrictions on their lowest fares. The difference between their fare structures and those of traditional airlines is that the differentiated fare structures of low-fare carriers are somewhat less restricted and narrower in terms of the ratio of fare levels offered.

Thus, it is not the differential pricing model that has changed in recent years, but its parameters. Figure 4.10 graphically illustrates how the major shifts described in this section can be described by the same differential pricing model introduced in Section 4.2. The overall decrease in business travel demand and a reduced WTP the highest fares are reflected by an inward shift of the demand curve, more so at its upper end. A greater willingness to accept restriction on lower fares is illustrated by the diversion of previous high-fare demand to lower fares. The reduction by airlines of the lowest fares is also shown, causing total loads to return to previous levels. However, the overall result is a lower total revenue, represented by the empty boxes under the demand curve. Figure 4.10 is a highly simplified representation of many changes, but it nonetheless illustrates how the fundamental concepts of the differentiated pricing model remain intact.

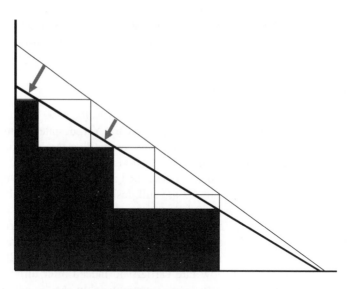

Figure 4.10   Shifts in differential pricing model

4.3   Airline Revenue Management

Although the discussion has focused on airline pricing thus far, there are in practice two different, yet related, components of airline revenue maximization:

- Differential pricing: As described, various "fare products" are offered at different prices with different characteristics for travel in the same O-D market.
- Yield management (YM): This process determines the number of seats to be made available to each "fare class" on a flight, by setting booking limits on low-fare seats.

Most airlines have implemented YM systems that routinely and systematically calculate the booking limits on each fare class (or booking class) for all future flight departures. Typically, YM systems take a set of differentiated prices/products, schedules and flight capacities as given. Under the assumption that the fixed operating costs associated with a committed flight schedule represent a very high proportion of total operating expenses in the short term, the objective of revenue maximization is effectively one of profit maximization for the airline.

YM became necessary when airlines began to realize that differential pricing alone is not enough to maximize revenues. Both leisure (discount) and business (full-fare) consumers typically prefer to travel at the same times, and compete for seats on the same flights (e.g., Friday afternoon and Sunday afternoon peak periods). Without *capacity controls* (booking limits) on discount fare seats, it is more likely that leisure travelers will displace business passengers on peak demand flights. This is due to the fact that leisure travelers tend to book before business travelers, a phenomenon made worse by advance purchase requirements on discount fares.

The main objective of YM is therefore to *protect seats* for later-booking, high-fare business passengers. This is accomplished by forecasting the expected future booking demand for higher fare classes and performing mathematical optimization to determine the number of seats that should be "protected" from (or not sold to) lower fare classes. In turn, any seats that are not protected for future high-fare demand are made available to lower fare class bookings.

YM involves the tactical control of an airline's seat inventory for each future flight departure. Given scheduled flights and capacities, as well as prices and products, YM is the airline's "last chance" to maximize revenue. Aircraft acquisition decisions will have been made years earlier, scheduling decisions will have been made six months or more prior to departure, and pricing decisions will have been made several months in advance. Setting booking limits on the different fare classes offered on a specific flight departure is a dynamic and tactical way for the airline to maximize total flight revenues, given the previous aircraft, scheduling and pricing decisions.

But too much emphasis on *yield* (passenger revenue per RPK, see Section 3.1) can lead to too much protection for high-yield passengers and overly severe limits on low fares, resulting in lower overall load factors for the airline. On the other hand, too many seats sold at lower fares will increase load factors but reduce yield and leave inadequate protection for late-booking high-yield passengers, with a potentially adverse affect on total airline revenues. As mentioned, *revenue* maximization is the proper goal for an airline. Achieving this goal requires a balance between load factor and yield. As a result, many airlines now refer to "revenue management" (RM) instead of "yield management" (YM). The two terms have come to be used interchangeably in the airline world.

To maximize revenue, RM systems try to fill each available seat on each future flight departure with the highest possible revenue. RM booking limits support the objectives of differential pricing, i.e., to make consumers with higher WTP purchase higher fares. On high-demand flights, RM systems will set booking limits on discount fare and group bookings, in order to protect seats for later-booking high-fare passengers. This can lead to slightly lower average load factors for the airline overall, but higher yields and increased total revenues. On low-demand flights with excess capacity, the proper RM principle is to sell the empty seats at almost any low fare by not setting stringent booking limits on low-fare classes. This can result in higher average load factors and lower yields for the airline, but higher total flight revenues.

Figure 4.11 provides an example of potential outcomes under RM strategies designed to maximize yield, load factor and revenue for a hypothetical flight with five booking classes. Under a yield maximization approach, the airline might decide to limit low-fare bookings too much, leading to higher yields but relatively low load factors. Under a load factor maximization approach, the airline takes a large proportion of low-fare traffic, but less high-fare traffic is carried. In this example (and as a general rule), an emphasis on load factors tends to lead to greater total revenues than a yield maximization emphasis – within reason, a seat filled with any revenue is more valuable to the airline than an empty seat.

The correct RM strategy, as mentioned, is to manage the seat inventory of each flight departure to maximize total flight revenues. As shown in Figure 4.11, a revenue emphasis leads to average load factors that are lower than under the load factor emphasis approach, and yields that are lower than under the yield emphasis approach. However, the total revenue for the flight is maximized.

EXAMPLE: 2100 km FLIGHT LEG   **CAPACITY = 200**

		NUMBER OF SEATS SOLD:		
FARE CLASS	AVERAGE REVENUE	YIELD EMPHASIS	LOAD FACTOR EMPHASIS	REVENUE EMPHASIS
Y	$420	20	10	17
B	$360	23	13	23
H	$230	22	14	19
V	$180	30	55	37
Q	$120	15	68	40
	TOTAL PASSENGERS	110	160	136
	LOAD FACTOR	55%	**80%**	68%
	TOTAL REVENUE	$28 940	$30 160	**$31 250**
	AVERAGE FARE	$263	$189	$230
	YIELD (CENTS/RPK)	**12.53**	8.98	10.94

Figure 4.11   Approaches to seat inventory control

4.3.1   Computerized Revenue Management Systems

The size and complexity of these airline seat inventory control problems require the use by airlines of computerized RM systems. Consider an airline that operates 500 flight legs per day, offers 15 booking classes in its reservations system and accepts reservations for each flight departure up to 330 days before the departure date. At any point in time, this airline's seat inventory includes almost 2.5 million booking limits that represent the airline's total revenue potential. Even with a large team of human RM analysts, monitoring and manipulating this large volume of inventory limits would be impractical.

Airline RM systems have thus evolved in both their computer database and mathematical modeling capabilities over the past 15–20 years. The first RM systems, developed in the early 1980s, were designed to collect and store data extracts from computerized reservation systems (CRSs). This capability provided a large amount of data to the human RM analyst, but did not provide any guidance as to what actions should be taken (in terms of booking limits) to improve airline revenues. The RM analysts were left to make their own judgments about inventory controls.

By the mid-1980s, several RM systems offered additional monitoring capabilities, as the systems could compare actual flight bookings to an expected or "threshold" booking curve for the flight, and then issue "exception reports" to RM analysts whenever a flight's bookings deviated from the expected booking profile. These exception reports identified flights that might require an RM analyst's attention, but did not provide the analyst with a recommendation as to what booking limit changes should be made. Although popular with RM analysts, these monitoring systems depended to a large extent on judgmental estimates of the "threshold curves" or, alternatively, on historical booking profiles to generate these curves. In either case, there was no guarantee that following the expected "threshold" curve would lead to revenue maximization for the flight.

By the late 1980s, the more advanced airlines and RM system vendors began to develop RM systems that could perform forecasting and optimization by booking class for each flight leg departure, in addition to having the same database and booking monitoring capabilities of previous systems. A typical "third-generation" RM system therefore includes the following capabilities:

- Collects and maintains historical booking data by flight and booking class.
- Forecasts future demand by flight departure date and booking class.
- Makes use of mathematical models to optimize total expected flight revenues, by determining both the optimal overbooking levels by aircraft compartment and the optimal booking class limits by booking class within each compartment (e.g., first, business, economy).
- Provides interactive decision support for RM analysts, allowing them to review, accept or reject the overbooking and booking limit recommendations.

The major components of a third-generation "automated booking limit system" are illustrated in Figure 4.12. Historical booking data for the same flight leg and day of week are combined with actual booking information for the future flight departure to generate a forecast of booking demand by booking class for the specific departure. These forecasts by booking class, together with estimates of the revenue value of each booking class, are then fed into an optimization model which calculates the recommended booking limits for each booking class on the flight departure in question. At the same time, the demand forecasts are fed into an overbooking model, which also makes use of historical information about passenger no-show rates for the same flight leg and day of week to calculate an optimal overbooking level for the future flight departure. Both the booking class limits and overbooking levels are calculated by the mathematical models and then summarized as recommendations to the RM analyst.

All third-generation RM systems revise their forecasts and booking limits at regular intervals during the flight booking process, as often as daily in some cases. Actual

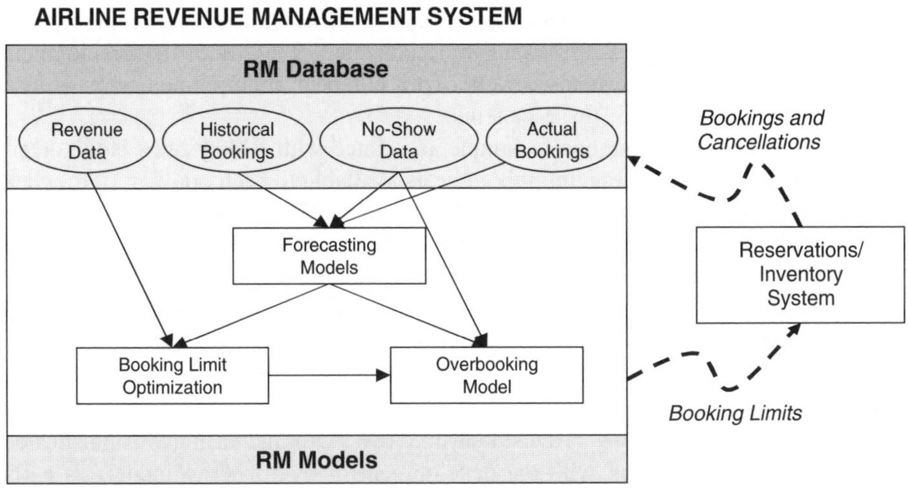

AIRLINE REVENUE MANAGEMENT SYSTEM

Figure 4.12   Components of typical third-generation RM system

bookings for the future flight are compared to the demand forecasts generated by the system. Should unexpected booking activity occur, the system reforecasts demand and reoptimizes its booking limit recommendations. A substantial proportion of the revenue gain attributable to fare mix optimization comes from this dynamic revision of booking limits. Human intervention to override the recommended booking limits is important in unusual circumstances (e.g., a surge in demand related to a special event), but most airlines tend to "overmanage" their booking limits based on human judgment, which can lead to reduced revenue gains from the RM system.

Based on a variety of empirical studies and simulation experiments performed by airlines and academics alike, it is now commonly accepted that proper use of an RM system can lead to airline revenue increases of 4 to 6% (Belobaba, 1989: Smith *et al.*, 1992). This revenue gain comes from better seat inventory management – both overbooking and fare class booking limits, with effectively no increase in flight operating costs.

Beyond the obvious incremental revenue benefits, the use of RM systems allows for better tactical matching of demand vs. supply by the airline. Booking limits on lower fare classes applied to high-demand flights can help to channel low-fare demand to empty flights, resulting in more even load factor distributions. By shifting low-fare demand to emptier flights, the RM system protects seats for highest-fare passengers on forecast full flights.

Finally, RM systems can play an important role in competitive airline pricing strategies. With RM capabilities, an airline can match or initiate almost any low fare that covers variable passenger carrying costs. Use of booking limits on the lower fare classes allows the airline to prevent revenue dilution, while maintaining a competitive pricing posture and its market share in the face of low-fare competition.

Computerized RM systems manage the airline's inventory of available seats by using mathematical models and computer databases to address three different problems:

- Overbooking: Airlines have been accepting reservations in excess of aircraft capacity for several decades, in an effort to reduce the revenue losses associated with unpredictable no-show behavior on the part of passengers (Rothstein, 1985). With the development of more sophisticated RM systems, overbooking has been incorporated into the seat inventory control functions of these systems.
- Fare class mix: The most common technique associated with RM systems is the determination of the revenue-maximizing mix of seats available to each booking (fare) class on each future flight leg departure. Virtually all airline RM systems were developed with the capability to optimize fare class mix as their primary objective.
- Origin–destination control: Currently limited to only the most advanced RM systems and airlines with large connecting hub networks, O-D control allows the airline to further distinguish between the seats it makes available to short-haul (one-leg) vs. long-haul (connecting) passengers, by fare product and price level.

Operations research (OR) has provided airline RM with increasingly sophisticated mathematical models, beginning with the development of overbooking models as early as the 1970s, followed by methods for forecasting fare class demand and optimizing fare class booking limits in the 1980s. In the 1990s, the focus shifted to application of various network optimization techniques to the airline O-D control RM problem. A comprehensive review of the applications of OR to airline (and other industry) RM problems is provided by McGill and Van Ryzin (1999). In the following sections, we provide an introductory-level overview of the types of models used in airline RM systems for overbooking, fare class mix optimization and O-D control.

4.3.2 Flight Overbooking

The objective of the airline flight overbooking component of RM is to determine the maximum number of bookings to accept for a future flight departure with a given physical capacity (in seats). Because the no-show behavior of passengers on future flights is unpredictable, there is an element of risk involved in accepting more reservations than physical capacity. If too many reservations are accepted and more passengers show up at departure time than there are physical seats, the airline must deal with the costs and customer service issues of *denied boardings* (DB). On the other hand, if not enough reservations are accepted for the flight and the no-show behavior of passengers is greater than expected by the airline, there are costs associated with the lost revenue from empty seats that could otherwise have been occupied, also known as *spoilage* (SP). The more specific objective of most airline overbooking models is therefore to minimize the total combined costs and risks of denied boardings and spoilage (lost revenue).

Why is overbooking even necessary? The simple answer is that airlines have historically allowed their passengers to make reservations (which removes seats from the airline's available inventory) and then to "no-show" with little or no penalty. In very few other service or manufacturing industries can the consumer "promise" to buy a product or service and then change his or her mind at the last minute with little or no penalty. The economic motivation for airline overbooking is substantial. In the USA, domestic airline no-show rates average 10–15% of final pre-departure bookings, and can exceed 20% during peak holiday periods. Although there are substantial regional differences, average no-show rates are almost as high throughout the rest of the world. Given that most airlines struggle to attain a consistent operating profit, the loss of 10–15% of potential revenues on fully booked flights (which would occur without overbooking) represents a major negative impact on profits.

As part of an RM system, effective overbooking has been shown to generate as much revenue gain as fare class mix optimization (Smith *et al.*, 1992). The intuitive explanation is that overbooking makes the difference between a full seat and an empty seat, while fare class mix makes trade-offs between passengers paying different fare levels.

For the purposes of discussing both the mechanics of the airline overbooking process and then the mathematical models that airlines use to calculate overbooking levels, there are important terms that need to be defined first. For each term below, an abbreviation is also introduced, for use in the models to be discussed later:

- Physical capacity (CAP): The actual number of seats on the flight or in the designated compartment that can be filled with passengers at departure. Usually, this is the maximum capacity of the aircraft (or compartment), unless the flight is constrained by weight limitations, for example.
- Authorized capacity (AU): The maximum number of bookings that the airline is willing to accept, given a physical capacity of CAP.

- Confirmed bookings (BKD): The total number of passenger reservations that have been accepted by the airline for a specific departure, counted just before the check-in process for the flight begins. Generally, we expect BKD to be less than or equal to AU.
- No-show rate (NSR): The mean proportion (percentage) of passengers with confirmed bookings that do not show up.

Mathematically, the airline overbooking problem is to determine an overbooking factor (OV) such that:

$$AU = CAP \times OV, \quad \text{where } OV > 1.00 \tag{4.1}$$

The major challenge in overbooking is the inability to predict with perfect accuracy the actual no-show rate for a future flight departure. In the following paragraphs, we review the evolution of airline overbooking approaches, from the simplest to the current state of the practice. To illustrate the differences between the overbooking factors of the different approaches, we use a simple example of a future flight departure with a physical capacity (CAP) of 100 seats, with a (correctly) forecasted mean no-show rate (NSR) of 20% and a standard deviation (STD) of the no-show rate equal to 8%.

The simplest approach used by airlines to determine an AU for a future flight departure relies on the judgment of human analysts. The analysts decide on overbooking factors based on their "market experience" and assessment of recent no-show history for the same flight (e.g., previous departures of the same flight leg, same day of week). Using strictly judgment, the tendency of most humans would be to choose $OV = 1 + NSR$ (or lower), because most airlines teach their analysts to try to avoid denied boardings when overbooking. In our flight example, a human analyst would therefore be most likely to apply an AU in the range of 115 to 120. Despite its simplicity and lack of scientific basis, many airlines around the world (primarily smaller and less sophisticated carriers) still rely heavily on judgmental overbooking.

The mathematically correct overbooking level for the above situation can be calculated using a simple "deterministic model," meaning the model assumes that the actual future no-show rate is known with certainty. Given the forecast NSR of 20%, and assuming that the flight does indeed book up to the point where BKD = AU, then the deterministic model is:

$$AU = CAP/(1 - NSR) \tag{4.2}$$

In our example, for CAP = 100 and NSR = 0.20, then $AU = 100/(1 - 0.20) = 125$. Therefore, the mathematically correct AU, assuming that we have perfect knowledge of the future flight's NSR = 0.20, is actually 125, higher than the intuitive (judgmental) answer for most human analysts.

If the deterministic model is applied in the real world where the actual no-show rate is uncertain (i.e., has STD greater than zero), then use of the deterministic overbooking model leads to a 50% probability of having DB should the flight book to AU. However, it also leads to a 50% probability of having spoiled seats (SP). The deterministic model provides an acceptable overbooking answer if the airline is *indifferent* (economically or otherwise) between denied boardings and spoiled seats. The major problem with the deterministic overbooking model is that it does assume that the future flight's no-show rate is known with certainty.

A more scientific approach to dealing with no-show uncertainty is with the use of a probabilistic or risk-based overbooking model. This model incorporates uncertainty about no-show rates for future flight departures, by representing the no-show forecast as a normal (Gaussian) probability distribution. The estimates of no-show behavior used by the overbooking model for a future departure now include both a (mean) NSR and a standard deviation, STD.

Using the properties of the normal distribution, the objective of the probabilistic/risk overbooking model is to find the AU that will keep DB to some airline-specified target level, in the event that the flight is fully booked (BKD = AU), with a specified level of confidence. The AU for a future flight departure given CAP and estimates of NSR and STD with an airline-specified objective (for example) of keeping DB = 0 with 95% confidence is:

$$AU = \frac{CAP}{1 - NSR + 1.645\ STD} \qquad (4.3)$$

The value 1.645 in the denominator is the "standardized normal value" (or "Z-score") for a one-tailed 95% confidence level under the assumption of a normal distribution of no-show rates. If the airline desired a lower level of confidence of keeping denied boardings equal to zero, e.g., 90%, this value would decrease (to 1.28). Note that if the airline decided that it only wanted to be 50% confident of having no DB, the appropriate Z-score value would be 0.00, and the probabilistic model result would revert back to that of the deterministic model.

Returning to our previous example with CAP = 100, NSR = 0.20 and now with STD = 0.08, and assuming that the airline wishes to limit DB = 0 with 95% confidence:

$$AU = \frac{100}{1 - 0.20 + 1.645 \times 0.08} = 107 \qquad (4.4)$$

The probabilistic model thus incorporates the uncertainty or variability in the distribution of no-show rates for the future flight departure. If recent observations of no-show rates suggest a larger STD, the denominator of the model will become larger and the optimal AU will be lower, due to increased uncertainty about no-shows. On the other hand, a lower STD will lead to a higher AU, due to lower uncertainty about future no-shows.

The probabilistic/risk model for overbooking lends itself to a variety of extensions that allow the airline greater flexibility in determining its own overbooking policy. The airline might wish to reduce the required level of confidence of exceeding the DB target, in order to overbook more aggressively. The Z-factor in the model's denominator will decrease, causing an increase in AU.

The airline can also decide to increase its DB tolerance to account for the likelihood that it will be able to find passengers willing to become "voluntary" denied boardings, at little or no cost to the airline. In essence, the "effective capacity" of the flight increases by the number of voluntary denied boardings (VOLDB) that the airline believes it can solicit at the time of departure, should it be necessary. The model's numerator becomes (CAP + VOLDB), increasing AU. It is also possible for the airline to include forecasted empty first-class or business-class cabin seats as being available for upgrading passengers from economy class, should more show up than the economy-class physical capacity. Again, this action increases the "effective capacity" of the flight.

A major extension of the probabilistic/risk overbooking approach is the cost-based overbooking model, which not only incorporates the uncertainty of future flight no-show behavior, but explicitly accounts for the actual costs associated with denied boardings and spoilage. The objective of the cost-based overbooking model is to find the optimal AU that minimizes the total combined costs of denied boardings and spoilage:

$$\text{MIN[Cost of DB} + \text{Cost of SP]} \tag{4.5}$$

Probability calculations of the normal distribution allow the following measures to be calculated for each value of AU, given CAP and estimates of NSR and STD:

$$\text{Total cost} = \$\text{DB} \times E[\text{DB}] + \$\text{SP} \times E[\text{SP}] \tag{4.6}$$

where: $\$\text{DB}$ and $\$\text{SP}$ = cost per DB and SP, respectively
   $E[\text{DB}]$ = expected number of DB, given AU
   $E[\text{SP}]$ = expected number of SP, given AU

Finding the optimal AU in this approach requires calculations of Gaussian probabilities and expected values, followed by a mathematical search over a reasonable range of AU values to find the minimum total cost. As shown in Figure 4.13 for our flight example, the total costs of denied boardings increase as the AU is increased above the physical capacity (CAP) of 100 seats. At the same time, each increase in AU results in a lower total spoilage cost. In the example shown, the cost per denied boarding is assumed to be $200 while the cost per spoiled seat is assumed to be $300. The minimum cost AU in Figure 4.13 is therefore 123. Some RM systems provide airline RM analysts with a graph

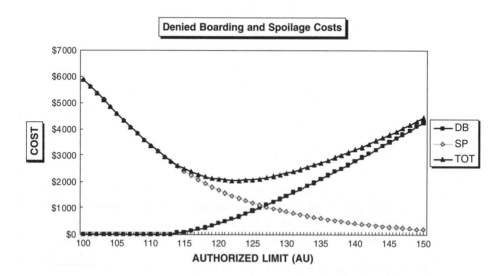

Figure 4.13   Cost-based overbooking model

very similar to that shown in Figure 4.13, allowing them to determine the sensitivity of total overbooking costs to their own judgmental modifications to the optimal AU.

The cost-based overbooking model has an explicit objective that matches the proper objective of airline overbooking (i.e., minimize the total costs of denied boardings and spoilage). However, the current "state of the practice" among most airlines does not reflect complete and widespread adoption of this model. Instead, many airlines continue to use the probabilistic/risk framework, due both to its relative simplicity and because the estimation of cost inputs required by the cost-based model can be difficult.

The costs of a denied boarding on a given flight can include a variety of elements, some of which are not readily quantifiable in monetary terms:

- Cash compensation paid to involuntary denied boardings.
- Free travel vouchers as incentives for voluntary denied boardings.
- Meal and hotel costs for displaced passengers.
- Space on other airlines to accommodate displaced passengers.
- Costs of lost passenger goodwill.

Many airlines have found it difficult to provide accurate cost estimates of the above elements. Furthermore, they realize that the costs of denied boardings will vary by length of flight (e.g., short-haul domestic versus long-haul international) and even by time of day on the same route (e.g., the costs of having involuntary denied boardings on the airline's last flight of the day are much higher than on earlier flights, due to hotel and meal costs).

Despite the development and use by many airlines of the mathematical overbooking models described here, some airlines continue to view aggressive overbooking strictly in negative terms. That is, they believe that denied boardings are associated with poor customer service and ultimately a loss of passenger goodwill. But a more objective and economic perspective on the overbooking problem suggests that the revenue loss of spoiled seats can actually be greater than DB costs. The rejection of a full-fare passenger on the day before departure because the AU level was not set high enough can mean lost revenue of many hundreds of and perhaps a thousand dollars or more.

On the other hand, it is possible to "take care of" a denied boarding at relatively low cost, by compensating the passenger with a free ticket, a future upgrade, and/or access to the airline's lounge while waiting for the next flight. A "voluntary DB program" is needed to control the costs and customer service issues associated with more aggressive overbooking, which then allows the airline to reduce what, for the vast majority of airlines, is a much larger revenue loss associated with inadequate overbooking.

With the help of voluntary DB programs, the largest US airlines have become extremely successful in managing DB and the associated costs, despite what is perceived to be very aggressive overbooking to reduce the revenue losses associated with spoilage. The involuntary DB rate among US major airlines in 2007 was only 1.12 per 10 000 passengers boarded (US DOT, 2008). Over 90% of all DB in the USA are volunteers, meaning that the total DB rate for US airlines was about 12 per 10 000 passengers boarded. This total is in line with world airline industry standards of 12 to 15 per 10 000. But the important point is that US airlines are able to report lower *involuntary* DB rates than most world airlines, thanks to effective voluntary DB programs.

4.3.3 EMSR Methods for Flight Leg Revenue Optimization

After overbooking, the second principal RM technique introduced above was "fare class mix" optimization, which involves determination of booking limits on different fare classes that share a common inventory of seats. Specifically, the fare class mix problem can be stated as follows: given a future flight departure with a known authorized booking capacity (AU), determine the number of seats that should be made available to each of the different fare classes that share the same inventory of seats in a physical compartment (typically the economy-class cabin) with the objective of maximizing total expected flight revenues.

To determine the number of seats to make available to each fare class, the airline's RM system must generate a forecast of future booking demand for each fare class on each future departure. Also, estimates of the revenue value associated with bookings in each fare class are required (see Figure 4.12). In this section, we describe the mathematical model used most commonly by airline RM systems to set booking limits on different fare classes. The "expected marginal seat revenue" or "EMSR" approach for setting RM booking limits was developed by Belobaba in 1987 ("EMSRa") and then refined to become the "EMSRb" model in 1992, which is described here (see Belobaba, 1987, 1989 and 1992).

All airline reservations systems include an "inventory" of available seats for future flight departures. The way in which this inventory is structured has an important impact on how booking limits should be determined by the RM system's optimization model. Because EMSRb was developed explicitly for use in "nested" inventory control systems, the concept of nested booking classes is described first. The vast majority of airline reservations systems now have inventory structures based on "serial nesting" of booking classes, as shown in Figure 4.14. Seats are not "allocated" to separate booking classes. Instead, seats are "protected" for higher fare classes and nested "booking limits" are applied to the lower fare classes.

For example, a future departure has an authorized capacity (AU) in the economy cabin of 120 and three fare classes (Y, B and M in decreasing revenue order) that must share this inventory. A "partitioned" booking class allocation might be as follows. A limit of

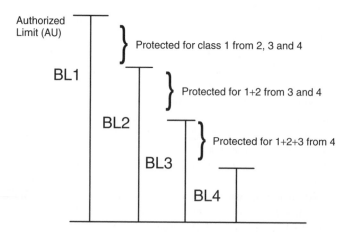

Figure 4.14 Nested seat protections and booking limits

30 bookings in Y class, 40 for B class and 50 for M. However, in a "serial nesting" class inventory structure, all 120 seats are made available to the highest Y class, in the (unlikely) event that the first 120 requests for bookings on the future departure are for the full-fare Y-class product. This ensures that the airline would never turn down a Y fare request, as long as there are any seats (Y, B or M) still available for the flight.

Then, 30 seats would be protected for the exclusive use of Y-class bookings, leaving a booking limit of 90 seats for B class. In turn, if 40 of these seats were not available to M class, and were instead protected for B-class demand (and for Y-class, should the need arise), the booking limit on M class would be 50. The *nested booking limits* for the flight would be 120 for Y, 90 for B and 50 for M. Note that these limits are nested – there are only 120 booking spaces available on the flight. If one M-class booking were accepted first, the remaining availability would be 119 Y seats, 89 B seats and 49 M seats.

If we were to assume that future booking demand for each fare class is deterministic (or known with certainty), it would be relatively simple to determine the fare class mix for a future flight departure. We would start with the highest fare class and protect exactly the number of seats as the "known" future demand for that class, and continue with the next lower fare class until the authorized capacity is reached.

In the real world, however, future booking demand is not known with certainty. The forecasts of future booking demand for each fare class on a given flight departure can be represented as being "probabilistic" or "stochastic," such that each forecast has both a mean and standard deviation to reflect uncertainty. In this case, the decision rule for seat protection is not as simple.

There are several modeling assumptions made in the EMSRb model for application to serially nested booking classes:

- Demand for each fare class is separate and independent of demand in other classes.
- Demand for each class is stochastic and can be represented by a probability distribution (usually normal).
- The lowest fare class books first, followed by the next lowest class, etc.

Because higher fare classes have access to unused lower-class seats in a nested booking class inventory structure, the problem is to find seat *protection levels* for higher classes, and *booking limits* on lower classes.

Given an estimated mean and standard deviation for fare class i, we can determine $P_i(S_i)$, the probability that actual demand for a fare class X_i will exceed any value of seats protected for that class, S_i, that we might choose. The EMSR of protecting the Sth seat for class i is defined as:

$$\text{EMSR}_i(S_i) = F_i \times P_i(S_i) \tag{4.7}$$

where F_i is the average revenue (or fare) for class i. That is, the expected revenue from protecting the Sth seat is the average fare of class i multiplied by the probability of actually receiving S or more requests for class i.

Then, the optimal seat protection level, π_{12}, for class 1 away from class 2 satisfies the condition:

$$\text{EMSR}_1(\pi_{12}) = F_1 \times P_1(\pi_{12}) = F_2 \tag{4.8}$$

The airline should protect seats incrementally for class 1 only to the point at which it is *indifferent* between selling that seat at the lower fare (in class 2) and keeping that seat protected for the possibility of selling it at the higher fare (in class 1). The above condition defines this point of indifference, as the airline will either receive the fare of the lower class 2 as revenue for the incremental seat, or have an equal expectation of filling that seat with the higher class 1 fare. Once π_{12} has been found, we set the booking limit (BL) on class 2 as:

$$BL_2 = AU - \pi_{12} \qquad (4.9)$$

In a nested booking class structure, $BL_1 = AU$.

The same methodology can be extended to many nested booking classes. At each step in the process, we must determine how many seats to protect jointly for classes 1 through n from class $n + 1$. Complete details of the EMSRb seat protection algorithm are provided in Belobaba and Weatherford (1996). Figure 4.15 shows an example of an EMSRb solution for a flight with six nested booking classes.

In the example of Figure 4.15, the future flight departure has an authorized cabin capacity (AU) of 135, with no bookings on hand. The airline offers six booking classes (Y, M, B, V, Q and L), each with associated fare products and average fare levels. The forecast means and standard deviations (sigma) of demand by booking class are shown. The EMSRb model protects eight seats for the exclusive use of Y class, meaning the nested booking limit for M class is 127. Note that the Y-class protection level is less than the mean forecast demand for that class, given the uncertainty of the forecast demand and the fact that the value of an M-class booking is 82% of the fare of Y class. Thus, the EMSRb model only protects seats for Y class that have a probability greater than 82% of being actually sold to Y-class passengers. In this case, eight seats meet this condition.

Similar calculations determine the nested booking limits of each lower class in Figure 4.15. For the top three booking classes, total combined mean demand is 39, and the model protects a total of 38 seats away from the three lowest classes. Note that, for the lowest L class, the EMSRb booking limit is 41, compared to a forecast L-class demand of 47. For this flight departure, even though the total mean demand is forecast

CABIN CAPACITY =		135				
AVAILABLE SEATS =		135				
BOOKING	AVERAGE	SEATS	FORECAST DEMAND		JOINT	BOOKING
CLASS	FARE	BOOKED	MEAN	SIGMA	PROTECT	LIMIT
Y	$ 670	0	12	4	8	135
M	$ 550	0	17	6	25	127
B	$ 420	0	10	5	38	110
V	$ 310	0	22	9	62	97
Q	$ 220	0	27	10	94	73
L	$ 140	0	47	12		41
	SUM	0	135			

Figure 4.15   Example of EMSRb solution for six nested classes

to be 135, exactly equal to the authorized capacity, the EMSRb limits will result in the rejection of some L-class demand, on average. The model calculations trade off the risk of not selling a seat with the expected revenue of protecting seats in higher classes, leading to higher total expected revenues for the flight.

4.3.4 Network Revenue Management

Airlines that have mastered the basics of RM with successful implementation of a third-generation system are developing and/or implementing the fourth generation of RM systems – those that provide "origin–destination" or "O-D" control in addition to all of the existing RM capabilities. O-D control represents a major step beyond the fare class mix capabilities of most third-generation RM systems, and is currently being pursued by the largest and more advanced airlines in the world. As its name implies, O-D control gives the airline the capability to manage its seat inventory by the revenue value of the passenger's O-D itinerary on the airline's network, not simply according to the fare class requested on a single flight leg.

In this section, we provide an overview of network revenue management, which is recognized as the fourth generation of RM systems development. The section begins with a description of how network RM overcomes the limitations of basic fare class control by flight leg. Then, we explore several inventory schemes and optimization approaches that have emerged as the most feasible for application to the network RM problem. Finally, the additional benefits of network RM above and beyond those of third-generation RM systems are discussed.

As described in the previous sections, the vast majority of world airlines still practice "fare class control" using third-generation leg-based RM systems. As shown in the example of a typical yield-based fare class structure provided earlier in Figure 4.4:

- High-yield ("full") fare types are associated with the top booking class (Y) on each flight leg.
- Lower-yield ("discount") fares are assigned to lower booking classes (B, M and Q), which allows the airline to limit sales at these lower-yield fares.

This approach to RM was in effect designed to maximize revenues (or more specifically, yield) on each flight leg, but not necessarily total revenues on the airline's network.

Under leg-based fare class control, seats for connecting itineraries involving more than one flight leg must be available in the same fare class across all flight legs requested by the passenger. The airline's RM and reservations systems cannot distinguish among itineraries in the same fare class. A Q-class seat can be taken by a passenger flying only one leg or a passenger on a long-haul connecting itinerary involving two or three legs. As a result, very high-demand, short-haul legs can create "bottlenecks" that prevent long-haul passengers with substantially higher network revenue contributions from obtaining a seat.

In the small-network example of Figure 4.16, we have a short-haul leg (LH100) from Nice (NCE) to Frankfurt (FRA), which connects to two different longer-haul flight legs – one is LH200 from FRA to Hong Kong (HKG) and the other is LH300 from FRA to New York (JFK). We can use this example to illustrate two limitations of leg-based

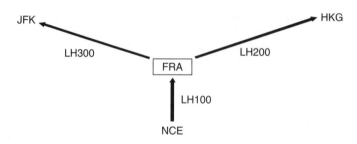

Figure 4.16   Connecting flight network example

fare class control that prevent this approach from maximizing total network revenues for the airline:

1. In situations where the short-haul NCE–FRA flight leg is expected to be booked very full while either of the connecting flight legs is expected to have many empty seats, limited fare class availability on the short-haul LH100 leg can block long-haul passengers willing to pay a higher total itinerary fare. The RM system might be willing to accept a "local" passenger in B class traveling only from NCE to FRA, yet cannot accept passengers willing to pay a much higher total fare for an M-class ticket from NCE to HKG. If LH200 has numerous available seats, this will clearly not maximize total network revenues.
2. The inability of fare class control to distinguish between passenger O-D itineraries, even within the same fare class, is another limitation of the leg-based RM approach. In our example, if B-class seats are available on both LH100 and LH300, the airline cannot control whether those seats are sold to connecting passengers or to pairs of "local" passengers NCE–FRA and FRA–JFK. Selling scarce seats to two "local" passengers, if possible, would generate substantially more network revenue, because the sum of two local fares is typically greater than a connecting itinerary fare of the same type (fare class).

Revenue maximization over a network of connecting flights thus requires two different strategies, under different conditions. The airline should try to increase availability to high-revenue, long-haul passengers, regardless of yield, when empty seats are expected on one or both of the connecting flights in an O-D itinerary. On the other hand, if both connecting flights are expected to be full, the airline should try to prevent lower-yield connecting passengers from displacing high-yield, short-haul or "local" passengers.

O-D Control Mechanisms

The term "O-D control" has been used to describe a variety of different approaches to network RM. Most simply, an RM system with O-D control provides the airline with the capability to respond to different O-D requests with different seat availability. O-D control can be implemented in a variety of ways, all of which depend to differing degrees on the concepts that will be described below. Each of these approaches to O-D control can increase network revenues for the airline, but each one has implementation trade-offs in terms of development costs and complexity.

Among the earliest approaches developed for O-D control is the use of "revenue value buckets" instead of fare classes for seat inventory management. Under the revenue value bucket concept, the fixed relationship between fare type and booking class is abandoned, such that booking classes ("buckets") are defined according to network revenue value, regardless of fare restrictions. Each O-D itinerary/fare-type combination (i.e., "ODF") is then assigned to a revenue value bucket on each flight leg. Then, the seat availability for a requested ODF depends on the availability of the corresponding revenue value bucket on each leg of the passenger's itinerary.

The revenue value bucket concept can be implemented through the development of "virtual" inventory classes. Pioneered by American Airlines (Smith and Penn, 1988), virtual class mapping requires the development of an entirely new "virtual" class booking inventory structure. As shown in the example of Figure 4.17, the airline now has 10 virtual booking classes on each flight leg (in this case NCE–FRA) which are not seen by anyone outside the airline. The Y-class full fare from NCE to FRA is now assigned to virtual class 6, whereas the Q-class discount fare from NCE to HKG is assigned to virtual class 5, giving the latter greater seat availability based on its higher total revenue value.

The "mapping" functions required to translate an incoming ODF request into specific virtual inventory class availability represent substantial implementation costs to the airline. Also, seamless availability links between the reservations systems (CRSs) of different

FARE VALUES BY ITINERARY

NCE/FRA		NCE/HKG (via FRA)		NCE/JFK (via FRA)	
CLASS	FARE (OW)	CLASS	FARE (OW)	CLASS	FARE (OW)
Y	$450	Y	$1415	Y	$950
B	$380	B	$975	B	$710
M	$225	M	$770	M	$550
Q	$165	Q	$590	Q	$425
V	$135	V	$499	V	$325

MAPPING OF ODFs ON NCE/FRA LEG TO VIRTUAL VALUE CLASSES

VIRTUAL CLASS	REVENUE RANGE	MAPPING OF O-D MARKETS/CLASSES	
1	1200 +	Y NCEHKG	
2	900–1199	B NCEHKG	Y NCEJFK
3	750–899	M NCEHKG	
4	600–749	B NCEJFK	
5	500–599	Q NCEHKG	M NCEJFK
6	430–499	V NCEHKG	Y NCEFRA
7	340–429	B NCEFRA	Q NCEJFK
8	200–339	V NCEJFK	M NCEFRA
9	150–199	Q NCEFRA	
10	0 – 149	V NCEFRA	

Figure 4.17   Virtual class mapping by fare value

airlines are required, because the virtual classes exist only within the inventory structure of the airline with this form of O-D control. All ODF requests must be transmitted to the airline for internal evaluation. Data collection and storage is performed by leg/value bucket, instead of by leg/fare class, meaning that demand forecasting and optimization can be performed by leg/value bucket, consistent with the structure of leg-based RM systems. And different ODF requests get different availability based on their network revenue value.

But such implementations of revenue value buckets also have their limitations. Each flight leg's revenue value buckets are still optimized separately and do not take into account demand and availability on other flight legs in the network. As a result, revenue value buckets based on total fare values can give too much preference to long-haul and connecting passengers. In fact, this implementation has come to be known as the "greedy" approach to O-D control, because the highest fare itineraries are given preference in terms of seat availability, even if it might make more sense to accept two local passengers with a higher sum of total fares.

Displacement Adjusted Virtual Nesting

The actual contribution of an ODF passenger on a given flight leg to the total airline network revenue must be less than or equal to the total ODF fare value of the passenger's itinerary. This is because connecting passengers accepted on a given flight leg can displace other passengers and revenue on down-line (or up-line) legs of their itinerary.

The mathematical problem is therefore how to determine the network revenue contribution of each ODF for O-D control purposes, in order to overcome the "greediness" of the revenue value bucket approach when it is based on total itinerary fares. The answers come from various mathematical models that have been applied to the airline network RM problem. Sophisticated network optimization techniques can be used to calculate the estimated revenue displacement cost on each flight leg, given fixed capacities and forecasted demand for each ODF on the network. Much simpler leg-based EMSR models can also be used to generate estimates of revenue displacement resulting from acceptance of connecting passengers (Belobaba, 1998).

Regardless of which mathematical approach is used to estimate down-line revenue displacement, it is possible to map ODFs to virtual classes based on their estimated *network* revenue value. The network revenue value of an ODF on leg 1 of a connecting itinerary is equal to the total ODF fare *minus* the sum of expected down-line leg displacement costs. When the connecting flight legs have high forecasted demand, the corresponding displacement costs will be higher, reducing the "displacement adjusted" network revenue value of the ODF. With a lower network value, this ODF would be mapped to a lower virtual class for seat inventory control purposes, meaning the seat availability for connecting passengers is reduced. At the same time, local passengers (with no down-line displacement costs) would get greater seat availability.

For our small-network scenario, Figure 4.18 shows an example of how the 15 ODFs that make use of LH100 from NCE to FRA could be mapped to 10 virtual classes based on total revenue value. The Q-class fare from NCE to HKG (i.e., QNCEHKG) is $770, and is initially mapped to virtual bucket 3. The "full" Y fare in the local NCE–FRA market, on the other hand, is mapped to virtual bucket 6, giving it less seat availability based strictly on its lower total fare value. If demand is forecast to be high on the connecting

FARE VALUES BY ITINERARY

NCE/FRA		NCE/HKG (via FRA)		NCE/JFK (via FRA)	
CLASS	FARE (OW)	CLASS	FARE (OW)	CLASS	FARE (OW)
Y	$450	Y	$1415	Y	$950
B	$380	B	$975	B	$710
M	$225	M	$770	M	$550
Q	$165	Q	$590	Q	$425
V	$135	V	$499	V	$325

MAPPING OF ODFs ON NCE/FRA LEG TO VIRTUAL VALUE CLASSES

VIRTUAL CLASS	REVENUE RANGE	MAPPING OF O-D MARKETS/CLASSES	
1	1200 +	Y NCEHKG	
2	900–1199	B NCEHKG Y NCEJFK	
3	750–899	M NCEHKG	
4	600–749	B NCEJFK	
5	500–599	Q NCEHKG M NCEJFK	Displacement Adjustment
6	430–499	V NCEHKG Y NCEFRA	
7	340–429	B NCEFRA Q NCEJFK	
8	200–339	V NCEJFK M NCEFRA	
9	150–199	Q NCEFRA	
10	0 – 149	V NCEFRA	

Figure 4.18 Displacement adjusted virtual nesting

flight leg LH200, the estimated displacement cost for that leg might be $400, for example. In this case, the network value contribution of QNCEHKG passengers on LH100 would be reduced to $370, and this ODF would be mapped to virtual bucket 7. The result would be lower seat availability for QNCEHKG, lower than the local full fare YNCEFRA.

"Displacement adjusted virtual nesting" or "DAVN" is therefore a mechanism that achieves both of the O-D control objectives introduced earlier. Virtual nesting based on total fare value increases availability to connecting and long-haul passengers, while adjustment for down-line displacement of revenues ensures that two local passengers will receive preference when two connecting flights are forecast to be fully booked.

Bid Price Control

Under the revenue value bucket control approach, the airline accepts an ODF request if its network value falls into an available bucket. For a virtual class bucket to be "available," it must be the case that the revenue value of the request must exceed the expected marginal seat revenue value of the "last" or lowest-valued seat remaining available on the flight leg. When we define the network value as the ODF total fare minus down-line revenue displacement, then we accept an ODF request if:

$$\text{ODF fare} - \text{Displacement cost} > \text{Value of last seat on leg}$$

With simple manipulation, the same acceptance rule can equivalently be expressed as:

ODF fare > Value of last seat + Displacement cost

or

ODF fare > Minimum acceptable "bid price" for ODF

The use of such an acceptance rule based on minimum acceptable fare values is known as "bid price control." Both bid price control and value buckets make use of the same O-D control concepts and mathematical models. They simply represent two different O-D control mechanisms.

Bid price control involves a much simpler inventory control mechanism than virtual nesting of revenue value buckets. The airline just needs to calculate and store a current bid price value for each future flight leg departure in its reservations system. At the time of an ODF request for seat availability, the ODF fare is evaluated against the sum of the leg bid prices over the itinerary being requested. However, the airline's RM system must reoptimize the leg bid prices frequently to prevent too many bookings of ODFs at current bid price levels. As was the case for estimates of revenue displacement associated with connecting passengers, leg bid prices can similarly be calculated either with network optimization tools or leg-based heuristics such as EMSR approaches.

Example: Bid Price Control

Consider a series of three flight legs from A to B to C to D, with the following calculated flight leg bid prices:

A–B : $50   B–C : $200   C–D : $150

The current seat availability for passenger ODF requests, with their associated total itinerary fares for travel on one leg from B to C, would be as follows:

B–C	Bid price = $200	Available?
Y	$440	Yes
M	$315	Yes
B	$225	Yes
Q	$190	No

ODF itinerary requests requiring more than one flight leg are evaluated by comparing the total ODF itinerary fare against the *sum* of the leg prices over the flight legs to be traversed:

A–C	Bid price = $250	Available?
Y	$500	Yes
M	$350	Yes
B	$260	Yes
Q	$230	No

A–D	Bid price = $400	Available?
Y	$580	Yes
M	$380	No
B	$300	No
Q	$260	No

Bid price control thus gives different ODF requests different seat availability responses. In this example, the current solution prevents low-yield connecting passengers wishing to travel from A to D from taking seats that can generate more revenue if sold to passengers on A–C and B–C itineraries.

As mentioned earlier, estimates of network displacement costs and bid prices can be derived using a variety of mathematical approaches. There are numerous applications of network optimization tools, both in the OR literature and offered by RM software vendors. However, most airlines still lack the detailed data required and face additional practical constraints in using network optimization models. Because most reservations systems and even third-generation systems were developed on the basis of leg/fare class data, most airlines do not have access to the detailed historical ODF booking data required by network optimization models.

Use of large-scale network optimization models also raises technical and computational issues related to the solution times and frequency of reoptimization. Even if an airline builds a database to capture detailed ODF historical data, there remain concerns about ODF demand forecasting accuracy. In a large network, the mean demand for any single ODF will be very small, with very high variance. Finally, it can be difficult for RM analysts to interact with network optimization solutions. For example, while an RM analyst might judge a Y-class allocation of eight seats to be "not enough," it is much less intuitive to judge a calculated leg bid price of $201 to be "too high" or "too low." As a result, relatively few airlines to date have implemented network optimization models for dynamic calculation of displacement costs and/or bid prices for O-D control.

Instead of moving immediately to advanced network optimization models for O-D control, several large airlines have implemented approximation models of network displacement costs and/or leg bid prices. These heuristic approaches to network RM generate estimates of displacement costs and/or bid prices based on leg/bucket EMSR calculations. Because the forecasting and optimization continues to use leg-based data, the airline can make use of its existing inventory structure, databases and RM system capabilities. And the leg-based bucket orientation is more compatible with current RM analyst work routines. Leg-based heuristic models represent a low-risk approach to O-D control, as an intermediate step to complete network optimization (Belobaba, 2002a).

Revenue Benefits of O-D Control

Simulations of O-D control benefits have been performed for a wide range of airline networks and characteristics, testing different optimization and control mechanisms (see Williamson, 1992; Belobaba, 2002a). They have found that the incremental revenue gains for an airline that moves to a fourth-generation RM system with O-D control capabilities, as compared to effective use of a third-generation leg-based RM system with fare class

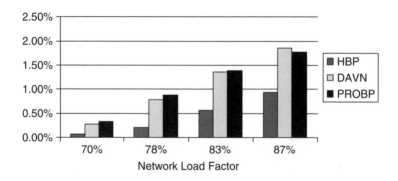

Figure 4.19   Incremental revenue gains of O-D control

control, are in the range of 1–2%. That is, O-D control can contribute an additional 1–2% in total revenue, above the 4–6% that can be realized through effective overbooking and fare class mix optimization.

Figure 4.19 provides an example of simulation results from the Passenger Origin–Destination Simulator (PODS), in which several O-D control methods are compared to leg-based EMSR fare class control in a large network with competing airline connecting hubs (Belobaba, 2002b). The revenue gains shown on the figure compare the performance of three O-D control methods to third-generation RM systems (i.e., control of fare classes by flight leg using EMSRb booking limits): leg-based heuristic bid price (HBP), displacement adjusted virtual nesting (DAVN) and probabilistic bid price (PROBP) control.

The simulation results show a range of potential O-D control revenue gains, from less than 0.5% to over 1.75%, depending on the average load factor of the hub network. That is, on days of the year when the average load factor of the flights into and out of the hub is around 70%, an airline can expect to increase its network revenue by about 0.3% with O-D control. On peak travel days when the average load factor at the hub reaches 87%, the revenue benefit of the most sophisticated O-D control schemes increases to almost 1.8%. Over the course of an entire year, the weighted average of gains for a large connecting hub airline would be well over 1% in incremental revenues, given average load factors that approached 80% in recent years. The simpler to implement heuristic bid price (HBP) approach provides about one-half of the incremental revenue gain of either DAVN or PROBP.

4.3.5 Revenue Management for Less Restricted Fare Structures

Airline RM systems were developed over the past 25 years to respond to the increasingly complex fare structures adopted by airlines using the concepts of differential pricing, as described in Section 4.2. Yet the heavily restricted fare structures designed to segment business and leisure passengers by WTP have recently been replaced in many markets by substantially simplified fare structures. With such dramatic changes in pricing practices among the world's airlines, it is fair to ask whether existing RM systems can manage airline seat inventories as effectively as they have in the past and, in fact, whether the benefits of effective RM are as great given the compressed and simplified fare structures.

Virtually all existing RM systems, whether developed in-house by airlines or purchased from software vendors, were designed to maximize airline revenues given the restricted fare structures of the 1980s and 1990s. The underlying forecasting and optimization models assumed that demands for each fare class were independent, given that restrictions prevented much diversion between fare products. With the move toward less restricted fare structures, consumers are more likely to buy the lowest fare available at the time of booking. The assumption of independent demands for each fare product was never completely correct, but it becomes much less valid with fewer fare restrictions.

Without significant modification to the forecasting algorithms used by RM systems, their ability to maximize revenues in less restricted fare structures is severely impaired. As more consumers take advantage of the lowest (less restricted) fares, fewer bookings are recorded by the RM historical database in the higher fare classes. The actual demand by consumers willing to pay the higher fares is then under-forecasted for future flights, allowing more seats to be sold at lower fares etc. This cycle of diversion is referred to as "spiral-down" in RM systems and, left unchecked, results in inadequate seat protection for higher fares, excessive sales at lower fares and substantial revenue losses for the airline.

The focus of much of RM research currently is on this problem of "spiral-down" in less restricted and even completely unrestricted fare structures. The goal of this research is to find methods for adapting existing RM forecasters to generate estimates of future demand that reflect the potential of consumers to "sell up" to higher fares. Alternatively, entirely new forecasting approaches based on consumer choice and forecasting by WTP might be required. While these new methods are being developed, many airlines are forced to use manual overrides of the recommendations of their existing RM systems to prevent excessive revenue loss.

The need for these modifications to RM systems becomes even more critical given that the revenue leverage of seat inventory control to limit the sales of low-fare seats is substantially greater in less restricted fare structures. Intuitively, without strong restrictions to impose demand segmentation by passengers' WTP, the only way an airline can force those with higher WTP to actually pay the higher fares is to limit seat availability of lower fare classes. Having lost the segmentation power of the most effective fare restrictions, airlines must rely to a greater extent on the remaining weaker restrictions, supported by RM booking limits on lower fares. The objectives of RM have not changed – the primary goal is to protect seats for later booking passengers assumed to be willing to pay a higher fare. The only difference is that, with less restricted fare structures, seat inventory control becomes more critical to maximizing revenues.

Simulation studies have confirmed that the percentage revenue gains of effective RM controls are greater under less restricted fare structures, for the reasons noted above. As shown in Figure 4.20, under traditional restricted fare structures, the revenue gains of basic leg-based RM as described in previous sections is approximately 8%, when compared to a scenario in which no booking limits are applied to lower fare classes ("No RM"). These percentage revenue benefits of basic RM are in line with previous studies for load factors of about 80%, and increase with higher load factors. With the removal of the Saturday night minimum stay restriction, the revenue gains of RM increase to 12% at comparable average load factors.

However, it must be emphasized that removal of fare restrictions with the goal of fare simplification still results in a net revenue loss for legacy carriers. The percentage revenue

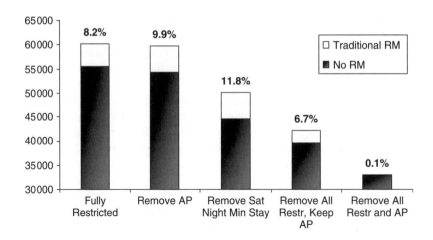

Figure 4.20   Gains of RM in less restricted fare structures (AP = Advance Purchase)

gains of effective RM compared to no RM controls or acceptance of all passengers as "first come, first served" (FCFS) are higher in less restricted fare structures. But the absolute revenues of the airline remain well below what they were under traditional restricted fare structures. At the extreme, under a completely undifferentiated fare structure with no restrictions on different fare classes, the airline will experience large revenue losses (compared to the traditional restricted fare structure) if no changes are made to the traditional forecasting and optimization models of the RM system. Under these conditions, traditional RM systems "spiral down" completely, allowing all passengers to purchase the lowest fare class. Of course, no airline would allow this progression to occur, but would intervene with manual RM practices to close down the lower fare classes, in the absence of more sophisticated models.

As more airline markets around the world shift to less restricted and, in some cases, even fully undifferentiated fare structures with multiple price levels, it is clear that RM systems will become more important for airlines that wish to maximize total revenues. This is true not only for legacy carriers facing new low-fare competitors, but for the low-fare operators themselves. As mentioned earlier, virtually every low-fare airline in the world practices some form of differential pricing, offering multiple price levels on the same flight. While the newest low-fare airlines might not currently use sophisticated RM systems, many of the largest and most successful low-fare carriers have long realized the importance of RM to revenue maximization.

As for the legacy airlines, greater sophistication in their RM capabilities is crucial for managing their seat inventories in markets in which they face direct low-fare competitors. It is also important in their efforts to manage seat inventories over their network of operations, particularly on flights that serve both low-fare markets with less restricted fare structures and connecting O-D markets that retain the more traditional restricted fare structures. Advanced network RM systems give network carriers an improved ability to provide greater seat availability to connecting passengers with higher revenue contribution to the total network.

References

Belobaba, P.P. (1987) "Air Travel Demand and Airline Seat Inventory Management," Ph.D. thesis, Massachusetts Institute of Technology.

Belobaba, P.P. (1989) "Application of a Probabilistic Decision Model to Airline Seat Inventory Control", *Operations Research*, Vol. 37, pp. 183–197.

Belobaba, P.P. (1992) "Optimal vs. Heuristic Methods for Nested Seat Allocation," Presentation to AGIFORS Reservations Control Study Group Meeting, Brussels, Belgium.

Belobaba, P.P. (1998) "The Evolution of Airline Yield Management: Fare Class to Origin-Destination Seat Inventory Control," *Handbook of Airline Marketing*, McGraw-Hill, New York, pp. 285–302.

Belobaba, P.P. (2002a) "Airline Network Revenue Management: Recent Developments and State of the Practice", *Handbook of Airline Marketing*, 2nd edition, McGraw-Hill, New York, pp. 141–156.

Belobaba, P.P. (2002b) "O&D Control: What Have We Learned?" Presentation to IATA Revenue Management and Pricing Conference, Toronto, Canada, October.

Belobaba, P.P. and Weatherford, L.R. (1996) "Comparing Decision Rules That Incorporate Customer Diversion in Revenue Management Situations," *Decision Sciences*, Spring, pp. 343–363.

Boeing Company (1988) "1988 Domestic Fare Survey," Seattle, WA.

Botimer, T.C. and Belobaba, P.P. (1999) "Airline Pricing and Fare Product Differentiation: A New Theoretical Framework," *Journal of the Operational Research Society*, Vol. 50, No. 11, pp. 1085–1097.

Dar, M. (2006) "Modeling the Performance of Revenue Management Systems in Different Competitive Environments," Master's thesis, Massachusetts Institute of Technology.

McGill, J.I. and Van Ryzin, G.J. (1999) "Revenue Management: Research Overview and Prospects", *Transportation Science*, Vol. 33, pp. 233–256.

Rothstein, M. (1985) "O.R. and the Airline Overbooking Problem," *Operations Research*, Vol. 33, pp. 237–248.

Simpson, R.W. and Belobaba, P.P. (1992) "The Prices for Air Transportation Services," Unpublished Notes for Air Transportation Economics Course 16.74, Massachusetts Institute of Technology.

Smith, B.C. and Penn, C.W. (1988) "Analysis of Alternative Origin-Destination Control Strategies," *AGIFORS Symposium Proceedings*, Vol. 28, New Seabury, MA.

Smith, B.C., Leimkuhler, J.F., and Darrow, R.M. (1992) "Yield Management at American Airlines," *Interfaces*, Vol. 22, pp. 8–31.

Tirole, J. (1988) *The Theory of Industrial Organization*, MIT Press, Cambridge, MA.

US DOT (2008) *Air Travel Consumer Report*, Office of Aviation Enforcement and Proceedings, Aviation Consumer Protection Division, Washington, DC, April.

Williamson, E.L. (1992) "Airline Network Seat Inventory Control: Methodologies and Revenue Impacts," Ph.D. thesis, Massachusetts Institute of Technology.

5

Airline Operating Costs and Measures of Productivity

Peter P. Belobaba

With airline deregulation and the spread of increased competition to airline markets around the world, control of operating costs and improved productivity have become critical to the profitability of airlines. The emergence and rapid growth of "low-cost" airlines is due in large part to their ability to deliver air transportation services at substantially lower costs and at higher levels of productivity than the traditional "legacy" airlines. In response, legacy airlines have had to find ways to reduce operating costs and improve the efficiency of how they utilize both their aircraft and employees.

This chapter is devoted to a discussion of airline operating costs and productivity measures. Section 5.1 provides an introduction to the source of cost and productivity data used in much of the discussion – the US Department of Transportation (DOT) Form 41 Traffic and Financial Statistics. It also describes some of the challenges in categorizing airline operating costs and explores alternative categorization schemes, including examples of differences with ICAO and European Joint Aviation Authority (JAA) cost categorization.

Section 5.2 then provides more detailed comparisons of airline operating costs, including the breakdown of total airline operating expenses by category, comparisons of operating costs reported by US legacy and low-cost airlines in different categories, as well as comparisons of operating expenses for different types of commercial aircraft. A brief discussion of the major characteristics of the "low-cost carrier (LCC) business model" is provided as background. Section 5.3 focuses on airline unit costs and their interpretation, with further comparisons of publicly available unit cost data for US, European and Asian airlines, both legacy and low-cost. Finally, Section 5.4 describes common measures of aircraft and employee productivity, as used in the airline industry, and further presents recent trends in the productivity of US airlines.

The Global Airline Industry   P. Belobaba, C. Barnhart and A. Odoni
© 2009 John Wiley & Sons, Ltd

5.1   Airline Cost Categorization

This section describes approaches for categorizing the operating costs incurred by airlines in providing air transportation services. For the purposes of this description, we rely primarily on cost data reported in the US Department of Transportation (DOT) Form 41 database (US DOT, 2007) – a detailed and comprehensive source of traffic, financial and operating cost data reported to the DOT by US airlines. The volume and detail of data compiled in this database, particularly data relating to airline operating costs, is unparalleled among publicly available airline data in the USA and around the world. In fact, no other country makes available such detailed operating cost data to the public (and, in turn, to competing air carriers). There do exist some public sources of world airline operating cost data at substantially greater level of aggregation, including the ICAO annual reports of "Series F" financial data (ICAO, 2007a), which we will use both to illustrate differences in cost categorization and to compare overall airline cost trends.

In the US DOT Form 41 database, data is reported by US airlines and published quarterly for most tables. The detail of reporting differs for different expense categories. For example, airlines are required to report aircraft operating expenses by specific aircraft type (e.g., Boeing 757-200) and region (e.g., Domestic, Latin, Atlantic and Pacific) of operation, for each calendar quarter. Other expenses that are more difficult to allocate by aircraft type, such as ground operating costs associated with processing passengers and baggage at the airport, are reported as system-wide totals.

Although the Form 41 reporting requirements have, from their inception in the early 1940s, attempted to impose a "uniform system of accounts" (Civil Aeronautics Board, 1942), airlines employ accounting methods and cost allocation schemes that may not always be completely "uniform." Consequently, inter-airline operating cost differences, or yearly cost trends, can in some cases be attributed as much to differences in cost accounting rules as to real differences in operating cost performance.

5.1.1   Administrative vs. Functional Cost Categories

One approach to airline cost categorization makes use of "administrative cost categories," which are typical of financial accounting statements used in many industries. Administrative cost categories in the Form 41 database include separate reporting of the following:

- Salaries and related fringe benefits for all personnel, including general management, flight personnel, maintenance labor, aircraft & traffic handling personnel, other personnel.
- Materials purchased, e.g., aircraft fuel & oil, maintenance materials, passenger food, other materials.
- Services purchased, such as advertising & promotions, communications, insurance, outside maintenance, commissions, other services.
- Additional categories for landing fees, rentals (including aircraft), depreciation (including aircraft), other expenses.

Administrative cost categorization is typical of financial statements, as it reports funds expended for labor (salaries), materials and services used as inputs for the "production" of the airline's output. Figure 5.1 shows an administrative categorization of airline costs,

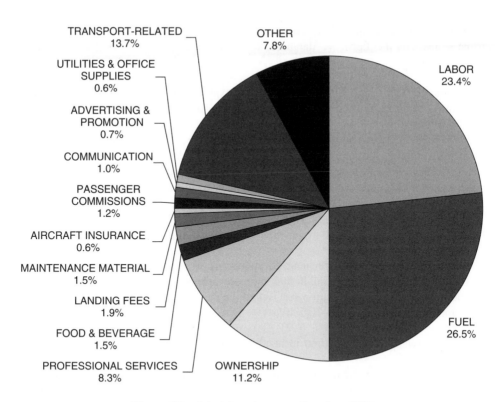

Figure 5.1   Administrative cost allocation, 2007
Data source: Air Transport Association (2008)

as reported by US airlines in the 2007 Form 41 data. Although consistent with general accounting principles, administrative cost categorization does not allow for more detailed analysis of the specific activities that comprise the airline operation and contribute to airline costs.

For example, the category "salaries and benefits" does not allow one to separate out important subsets of this category, most notably aircraft crew costs. In practice, total aircraft operating costs include elements of salaries (pilots, maintenance personnel), materials (fuel, spare parts) and services (insurance). Under the administrative cost categorization, it is difficult to separate out the components of salaries, materials and services that are explicitly associated with operating the aircraft, as opposed to ground operations, for example.

An alternative approach to the categorization of airline operating costs is to define "functional" cost categories, in a way that allocates costs to different functions within the airline's operation. Specifically, the three major functional cost categories for airlines are flight operating costs, ground operating costs and system operating costs (Simpson and Belobaba, 2000).

"Flight operating costs" include all expenses associated with operating aircraft, and are also referred to as "direct operating costs" (DOC) or "aircraft operating costs." Flight operating costs represent the largest proportion of an airline's operating expenses (typically about half) and are usually allocated against the number of block hours operated by the airline's fleet. In the Form 41 database, the following cost items contribute to flight operating costs:

- *Flying operations*: This function consists of "expenses incurred directly in the in-flight operation of aircraft" (US DOT, 2007), including all costs associated with flight crew and fuel costs.
- *Maintenance*: Maintenance expenses are "all expenses, both direct and indirect, specifically identifiable with the repair and upkeep" of aircraft and equipment (US DOT, 2007). This element includes both routine maintenance and more extensive major checks, with costs driven in large part by extensive use of labor and the consumption of spare parts.
- *Depreciation & amortization:*: This function spreads the capital cost of the airline's assets – specifically, aircraft – over their expected lifetime.

"Ground operating costs" are incurred at the airport stations in handling passengers, cargo and aircraft or by the airline in making reservations and ticket sales, and are directly incurred in providing transportation services to the customer. The three major components of ground operating costs are:

- *Aircraft servicing* costs incurred in handling aircraft on the ground, including landing fees.
- *Traffic servicing* costs of processing passengers, baggage and cargo at airports.
- *Promotion and sales* costs associated with airline reservations centers and ticket offices, including travel agency commissions and distribution system fees.

"System operating costs" are the indirect operating costs remaining after ground operating costs are accounted for. They are not directly associated with supplying the transportation service, but are more of a corporate overhead expense. For example, advertising costs are those spent to increase system revenues, while on-board passenger service expenses include food, entertainment and cabin crew costs. Administrative expenses are those of a general management or corporate nature for the complete airline system (except maintenance administration). The major components of system operating costs can be summarized as follows:

- *Passenger service* costs, including meals, flight attendants and in-flight services.
- *Advertising* and publicity expenses.
- *General and administrative* expenses that are truly general to the airline or those that cannot be associated to a particular activity.
- *Transport-related* expense items are costs associated with "the generation of transport related revenues" (US DOT, 2007). They include fees paid to regional airline partners for providing regional air service, extra baggage expenses and other miscellaneous overheads.

Figure 5.2 shows the functional categorization of US airline operating costs reported for 2007, as a direct comparison to the administrative categorization of Figure 5.1. To summarize, administrative and functional cost categorization schemes reflect two different

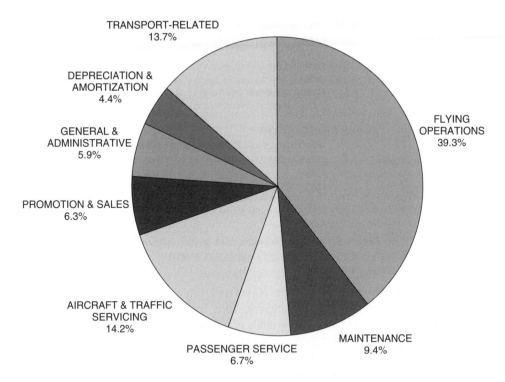

Figure 5.2 Functional cost allocation, 2007
Data source: Air Transport Association (2008)

approaches for partitioning airline operating costs. While the administrative approach is useful in financial reports and related analyses, it is the functional categorization that allows one to perform more detailed cost comparisons across airlines and even among different aircraft types.

The distribution of airline operating expenses by functional cost category depends, of course, on the specific definitions of the cost elements to be included in each category. There can be differences between the cost categorizations used by different entities, depending on their perspectives and history of cost analysis. In many respects, ICAO has historically provided the closest to what can be considered a worldwide standard of airline cost categorization, as it requires its Member States to submit annual operating cost data for their airlines in a standardized form (in addition to various traffic and financial data). The ICAO cost categories are summarized in Figure 5.3, and are in many respects similar in structure to both the US Form 41 functional categories and to allocation schemes used by airlines and government authorities around the world.

There are nonetheless some differences. Most relevant to our ability to perform detailed comparisons with US airline operating costs reported in Form 41 is the greater level of aggregation of the ICAO categories. Within the functional categories described above for US Form 41 there are, in many cases, even more detailed allocations that must be reported by airlines. For example, the Form 41 data breaks its "maintenance" category into "direct airframe maintenance," "direct engine maintenance" and "maintenance burden" (overhead)

Direct Aircraft Operating Costs
   Flight Operations (Total)
   Flight Crew
   Fuel and Oil
   Other
   Maintenance and Overhaul
   Depreciation and Amortization

Indirect Operating Costs
   User charges and station expenses (Total)
   Landing and associated airport charges
   Other
   Passenger services
   Ticketing, sales and promotion
   General, administrative and other

Figure 5.3   ICAO airline operating cost categories
Source: Adapted from ICAO (2007b)

subcategories. This is not the case for the ICAO data, which groups all "maintenance and overhaul" expenses into a single cost category. Thus, any comparisons of US and non-US airlines' maintenance costs can only be made at the more aggregate level. Overall, however, the "direct aircraft operating costs" of the ICAO scheme are reasonably comparable to the "flight operating costs" of the functional cost categorization described above.

Looking at the ICAO "indirect operating cost" categories, there are some further differences. Again, the ICAO approach is more aggregate, as it does not separate "station expenses" into the "aircraft servicing" and "traffic servicing" categories of ground operating costs described above. ICAO includes landing fees and airport charges in the same category as station expenses, similar to the categorization by the US DOT of landing fees as aircraft servicing expenses. The allocation of landing fees to the "aircraft servicing" category of ground operating costs (as is done in the US DOT categorization) might make sense to some, but others (e.g., some large European airlines) include landing fees in flight operating costs, since an aircraft cannot operate a flight without landing at an airport.

The ICAO scheme also aggregates all "ticketing, sales and promotion" into a single category, whereas the US Form 41 data separates out the advertising components from the remaining distribution costs (reservations, ticketing, distribution system fees). Under "passenger services," both approaches include cabin crew (flight attendant) costs. Thus, neither the US DOT nor ICAO include flight attendant expenses in flight operating costs with the rationale that different airlines will operate the same aircraft type with different numbers of flight attendants (above the minimum requirement for safety), for passenger service reasons. The logic of not including flight attendant expenses in flight operating costs is that doing so would distort potential comparisons of aircraft-related operating expenses. On the other hand, the European JAA has in the past categorized all flight attendant expenses as part of flight operating costs, despite the fact that as many as 50% of the flight attendants on an international flight can represent staffing above the minimum safety requirement.

These examples illustrate the difficulties of operating cost categorization, and the reality that there exists no perfectly "clean" or fully defensible definition of airline cost categories.

In the airline industry, much of the rationale for existing operating cost categories and the way in which costs are allocated and reported is historical. Fortunately, many of the differences between the US Form 41 approach that we rely on for detailed cost analysis and the ICAO standard for world airline costs are in the level of aggregation and in the allocation of a few minor expenses to different categories. Overall, the structure of the cost allocation schemes is very similar.

5.1.2 Cost Drivers by Functional Category

The reporting of operating costs for each functional category is typically based on various cost "drivers" or specific activities that generate the expenses for the airline in each category. These cost drivers are based largely on historical precedent, but in most cases they also make intuitive and economic sense:

- *Flight operating costs* (FOC) and all of the components of this category are reported "per block hour," given that the large majority of aircraft operating expenses are directly correlated with the amount of time the aircraft is being utilized.
- *Aircraft servicing costs* are reported per aircraft departure, given that these expenses are incurred by preparing the aircraft for each departure (cleaning, fueling, marshalling the aircraft).
- *Traffic servicing costs* are reported per enplaned passenger for passenger airlines, since these expenses involve the processing of passengers and their baggage at airports
- *Passenger service costs* are reported per revenue passenger kilometer (RPK), reflecting the fact that on-board services for a particular passenger trip will cost more for longer distances.
- *Promotion and sales costs* are reported as a percentage of revenues, given that this cost category is directly responsible for the generation of revenues for the airline.
- *Other indirect and system overhead costs* are reported as a percentage of total operating expenses, given the difficulties of more specific functional allocation.

Although these are the most common ways in which the different cost categories are reported and compared, more aggregate comparisons can involve measures of total operating cost per available seat mile – also known as "unit cost" or "CASK" (i.e., cost per ASK). Other comparisons might be based on total operating expense per passenger enplaned, or per RPK. The use of different bases for airline operating cost measures, however, can lead to misleading conclusions when different airlines with different networks and operating patterns are compared. In the following sections, some of the potential distortions embedded in various aggregate cost measures are described.

5.2 Operating Expense Comparisons

In this section, we make use of the functional cost categories and typical cost drivers introduced in the previous section to perform several types of airline operating cost comparisons. We begin with a closer look at the distribution of airline operating costs by functional cost category, comparing the changes in this distribution in recent years, including a comparison of US and world airlines. We then present a comparison of operating costs by category, as reported by US "legacy" and "low-cost" airlines, and examine recent trends and differences between them. Finally, we examine the flight operating cost categories in even greater detail, with comparisons across different types of aircraft. The purpose of this section is not simply to describe these comparisons and recent trends, but also to provide some insight into the potential pitfalls of performing such analyses without ensuring a valid basis for comparison.

5.2.1 Percentage Breakdown of Operating Expenses

Based on the functional cost categorization described in Section 5.1, we can use the breakdown of US airline operating costs for 2007 shown in Figure 5.2 to examine the three major categories of airline costs introduced earlier:

- Flight operating costs (FOC) comprise 53.1% of total expenses, if we sum the flying operations, maintenance, and depreciation and amortization categories in Figure 5.2. Historically, FOC have accounted for approximately one-half of total airline operating expenses, with variations over the years driven in large part by fluctuations in fuel prices.
- Ground operating costs represent another 20.5% of total operating expenses, divided into the major functional components: aircraft and traffic servicing (14.2%) and promotion and sales (6.3%). While this category has historically represented up to 30% of total operating expenses, major reductions in promotion and sales costs (as well as increases in other categories) over the past decade have reduced this proportion, reflecting the imposition of travel agency commission caps and the lower costs associated with Internet distribution of airline tickets.
- System operating costs account for the remaining 26.4%, including passenger service (6.7%), general and administrative (5.9%) and transport-related expenses (13.7%). This category has grown in the recent past, from its historical average of approximately 20% of total operating expenses. Since 2003, US DOT has required airlines to report the amount they spend on capacity purchases from regional airlines as "transport-related" costs. The growth in the proportion of system operating costs is therefore attributable both to a change in reporting rules and the recent trend of US network carriers to increase their use of regional partners as feeder airlines.

An illustration of how this percentage breakdown can change, even over a relatively short period, is provided by Figure 5.4. It shows the breakdown by category of the total operating expenses for both 2000 and 2004, as a comparison to the 2007 proportions shown in Figure 5.2. The most obvious change in cost proportions since 2000 is the substantial increase in the percentage of transport-related costs, which increased from 3.6% in 2000 to 13.0% in 2004, and reached 13.7% in 2007, for the reasons mentioned above. Also worth noting is the large drop in the promotion and sales cost category, from 13.4% in 2000 to 6.3% in 2007. Maintenance, passenger service, and aircraft and traffic service proportions also decreased, due in large part to the large increase in the transport-related cost proportion. The flying operations proportion also increased, reflecting an increase in fuel prices between 2000 and 2004. The flying operations category increased even more as a proportion of total costs from 32.7% in 2004 to 39.3% in 2007, as an unprecedented surge in fuel prices to historically high levels

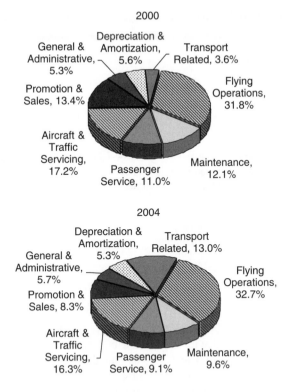

Figure 5.4 US airline operating cost breakdown, 2000 and 2004
Source: Strina (2006)

overwhelmed the cost reductions achieved by airlines in other categories (e.g., lower labor costs).

Changes in the breakdown of operating expenses by functional category provide some insight into shifts in airline operating costs, but it is important to remember that any significant change in the share of one category will affect all other categories, in the opposite direction. This phenomenon is well illustrated by Figures 5.2 and 5.4, in which dramatic increases in the percentages of transport-related and flying operations cost proportions reduce the apparent importance of many other categories.

Another example of the changes in the distribution of airline operating costs is provided by Figure 5.5, which shows the percentages based on the ICAO cost categorization scheme described in Section 5.1. Although the specific categories differ slightly from those of Form 41, the general trends in world airline cost proportions are nonetheless similar to those described above for US airlines. Direct operating costs are shown in Figure 5.5 to have increased from 44% in 1992 to 49.1% in 2002 and to 54% in 2005. Recall that the equivalent FOC comprised 50% of US airline costs in 2000 and 48% in 2004 (Figure 5.4), so these proportions are quite similar, averaging about 50% of total operating expenses.

The proportions of total operating expenses associated with most of the other (aggregate) cost categories are also very similar between US and world airlines. The "aircraft and traffic servicing" category comprises 16–17% of US airline costs, very similar to

ICAO OPERATING COST CATEGORIES	1992	2002	2005
Direct Aircraft Operating Costs	**44.0**	**49.1**	**54.0**
Flight Operations (Total)	<u>26.1</u>	<u>30.7</u>	<u>37.7</u>
Flight Crew	7.2	9.0	7.8
Fuel and Oil	12.2	13.0	21.9
Other	6.7	8.7	8.0
Maintenance and Overhaul	10.9	11.3	10.2
Depreciation and Amortization	7.0	7.1	6.1
Indirect Operating Costs	**56.0**	**50.9**	**46.0**
User charges and station expenses (Total)	<u>17.2</u>	<u>17.0</u>	<u>16.2</u>
Landing and associated airport charges	3.9	4.0	3.8
Other	13.3	13.0	12.4
Passenger services	10.8	10.3	9.3
Ticketing, sales and promotion	16.4	10.7	9.1
General, administrative and other	11.6	12.9	11.4

Figure 5.5 World airline cost distributions, 1992–2005
Data source: ICAO (2007b)

world airline costs. "Passenger services" average approximately 10% for both sets of airlines. The categories associated with sales, distribution and promotion show reductions over the years in both Figures 5.4 and 5.5, having dropped to 9.1% for world airlines by 2005, and to 8.3% for US airlines by 2004. Perhaps the largest difference in cost proportions is with respect to general and administrative costs, which still appear to be significantly higher for world airlines than for US airlines.

5.2.2   Legacy vs. Low-Cost Airlines

In this section, we explore some recent trends in operating costs by category, with the goal of comparing the reported cost measures for "network legacy carriers" (NLCs) as opposed to "low-cost carriers" (LCCs). NLCs are typically more traditional airlines that operate large hub-and-spoke networks consisting of regional, domestic and international services. LCCs, on the other hand, tend to operate smaller networks that tend to include a higher proportion of "point-to-point" or non-hub services – although many LCCs around the world operate networks that include "focus cities" as connecting points for passengers. LCCs typically offer reduced levels of service and low fares, although the wide range of LCC characteristics makes it difficult to identify a single representative LCC "business model."

The LCC "Business Model"

There are many characteristics and operating strategies commonly assumed to be shared by all, or at least most, LCCs. The following list reflects the most important characteristics that can contribute to the ability of an LCC to compete in a market with fares substantially lower than those offered by legacy airlines before the entry of the LCC. Not surprisingly, all of these characteristics are associated with productivity efficiencies and lower operating costs for the LCC:

- Use of a single aircraft type or interchangeable family of aircraft: as will be discussed in Section 6.1, "fleet commonality" reduces the costs of spare parts, maintenance and crew training.
- Operation of "point-to-point" instead of connecting hub networks: focusing on non-stop flights serving only local passengers flying from point A to point B reduces costs associated with the handling of connecting passengers, and improves the productivity of both aircraft and crews (see Section 6.2).
- No labor unions and lower wage rates for employees: keeping their employees non-unionized allows LCCs to pay lower salaries and achieve higher productivity due to less restrictive work rules.
- Single cabin service, with no premium classes offered: given that LCCs focus on very low fares aimed at price-sensitive leisure travelers, multiple product offerings would both increase complexity and costs.
- No seat assignments: open seating means less time spent on processing passengers at the airport and no need to print boarding passes, again improving productivity and reducing costs.
- Reduced "frills" and seating space on board: elimination of food and beverages reduces passenger service costs, while reduced seating space increases the ASK produced by each flight, in turn lowering its unit costs.
- No frequent-flyer loyalty programs: such programs incur administrative costs that can be avoided.
- Avoid use of traditional distribution channels: travel agencies still receive commission fees in many countries, and make use of computer reservations systems called "global distribution systems (GDS)", as will be discussed in Chapter 16. GDS charge fees for booking ands ticketing, which can be avoided if the LCC limits its passengers to making reservations and buying tickets directly from the airline, either on its website or by telephone.

Although not exhaustive, this list of characteristics captures what most would agree are the basic elements of the LCC business model, as it has evolved and been adopted by airlines in this general category around the world. The only problem is that many of these characteristics do not accurately represent the actual strategies employed by some of the largest and most successful LCCs. Figure 5.6 presents a comparison of six well-known and very successful LCCs in the USA, Canada and Europe, showing the extent to which each airline meets the above "typical" LCC characteristics:

- Southwest Airlines (USA) is the oldest, most successful and most studied LCC in the world. Many believe that it still maintains all of the typical LCC characteristics. South-west does indeed operate a single aircraft family (Boeing 737), offers a single cabin service and no seat assignments. On the other hand, Southwest does offer complimentary snacks and beverages, its aircraft have leather seats with legroom similar to that of legacy carriers. While the airline's network is not a classical hub structure, it has many

	Southwest	JetBlue	AirTran	WestJet	easyJet	Ryanair
Single aircraft type or single family of aircraft	✔	✗	✗	✔	✗	✔
Point-to-point ticketing, no connecting hubs	✗	✗	✗	✗	✔	✔
No labor unions, lower wage rates	✗	✔	✗	✔	✗	✔
Single cabin service, no premium class	✔	✔	✗	✔	✔	✔
No seat assignments	✔	✗	✗	✗	✔	✔
Reduced frills for on-board service (vs. legacy)	✗	✗	✗	✗	✔	✔
No frequent-flyer loyalty program	✗	✗	✗	✗	✔	✔
Avoid global distribution systems (GDS)	?	✗	✗	✗	✔	✔

Figure 5.6   Comparison of major LCC characteristics

"focus cities" at which a large proportion of its passengers make connections. And, perhaps most surprisingly, Southwest is the *most heavily unionized* airline in the USA; its employees' salaries are very similar to those paid by US legacy carriers (Gittell, 2003).

- JetBlue (USA) was launched in 2000, and has grown very rapidly ever since. While a successful LCC, it has even fewer of the typical LCC characteristics than Southwest. Unlike Southwest, it has avoided employee unionization to date, making it more "typical" in this respect. On the other hand, JetBlue operates two very different aircraft types (the Airbus 320 and Embraer 190), and its focus on flights to and from its home base at JFK Airport in New York has given it a natural connecting hub there. JetBlue also offers advance seat assignments to all passengers, on-board service (and live television) that is perceived by consumers to be superior to that of legacy airlines, a frequent-flyer program, and in 2007 made its flights available through traditional distribution channels (GDS).
- The third largest US LCC is AirTran and, as shown in Figure 5.6, it has none of the typical characteristics shown, but it is a successful and low-cost operator. Some of AirTran's employees are unionized, and the airline operates a traditional connecting hub at Atlanta. It is the only LCC shown that also offers a business-class product with more seating space and enhanced on-board passenger service.
- WestJet (Canada) was launched as an imitator of the "Southwest" model, and has grown to be the second-largest airline in Canada and a formidable competitor for Air Canada in domestic and transborder (between Canada and the USA) markets. WestJet employees are not unionized, and it operates a single aircraft family (Boeing 737) with a single cabin. Otherwise, it fails to meet all of the other criteria of a "typical" LCC – it has focus cities such as Toronto and Calgary where passengers make connections, it

participates in both a loyalty program and traditional distribution channels, and offers a level of on-board passenger service that is very competitive with legacy airlines.

- One of the largest European LCCs is easyJet (UK), based at Luton Airport outside of London. This airline has more of the typical LCC characteristics than any of the North American LCCs discussed above. However, some of its employee groups are unionized (e.g., pilots) and it now operates two different aircraft families, the Boeing 737 and Airbus 319 types.
- The other large European LCC is Ryanair (Ireland), and it is the only airline in this comparison that has all of the "typical" characteristics of an LCC business model. Ryanair has indeed been both profitable and successful, and it remains the largest (but only one of a few) LCCs that meet all six of the criteria of Figure 5.6.

As LCCs around the world continue to grow and mature, some might choose to adhere strictly to the typical LCC characteristics discussed here. The vast majority of LCCs, however, will develop different strategies in order to compete and survive against legacy airlines, as demonstrated by this comparison of six well-known and large LCCs. In fact, there appears to be some blurring of the dividing line between LCCs and legacy airlines, as LCCs adopt effective distribution and loyalty strategies developed by legacy carriers. At the same time, legacy airlines have adopted some of the cost efficiency and productivity strategies of LCCs, leading to partial convergence in terms of both business models and operating costs, as detailed in the next section.

Recent Trends in Operating Costs – Legacy vs. Low-Cost Airlines

The initial focus of our cost comparisons between NLCs and LCCs is on US airlines, due once again to the availability of detailed operating cost data in the Form 41 database and the lack of similarly detailed information in the public domain for non-US carriers. We first compare various cost measures by category based on the available US data. We then include more aggregate cost measures for NLCs and LCCs in Europe.

The classification of US airlines as NLCs as opposed to LCCs is not entirely straightforward. Although, at least from a historical perspective, most would agree that the NLC group should include the "Big Six" US network carriers (American, United, Delta, Northwest, Continental and US Airways), the recent merger of US Airways and America West raises questions about the classification of the "new" US Airways. On the LCC side, few would argue that Southwest, JetBlue and AirTran clearly belong in this group. However, while Alaska has intermittently pursued a "low-fare" market position, its costs are not in the same range as those of others in this group.

With some recognition of its limitations, for the purposes of this discussion, we will use the following classification of large US airlines, as we examine the evolution of cost measures by category over the period 1999–2007:

- NLCs: American, United, Delta, Northwest, Continental, US Airways/America West
- LCCs: Southwest, AirTran, Frontier, American Trans Air (ATA), JetBlue, Spirit.

In the following paragraphs, recent trends in operating costs by functional category are examined, for the NLC and LCC subgroups separately. Note that in this section and throughout the chapter, operating costs are expressed in current dollars (as opposed to constant dollars adjusted for inflation). The reason is that our emphasis is on the *comparison* of time streams of operating costs (e.g., cost per block hour for the NLC group vs. cost per block hour for the LCC group). Current dollars are preferable in this context, since they provide a comparison of the actual differences in costs at each point in time.

Flying Operations

As mentioned in Section 5.1, since the expenses categorized in this function are incurred during flight, they are typically allocated against block hours. The change in flying operations expenses per block hour between 1995 and 2007 is illustrated in Figure 5.7 for each group of airlines. Flying operations expense per block hour effectively doubled during this period regardless of the type of airline. The steepest increases occurred in 2000 – with a jump of about 20% – and then during the years after 2003, an increase of over 30%. In 2000, part of this increase was related to higher flight crew (pilot) wages resulting from new union contracts put into place at several of the largest NLC airlines. After 2003, steep increases in fuel prices contributed to this increase for both groups.

Figure 5.7 also shows that flying operations expenses per block hour for NLCs have remained approximately 50% higher than for the LCCs. Part of this difference can be attributed to lower wage rates and higher levels of productivity among LCCs – the differences that make them "low-cost" operators. However, in this case, a large part of the higher flying operations expense per block hour for the NLC group is the result of a much simpler explanation – legacy airlines typically operate substantially larger aircraft that consume more fuel per block hour, cost more to acquire and maintain, and have more senior (and higher-paid) pilots in the cockpit. Comparisons of flying operations expense per block hour across airlines (or airline groups) thus carry the risk of misleading conclusions if aircraft size, among other factors, is not accounted for. One way to account for aircraft size is to make such comparisons on a "unit cost" basis, per available seat or available seat mile (ASK), as will be discussed.

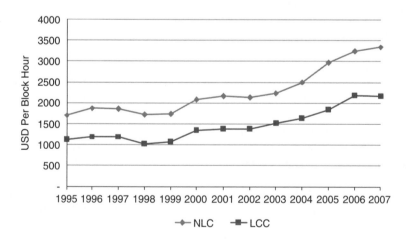

Figure 5.7   Flying operations expense per block hour, 1995–2007
Data source: US DOT Form 41

Maintenance

Figure 5.8 illustrates the maintenance expense per block hour for the two groups of US airlines. Between 2000 and 2005, legacy carriers spent about 60% more (peaking at 80% more in 2002) for maintenance per block hour than their low-cost competitors. By 2007, this difference had increased to 100% (i.e., legacy airlines spent twice as much per block hour for aircraft maintenance than LCCs). One reason why LCCs have lower maintenance costs per block hour is that their fleet is usually younger than that of the legacy airlines. Figure 5.9 shows a positive relationship between the average age of an airline's fleet and its maintenance cost per block hour, based on an analysis of 2004 data.

It is also true that maintenance expenses per block hour will be higher for larger aircraft (with more engines, and more airframe to maintain). Legacy airlines that operate international networks tend to have aircraft that are not only older, but also larger, wide-body types that are commonly used for longer-haul international routes. Many other factors can contribute to higher maintenance cost per block hour, among them fleet diversity and differences in wage rates paid to maintenance personnel. The use of outsourcing for maintenance has been more common among LCCs in the USA, further increasing the differences in reported maintenance costs (as payments made to external maintenance suppliers are not uniformly included in the "maintenance" cost category).

Passenger Service

It is in this category of costs that we observe the largest difference between the NLC and LCC groups. Recall that passenger service costs are most typically reported on a per RPM

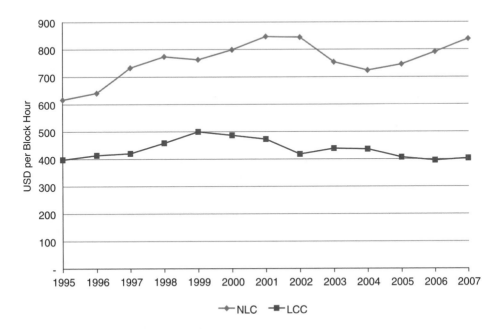

Figure 5.8   Maintenance expense per block hour, 1995–2007
Data source: US DOT Form 41

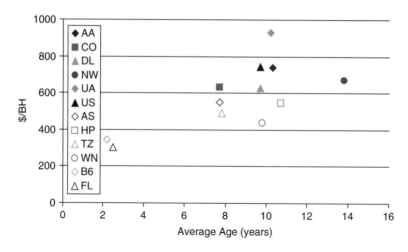

Figure 5.9 Maintenance expense per block hour vs. average fleet age, 2004
Source: Strina (2006)

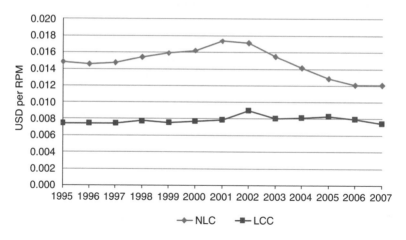

Figure 5.10 Passenger service expense per RPM, 1995–2007
Data source: US DOT Form 41

(or RPK) basis. Figure 5.10 shows that legacy carriers spent on average two times more per RPM for passenger service than the LCC group ($0.016/RPM vs. $0.008/RPM), during the period up until 2001, but this gap has narrowed noticeably since 2002. By 2007, legacy airline passenger service costs had dropped to $0.012/RPM, or 50% higher than LCCs.

Passenger service expenses are an easy target for cost cutting, and legacy airlines were able to reduce this cost per RPM by about 30% between 2001 and 2006. For the LCC group these expenses dropped only slightly between 2000 and 2006, since their passenger service costs were already quite low. Traditionally, LCCs offered less in the way of in-flight amenities and were thus able to limit their expenses in this category. More recently, many NLCs have cut passenger service costs by eliminating meals, pillows and other amenities, while some of the LCCs have actually improved their in-flight services, leading to higher passenger service expenses.

Aircraft and Traffic Servicing

Both of these expense categories are incurred on the ground, and there are again substantial differences in these costs between NLCs and LCCs. The recent trends in aircraft servicing expenses per departure are shown in Figure 5.11. Both groups of airlines have seen a steady increase in their reported aircraft servicing expenses per departure over the period shown. A good portion of this increase occurred in 2001 and subsequent years, due to the increase of security-related expenses for both passengers and baggage after 9/11.

More striking is the absolute difference in aircraft servicing costs per departure, when NLCs and LCCs are compared. Legacy airlines reported aircraft servicing expenses per departure on average twice those of low-cost airlines. A major reason for this difference is again simply the fact that NLCs typically operate larger aircraft and fly routes that involve longer distances. Both of these characteristics help explain the much higher aircraft servicing costs – larger aircraft require more resources (labor) to be serviced between flights, and longer flights also require more aircraft preparation, including fueling and cleaning. Also contributing to the difference are the rental fees and user charges paid by airlines to airports. LCCs are more likely to operate at secondary airports that have lower fees. And, because landing fees and various other airport charges depend on the weight (e.g., the size) of the aircraft, low-cost airlines operating smaller aircraft will further see lower per departure costs for aircraft servicing.

The comparisons and explanations for the differences between NLCs and LCCs in terms of traffic servicing expenses per passenger enplaned are very similar (Figure 5.12). The legacy airlines report traffic servicing expenses per enplanement that are two to three times those of LCCs. Again, differences in the network and operations of these two airline groups and, in turn, the characteristics of their passengers explain much of this

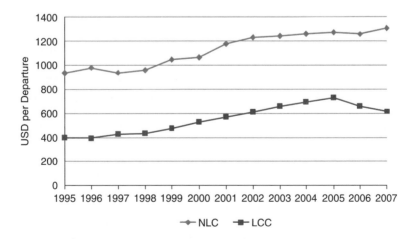

Figure 5.11   Aircraft servicing expense per departure, 1995–2007
Data source: US DOT Form 41

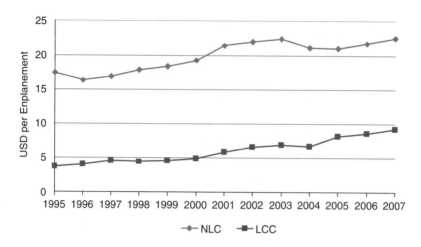

Figure 5.12   Traffic servicing expense per enplanement, 1995–2007
Data source: US DOT Form 41

cost difference. Legacy airlines are more likely to be carrying passengers on longer-haul itineraries, involving connecting flights and perhaps international travel. Processing passengers at the airport for international trips requires substantially more effort (passport checks, security controls) than domestic passengers. Passengers on longer itineraries, domestic or international, are also much more likely to check more baggage, which incurs a higher traffic servicing cost. And NLCs that rely on the hub-and-spoke network model transport many more connecting passengers (and their baggage), who require additional processing at their connecting points.

Promotion and Sales

Promotion and sales expenses have been falling since the first introduction of travel agency commission caps in 1994, and continued to decline over the years 2000–2005 (see Chapter 16). This was evident in our previous discussion of the percentage breakdown of total operating expenses, and is equally apparent in the promotion and sales expense per RPM shown in Figure 5.13. Between 1995 and 2002, this cost measure dropped by about 40% (from $0.025 to $0.014 per RPM) for the legacy airline group, and by about 25% for the LCCs. With the continued expansion of electronic ticketing and the rapidly growing use of the Internet to distribute tickets, the industry has been able to reduce drastically its distribution costs.

The difference between NLCs and LCCs in this cost category has decreased in absolute terms per RPM over the period shown in Figure 5.13. Still, legacy airlines continue to spend more than twice as much as low-cost airlines per passenger enplaned in the promotion and sales category, as shown in Figure 5.14. This persistent difference can be explained once more to a large extent by the longer and more complicated itineraries flown by legacy airline passengers. From confirming reservations to issuing tickets, it is more complicated and costly for the legacy airline to process international and long-haul passengers with many connecting flights (especially those with "inter-airline" connections

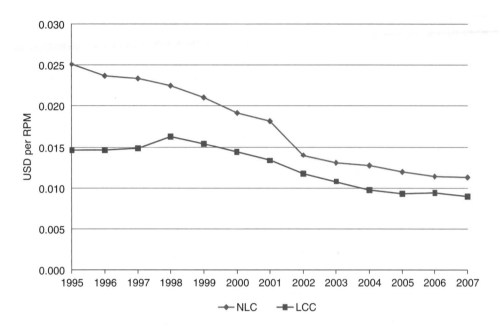

Figure 5.13 Promotion and sales expense per RPM, 1995–2007
Data source: US DOT Form 41

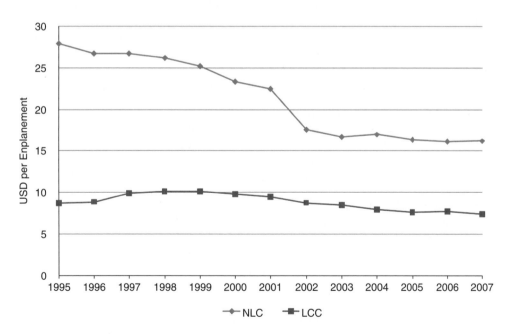

Figure 5.14 Promotion and sales expense per passenger, 1995–2007
Data source: US DOT Form 41

involving multiple airlines). Moreover, these passengers are less likely to be willing (or able) to use the Internet to make bookings and purchase tickets.

These comparisons by functional category between US NLCs and LCCs illustrate some of the differences between the two groups in terms of operating costs. In each cost category examined, the LCC group has consistently reported substantially lower costs than the NLC group. Part of this cost gap is clearly due to the differences in underlying cost structures, wage rates and overall productivity of the two groups, but it is also due in large part to significant differences in the complexity of the services provided by the carriers, as well as the characteristics of their networks, size and age of their aircraft, and conditions of their labor agreements. Our analysis of recent trends in these cost measures has shown that both carrier groups are affected by the same exogenous forces that impact costs (e.g., fuel prices). And, in many of the cost categories examined, the gap between the NLC and LCC groups appears to be narrowing.

5.2.3 Flight Operating Cost Comparisons

In this section, we will focus on flight operating costs (FOC) and how they can differ across both different aircraft types and different airlines using the same type of aircraft. For these more detailed comparisons, we will use the set of definitions of the functional cost components introduced in Section 5.1. These categories are based heavily on the Form 41 functional categories, with FOC representing all costs related to aircraft flying operations. As noted in Section 5.1, the total FOC is the sum of the following reported functional cost categories in the Form 41 data: flying operations, maintenance, and depreciation and amortization. FOC also include an allocation of maintenance burden (overhead) in addition to direct maintenance costs.

FOC by aircraft type reflect an average allocation of system-wide costs per block hour, as reported to the US DOT Form 41 database by airlines for each aircraft type. These average costs per block hour can be affected by specific airline network and operational patterns, as will be discussed below. The four components of FOC are as follows:

- *Crew*: All wages and benefits paid to pilots (flight attendant costs are not included).
- *Fuel*: The easiest to allocate and most clearly variable cost element of FOC.
- *Maintenance*: Direct airframe and engine maintenance cost, plus "burden" or overhead (hangars and spare parts inventory).
- *Ownership*: Depreciation allowances, leasing costs, taxes and insurance.

For example, the total FOC per block hour, as reported by US airlines for the operation of the Airbus A320-200 aircraft type (average capacity 150 seats) for the year 2005, were as follows:

Crew	$470
Fuel	$1327
Maintenance	$524
Ownership	$570
Total FOC	**$2891 per block hour**

The reported FOC of $2891 per block hour for this aircraft type is an industry-wide average based on an 1139 mile (1834 km) average stage length and 11.6 block hours

Airline	Crew	Fuel	Maintenance	Ownership	Total
Northwest	$750	$1256	$597	$613	$3214
United	$478	$1241	$539	$436	$2694
JetBlue	$362	$1159	$358	$486	$2365

Figure 5.15   Airbus 320 flight operating costs per block hour, 2005
Data source: US DOT Form 41

of daily utilization, an average for all US major airlines operating the A320. Different stage lengths and aircraft utilization levels by different airlines can result in substantial variations in block hour costs among airlines for the same aircraft type. Also, the FOC per block hour can vary across airlines due to differences in crew costs (wage rates paid due to union contracts, seniority of pilots flying this aircraft type), maintenance costs (wage rates, age of each airline's aircraft of this type) and ownership costs (age of aircraft, specifics of leasing arrangements).

Figure 5.15 presents a comparison of FOC per block hour, by cost category, for the same A320-200 aircraft type, across three of the US airlines that operated this particular type of aircraft in 2005. While the average FOC of all US airlines operating the A320 was $2891 per block hour, Figure 5.15 shows that the FOC for individual operators ranged from a high of $3214 at Northwest (an NLC) to a low of $2365 at JetBlue (an LCC). For the same aircraft type, Northwest spent 35% more than JetBlue in FOC for each block hour of operation in 2005.

We can explain this difference by comparing each of the four components of FOC. Fuel costs per hour are slightly lower for JetBlue, perhaps due to its use of fuel hedging to lower the price of fuel it paid per gallon in 2005 or due to its younger A320 aircraft with slightly more fuel-efficient engines. Note that, overall, there is little difference in fuel cost per block hour across these airlines operating the A320. On the other hand, there are large differences in crew costs per block hour among the airlines, due to differences in union contracts (or lack thereof), wage rates and seniority of the pilots that fly this aircraft type. In a comparison of Northwest and JetBlue, Northwest's crew costs per block hour for the A320 were more than twice those of JetBlue, representing almost $400 of the difference in total FOC per block hour.

The other two categories of FOC – maintenance and ownership – also reflect differences across the airlines. Differences in maintenance costs per block hour could be due to different wage rates paid to maintenance personnel and/or aircraft age (an airline will require less maintenance for newer A320 aircraft). Maintenance cost differences could also reflect differences in the use of maintenance outsourcing and how those costs are allocated. As for ownership, these costs will be affected by the particulars of each airline's negotiated purchase price or leasing rate for the aircraft, as well as by the age of each aircraft (which can affect depreciation expenses).

There is another important factor that has a tremendous impact on these comparisons of FOC per block hour reported by Northwest and JetBlue for 2005 – the daily utilization of the A320 aircraft (block hours per day). For example, although JetBlue has younger A320s that might be expected to have higher daily ownership costs, the ownership cost per block hour is much lower at JetBlue because this airline achieved a utilization of 13.7 block hours per day for this aircraft type. Northwest's utilization for its A320 fleet was only 9.3 hours per day. Thus, JetBlue was able to spread its ownership costs (along with other fixed cost components of maintenance and even crew benefits) over 47% more block hours, resulting in substantially lower reported costs per block hour.

This comparison of FOC across different airlines operating the same type of aircraft illustrates the many factors that can result in substantial cost differences per block hour. Some of these factors reflect genuine differences in the costs of inputs used to generate air transportation services – for example, one airline pays its pilots a higher wage rate to fly the same type of aircraft. Many other factors, however, are "structural" in that the FOC differences are driven in large part by the characteristics of each airline's network and schedule – for example, JetBlue achieves high aircraft utilization with transcontinental point-to-point flights that include many "red-eyes" that operate overnight, whereas Northwest's utilization is much lower, due to its hub-and-spoke pattern of operations.

The airline-specific comparisons of FOC for the A320 aircraft type shown in Figure 5.15 reflect data reported for 2005, a period before the tremendous surge in fuel prices and before the post-bankruptcy restructuring efforts of Northwest. As an indication of how FOC per block hour can change in just two years, both in absolute terms and in relative terms across airlines, Figure 5.16 shows the comparable A320 FOC data for 2007. The increase in fuel costs is most obvious, climbing by 33–45% per block hour and contributing to an increase in A320 FOC per block hour for all three airlines shown.

Among the other cost components, the three airlines experienced very different changes between 2005 and 2007, in both direction and magnitude. Northwest reported a 28% decrease in crew costs per block hour for this aircraft type, due in large part to wage concessions the airline obtained as part of its post-bankruptcy restructuring in 2006. These crew cost reductions, combined with smaller decreases in maintenance and ownership cost components, allowed Northwest to report only a 3% increase in FOC per block hour for its

Airline	Crew	Fuel	Maintenance	Ownership	Total
Northwest	$540 (−28%)*	$1659 (+32%)	$564 (−6%)	$554 (−10%)	$3317 (+3%)
United	$435 (−9%)	$1740 (+40%)	$569 (+6%)	$576 (+32%)	$3555 (+32%)
JetBlue	$428 (+18%)	$1676 (+45%)	$338 (−6%)	$431 (−11%)	$2872 (+21%)

*(change in % from same figure in 2005)

Figure 5.16   Airbus 320 flight operating costs per block hour, 2007
Data source: US DOT Form 41

A320 fleet. During the same period, United's FOC per block hour for the A320 increased by 32%, driven in large part by an increase in reported ownership costs. (United's much lower reported ownership costs in the 2005 data could well have been a result of its then bankruptcy and reduced payment obligations to creditors.) The LCC JetBlue also reported 21% higher FOC per block hour in 2007, driven by both higher fuel prices and higher crew costs paid to its increasingly senior pilots.

These comparisons of FOC per block hour for a single aircraft type are intended to illustrate how different operating cost components contribute to differences in the reported FOC. The example also shows how factors not readily apparent in the reported cost data (bankruptcy effects, labor agreements, pilot seniority, and many others) can have an important and sometimes distorting impact on cost comparisons across airlines, even if we limit the comparisons to the same type of aircraft.

Comparison of Flight Operating Cost across Aircraft Types

In this section, we explore some comparisons of FOC per block hour across different types of aircraft operated by US airlines. As was the case in the previous section, many of the same "structural" factors such as labor unions, network configuration and fleet characteristics will affect the expected cost relationships across aircraft types.

All else being equal, we expect that larger aircraft should have higher flight operating cost per hour, but a lower unit cost per seat or per ASK. There exist some expected economies of scale due to aircraft size – for example, two pilots can fly a 100-seat or a 400-seat aircraft (even though they might be paid at different rates), so we would expect the crew cost per seat to be lower for larger aircraft. Also there exist some economies due to stage length, as the fixed costs associated with taxi, takeoff and landing are spread over longer flight distances.

But many other factors can distort cost comparisons across aircraft types and result in what appear to be counterintuitive relationships. For example, airline pilots are paid more per hour on larger aircraft that fly international routes, and these longer-haul routes are typically flown by the most senior pilots at each airline, who also happen to receive the highest wage rates. As well, newer technology engines are more efficient, even on smaller aircraft, so that smaller aircraft with newer technology engines may report lower fuel costs per seat than much larger aircraft with older technology engines. Most important, aircraft utilization rates can have a large impact on the allocation of FOC per block hour, as discussed.

The FOC comparisons in Figure 5.17 provide several examples of how different aircraft characteristics and their pattern of operations at US airlines affect the reported costs per block hour, based on 2005 data. The largest B747-400 aircraft has the highest total FOC per block hour, while the smallest DC9-30 has the highest FOC per seat hour, as might be expected. However, of the aircraft shown in Figure 5.17, the lowest costs per seat hour (and in turn per ASK) are reported for the newer technology A330 (100 seats smaller than the 747-400). Similarly, the B737-300 shows a lower cost per seat hour than the B757-200, which offers 50 more seats.

In both examples above, structural factors prevent us from making an "all else equal" theoretical comparison. For example, the B747-400 costs per seat hour are higher than those of the A330 due to factors that include higher wage rates paid to senior pilots at the airlines that operate the B747, as well as the older age and technology of the B747,

Aircraft Type	Seats	FOC/ Block-hr	FOC/ Seat-hr	Average Stage (mi.)	Utilization (block-hrs/day)
DC9-30	100	$3217	$32.17	473	6.9
A319	122	$2712	$22.23	943	10.5
B737-300	131	$2690	$20.53	568	9.6
B757-200	184	$4003	$21.75	1330	10.1
A330-200	270	$5382	$19.93	3668	13.8
B747-400	367	$9688	$26.39	4393	12.4

Figure 5.17   Flight operating costs, selected aircraft types, 2005
Data source: US DOT Form 41

which results in higher maintenance and fuel costs, even on a per seat basis. Figure 5.17 shows data reported for 2005 but, if we were to examine 2007 data, the distortions to the expected relationships would be even greater, caused by the disproportionate impact of high fuel costs on the older technology aircraft.

This section first examined the differences and recent trends in operating cost measures by functional category between US NLCs and LCCs. Then, more detailed comparisons of flight operating costs (FOC) were discussed, both among airlines that operate the same type of aircraft and among different aircraft types. Throughout this discussion, some of the challenges of making valid and consistent airline cost comparisons were identified. All of the cost measures described can be affected by structural differences among airlines, their networks, fleets and the type of air transportation service each operates.

5.3   Comparisons of Airline Unit Costs

The challenges of making fair comparisons across airlines and aircraft types cannot be escaped, even as we introduce some of the most commonly used airline measures of unit cost. In this section, we first discuss the relationships between total operating costs and unit costs, and the factors that affect them. We then examine the recent evolution of unit costs for US NLCs and LCCs, once again with an eye to the differences in measures such as aircraft utilization and average stage length that help to explain the unit cost gap between the two groups. Based on much more limited available data, we also present comparisons of legacy and LCC unit costs for a sample of European carriers, and conclude this section with a discussion of unit cost comparisons among the largest of the world's airlines.

5.3.1   Total Operating Costs vs. Unit Costs

The total operating cost for an airline is the sum of all of its operating expenses incurred during a particular period (e.g., a year). Total operating expenses will clearly be greater for larger airlines with more employees, operating more aircraft and carrying more traffic. Total operating expenses will also be higher for larger aircraft with greater capacity, as larger aircraft have higher FOC both per block hour and for any given flight than a smaller-capacity aircraft. Also, total operating costs will be greater for a longer stage length (flight distance), as more fuel is burned and more labor hours (pilots and flight attendants) are required. Thus, we expect that total operating costs for an airline will be positively correlated with the size of its network and fleet, the average capacity of its aircraft and its average stage length.

Unit cost is the ratio of the airline's total operating expenses to ASK produced, or ATK produced in the case of cargo airlines. For passenger airlines, unit cost is also known as "CASK," meaning "Cost per ASK" (the equivalent measure in miles is "CASM"). Unlike total operating costs, the relationships between unit costs on the one hand and airline size, average aircraft capacity and average stage length on the other are all expected to be negatively correlated, in theory at least. That is, a large airline is expected to see some economies of scale (reduction in unit costs with increased output), as its fixed costs are spread over a larger output of ASK. A larger-capacity aircraft is also expected to show some economies of aircraft size, as the fixed costs are spread over more seats for any given flight, resulting in lower costs per seat hour and per seat kilometer, as discussed above. And longer stage lengths mean that the relatively fixed costs of aircraft servicing and traffic servicing, for example, can be spread over more block hours of utilization and, in turn, more ASK produced. Even fuel consumption per ASK should be lower for longer stage lengths, as the higher fuel consumption associated with takeoff and descent are outweighed by longer periods of more efficient fuel consumption at cruise altitude.

With all else equal, then, the theoretical expectation is that the largest airlines flying the largest aircraft over the longest average stage lengths should report the lowest unit costs per ASK. This expectation underlies the worldwide trend toward airline mergers and consolidation. One of the primary objectives of most airline mergers is the desire to achieve lower unit costs with a larger scale of operations. If this theoretical expectation held true in practice, the largest NLCs both in the USA and around the world would consistently report lower unit costs (and correspondingly higher operating profits) than their LCC counterparts. NLCs are much larger in their scale of operations, and typically operate larger aircraft on longer stage lengths, as compared to most LCCs.

The reality does not match our theoretical expectations, as LCCs consistently report significantly lower unit costs than legacy carriers, despite being smaller in scale and flying smaller aircraft on shorter stage length flights. For example, Figure 5.18 shows 2003 data for 12 major US airlines, in which their unit costs are plotted versus average stage length. At first glance, there is little evidence of the expected inverse relationship between unit cost and stage length. But if the legacy group is separated from the low-cost group (by the ovals drawn on the graph), we observe the expected inverse relationship, but only within each group. That is, among the legacy carriers, it is clear that those with the longest average stage length have the lowest unit costs and vice versa. The same pattern is also evident within the low-cost group.

There are two lessons to be learned from Figure 5.18. First, unit cost comparisons should not be made between airlines unless differences in their average stage length are taken into account. US Airways shows a much higher unit cost than Continental, but it also operates flights with much shorter stage lengths. It could well be that US Airways is

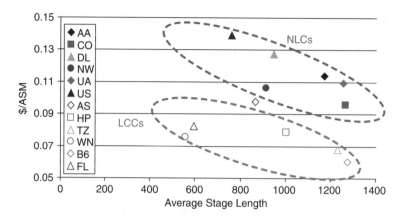

Figure 5.18   US airline unit cost vs. stage length, 2003
Data source: US DOT Form 41

a higher-cost producer than Continental, but unit cost comparisons alone cannot be used to make this determination. Second, we see that the differences in unit costs between legacy and low-cost airlines in 2003 were significant, even when we account for stage length. Low-cost airlines reported a much lower unit cost than legacy carriers across all ranges of average stage length.

5.3.2   Legacy vs. Low-Cost Airline Unit Costs

In this section, we examine the evolution of unit costs and provide comparisons between legacy and low-cost carriers. We begin with a comparison of US NLCs and LCCs (as defined in Section 5.2) made possible by the available Form 41 data, followed by a similar comparison of NLC and LCC unit costs for a sample of European airlines.

Figure 5.18 compared total unit costs as reported by US airlines for the year 2003, due in large part to the fact that 2003 is effectively the last year for which it makes sense to compare total unit costs reported by US airlines. In the Form 41 data, total unit costs include all operating expenses, including a category called "transport related expenses" which consists largely of payments made by airlines to regional carriers to provide connecting services on their behalf. These connecting services provide the paying airline (usually a network legacy airline) with incremental connecting traffic and revenue. However, these payments are not actual "operating expenses" incurred in the production of the capacity (ASK) output of the mainline carrier. Since the ASK units of the regional carrier are not included in the denominator of unit costs for the mainline carrier, transport-related payments should not be included in the numerator for the purposes of unit cost comparisons across airlines (Tsoukalas *et al.*, 2008).

This adjustment to Form 41 cost data is especially important when comparing the unit costs of US NLCs and LCCs, given that LCCs do not typically rely on regional partners for connecting traffic feed. Retaining the transport-related expenses in calculating NLC CASM can therefore lead to distortions in unit cost comparisons that make the gap between NLC and LCC unit costs appear to be greater than it really is.

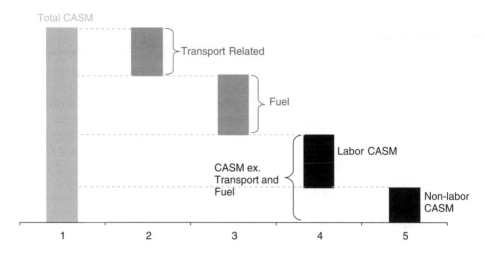

Figure 5.19   Major components of unit costs (CASM)
Source: Tsoukalas *et al.* (2008)

Figure 5.19 illustrates the four reported components used in the comparisons of the evolution of unit costs between US NLC and LCC groups, over the period 1995–2007, which are presented next. Transport-related expenses are reported but, as explained above, are excluded from all of the comparisons. The remaining three cost components of interest are fuel expenses, labor costs and non-labor costs.

Fuel expenses may be included in unit cost comparisons, under the historical assumption that all airlines are subject to the same fuel price environment. However, recent growth in the use of financial hedging instruments has given airlines a way to manage fuel costs in different ways, making direct comparisons less valid. Labor costs include total salaries, benefits and other costs paid by airlines to employees, providing an indication of the use and cost efficiency of labor inputs in an airline's total cost structure. Non-labor costs include all other operating expenses not included in the transport-related, fuel or labor-related cost components. This last category includes cost items that represent the "structural" costs of the airline over which management can exert influence and are therefore a good gauge of how management strategies affect "controllable costs" not related to fuel or labor inputs.

Figure 5.20 compares US NLC and LCC airline unit costs, with transport-related expenses removed. If we exclude this expense category, and focus on the costs of actually operating the airline and generating capacity in ASM form, the NLC group shows unit costs that have consistently been approximately 2 cents (USD) per ASM higher than the LCC group. In percentage terms, the unit costs of the two airline groups appear to be converging. LCCs still have a clear unit cost advantage in 2007, but their unit costs relative to NLCs were 18% lower in 2007 compared to 30% lower in 2001.

While this measure of total unit costs (excluding transport-related expenses) increased overall between 1995 and 2007 for both groups, the pattern of increase and the underlying factors driving the evolution of unit costs differs. Both groups experienced a drop in unit

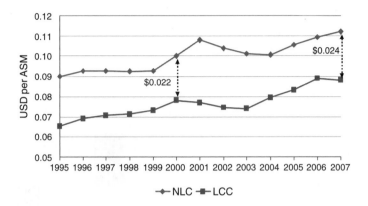

Figure 5.20   Unit cost excluding transport-related expenses, 1995–2007
Data source: US DOT Form 41

costs after 2001. For the legacy airlines, the sudden decline reflects cost-cutting strategies put into place after 9/11, including employee layoffs and passenger service cutbacks. However, this decrease did not last, as NLC unit costs began to climb again from 2004 through 2007, due in large part to soaring fuel prices. LCC airlines saw their unit cost follow a similar pattern, but the unit cost increase for LCCs was tempered by their rapid capacity growth during the same period. The expansion of LCCs' capacity (ASK) relative to their operating expenses has kept the unit costs of this group more stable, despite the rise in fuel prices. With a substantial proportion of fixed operating costs, airlines typically see lower unit costs with capacity expansion.

If both transport-related and fuel expenses are removed from the reported cost data, the remaining cost components provide an even more consistent unit cost comparison between the NLC and LCC groups, shown in Figure 5.21. The results for this modified

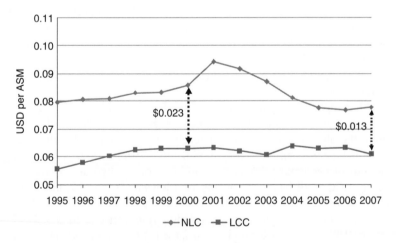

Figure 5.21   Unit costs excluding transport-related and fuel expenses
Data source: US DOT Form 41

unit cost measure ("CASM$_{exTF}$") present a very different picture. NLC CASM$_{exTF}$ has decreased by 17% from its peak in 2001 while it has decreased very slightly for LCCs. Between 2001 and 2006, the NLC CASM$_{exTF}$ dropped from 9.4 cents to 7.8 cents, while LCC CASM$_{exTF}$ remained flat at around 6.1 cents per ASM. As a result the difference (or "unit cost gap") between the carrier groups decreased from 3.1 cents to 1.3 cents per ASM. These results suggest that fuel expenses have been the major driver of increased unit costs for both NLCs and LCCs since 2000. Once we remove fuel and transport-related expenses, we see significant cost convergence in CASM$_{exTF}$ between 2000 and 2006.

To better understand the reasons for this cost convergence, we can further break this unit cost measure into its labor and non-labor components. Figure 5.22 shows that the non-labor unit cost component of CASM$_{exTF}$ remained relatively unchanged from 1995 to 2007 for both groups, despite fluctuations during the period. The non-labor unit cost gap between the groups has remained approximately 1 cent per ASM. The non-labor CASM$_{exTF}$ category reflects the airline's structural costs that are driven by a variety of factors such as network structure, fleet type and outsourcing activity, to name a few. This comparison of NLCs and LCCs suggests, therefore, that the NLCs have certain structural costs (hub operations, international flights, lounges and other services) that result in an inherent and consistent non-labor unit cost gap that appears to be about 1 cent per ASM.

If a comparison of non-labor costs shows a consistent unit cost gap, then it must be the case that the major factor driving unit cost convergence between NLCs and LCCs is labor unit costs. Figure 5.23 confirms that NLC labor unit costs dropped by almost 30% from 2002 to 2006, from 4.5 cents to 3.3 cents per ASM. On the other hand, the labor unit costs of the LCC group have increased steadily since 1995, from 2.2 cents in 1995 to 3 cents in 2006, with a slight decrease in 2007. The labor unit cost gap was reduced dramatically from its peak in 2002, to a difference of only 0.3 cents by 2006.

The large decrease in labor unit costs for NLCs is a direct result of their cost-cutting strategies during the bankruptcies and the threat of bankruptcies that occurred between 2002 and 2005. The financial crisis of that period allowed NLCs to cut employment levels,

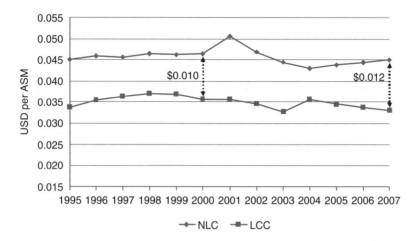

Figure 5.22 Non-labor unit costs
Data source: US DOT Form 41

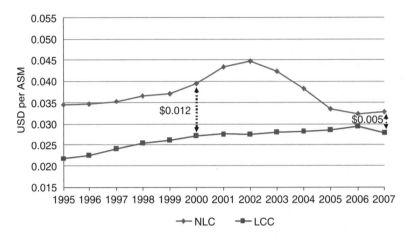

Figure 5.23 Labor unit costs
Data source: US DOT Form 41

negotiate lower overall compensation and increase productivity. On the other hand, LCCs had to deal with increasing seniority and some slowing in their growth rates, both of which contributed to slightly higher unit labor costs. In the US airline industry, the LCC advantage can no longer be simply attributed to lower labor costs.

5.3.3   Unit Cost Comparisons in Europe, Asia and Worldwide

Similarly detailed comparisons of the unit costs of NLCs and LCCs operating in other parts of the world are much more difficult, if not impossible. Once again, the operating cost data for non-US carriers is typically not available at the level of detail provided in the US Form 41 database, precluding comparisons of individual cost categories (e.g., aircraft servicing) across airlines or of the major unit cost components (labor vs. non-labor) presented in the previous section. Even if individual airline data can be obtained from public sources, e.g., annual financial reports, the categorization of operating costs included in those reports can preclude head-to-head comparisons between airlines and with the US airline data presented above.

As mentioned in Section 5.1, the ICAO air transport financial reports provide a compilation of annual operating expenses reported by ICAO Contracting States on behalf of their scheduled airlines, broken down by the categories shown in Figure 5.3. That categorization makes it impossible to separate labor and non-labor cost components. For example, categories such as "maintenance and overhaul" include both labor and non-labor expenses. Nonetheless, this section presents some unit cost comparisons of world airlines based on publicly available data.

Figure 5.24 shows a plot of unit costs (in US dollars per ASM) for European NLCs and LCCs, between 2000 and 2006. Although the use of ASM as a standardized measure of capacity allows for direct comparisons of airlines operating in different world regions, the use of US dollars can create some distortions due to currency fluctuations. Notably, the surge in the value of the euro between 2004 and 2006 complicates unit cost comparisons

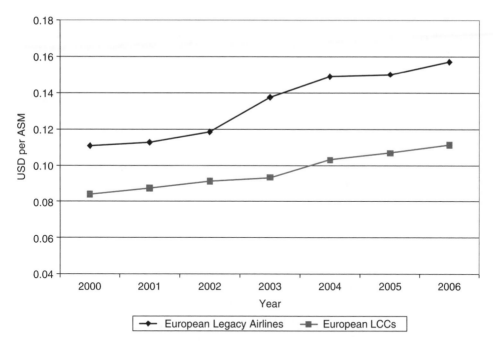

Figure 5.24   European NLC vs. LCC unit costs
Data source: ICAO (2007a)

between European and US airlines, for example. On the other hand, comparisons of NLC and LCC unit costs within regions remain valid.

European NLCs reported consistently higher unit costs than LCCs, and the difference in unit costs has increased since 2000. Among European carriers, the unit cost gap has increased over the period shown, from approximately $0.03 per ASM in 2000 to more than $0.04 per ASM. Recall that the comparison of US airline unit costs in Figure 5.20 showed a gap of about $0.02 between NLCs and LCCs. In Europe, however, the ICAO data suggests that NLC unit costs have been increasing faster than LCC unit costs, particularly since 2003, so that NLCs have unit costs over 40% higher than their LCC competitors. Again, this NLC cost disadvantage persists despite the fact that LCCs operate smaller aircraft on shorter stage length routes.

A similar comparison of NLC and LCC unit costs is provided for Asian airlines in Figure 5.25, based on data from the same ICAO source. As would be expected, NLC unit costs are substantially greater than LCC unit costs, but in Asia this gap has widened even more dramatically than in Europe, over the same period. Whereas the unit cost gap was less than $0.02 per ASM in 2002, it had grown to more than $0.06 per ASM by 2006. NLC unit costs increased by over 50% while LCC unit costs actually decreased slightly.

The evolution of unit costs among Asian carriers and the differences with their European counterparts suggest some interesting insights. First, Asian NLCs have seen their unit costs increase more dramatically than European NLCs (in percentage terms). Both groups have experienced higher labor and, especially, fuel costs, but these impacts have affected

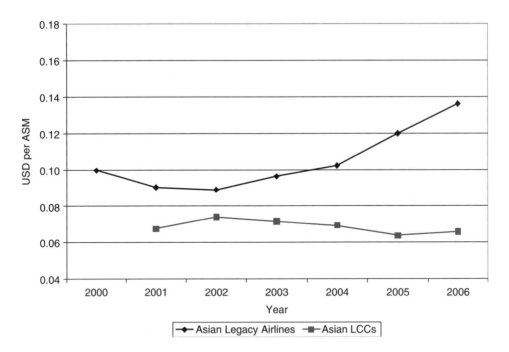

Figure 5.25   Asian NLC vs. LCC unit costs
Data source: ICAO (2007a)

Asian NLCs to a greater extent. Given historically higher labor costs in Europe and recent efforts by European NLCs to improve cost efficiency and productivity, the increases in labor costs at Asian NLCs have been proportionately greater. And the growing strength of European currencies between 2004 and 2006 reduced the impacts of surging fuel costs on European NLCs, compared to Asian NLCs (airline fuel prices are tied to the US dollar).

At the same time, Asian LCC unit costs dropped slightly during this period, while European LCC unit costs climbed by over 30%. This difference in cost evolution is attributable simply to the relative maturity of LCCs in Europe versus Asia. European LCC data is dominated by well-established large LCCs such as easyJet and Ryanair. Although these carriers clearly still have low costs, their unit costs continue to increase as they mature – employees accumulate seniority and demand higher wages, and formerly new aircraft become older and require more maintenance, for example. The Asian LCCs are much younger and much smaller than those in Europe (and, in turn, those in the USA). They are still experiencing very rapid growth which, along with somewhat longer stage lengths, allows them to report stable (or slightly decreasing) unit costs.

These comparisons of European and Asian airline unit costs, when combined with the US unit cost comparisons in the previous section, lead to several overall cost trends that reflect the spread of deregulation, growth of LCCs and restructuring of NLCs in response. In the USA, NLC unit costs have remained relatively stable since 2000 (despite surging fuel prices) as NLCs have restructured and dramatically reduced labor costs. LCCs in the USA are reaching maturity and their unit costs are increasing, such that the overall unit cost gap between the two groups remains at about $0.02 per ASM. In Europe, LCCs are also maturing and reporting higher unit costs. But European NLCs have not undergone the same restructuring or reduction of labor costs as US NLCs, resulting in a unit cost gap of well over $0.04 per ASM in 2006. In Asia, LCCs have not reached maturity and NLCs have made even less progress in restructuring and cutting costs, leading to an even higher unit cost gap of over $0.06 per ASM in 2006.

These differences in total unit costs among airlines around the world are reflected in one final comparison of the top 20 passenger airlines (ranked in terms of RPK carried), shown in Figure 5.26. On this graph, the 2006 total unit costs are plotted versus average passenger trip length (as a proxy for average stage length) for these largest airlines, as reported in the Airline Business financial database (*Airline Business*, 2007). Several of the most important concepts of this section are illustrated by Figure 5.26. First, among the non-US airlines, there exists an inverse relationship between total unit cost and average trip length. Singapore, Qantas and Emirates have the longest average trip lengths, and report relatively low unit costs (especially in view of their high levels of service quality). Lufthansa, Japan Air Lines (JAL), Air France/KLM and British Airways all have shorter average trip lengths, and report correspondingly higher unit costs.

A second important concept illustrated by Figure 5.26 is that expected theoretical relationships between airline costs and other explanatory factors do not always hold true. Specifically, all of the US NLCs reported unit costs well below those of the non-US airlines in the figure, despite having substantially shorter average trip lengths. The dramatic cuts and restructuring efforts of US legacy airlines have made them very competitive with their largest international counterparts, in terms of unit costs if not overall service quality. And, of course, Southwest Airlines continues to confound airline economic theory

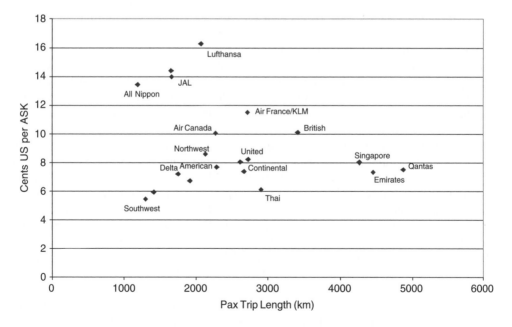

Figure 5.26   Unit cost vs. trip length, top 20 world airlines, 2006
Data source: *Airline Business* (2007)

by appearing on this chart with the lowest unit cost and the next-to-shortest average passenger trip length.

5.4   Measures of Airline Productivity

In most industries, productivity is typically measured as the amount of output created per unit of input: that is, a ratio of output over input. From an airline point of view, outputs are best defined by the capacity created. Because we have focused in this chapter primarily on passenger airlines, the appropriate measure of capacity for productivity measures is ASK (and "available tonne kilometers" for cargo airlines). On the input side, the most important productive inputs for airlines are capital (aircraft) and labor (employees). In this section, we describe the most common measures of both aircraft and employee productivity used in the airline industry, and further explore some recent trends in these measures, once again comparing the performance of NLCs and LCCs based on the detailed Form 41 data available for US airlines.

5.4.1   Aircraft Productivity

The most commonly used metric in evaluations of aircraft productivity is aircraft "utilization." Aircraft utilization is measured in block hours per day per aircraft. Block hours refer to the period during which an aircraft is in use, beginning at door close (blocks taken away from wheels) to door open (blocks placed under wheels). Block hours are thus the "gate-to-gate" times, including ground taxi times as well as flight times.

The ability of an airline to achieve a certain level of aircraft utilization depends on the characteristics of its network, its schedule and its efficiency in turning an aircraft around on the ground between one arrival and the next departure. The longer the "turn times", the less time there is for increasing block hours given a limited number of feasible operating hours during the course of each day. Differences in aircraft turn times can be substantial – Southwest Airlines has made turn times that average 20–30 minutes a focal point of their low-cost strategy and success (Gittell, 2003), while network legacy airlines can have aircraft sit on the ground at a connecting hub for as long as 1.5 to 2 hours to ensure connections for passengers and their baggage. Larger aircraft also require more turnaround time than smaller aircraft, as do aircraft arriving or departing on international services (due to additional customs and security requirements).

Like unit cost, aircraft utilization is also affected by an airline's average stage length, at least in theory. Longer flights mean less time spent by the aircraft on the ground during the day, and the turn time for a longer-haul flight, although longer than for a short flight, does not increase proportionally with stage length. Thus, we would expect airlines with longer average stage lengths to consistently report higher utilization rates. Yet this is not the case when we compare the US NLC and LCC airline groups, as we have done in the previous sections of this chapter. Figure 5.27 shows the trend in aircraft utilization for the two airline groups between 1996 and 2007.

Figure 5.27 shows that both airline groups saw their aircraft utilization drop in 2001 – in the aftermath of 9/11, most airlines cut back their schedules in response to lower demand for air travel. However, aircraft utilization rates at NLCs continued to decline through 2003, before showing a substantial rebound. The LCC group, on the

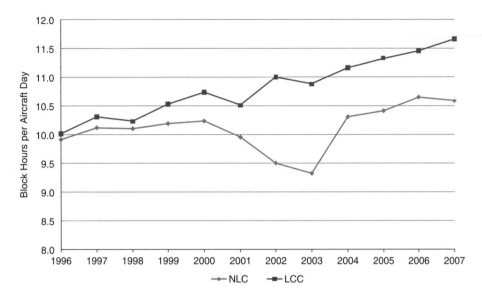

Figure 5.27   Aircraft utilization, 1996–2007
Data source: US DOT Form 41

other hand, saw its utilization rates increase again beginning in 2002, with improvements in this measure of aircraft productivity continuing through 2007. Throughout the period shown in Figure 5.27, LCCs had consistently higher aircraft utilization rates than the NLC group, and the gap between them has actually increased significantly since 1996, despite substantial improvements in NLC utilization rates since 2003. Given that LCCs operate shorter stage lengths on average, the difference between the two groups in terms of aircraft utilization is all the more impressive.

Another measure of aircraft productivity is that of ASK (or ASM as reported in the US Form 41 data) generated per aircraft per day, calculated as the product of the number of departures per day per aircraft, the average stage length of these departures and the number of seats on the aircraft:

$$\text{ASM per aircraft day} = \text{No. of departures} \times \text{Average stage length} \times \text{No. of seats} \quad (5.1)$$

The three factors contributing to this measure illustrate how an airline can increase its aircraft productivity by pursuing one or more of the following strategies:

- Increase the number of flight departures per day with the existing aircraft fleet, by reducing turn times and/or by increasing the operation of flights at off-peak departure times.
- Increase the average stage length for the aircraft fleet, by choosing to fly longer-distance routes and reducing the number of flights operated on short-haul routes. Longer stage lengths can increase both aircraft productivity and aircraft utilization (block hours per day).
- Increase the number of seats on each aircraft, by removing first- or business-class seats in favor of more economy-class seats and/or by reducing the "seat pitch" or distance between adjacent rows of seats.

Airline	Flights per Day	Block Hours	Stage Length	Seats	ASM
Northwest	3.6	9.3	957	148	513513 (Base)
United	3.9	11.2	1116	146	639801 (+25%)
JetBlue	4.0	13.7	1358	157	861627 (+68%)

Figure 5.28   Airbus 320 productivity comparison, 2005
Data source: US DOT Form 41

Each of these strategies for increasing aircraft productivity is reflected in the comparisons shown in Figure 5.28. The 2005 data shown is for the A320 aircraft type operated by Northwest, United and JetBlue, the same aircraft and set of airlines used in the comparisons of FOC per block hour in Section 5.2. The differences in aircraft productivity (ASM per aircraft per day) are dramatic – JetBlue in 2005 was able to generate 68% more ASM per day with its A320 aircraft than Northwest and 35% more than United. JetBlue's A320 aircraft operated more departures per day, on a longer average stage length, with more seats than the two legacy airlines shown. The much higher productivity of JetBlue aircraft is a major reason for its much lower FOC per block hour, as discussed previously and shown in Figure 5.15.

5.4.2 Labor Productivity

During the past several decades, labor-related expenses have accounted for over a third of the total operating expenses for US passenger airlines. More recently, this proportion has been reduced, both by airline efforts to reduce labor costs and by the growth of other expense categories such as fuel. In 2007, labor expenses accounted for 23.4% of total US airline operating expenses (see Figure 5.1), second only to fuel costs. In light of increased competition from low-cost airlines, legacy carriers have focused on these expenses in their efforts to cut their unit operating costs, as discussed in Section 5.2.

One way in which legacy carriers have cut total labor expenses is by reducing their total number of employees. Figure 5.29 shows the trend in total employment for the US airline industry, and by airline group, from 1999 to 2006. NLC employment decreased by over 30% between its peak in 2000 and 2006, representing a total loss of over 125 000 jobs. At the same time, LCC employment increased with their continued rapid growth, but clearly did not make up for the dramatic cutbacks at legacy airlines.

Similar to aircraft productivity, labor productivity in the airline industry is most typically measured in terms of output (ASK or ASM) per period, but on a per employee basis. This measure is thus the ratio of ASM to employees, irrespective of the type of employee.

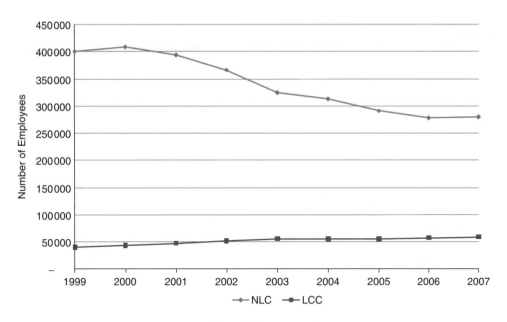

Figure 5.29   US airline employees, 1999–2007
Data source: US DOT Form 41

As with aircraft productivity, we expect that employee productivity should be higher for airlines with:

- Longer stage lengths, since the amount of labor required for aircraft and traffic servicing for each flight departure is not proportional to stage length.
- Larger aircraft sizes, given that there should be economies of scale in the labor required per seat for each flight departure.
- Higher levels of aircraft productivity due to shorter turnaround times, as more ASM generated by the aircraft contributes to higher employee productivity measures.

Yet legacy airlines with longer average stage lengths and larger aircraft sizes have historically reported lower employee productivity rates than LCCs operating smaller aircraft on shorter stage lengths on average.

As shown in Figure 5.30, the employee productivity of the LCC group of US airlines (as defined previously) continues to be about 10% higher than that of the NLC group, even as both groups have increased ASM per employee by more than 35% in recent years. Legacy airlines have achieved this increase in labor productivity through reductions in workforce and concomitant relaxation of restrictive work rules for different employee groups, providing greater flexibility in employee assignments and coordination of labor functions (to be discussed in detail in Chapter 10). And both NLCs and LCCs have continued to increase this measure of employee productivity simply by eliminating the need for humans to perform many customer service duties – making reservations, buying

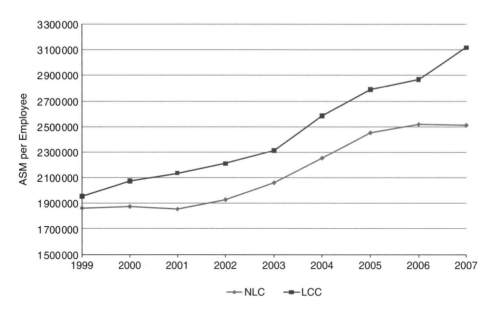

Figure 5.30   Labor productivity at US airlines
Data source: US DOT Form 41

tickets, checking in and printing boarding passes can now all be done by the passengers themselves instead of airline employees.

We have shown in this chapter that, like virtually all other measures of airline operating costs, unit costs can be affected by a variety of operational and network characteristics of each airline, e.g., average stage length. Our comparisons of unit costs reported by legacy and low-cost airlines indicate that the LCC group has lower unit costs despite operating smaller aircraft and shorter stage lengths on average than its legacy counterparts. These differences in unit costs are driven in large part by differences in the productivity of inputs, specifically aircraft and labor. With the rapid growth in the USA and around the world of LCCs as effective competitors, reduced unit costs and improvements in both employee and aircraft productivity have become critical to the profitability of all airlines.

References

Airline Business (2007) "Scheduled Passenger Traffic Rankings 2007," Annual Surveys, Airline Business Database, http://www.flightglobal.com/staticpages/premiumdata.html.

Air Transport Association (2008) "US Passenger Airline Cost Index," www.airlines.org.

Civil Aeronautics Board (1942) "Uniform System of Accounts for Domestic Air Carriers," CAB Form 2780, Manual 1-1-42, US Printing Office, Washington, DC.

Gittell, J.H. 2003. *The Southwest Airlines Way*, McGraw-Hill, New York.

International Civil Aviation Organization (ICAO) (2007a) *Digest of Financial Statistics, Financial Data 2007, Series F*, Montreal.

International Civil Aviation Organization (ICAO) (2007b) *Outlook for Air Transport to the Year 2025*, Montreal, September.

Simpson, R. and Belobaba, P. (2000) "Air Transportation Operating Costs," Unpublished Course Notes for Air Transportation Economics, Massachusetts Institute of Technology.

Strina, F. (2006) "US Airline Industry Trends and Performance 1999-2004: Analysis of Form 41 Data," Master's thesis, Massachusetts Institute of Technology.

Tsoukalas, G., Belobaba, P., and Swelbar, W. (2008) "Cost Convergence in the US Airline Industry: An Analysis of Unit Costs 1995-2006," *Journal of Air Transport Management*, Vol. 14, pp. 179–187.

US Department of Transportation (US DOT) (2007) "Air Carrier Financial Reports (Form 41 Financial Data)," Bureau of Transportation Statistics, Washington, DC, http://www.transtats.bts.gov.

6

The Airline Planning Process

Peter P. Belobaba

The objective of this chapter is to provide an overview of the airline planning process, from the longest-range strategic decisions involving aircraft acquisition to medium-term decisions related to route planning and scheduling. The principal characteristics of each step in the airline planning process are introduced, along with a brief mention of the decision support models and systems used by airlines for each decision.

The most important planning decisions faced by airline managers can be categorized as follows:

- *Fleet planning*: What type of aircraft to acquire, when and how many of each?
- *Route planning*: Where to fly the aircraft profitably, subject to fleet availability constraints?
- *Schedule development*: How frequently and at what times on each route should flights be operated, subject to operational and aircraft limitations?

More tactical decisions related to marketing and distribution are also required closer to flight departure, involving pricing and revenue management, as discussed in Chapter 4.

This chapter introduces the fundamental decisions required in the airline planning process, including the major trade-offs involved, the interrelationships between these decisions, and the modeling approaches used to support the planning process. Each of the above steps is examined in greater detail throughout the remaining sections of this chapter.

6.1 Fleet Planning

Fleet composition is among the most important long-term strategic decisions for an airline, in terms of both its planning process and, ultimately, its operations. An airline's fleet is described by the total number of aircraft that an airline operates at any given time, as well as by the specific aircraft types that comprise the total fleet. Each aircraft type has different characteristics related to technical performance, the most important

The Global Airline Industry   P. Belobaba, C. Barnhart and A. Odoni
© 2009 John Wiley & Sons, Ltd

of which determine its capacity to carry payload over a maximum flight distance, or "range."

Decisions made by an airline to acquire new aircraft or retire existing aircraft in its fleet have direct impacts on the airline's overall financial position, operating costs and especially its ability to serve specific routes in a profitable manner. The decision to acquire a new aircraft by an airline represents a huge capital investment with a long-term operational and economic horizon. In 2008, prices for a typical twin-engine, narrow-body, 150-seat aircraft that might be used for short- to medium-haul domestic services ranged from $50 to 60 million per unit. The purchase price of a long-range, wide-body aircraft such as the Boeing 747-400 with over 400 seats exceeds $225 million per unit, and the Airbus 380 aircraft, which can seat up to 600 passengers on long-haul flights, has a list price of over $300 million per unit (Airbus, 2008), although most airlines pay much less than the list price due to discounting.

The impacts on an airline's financial position of such an investment include depreciation costs that typically are incurred for 10–15 years, as well as increases in long-term debt and associated interest expenses. From an operational perspective, the decision to acquire a specific aircraft type can have an even longer impact, as some commercial aircraft have been operated economically for more than 30 years. For example, early versions of the McDonnell Douglas DC-9 were introduced in the late 1960s and are today still operated (with proper maintenance and refurbishment) by many airlines around the world.

It is therefore somewhat surprising that the decision support tools used to make these very important long-term decisions are not as sophisticated as one would expect (or as sophisticated as some of the tools available to airlines for more tactical decisions like scheduling and revenue management). The highly uncertain nature of conditions 10–20 years into the future has limited the development and use of detailed optimization models for airline fleet planning. Instead, most airlines rely primarily on (relatively sophisticated) spreadsheet-based financial models to make fleet planning decisions.

Section 6.1.1 first summarizes the many different issues faced by an airline in making fleet planning decisions. The most important characteristics of different aircraft types are introduced, and the various selection criteria that affect airlines' ultimate choice of aircraft types are discussed. In Section 6.1.2, the role of fleet planning in the overall airline planning process is then discussed in greater detail, and a framework for evaluating the economic and financial impacts of alternative fleet acquisition options is described. Two theoretical approaches for assessing the impacts of different fleet acquisition strategies are then compared – "top-down" as opposed to "bottom-up" fleet planning.

6.1.1 Airline Fleet Decisions

The fleet planning problem facing an airline is in fact an optimal staging problem. An airline's fleet composition is defined for a particular point in time, but it changes with every additional aircraft acquired and every existing aircraft that is removed from the fleet. An airline's fleet plan should therefore reflect a strategy for multiple periods into the future, including the number of aircraft required by aircraft type, the timing of future deliveries and retirement of existing fleet, as well as contingency plans to allow for flexibility in the fleet plan given the tremendous uncertainty about future market conditions. The definition of such a multi-stage fleet plan must also recognize constraints imposed by the existing fleet, the ability to dispose of older aircraft, and the availability of future delivery slots (i.e., planned delivery times) from aircraft manufacturers and/or leasing companies.

In this section, we describe some of the most important factors that can affect the airline's choice of aircraft types. First, an overview of commercial aircraft categories and their characteristics is presented. Then, the specific aircraft selection criteria that have the greatest influence on typical airline fleet decisions are discussed, including technical and performance characteristics, the economics of operations and revenue generation, marketing and environmental issues, and, finally, political and international trade concerns.

Commercial Aircraft Categories and Characteristics

The major categories of commercial aircraft in use today and available for airline acquisition are most commonly defined by the aircraft type's *range* and *size*. The "range" of an aircraft refers to the maximum distance that it can fly without stopping for additional fuel, while still carrying a reasonable payload of passengers and/or cargo. The "size" of an aircraft can be represented by measures such as its weight, its seating or cargo capacity, as indicators of the amount of payload that it can carry. Thus, broad categories such as "small, short-haul" or "large, long-haul" aircraft can include several different aircraft types, perhaps built by different manufacturers. Because aircraft types within each category can provide similar capabilities to airlines, they are regarded as "competitors" in the airline's fleet planning evaluations. For example, the Airbus 320 and Boeing 737–800 are competing aircraft types, as they are both single-aisle, twin-engine aircraft with approximately 150 seats, each with similar range capabilities.

Figure 6.1 is a representation of the size and range characteristics of different commercial aircraft types available to airlines in 2008. Historically, it was generally the case that the largest aircraft were designed for routes with the longest flight distances. The relationship between aircraft size and range in the 1970s was almost linear, in that an airline wishing to serve a very long-haul non-stop route had little choice but to acquire the largest Boeing 747 aircraft type. Over the past 30 years, the number of range/size product options made available by the principal aircraft manufacturers has increased substantially.

With the emergence of greater competition among both airlines and especially aircraft manufacturers, airlines now have a much wider choice of products by range and capacity in each category. The range capability of new aircraft in the "small" category (100–150 seats) has increased dramatically, allowing US transcontinental routes to be flown with Boeing 737 and Airbus 320 series aircraft, for example. The sizes of newer "long-range" aircraft have decreased, allowing airlines to serve certain low-demand, long-haul international routes non-stop. Examples of such operations include Boeing 757 services (180 seats) from the northeast USA to some European destinations (e.g., Newark to Lisbon) and from the US west coast to Hawaii (e.g., Los Angeles to Maui). In addition, a medium-size/medium-range category of new aircraft types has emerged, as airlines find new "missions" for aircraft with intermediate combinations of range and capacity.

The two remaining dominant aircraft manufacturers (Boeing and Airbus) both continue to expand their aircraft product "families" in order to offer airlines as many size/range combinations as possible. These families of aircraft type allow each manufacturer to be competitive in as many aircraft categories as possible, matching the specific performance characteristics of each airline's fleet requirements. Aircraft families also have the appeal to airlines of the advantages of "fleet commonality," as will be discussed.

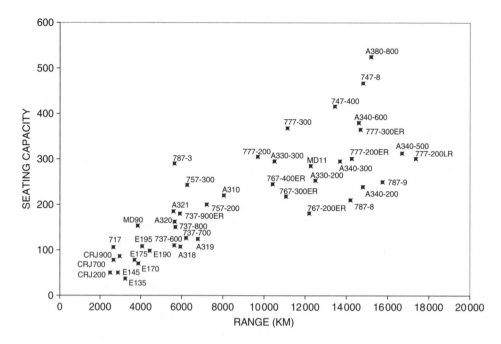

Figure 6.1 Commercial passenger aircraft

Data sources: Manufacturers' websites – www.airbus.com, www.boeing.com, www.embraer.com, www.bombardier.com

Technical and Performance Characteristics

As reflected in the above description of commercial aircraft categories by size and range, perhaps the most important technical and performance characteristic that determines the airline's choice of aircraft type is the "payload–range curve." As shown in the example provided for a Boeing 767 aircraft type in Figure 6.2, the payload–range curve defines the technical capability of each aircraft type to carry a payload of passengers and/or cargo over a maximum flight distance. Each specific aircraft model has its own payload–range curve, with different curves depending on which particular engine type is to be attached to the airframe. The payload–range curves shown in Figure 6.2 are for the Boeing 767-300ER aircraft types, using General Electric engines.

The specific shape of an aircraft's payload–range curve is affected by its aerodynamic design, engine technology, fuel capacity and typical passenger/cargo configuration. For all aircraft, the typical shape of the curve is such that the aircraft is able to carry a maximum payload over a certain distance, while longer distances can be flown if the operator is willing to reduce its flight payload in exchange for extra fuel. This trade-off continues until a maximum operational range is reached.

In addition to the payload–range curve, other important technical and performance characteristics of each aircraft type include a wide variety of factors related to both airline operational and airport constraints. For example, each type has its own maximum take-off and landing weights that determine minimum runway length requirements and, in turn, feasible airports for operating the aircraft. Similarly, limitations on taxiways and

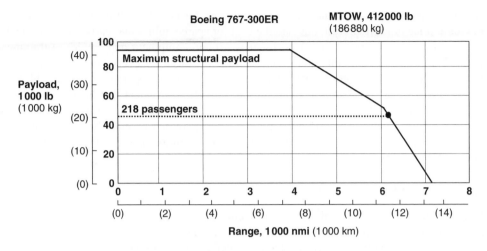

Figure 6.2   Example of payload–range curve
Data source: Boeing (1998)

gate space and even ground equipment at different airports can impose constraints on the airline's ultimate choice of aircraft type.

As mentioned above, fleet commonality with the airline's existing (or planned) fleet is a particularly important issue, as it can significantly reduce the costs associated with training of pilots and mechanics, as well as the need for new equipment and spare parts inventory for new aircraft types not previously in the airline's fleet. Fleet commonality refers not only to having the same exact aircraft type in an airline's fleet, but also to having closely related aircraft types made by the same manufacturer, as such aircraft types will have similar or identical cockpit layouts, and maintenance and spare parts requirements. For example, the Airbus A318 (approximately 110 seats), A319 (130 seats), A320 (150 seats) and A321 (170 seats) aircraft are virtually identical in every respect except their seating capacity. All aircraft in this "family" have the same cockpit crew requirements, allowing crews qualified to operate one aircraft type in the family to operate all types in the family. This provides the airline much greater flexibility in crew scheduling, and leads to reduced crew costs (as will be discussed in Chapter 7).

Financial and Economic Issues

The acquisition of new aircraft by an airline depends on the availability of required financing from internal or external sources. Full payment to the manufacturer is generally required at the time of aircraft delivery to the airline, and this payment can come from cash on hand, retained earnings, debt (loans) or equity (stocks) for aircraft purchases. Aircraft leasing is another option for many airlines. Leasing can appear to be more expensive in terms of monthly lease payments, but many airlines now lease many of their aircraft because it is also more flexible, allowing for more frequent fleet renewal and requiring less up-front capital investment.

As will be described in greater detail in Section 6.1.2, an economic and financial evaluation of alternative aircraft types is performed by the airline to determine cost and revenue impacts of each alternative. The initial costs for new aircraft include not only the purchase price of each aircraft, but also the cost for spare engines and parts inventory, and if the aircraft type is new to its fleet, for new ground equipment and employee training costs. The principal economic and financial trade-off faced by every airline considering the acquisition of new aircraft is between the promise of lower operating costs and higher ownership costs, as compared to older aircraft already in the fleet that have higher fuel and maintenance costs but which have been fully depreciated. Included in the economic and financial evaluation should be the potential for increased revenue from new aircraft with greater payload (seating) capacity, and possibly from the marketing appeal of newer aircraft to passengers.

Other Aircraft Selection Criteria

While economic/financial evaluations and technical/performance characteristics of alternative aircraft types tend to dominate the fleet planning decision process at most airlines, there are several additional aircraft selection criteria that cannot be overlooked, including environmental, marketing and political issues.

Environmental concerns and regulations either imposed or under consideration by governments around the world are having a growing impact on airline fleet decisions. The noise performance of commercial jet aircraft has become a major issue for airports and the communities that surround them. Many airports now have regulations and/or curfews that limit or prevent the operation of older aircraft types with engines that exceed specified noise levels. Similarly, there is a growing trend toward imposition of air pollution regulations designed to limit the type and quantity of aircraft emissions around airports. These environmental regulations provide further incentives to airlines to update their aging fleets with newer technology aircraft that are both quieter and cleaner in terms of emissions, but at a higher capital cost to the airlines. (See Chapter 14 for a more complete discussion of environmental issues related to aviation.)

Aircraft manufacturers tend to overstate the marketing advantages of newer aircraft in terms of passenger preference and their impact on generating incremental market share and revenues for the airline. Typically, most consumers have little aircraft preference, and this has become even more the case with the use of websites emphasizing flights with the lowest fares and with the proliferation of airline code-sharing practices. In both situations, passengers are less likely to choose (or even be aware of) different aircraft types involved in a given itinerary. However, there is also evidence that the first airline to operate the newest aircraft type or the airline with the youngest fleet (with proper advertising of these facts) can generate incremental revenues. For example, the introduction in 2008 of the A380 super-jumbo aircraft by Singapore Airlines generated a great deal of demand specifically for flights operated with that aircraft type, allowing the airline to charge higher fares on A380 flights than on flights operated with other aircraft types on the same routes.

Finally, despite detailed financial and economic evaluations and careful consideration of the technical, performance and other selection criteria described above, it is also true that political influences and international trade issues can dominate airline fleet decisions. There are numerous well-documented examples of political pressure for an airline to purchase from a particular manufacturer or country, especially at government-owned national airlines. A vivid example of such influence is the large purchase of US-made Boeing and McDonnell Douglas aircraft by Saudia, the national airline of Saudi Arabia, in the aftermath of the Gulf War in the early 1990s. Saudia was effectively instructed by its government to commit to the US-made aircraft, after President Clinton and his administration applied direct political pressure to the government of Saudi Arabia (Lynn, 1995).

6.1.2 Fleet Planning Methods

Airline decisions related to fleet planning depend primarily on an evaluation of the expected impacts of new aircraft on the airline's economic and financial performance. Although many technical and other characteristics of different aircraft types (discussed in Section 6.1.1) must also be incorporated into the decision of the type of aircraft to acquire and operate, the focus of fleet planning "models" is on assessment of the more quantifiable economic and financial impacts.

A representation of an economic evaluation process for fleet planning is shown in Figure 6.3. The evaluation process illustrated is representative of the analysis approach employed by most airlines to make fleet decisions, albeit at different levels of sophistication and detail. The major steps in the process shown can be summarized as follows:

- The most important driver of the evaluation is a forecast of expected traffic, expressed in RPK, for the network, subnetwork, or set of routes under consideration for the fleet acquisition decision at hand.
- Given a forecast of future RPK by period, a "target average load factor" is used to determine the future ASK required in each period to accommodate the forecasted demand at a reasonable average load factor. That is, the forecasted RPK is divided by the target average load factor to calculate the ASK required.
- Based on assumptions about the productivity of an aircraft type measured in terms of its ability to generate ASK per day (or per month, per year), the number of aircraft required to fulfill the future ASK requirement can then be calculated.

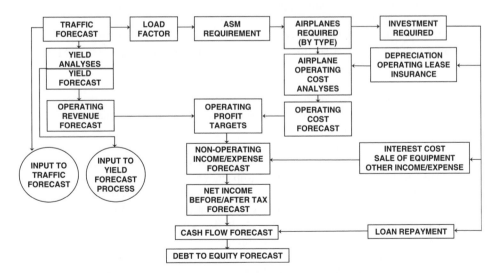

Figure 6.3 Fleet planning economic evaluation process
Source: Adapted from McDonnell Douglas (1981)

- The number of aircraft to be acquired has financial impacts in terms of investment funding, depreciation and interest expenses. At the same time, analysis of expected operating costs for the aircraft type leads to a combined forecast of operating costs and financial impacts.
- The estimation of revenues that will be generated by the new aircraft is based on both the traffic forecasts used initially to determine the number of aircraft required and forecasts of the yield (per RPK) that the airline can expect from operating this aircraft type on the routes or system that it has under consideration.
- Combining these revenue forecasts with the operating cost forecasts allows the airline to establish operating profit estimates or targets for the aircraft type, and these estimates can be used to predict effects on the airline's balance sheet, cash flow and debt load.

This fleet planning evaluation process is ideally an ongoing effort that requires input from many sources within the airline. The most critical inputs include the initial traffic forecasts, yield forecasts, estimated aircraft productivity and operating costs, all of which combine to determine the outcome of the evaluation. At most airlines, these estimates must be obtained from different databases managed by different departments, which can make this idealized airline evaluation process difficult to implement in practice.

Approaches to Fleet Planning Evaluations

The airline evaluation process introduced above can be applied at a variety of levels of analysis, depending on the specific context of the fleet planning decision at hand. For example, an airline wishing to acquire a dozen aircraft to replace its current fleet of short-haul domestic aircraft could define the relevant "system" to be its complete set of current domestic routes and planned routes in the future that will utilize the new aircraft. On the other hand, an airline wishing to acquire only one or two aircraft to serve a particular new international route opportunity would limit its evaluation to the specific route under consideration.

In either case, the evaluation process can be supported by one of two approaches to economic and financial assessment of the available aircraft alternatives (McDonnell Douglas, 1981):

- A "top-down" or "macro" approach based on relatively high-level aggregate analysis.
- A "bottom-up" or "micro" approach based on much more detailed analysis of data and forecasts by flight and route.

In the "top-down" approach to fleet planning evaluation, aggregate demand and cost spreadsheets are used to evaluate financial impacts of aircraft options for a defined sub-system, region or route under consideration. The evaluation process described above can accommodate these aggregate inputs without much modification. As described, a target average load factor (ALF) establishes the ASK needed to accommodate forecast RPK growth (e.g., 75% planned ALF) for future time periods.

As shown in Figure 6.4, the "capacity gap" is defined as the difference between required future ASK and existing ASK minus planned retirements, calculated for each period (year) in the fleet planning horizon. Aircraft with different technical characteristics (e.g., payload–range capabilities) might be required to provide service to various routes with

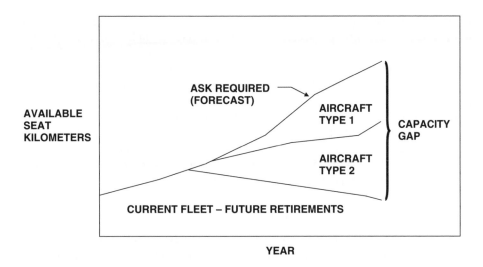

Figure 6.4 Capacity gap analysis
Source: Adapted from McDonnell Douglas (1981)

different stage lengths and demand estimates, in which case the airline's future capacity gap can be met with aircraft of several types, as shown. Assumptions about average aircraft stage length and daily utilization for each aircraft type being considered determine its "aircraft productivity" in units of ASK per day (see Section 5.4), which is then used to calculate the number of aircraft required. Estimates of aircraft operating costs from historical data or manufacturers' forecasts can then be used to compare the economic performance of different aircraft types.

The "top-down" approach is best described as a model that can be entirely contained in a relatively simple spreadsheet of traffic forecasts, aircraft characteristics, operating costs, and estimates of operating profit for different aircraft types. The inputs required are aggregate or average estimates for the entire subnetwork or set of routes under consideration, and the outputs are the relative operating profit impacts of different aircraft types.

The alternative "bottom-up" or "micro" approach involves a much more detailed evaluation of specific route characteristics and aircraft requirements, and correspondingly requires much more detailed descriptions of future scenarios for the subnetwork or routes under consideration. Detailed forecasts of future demand by origin-destination market as well as expectations concerning future routes and schedules must be generated for the airline sub-network. The airline's own share of total demand in each origin-destination market must be estimated with a market share model, given the routes and schedules in the scenario, taking into account the possible route networks and schedules of competitors. Forecasts of demand and revenues by origin–destination (O-D) market are then allocated to each future flight in the airline's schedule using a traffic allocation model.

With more detailed inputs, the bottom-up approach generates much more detailed outputs that can include individual aircraft tail assignments and operating statistics by route, flight and even aircraft tail number. The outputs of bottom-up evaluations can be used to develop projections of financial results under different fleet plans. In theory, a comprehensive bottom-up model would provide the airline with a complete representation of its network and operations under different fleet alternatives for a range of time periods into the future. Many different scenarios could be developed under different demand and operating cost assumptions, allowing the airline to perform extensive "what if" analyses. Such a detailed representation could then be integrated into subsequent decisions regarding route planning and scheduling, because specific routes and schedules are explicitly part of a bottom-up model.

The principal difference between the top-down and bottom-up models, then, is that the bottom-up approach is substantially more detailed in its representation of future scenarios, including demand forecasts by O-D market, route networks, schedules and operating cost estimates by individual flight. The top-down approach does not explicitly include details of specific routes, O-D market demands, or schedules.

Top-Down vs. Bottom-Up Fleet Planning Models

The top-down approach allows for rapid evaluation of new aircraft types using relatively simple spreadsheet models, given high-level assumptions about changes in traffic forecasts and/or operating costs (e.g., fuel price) at the level of an airline subnetwork. Airline network structural changes can be represented with aggregate measures (e.g., average stage length of flights).

The bottom-up approach both requires and provides substantially more detail about the demand, revenue, operating cost, schedule and competitive characteristics of the specific route network the airline expects to operate in the future. This allows the airline to evaluate the impacts on the best aircraft alternative of changes to individual route characteristics, for example. In practice, however, it is very difficult to incorporate future competitors' strategies, given that uncertainty about future competitive conditions is even greater than uncertainty about an airline's own future route and schedule strategies.

As a result, the simpler top-down approach is more commonly used for fleet planning evaluations, given that detailed future scenarios over 10–15 years are highly speculative. Demand and costs estimates are quite likely to be inaccurate in face of changing market conditions, putting into question the value of the enormous effort required to develop the detailed scenarios for the bottom-up approach. And in many airline fleet decisions, political decisions can overrule even the "best" analysis of options, making the bottom-up approach an ineffective use of effort and resources.

6.2 Route Planning

Given the airline's choice of aircraft and a fleet plan that determines the availability of aircraft with different capacity and range characteristics, the next step in the airline planning process is to determine the specific routes to be flown. In some cases, the sequence of these decisions is reversed, in that the identification of a profitable route opportunity might require the acquisition of a new aircraft type not currently in the airline's fleet.

Economic considerations and expected profitability drive route evaluations for most airlines. Route profitability estimates require demand and revenue forecasts for the period under consideration. In large airline networks, traffic flow support from connecting flights can be critical for route profitability. With the evolution of connecting hub networks around the world, very few flights operated by network airlines on a route carry only local O-D passengers (i.e., demand originating at A and destined for B on a flight from A to B). As explained in Section 3.3, the vast majority of flights operated in airline hub networks provide a joint supply of seats to numerous O-D markets simultaneously, a reality that makes the estimation of route profitability even more complicated.

In this section, the process of route evaluation is described, beginning with a discussion of the fundamental economics of hub network structures and their importance in route planning decisions. The process of route evaluation is then described, including the use of route profitability models designed to provide the airline with estimates of the economic impacts of adding a new route to its network.

6.2.1 Hub Economics and Network Structure

Hub-and-spoke network structures allow airlines to serve many O-D markets with fewer flight departures, requiring fewer aircraft generating fewer ASK at lower total operating costs than in a complete point-to-point route network. Consider a simple connecting hub network with 20 flights into and 20 flights out of a single "connecting bank" at a hub airport, as shown in Figure 6.5. A "connecting bank" refers to a hub operation in which many aircraft arrive at the hub airport, passengers and baggage are moved between connecting flights, and the aircraft then depart with the connecting passengers and baggage on board. Connecting banks last from approximately 1 hour in smaller domestic hub networks to 2–3 hours in larger international hub networks.

In this example of a connecting bank with 20 arriving flights followed by 20 departing flights, each flight leg arriving or departing the hub simultaneously serves 21 O-D markets – one "local" market between the hub and the spoke, plus 20 additional "connecting" markets, if we assume a single direction of passenger flow. This single connecting

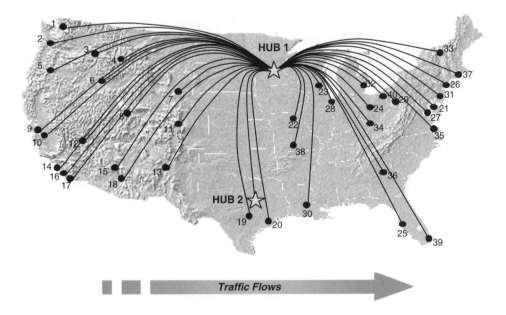

Figure 6.5   Connecting hub network example

bank thus provides service to a total of 440 O-D markets with only 40 flight legs and as few as 20 aircraft flying through the hub. In contrast, a complete "point-to-point" network providing non-stop service to each market would require 440 flight legs and hundreds of aircraft, depending on scheduling requirements. Routing both flights and passengers through a connecting hub is more profitable for the airline if the *cost savings* from operating fewer flights with larger aircraft and more passengers per flight are greater than the *revenue loss* from passengers who reject connecting service and choose a non-stop flight instead, if one exists (Morrison and Winston, 1986).

The hub airline's ability to consolidate traffic from many different O-D markets on each flight leg into and out of the hub allows it to provide connecting service even to low-demand O-D markets that cannot otherwise support non-stop flights. Consolidation of O-D market demands further allows the hub airline to provide increased frequency of connecting departures, as it likely operates several connecting banks per day in each direction at its hub airport. In fact, several connecting departures per day (via the hub) in these O-D markets may be more convenient for travelers than a single daily non-stop flight; that is, "total trip time" is lower, when schedule displacement time is included (see Section 3.3).

The use of "total trip time" to evaluate the impacts that hub networks have on passenger demand suggests that passengers trade off the shorter flight time of one or two daily non-stop flights against the reduction in schedule displacement ("wait time") offered by multiple daily connecting departures through a hub. For example, in the large US domestic multiple hub network operated by Delta, the airline is able to provide over a dozen daily connecting departures between Boston and San Diego. If a new entrant airline were to initiate one non-stop flight per day in this market, it might find it difficult to gain substantial market share given the connecting hub competition from Delta and most of the other large network airlines that operate hub networks.

Large hub networks result in market share advantages that translate into increased revenue for the airline (in addition to the reduced operating costs described above). With the potential for the airline to offer greater (connecting) departure frequency in many O-D markets, more convenient schedules (less schedule displacement) can lead to higher market shares against competitors. On-line connections (i.e., between two flights operated by the same airline) at the hub improve passenger convenience, compared to inter-airline connections (between flights operated by two different airlines). With larger hub networks, the airline can offer greater frequent-flyer program earning and reward options for passengers given greater network coverage and online service to many O-D markets. At the same time, schedule dominance of "local" markets into and out of the hub may lead to pricing and revenue advantages for the hub airline in those markets.

Hub-and-spoke airline networks are by no means limited to the US airline environment. Over the past 25 years, the vast majority of large international airlines have also developed hub networks, including Lufthansa at Frankfurt, KLM at Amsterdam, Cathay Pacific at Hong Kong and Singapore Airlines at Singapore, to name but a few examples. Large airlines that operate hub networks depend heavily on the revenues generated by connecting traffic, especially at hubs located in cities with low levels of local O-D demand. In fact, international airlines such as KLM (Amsterdam hub) and Singapore Airlines (Singapore hub) with relatively low populations around their home bases would not have been able to grow to their current level of operations without focusing to a large extent on connecting passengers (also known as "Sixth Freedom traffic") through their hubs.

Operational Advantages and Incremental Costs of Hubs

The consolidation of an airline's operations at a large hub airport has several operational and cost advantages. The airline will generally require fewer "base" locations for its aircraft maintenance and crew domiciles, resulting in reduced crew and maintenance expenses. There are fewer locations where passengers or bags can misconnect, and multiple connecting banks each day can reduce the delays associated with such missed connections. The large volume of operations at the hub airport leads to economies of scale in terms of aircraft maintenance operations, catering facilities and airport ground handling.

Hub networks also offer some potential aircraft and crew scheduling advantages for the airline. The establishment of fixed connecting bank times at the hub allows for simplified aircraft and crew scheduling, in that the "best" arrival and departure times at the hub airport are in essence predetermined by the connecting banks. It also provides more opportunities for "swapping" aircraft in response to delays, cancellations and irregular operations, given that a large number of aircraft are on the ground simultaneously during a connecting bank. To the extent that many of the aircraft are likely to be of the same fleet type, this further increases flexibility for the airline to exchange aircraft from one flight to another, as required.

Advance planning for aircraft swaps at the hub in response to changing demand for different flights, also known as "demand-driven dispatch" (Berge and Hopperstad, 1993), is another advantage of a hub operation. Different-size aircraft from the same fleet family (e.g., A319, A320 and A321) can be assigned to specific flight departures as little as 2–3 days before departure, to better match seat capacity to booking demand. In a hub network, the fact that most aircraft fly routings "out and back" from the hub to the spoke allows for greater opportunities for such aircraft interchanges.

On the other hand, there are incremental costs to the airline associated with a hub network. For example, hub operations can lead to reduced aircraft and crew utilization, compared to point-to-point networks. Although aircraft and crew scheduling may be simplified, hub operations can lead to reduced flexibility in scheduling of departures, and aircraft rotations due to the fixed timing of connecting banks at the hub. An aircraft that serves a spoke city closer to the hub usually has to sit longer on the ground at the spoke, awaiting the next connecting bank (in the other direction) at the hub. Because the revenue benefits of hub networks depend on load consolidation and feeding of connecting banks, it makes little sense to fly the aircraft back to the hub during a non-bank time. As a result, the average aircraft utilization (block hours per day) can be reduced by this effect.

Another negative impact on aircraft utilization related to hub networks is the need for increased ground ("turnaround") times for aircraft transiting the hub, to accommodate passenger and baggage connections. A connecting bank consists of a closely scheduled set of arriving flight legs followed, after a short period of time, by a set of closely scheduled departing flight legs. The rationale for this structure is that it allows arriving passengers (with many different origins) to connect with little waiting time to departing flights, to their destinations. Providing this short connection time service, however, has its costs. Aircraft arriving at a hub must stay on the ground, typically beyond the time at which the aircraft is ready to depart, to allow all connecting passengers adequate time to disembark their arriving flights and walk to the gate at which the aircraft is parked. While a point-to-point airline like Southwest can turn around a B737 in 20–30 minutes at most of its airports, a hub airline like Continental will keep the same type of B737 on the ground at its Houston hub for 60 minutes or more, to accommodate connecting traffic. Moreover, banking results in uneven use of resources (such as gates and runway capacity) at the airports, and uneven utilization of airline resources and personnel, with large numbers of resources needed to operate during the banks and much reduced numbers needed at non-peak times.

Some hub airlines have attempted to reduce the negative impacts on fleet and crew utilization of fixed connecting banks by shifting to "continuous hub" or "rolling hub" operations. In a continuous hub operation, the fixed connecting banks are eliminated. A "de-banked" hub schedule has a more even distribution of flight arrivals and departures to achieve a more balanced utilization of physical infrastructure, ground resources, air crews and aircraft. Flights are scheduled to arrive at the hub and then depart as soon as they can be turned around, without waiting for a fixed time to accommodate connections. This type of hub operation attempts to replicate the aircraft utilization success of low-cost carriers that typically operate more point-to-point flights and/or do not typically operate fixed connecting banks at their "focus cities" (hubs). American Airlines was the first to introduce the continuous hub concept at its Dallas–Fort Worth (DFW) hub in 2002, and was followed thereafter by other hub network airlines, including Delta and Lufthansa.

While the continuous hub concept can increase aircraft and crew utilization, thereby reducing unit costs for the airline, it can also have some disadvantages. Perhaps its greatest potential impact is on the timing and reliability of passenger (and baggage) connections. Unless the airline's hub operation is very large, with numerous flights between most spoke cities and the hub, the move from fixed to less defined connecting banks will both eliminate some previously offered connecting possibilities and make others much longer. Both impacts might reduce the attractiveness of the airline's connecting flights relative to other competing hubs, leading to a decrease in connecting traffic, market share and revenues. On the other hand, longer connection times at "non-banked" hubs can result in fewer passenger misconnections and more reliable service quality. With proper schedule design (see Chapter 7), the average increase in connection time can be kept to a minimum and key markets can maintain their shorter connections.

There are also potential congestion and delay costs at the hub airport. The fixed timing of connecting banks and surges in passenger activity can require peak airport staffing during several busy periods of the day, and result in underutilized staff resources during off-peak periods. Outside the hub airport terminals, the number of scheduled departures and arrivals during connecting banks can exceed the airport's runway capacity, leading to delays in peak periods and unused capacity in off-peak periods. Operationally, weather delays at the hub airport will affect the airline's entire network of operations. For example, should a major snowstorm close an airline's only hub, it effectively forces the airline to cancel all of its hub-to-spoke and spoke-to-hub flights. Even weather delays of an hour or less affecting the airline's hub airport can have severe impacts on the ability of passengers to successfully make their planned connections at the hub.

Recent Trends: Hub Strengthening

Despite repeated forecasts of more non-stop flights designed to bypass congested hubs, the trend toward development of bigger and stronger hubs continues, especially during slow economic times and/or high fuel costs. After the terror attacks of 2001, the largest US and European airlines responded to the drop in demand for air travel by eliminating from their networks virtually all flights that did not originate or terminate at their hubs. A very similar pattern emerged in 2008 as legacy airlines struggled with record-high fuel costs and weakening demand. Most US legacy airlines trimmed total capacity by 10–15%, with most of the schedule reductions affecting point-to-point rather than hub–spoke flights, particularly in lower-yield leisure markets.

There are several factors that have reinforced hub growth around the world. Liberalized bilateral agreements between countries have allowed airlines to fly what once were thought to be low-density international non-stop routes from their hubs (e.g., Cincinnati–Paris by Delta, Frankfurt–Portland, Oregon by Lufthansa, and Dubai–Hamburg by Emirates). At the same time, smaller regional jets are being used by hub airlines and their commuter partners to increase the frequency of service from the hub to small spoke cities, *not* to overfly the hub with new non-stop services. The vast majority of regional jet flights in both North America and Europe operate between existing large hub-and-spoke cities (Mozdzanowska, 2004).

The development of global airline alliance networks has also focused on increasing linkages between hub networks, specifically by increasing flights between the major hubs of the airline partners in the alliance. For example, in the oldest such alliance between Northwest and KLM, non-stop flights between Northwest's US hubs and KLM's Amsterdam hub increased substantially in the years after the alliance was launched (de la Torre, 1996). In their combined networks today, there are very few (if any) transatlantic flights offered that do not originate and/or terminate at one of the airlines' hubs.

Overall, the economic advantages of hub network operations have consistently outweighed their operational disadvantages, and continue to be the driving force in the evolution of large airline networks through good and bad economic times. Especially during weaker periods of demand and, in turn, less profitable operations, airlines rely heavily on the economic benefits of load consolidation offered by large connecting hubs.

Hub Impacts on Route Planning

From a route planning perspective, a hub-and-spoke network structure affects how the economics of new services are evaluated by the airline. New routes to smaller spoke cities become easier to justify in an established hub network. In the hypothetical hub network of Figure 6.5, the airline needs only two or three passengers per flight out of a new spoke city to each of 20 connecting destinations to make the operation of that flight with a 100-seat aircraft profitable. Even if the "local" spoke-to-hub O-D market demand is too small to justify the new service on its own, the new connecting passengers carried by the flight can make a positive contribution to the airline's total network revenues that exceeds the operating costs of the new service.

However, such incremental analysis can also lead to a tendency on the part of airline planners to overlook potential displacement of other traffic on connecting legs. That is, for the overall network revenue contribution of the new connecting passengers to be positive, they must not displace existing passengers on connecting flight legs beyond the hub, nor should they represent traffic "cannibalized" from alternative routings already offered by the airline (e.g., via another hub).

The estimation of the network contribution of passengers on a flight between a spoke city and a hub is a particularly difficult task, given the need for numerous, sometimes arbitrary assumptions about how much of the total ticket revenue of connecting passengers should be allocated to the flight leg in question. These revenue allocation assumptions, when combined with assumptions about displacement costs and even the allocation of different components of operating costs to the flight, can result in very different answers to the seemingly simple question, "is this flight leg profitable?" (Baldanza, 1999).

The same "incremental" logic for evaluating the economics of a route makes it more difficult to eliminate service to a potentially unprofitable destination, which provides connecting traffic support to other flights. That is, a flight leg from a spoke city to a hub might have a lower load factor than other flights, but if it typically carries a large number of, for example, international connecting passengers who make a substantial contribution to the total network revenues, it might be judged to be economically viable by the airline. Once again, the many assumptions made by the airline planner about the allocation of network revenue contribution to the flight in question can lead to very different estimates of its profitability.

For many of the same reasons, it can also be difficult for an airline to justify a new non-stop service to bypass its hub, as the non-stop flights might simply take traffic away from existing hub flights. However, a large number of connecting departures in an O-D market can allow the airline to build market share and perhaps allow it to introduce a non-stop flight, which would be supported in terms of frequency of departures by the existing connecting opportunities. As mentioned above, the introduction of new non-stop flights is limited to larger O-D markets in which a non-stop flight is able to capture adequate passenger share given the existence of many connecting alternatives.

6.2.2   Route Planning and Evaluation

Given a fleet plan, the process of route planning and evaluation involves the selection by the airline of which routes should be flown. The route selection decision is both strategic and tactical. It is an essential component of an integrated network strategy or "vision" for the airline, which must decide whether to focus on short-haul or long-haul services, domestic or international operations. At the same time, the characteristics of the selected routes will affect the types of "products" the airline offers to travelers. For example, an international route network will likely lead to a decision that business- and first-class products should be offered in order to be competitive.

The distance or "stage length" of the selected routes will also affect the airline's cost structure, as longer routes will likely be flown with bigger aircraft that have lower unit costs per seat and per ASK. Although fleet planning was introduced first in this chapter, it is important to recognize that the aircraft performance requirements for specific routes can and do provide a feedback loop to fleet planning decisions.

Route planning can also be a much shorter-term tactical process, as unexpected route opportunities often present themselves to the airline with changes to the market environment. For example, the bankruptcy of another airline, a withdrawal from a route by a competitor, or a newly negotiated bilateral agreement with another country can lead to new route opportunities that must be acted upon within months or even weeks.

Route Evaluation Issues

As was the case with fleet planning evaluations, economic considerations dominate route evaluation, especially for airlines operating in competitive environments with a profit maximization objective. These evaluations of the potential for a route to be profitable can be performed at a very high level (like "top-down" fleet planning) or at a more detailed level of assessment. Perhaps the most important inputs to any route evaluation are forecasts of potential passenger and cargo demand (as well as expected revenues) for the proposed route. For a given route, O-D market demand is likely to be the primary source of demand and revenues, but far from the only source. In airline hub networks, traffic flow support to the new route from connecting flights and other (non-local) O-D markets can make the difference between expected route profitability and loss.

Once a forecast of the *total* O-D market demand (per period) for the route in question has been generated, an equally important step is the estimation of the *market share* that the airline can expect of this total demand. The airline's own market share of total forecast demand will depend on its frequency share in the market, the path quality of its planned services (non-stop versus connecting flights), as well as its planned departure times. To the extent that the competitive marketplace will allow differences to be maintained, relative prices and service quality can also have a significant impact on expected market share of total demand.

The fundamental economic criterion for evaluation of a planned route is its potential for *incremental* profitability in the short run, given the opportunity cost of taking aircraft from another route. However, in route planning (as in fleet planning) a number of practical considerations can be just as important as the outcome of the economic evaluation. The technical capability to serve a new route depends on availability of aircraft with adequate range and proper capacity. The performance and operating cost characteristics of available aircraft in the airline's fleet will in turn have a substantial impact on the economic profitability of the proposed route. If the route involves a new destination, there will be additional costs of establishing the required airport facilities and sales offices, along with staff relocation.

Regulations, bilateral agreements and limited airport slots can also impose constraints on new route operations, to the point of non-profitability. For example, while a new route to Hong Kong might appear to be a viable option for the airline, the availability of landing slots only at undesirable arrival and departure times will have a negative impact on the airline's market share and route profitability.

In some cases, strategic considerations might be used by an airline to proceed with the initiation of service on a route despite a negative outcome from the economic evaluation of expected route profitability. Given the political and other uncertainties of international bilateral agreements and route opportunities, the airline might focus on the longer-term competitive and market presence benefits of entering a new route even if it is expected to be unprofitable in the short run.

Route Planning Models

Typically, the economic assessment of a proposed route involves a detailed evaluation approach, much along the lines of "bottom-up" fleet planning. Demand, operating cost and revenue forecasts are required for the specific route under consideration, perhaps for multiple years into the future. An estimate of the airline's own market share of total demand is also critical, based on models of passenger choice of different airline and schedule options. Both the forecasts of future demand and market share will depend to a large extent on the presence and *expected response* of competitors to the planned route entry.

"Route profitability models" have been developed both by some airlines and by software vendors for purchase by airlines. These are computer models designed to perform such route evaluations based on detailed inputs that include demand forecasts, operating cost estimates, and planned frequency and schedule of operations for the airline using the model. The objective such models is to allow airlines to select routes to maximize total airline profits, given a set of candidate routes and estimated demands, subject to fleet and capacity constraints.

These models have proven to be useful in comparing alternatives, but airlines must recognize that the profit estimates generated by the models are entirely dependent on the accuracy of future demand and revenue estimates, the allocation of operating costs to each route, and assumptions concerning expected market shares. A major shortcoming of even the most sophisticated airline profitability models is a very limited ability to integrate competitive effects. Regardless of the sophistication of the route planning models available to the airline, it is important to remember that the quality of the profit estimates generated by these models is entirely dependent on the accuracy of the inputs and the nature of the many assumptions used in the models.

Example: Route Profitability Analysis

As introduced above, the evaluation of the potential for profitability of a new route is most typically based on spreadsheet analysis of the forecasted demands, market share, revenues and operating costs for the flight(s) in question, over one or more periods in the future. In this section, we describe a hypothetical example of such an analysis for a Montreal (YUL) to Rome (ROM) non-stop route proposal. The example is hypothetical, but makes use of realistic operating and cost data, and reflects the most important components and assumptions of typical route profitability evaluations.

Figures 6.6 and 6.7 illustrate the revenue and operating cost calculations, respectively, required for an estimate of annual operating profit of a daily new non-stop service on the Montreal–Rome route. The airline is planning to operate a Boeing 767-300ER aircraft with 210 seats in a mixed configuration of business-class and economy-class seats. The calculations shown are for an entire year of operations, leading to an estimated $2.4 million annual operating profit for this route, representing a 4.8% operating margin. A closer look at both the revenue and cost calculations, however, will reveal that this estimate is subject to a large number of forecasts, estimates and assumptions required in the evaluation.

Figure 6.6 provides the details of the demand and revenue calculations. The demand forecasts comprise annual estimates for several categories of traffic expected to be captured by the airline operating this new service. The local O-D demand forecast is for 102 000 one-way passenger trips (summed over both directions) per year. The airline expects that it can capture 70% of this local O-D market demand, at an expected average fare of $450 per one-way passenger trip. This relatively high estimate of market share is based on the

DEMAND AND FARE ESTIMATES FOR YEAR	ANNUAL DEMAND	Prorated Average One-Way Revenue	TOTAL REVENUE ($)
Total YUL–ROM local O-D passengers (both directions)	102 000		
Expected market share for one daily flight	70.00%		
Local YUL–ROM passengers on new flight	71 400	$450	32 130 000
Additional Traffic			
Connections North American cities behind YUL to/from ROM	24 000	$425	10 200 000
Connections to/from YUL beyond ROM	12 000	$400	4 800 000
Connections behind YUL to/from destinations beyond ROM	4 500	$375	1 687 500
Total passengers (both directions)	**111 900**		48 817 500
Additional cargo revenue	10% of passenger revenue		4 881 750
		TOTAL REVENUES	**53 699 250**

Figure 6.6 Revenue calculations for route example

INPUTS AND ASSUMPTIONS

Aircraft Type	B767-300ER	
Number of Seats	210	
Total Annual Flights (each direction)		358
(Reflects 98% completion of daily schedule)		
Block Hours YUL to ROM		08:00
Block Hours ROM to YUL		09:00
Non-stop miles YUL/ROM		4086

Aircraft Operating Costs per Block Hour:

Crew Cost	$	890
Fuel/Oil	$	3,280
Ownership	$	870
Maintenance	$	710
Total per Block Hour	**$**	**5,750**

Indirect Operating Costs

Passenger Service	0.015	per RPM
Traffic Servicing	$22	per Enplanement
Aircraft Servicing	$1800	per Departure
Promotion and Sales	9.00%	of Passenger Revenues
General and Administrative	$0.002	per ASM

CALCULATED MEASURES (ANNUAL)

Annual Flights	716
Block Hours	6086
RPMs	457 223 400
Passenger Yield	0.1068
ASM	614 370 960
Seat Departures	150 360
Passengers Enplaned	111 900
Average Load Factor	74.42%
DIRECT OP COSTS	$ 34 994 500
PAX SERVICE	$ 6 858 351
TRAFFIC SERVICE	$ 2 461 800
AIRCRAFT SERVICE	$ 1 288 800
PROMOTION/SALES	$ 4 393 575
GEN ADMINISTRN	$ 1 228 742
OPERATING COSTS	**$ 51 225 768**
OPERATING PROFIT	**$ 2 473 482**
OPERATING MARGIN	**4.8%**

Figure 6.7 Operating cost and profit calculations

fact that the airline will be providing the only non-stop service in this O-D market. Note that 30% of the Montreal–Rome local O-D demand is expected to continue to choose other flight options, specifically connecting itineraries on other airlines via a variety of alternative hubs (e.g., New York, Philadelphia, London, Amsterdam, Frankfurt, etc.)

The local O-D market represents 64% of the total expected traffic on the planned new flight in this example. The remaining demand is forecast for several categories of connecting passengers that will also share the supply of seats on this new non-stop service. Their "prorated" fares as allocated to this non-stop service are lower than the average fares of the local O-D market, as the airline will inevitably receive a smaller per mile revenue for these connections. The total passenger revenues are estimated to be $48.8 million per year, and then additional cargo revenues are assumed to be 10% of the passenger revenues, leading to total annual revenues for this new service of \$53.7 million. Such an estimate of cargo revenue is admittedly simplistic, but is typical in preliminary route profitability evaluations, and would be based on percentage estimates from other similar (transatlantic) routes.

Figure 6.7 summarizes both the inputs and calculated measures for this example as they relate to the operating costs and, in turn, estimated annual profit of this route. The planned block times in each direction are used to calculate total aircraft operating costs (crew, fuel, maintenance and ownership) for the new route. The aircraft operating costs per block hour are estimated by the airline based on the operation of its existing 767-300ER aircraft on other comparable routes, or from system-wide cost data for the aircraft type. In this example, the cost inputs are based on actual reported 2007 Form 41 data for US airlines.

In addition to the aircraft operating costs, several categories of indirect operating costs (as introduced in Chapter 5) also represent important inputs to the calculation of total estimated costs for the new flights. Again, the values shown in Figure 6.7 are based on data reported in Form 41, whereas an airline would use its own estimates of these indirect cost rates for such route evaluations.

The right side of Figure 6.7 summarizes the calculated measures, annual costs and estimated route profit for this example. Based on the inputs and assumptions made in the analysis, this route is estimated to operate at a 74.4% average load factor and generate a 4.8% operating margin. It should be clear from even this hypothetical example that these profitability estimates depend heavily on (and are quite sensitive to) the accuracy of virtually every input variable. The demand forecasts, estimated airline market share, expected fares, and all of the operating cost estimates can have a critical impact on the profit estimate.

This type of spreadsheet-based route evaluation allows the sensitivity of different inputs to be tested, holding all other inputs constant. For example, this new route would become unprofitable for this airline if its actual share of local Montreal–Rome O-D demand were to drop below 64%. Similar sensitivity analyses of the various demand forecasts, average fares and even operating cost components such as fuel costs can be performed to give the airline a better idea of the overall potential for route profit under uncertain future conditions.

This route evaluation example incorporates variable flight costs, aircraft ownership costs, variable passenger costs and allocated overhead and non-operating costs, reflecting "fully allocated flight costs" (Baldanza, 1999). On the revenue side, the fares have been fully prorated to the Montreal–Rome flight legs, so the end result is an estimate of "fully allocated segment profitability." This approach to route evaluation is commonly used by airline planners in assessing profitability over the long term (i.e., several years of operations). In the long term, it is appropriate to attribute a complete allocation of airline costs to the route under consideration. And a proration of revenues to the route is consistent with an accounting approach in which the estimated profits for all routes sum to a total operating profit for the airline.

Alternative approaches to the evaluation of route profitability can exclude certain components of operating costs and include additional components of network revenue contribution. These approaches are more suitable for evaluations of route profit in the shorter term. For example, if the Montreal–Rome flights were only to be operated for several months, it could be argued that general and administrative costs should not be included (under the assumption that these costs would not increase for such an incremental addition of service). Furthermore, if the route decision involved the short-term use of an existing aircraft that otherwise would not be operated by the airline, it might choose to exclude aircraft ownership costs from the evaluation.

On the revenue side, the revenue estimates could be adjusted to better reflect the "network contribution" of this new service, as introduced earlier. Including network contribution would increase the estimates of total revenues generated by this new service and, in turn, the estimated profit. The network revenue contribution could include the total fares paid by passengers flying on connecting flights operated by the same airline, with reductions for variable passenger carrying costs and the estimated displacement of revenues on those connecting flights.

The use of network revenue contribution in route evaluations can become quite complicated, potentially requiring estimates not only of network displacement costs on flights connecting to the Montreal–Rome flights, but of potential "cannibalization" of existing traffic on other routings that serve the same market(s). For example, the airline might already provide connecting service Montreal–Rome via another hub, either its own or that of an alliance partner.

Finally, the revised estimation of route profitability to include network revenue contribution raises the risk of "double counting" revenues and profits over the medium to long term (Tretheway, 2004). That is, while the allocation of more than the prorated revenue of connecting passengers can be helpful in measuring their network contribution on a given flight leg, the resulting estimates of route "profit" include portions of connecting passenger revenue that contribute to the other connecting legs as well. Summation of flight leg estimates of "profit" that include network contribution would result in the double counting of the same connecting passenger revenues on different flight legs. The network contribution approach to route evaluation is therefore most appropriate for short-term and incremental route decisions, not for longer-term evaluations of hub or network profitability (Baldanza, 1999).

6.3   Airline Schedule Development

The next major step in the airline planning process, as introduced at the beginning of this chapter, is that of schedule development. Given a set of routes to be operated in an airline network, and a fleet of aircraft, schedule development involves four different but interrelated tasks:

- *Frequency planning*: How often should the airline operate flights on the selected route(s)?
- *Timetable development*: At what times should flight departures be scheduled?
- *Fleet assignment*: What type of aircraft should be used for each departure time?
- *Aircraft rotation planning*: How should each aircraft type be flown over the airline's network in order to ensure an overall balance of aircraft arrivals and departures at each airport?

Each of these tasks is part of the schedule development process, and will be described in greater detail in this section.

In general, the schedule development process begins a year or more in advance of flight departure, and can continue right up until actual departure time. Frequency plans are established first, based both on route evaluations performed a year or more in advance and on aircraft availability resulting from fleet planning decisions typically made 2–5 years earlier. Specific timetables of departure times and aircraft rotation plans are developed up to 1 year in advance, and finalized 2–6 months before departure. Final revisions to the planned timetable and aircraft rotations are made as necessary closer to departure, while unexpected operational constraints (maintenance, weather) can necessitate schedule changes and "irregular operations" planning until the flight actually departs.

The trade-offs inherent in airline schedule development involve the sometimes conflicting objectives of maximizing passenger convenience (which in turn leads to increased revenues for the airline) and meeting numerous operational constraints that might not always maximize convenience for each passenger. Recall that the frequency of service and schedule of flight departure times on each route served are perhaps the most important determinants of airline market share, particularly on routes with a large proportion of business demand. As introduced in Chapter 3, passengers will choose the combination of flight schedules, prices and product quality that minimizes their overall disutility of air travel.

6.3.1 Frequency Planning

Increases in the frequency of departures on a route improve the convenience of air travel for passengers. The airline can benefit from higher traffic and revenues associated with this increased frequency, in terms of both more passengers willing to fly given the more convenient air travel services (i.e., demand stimulation due to higher frequency) and increased market share at the expense of its competitors.

Peak departure times (early morning and late afternoon) are most attractive to a large proportion of travelers in many markets, as "time of day" distributions of desired departure times tend to be clustered around 08:00–09:00 and 17:00–18:00. More frequent flight departures further reduce the schedule displacement or "wait time" between flights, reducing travel inconvenience for more passengers. Recall that frequency is much more important in short-haul markets than for long-haul routes where actual flight time dominates "wait time." In competitive markets, airline frequency share is most important for capturing time-sensitive business travelers. Frequency of departures can be as important as path quality (non-stop vs. connection) in many cases, as demonstrated in our discussion of hub network effects.

Although the determination of the "optimal" frequency of service on a route does involve elements of competitive strategy, in most cases an analysis of demand forecasts and expected market share can be used to establish a "baseline" frequency of flights on a route that can be operated profitably by the airline. Based on the outputs of the route evaluation process, the airline should already have estimates of the total "local" demand between the origin and destination served by the route under consideration. The airline's expected market share of this total demand is then determined by its frequency share and specific timetable of flight departures, relative to its competitors on the same route.

Not to be forgotten in most airline hub and/or connecting networks is the potential for additional traffic on the route under consideration from connecting flights. If connecting passengers represent a substantial proportion of expected traffic flow on the route, the airline can decide to offer even greater frequency of departures (further increasing its own market share of the local O-D market) and/or operate larger aircraft with lower costs per seat kilometer on the route.

Although frequency planning is introduced here as being separate from the choice of aircraft type for each flight on the route (i.e., "fleet assignment"), it should be clear that the two decisions are interrelated. The airline's supply decision for a route in fact consists of two simultaneous choices – the number of departures per day and the number of seats to be offered on each departure. If the demand estimates for a route suggest that 400 seats per day are required to accommodate the expected traffic, the airline can choose to operate one frequency per day with a 400-seat aircraft or four frequencies per day with a 100-seat aircraft. Although both choices supply 400 seats per day to the route, the outcome in terms of this airline's market share, passengers carried and revenues generated will be very different, especially if there exists a competitor on the route that operates four flights per day.

In this example, a decision to offer one flight with 400 seats will almost certainly give the airline a relatively low market share due to its low-frequency (20%) share, with prices and quality of service equal. Such a low market share would not likely allow the airline to fill its 400-seat aircraft to profitable levels. Although hypothetical, this example illustrates how frequency planning can be driven as much by the presence and frequency of one or more competitors as by an assessment of the supply of seats required to carry expected demand on a route. This example also illustrates how airlines have little choice but to operate smaller-capacity aircraft with higher operating costs per seat and per seat kilometer in competitive markets, especially those involving relatively short-haul routes.

The objective of "load consolidation" also affects frequency and aircraft size decisions. A single flight with multiple stops provides service to several O-D markets at the same time. Consolidation of traffic from multiple O-D markets onto one aircraft can allow that airline to operate higher frequency on the route (increasing its market share) and/or larger aircraft (reducing its unit operating costs). As discussed in Section 6.2.1, this ability to consolidate loads is a fundamental reason for the economic success of airline hubs.

6.3.2 Timetable Development

Given a chosen frequency of service on each route, the next step in the schedule development process is to generate a specific timetable of flight departures. As introduced earlier, an initial goal for the airline is to provide departures at peak periods (09:00 and 17:00), especially on routes that serve business demand. However, not all departures can be at peak periods on all possible routes, given aircraft fleet and rotation considerations. That is, most airlines do not have enough aircraft to allow them to schedule departures on each route in their network at 09:00, for example.

Establishing a timetable of flight departures requires a trade-off between maximization of aircraft utilization (block hours per day) and schedule convenience for the passengers. The timetable must incorporate minimum "turnaround" times required at each airport to deplane and enplane passengers, refuel and clean aircraft. Minimum turnaround times will vary by aircraft type and the characteristics of the flights involved. A 100-seat aircraft in domestic service can be turned around in as little as 20–30 minutes, while a 400-seat aircraft used for international flights can take 2 hours or more to turn around.

Even if minimum turnaround times are met, the earliest feasible departure time for the next flight to be operated by the same aircraft might not be desirable in terms of schedule quality. For example, an 09:00 departure from city A with an 11:00 arrival at B and a 60-minute minimum turnaround time results in a possible departure time for the aircraft from B at 12:00. If this aircraft is to return to A, a 12:00 departure will be off peak and have potentially lower demand, but keeping the aircraft on the ground until the next peak period reduces aircraft utilization and increases unit operating costs (as fixed costs are spread over fewer seat kilometers).

Given the trade-off described above, most airlines choose to maximize aircraft utilization in establishing a timetable of scheduled flight departures. The overall tendency of timetable planners is to keep ground "turnaround" times to a minimum, to make as much use of the expensive aircraft resources as possible. For this reason, airlines will schedule off-peak flights with relatively low load factors to maintain frequency share and to position aircraft for peak flights at other cities. Taken to the extreme, this philosophy of maximizing aircraft utilization in the timetable development process leaves little buffer time for maintenance and weather delays, and can also result in many low load factor flights and reduced airline profitability.

In addition to the constraints described, numerous additional factors constrain an airline's timetable development. As described in Section 6.2, connecting hub networks that operate on a "fixed bank" basis require that flights arrive from spoke cities within a prescribed time range, to facilitate passenger connections. This requirement leaves relatively little flexibility for scheduling departures from the spoke city to the hub, if the flights are expected to provide service to both local and connecting passengers.

Time zone differences also limit feasible departure times, especially on long-haul routes. For example, flights from eastern cities in North America to Europe typically do not depart before 16:00, as passengers do not want to arrive at their European destination in the middle of the night. To the limited extent that there exist "daytime" flights eastbound from North America to Europe, they must depart by 09:00 to arrive in Europe no later than 22:00 or 23:00. Regulatory constraints such as airport arrival and departure slot times, as well as noise curfews, can further limit the scheduling flexibility for an airline.

Finally, crew scheduling and routine maintenance requirements also have substantial impacts on timetable development. One example: if an airline operates a single daily flight into an airport, arriving late in the evening (say, 22:00), the subsequent departure of the same aircraft and crew cannot be scheduled for the next morning until minimum crew rest requirements have been met. The alternative for the airline is to schedule the aircraft to depart very early the next morning with a new crew, but this requires that each crew be given a 30-hour layover, increasing crew costs.

The complexity and size of a typical airline's timetable development problem makes most schedule changes incremental. A single change in departure time of a flight can have major impacts on down-line arrival times for the same aircraft, feasible passenger connections, aircraft rotations and even the number of aircraft required to operate the revised schedule. The problem is further complicated by crew and maintenance schedule constraints, requiring coordination between the timetable planners and several other airline operational departments.

Computer models that can determine an "optimal" timetable for an airline network do not yet exist, given the huge combination of possible departure/arrival times at a large set of airports, demand and market share variability by time of day, and thousands of operational constraints. Even with advances in OR and computational speeds, there are no tractable models that can fully capture the complexity of the problem, including the competitive dynamics and inherent uncertainty of passenger demand and choice. Hence, most models developed and used today are incremental in nature, creating new airline schedules by making network-wide trade-offs to identify routing, fleeting and scheduling changes to an existing airline schedule. However, interactive computer scheduling databases and decision support tools are used by the more advanced network airlines to allow for rapid "what if" analysis under different timetable scenarios. There has also been substantial decision support progress in the use of mathematical models for fleet assignment and aircraft rotation optimization, as described in the next section.

Example of a Schedule Map

The timetable and aircraft rotation plan in an airline's schedule can be represented graphically by a "schedule map," in which the horizontal axis reflects the movement of aircraft from one airport to another and the vertical axis reflects movement in time. Figure 6.8 provides a simple example of a schedule map, for an airline that operates two aircraft of the same type. This schedule map shows a timetable with a balanced aircraft rotation plan. One aircraft begins its day at Stockholm (STO) and flies to Frankfurt (FRA), then to Madrid (MAD), back to FRA, then to Amsterdam (AMS), and back to FRA and finally MAD. This aircraft then overnights at MAD, before flying a second day in this rotation, from MAD to FRA to AMS to FRA and back to STO.

The schedule map shown in Figure 6.8 therefore illustrates a two-day aircraft rotation that begins and ends in STO – the same aircraft ("tail number") will operate the 07:00 departure out of STO every second day. Therefore, for the airline to operate each of

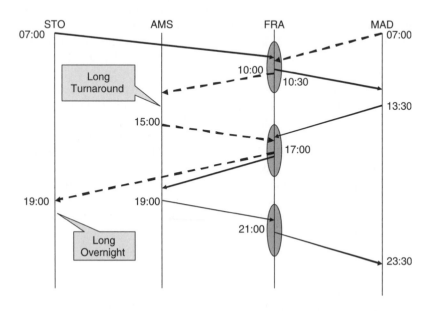

Figure 6.8   Example of a schedule map

the flights on the schedule map daily, it needs two aircraft of the same type. (Further discussion of this topic can be found in Chapter 7.)

This hypothetical example also illustrates several of the challenges in the development of a feasible schedule, given operational and marketing constraints. In this example, we have assumed that the airline must deliver and receive connecting passengers traveling to many other destinations at the FRA hub. The connecting banks at FRA are fixed, as indicated on Figure 6.8 by the tinted ovals at FRA, so that our hypothetical airline must have its feeder flights arrive and depart at FRA only during the connecting bank periods. Other constraints faced by the airline schedulers include minimum turnaround times at each station, which we have assumed to be 60 minutes.

During the first day of the aircraft rotation plan shown on the schedule map, the aircraft that departs STO at 07:00 traverses the FRA hub three times during designated connecting bank periods, operates six flight legs, and accumulates 11.5 block hours of utilization. All turnaround times are at the minimum 60-minute constraint. On the second day of this rotation plan, however, the aircraft operates only four flight legs for 6.5 block hours, as scheduling constraints limit its utilization. First, the aircraft arrives in AMS from FRA at 11:30 but must sit on the ground until 15:00 so that it can feed the next connecting bank at FRA. This is a good illustration of the utilization penalties on short hub–spoke services discussed earlier. The aircraft spends 3.5 hours on the ground at AMS, well above the minimum 60-minute turn time. Second, the aircraft departs from the FRA hub at 17:00 and arrives at STO at 19:00 to be in position for the next morning's 07:00 departure, as required to ensure a balanced schedule. By ending its productive day at 19:00, the aircraft sits idle for several hours in the evening that might have been used to generate additional utilization. On the other hand, this "early" end to the aircraft's day at STO might allow for overnight maintenance to be performed. It might also give the inbound crew enough rest time to depart the next morning on the 07:00 flight, thereby reducing crew costs.

In the schedule map shown, the airline's fleet of two aircraft operates on average five flight legs per day with an average daily utilization of 9.0 hours. It might not be the "best" solution to this scheduling problem and its associated constraints, and it is certainly not the only feasible solution (the reader is encouraged to develop an alternative feasible schedule as an exercise).

Schedule maps are a visual representation of flight legs that incorporate aircraft movement in terms of both geography and time. In the early days of airline planning, schedule maps were large paper charts on the walls of scheduling departments. Today, schedule maps are presented to airline scheduling analysts on computer screens that are linked to extensive fleet, route and timetable databases, and which allow the analysts to "point and click" on specific flight legs to make changes to the timetable. Each proposed change to a flight's departure or arrival time can trigger a recalculation of the minimum number of aircraft required to operate the new schedule and can inform the analyst of any constraints (crew rules, maintenance requirements, passenger connection times) that might be broken by the proposed shift of times.

6.3.3 Fleet Assignment and Aircraft Rotations

The fleet assignment problem is to determine the type of aircraft to be flown on each flight leg departure, given a planned network of routes and specified timetable of flights. "Fleet assignment" should not be confused with "fleet planning," as described in Section 6.1.

Fleet assignment is a tactical decision made after an initial timetable has been developed, and well after longer-term decisions about fleet acquisition and route networks have been made. Fleet assignment is limited to a choice of aircraft types from the airline's existing fleet.

The objective of fleet assignment is to minimize the combined costs of "spill" (rejected demand) and aircraft operating costs. Spill occurs when the aircraft assigned to a flight departure is too small and potential demand and revenues are lost to the airline. Spill is the loss of bookings due to the fact that the flight has been fully booked to capacity. The magnitude of spill is difficult to measure, but can be estimated from observed loads and full flight data using "spill models" (Boeing, 1978). Spill can be reduced (or eliminated) by assigning a large enough aircraft to accommodate all possible peak day demands for the flight in question. However, larger aircraft have higher operating costs and will fly with many empty seats on most non-peak days.

Many airlines have fleet assignment models based on large-scale mathematical network optimization methods (e.g., Subramanian *et al.*, 1994; Hane *et al.*, 1995). These computer models optimally assign aircraft so as to minimize spill costs plus operating costs (or equivalently, maximize profitability), subject to constraints such as minimum ground times, maintenance requirements, number of aircraft by type available in the airline's fleet. Aircraft rotation constraints in the fleet assignment model formulations ensure feasible aircraft cycles and balance of aircraft inflow/outflow at each airport.

Even with fleet assignment optimization models, it is important to recognize that it is not possible to achieve a perfect match of seats with demand on each flight in the timetable, given the need to balance the flow of aircraft over the entire airline network. The "optimal" aircraft assignment to a given flight might in fact be too big or small for the demand on that flight, but it can lead to maximum operating profit over the entire network.

Consider this simple example: assigning a 100-seat aircraft to a peak 09:00 departure from A to B and to the subsequent non-peak departure from B to A might result in spill from A to B on certain high-demand days, but the load factor on the return flight from B to A at the off-peak time might be very low. In this case, reducing spill on the first flight by operating a larger aircraft will only worsen the load factor performance of the second (return) flight. The "optimal" aircraft size must consider two (or more) flights in the same aircraft rotation.

Fleet assignment models also include fleet size and aircraft balance constraints. Aircraft routing models are used to assign specific aircraft "tail numbers" to each flight, creating rotations that satisfy aircraft maintenance requirements. A detailed description of airline fleet assignment and maintenance routing optimization models is provided in Chapter 7. The solutions to such optimization problems have allowed airlines to achieve higher aircraft utilization rates. However, as mentioned earlier, higher aircraft utilization achieved by minimizing turnaround times can also lead to less "buffer" time for recovering from unexpected maintenance and weather delays.

The handling of "irregular operations" involves dynamic revisions to the planned schedule right up until the flight departs or is cancelled. In Chapter 8, the operations control process is described in more detail, while Chapter 9 is devoted to the problems of schedule recovery from such irregular operations. At this point, it is important to note that flight cancellations and deviations from the planned timetable of operations are decisions that are not made lightly by the airline. A single cancelled flight can seriously disrupt aircraft rotations, crew schedules and maintenance plans, not to mention passenger trips. Under conditions of disruptions and/or flight cancellations, the primary objective for the airline is to return to "normal" operations as quickly as possible. In this effort to get the airline "back on plan" with respect to the planned timetable, flight cancellations or aircraft rerouting sometimes take precedence over passenger convenience.

6.4 The Future: Integrated Airline Planning

As described in this chapter, the current airline planning practice is to make decisions concerning fleet planning, route evaluation and schedule development *sequentially*. The sequential nature of the process is driven in large part by the different time horizons (prior to departure date) involved for fleet planning (2–5 years), route evaluation (1–2 years) and schedule development (2–6 months). After these airline planning decisions have been made, more tactical decisions concerning pricing and revenue management (as described in Chapter 4) are then made even closer to departure. In the future, the development of integrated models will allow for a *joint optimization* of schedules, capacity, prices and seat inventories.

It should be clear that a joint decision approach can have substantial benefits for the airline. For example, better feedback from pricing and revenue management systems can affect the optimal choice of schedule and aircraft capacity. At the same time, improved decisions related to schedule and capacity can reduce the need for excessive discounting and "fare wars."

Joint optimization and planning is a big challenge, both theoretically and practically, for the following reasons. Despite having an enormous amount of detailed booking, revenue and operations data, few airlines have "corporate databases" that provide a single source of consistent and detailed demand/cost data, as required for joint decisions. Furthermore, a great deal of research is still required to identify and calibrate models that can capture market dynamics and competitive behaviors that are necessary for joint optimization of airline planning decisions. Finally, there are obstacles within many airlines in terms of organizational coordination and a willingness to accept a large-scale decision tool that claims to "solve" several airline planning problems simultaneously. In fact, it might never be possible to integrate all the subtleties of airline planning decisions into a single optimization tool.

There have been many recent developments in decision support models for airline planning, but for many airlines, these developments are outpacing their capabilities to use them effectively. The more immediate challenge is for both users and developers of airline planning decision support tools to recognize the potential of these tools, as well as their limitations. Critical to the long-term success in adoption of more sophisticated decision support tools is improved education about the underlying models, their assumptions and how users should interpret results to achieve benefits.

References

Airbus (2008) "Aircraft List Prices 2008," http://www.airbus.com/en/presscentre/, April 22.

Baldanza, B. (1999) "Measuring Airline Profitability", *Handbook of Airline Operations*, McGraw-Hill, New York, pp. 147–159.

Berge, M.E. and Hopperstad, C.A. (1993) "Demand Driven Dispatch: A Method for Dynamic Aircraft Capacity Assignment, Models and Algorithms," *Operations Research*, Vol. 41, No. 1, pp. 153–168.

Boeing Commercial Airplane Company (1978) "Load Factor Analysis: The Relationship between Flight Load and Passenger Turnaway," Seattle, WA.

Boeing Commercial Airplane Company (1998) "Product Reviews on CD ROM," Seattle, WA, May.

de la Torre, P.E. (1996) "Airline Alliances: The Airline Perspective," Master's thesis, MIT Department of Aeronautics and Astronautics.

Hane, C.A., Barnhart, C., Johnson, E.L., Marsten, R.E., Nemhauser, G.L., and Sigismondi, G. (1995) "The Fleet Assignment Problem: Solving a Large-Scale Integer Program," *Mathematical Programming*, Vol. 70, pp. 211–232.

Lynn, M. (1995) *Birds of Prey: Boeing vs. Airbus*, Heinemann, London, pp. 1–10.

McDonnell Douglas Aircraft Company (1981) "Airline Fleet Planning," Unpublished Presentation, Long Beach, CA.

Morrison, S. and Winston, C. (1986) *The Economic Effects of Airline Deregulation*, The Brookings Institution, Washington, DC, pp. 4–10.

Mozdzanowska, A. (2004) "Evaluation of Regional Jet Operating Patterns in the Continental United States," Master's thesis, MIT Department of Aeronautics and Astronautics.

Subramanian, R., Scheff, R.P., Quillinan, J.D., Wiper, D.S., and Marsten, R.E. (1994) "Coldstart: Fleet Assignment at Delta Air Lines," *Interfaces*, Vol. 24, No. 1, pp. 104–120.

Tretheway, M. (2004) "Distortions of Airline Revenues: Why the Network Airline Business Model is Broken," *Journal of Air Transport Management*, Vol. 10, No. 1, pp. 3–14.

7

Airline Schedule Optimization

Cynthia Barnhart

An airline's schedule is perhaps the single most important indicator of its business strategy and competitive position. In deciding which markets to serve with what frequency, and how to schedule flights to meet these frequencies, an airline determines how and where to operate and compete. These important strategic decisions are limited by many practical considerations facing the airline, including fleet composition, the locations of crew and maintenance bases, constraints on access to certain airports due to gate or landing slot restrictions, and, for international flights, bilateral agreements and government allocations that specify where an airline may fly.

Chapter 6 provided an overview of the overall airline planning process, including three major components that involve fleet planning and route evaluation, followed by airline schedule development. In this chapter, we examine the use of optimization models to develop airline flight schedules.

Airline schedule optimization involves the design of future schedules for aircraft and crews. The objective is to identify profit-maximizing schedules that are consistent with a large set of operational, marketing and strategic goals. This is a complex task, involving a host of expensive and diverse elements, including:

1. numerous airport locations with various arrival and departure restrictions, and each with gates, ground personnel and equipment to manage;
2. multiple aircraft types with different operating characteristics, costs, crew requirements, seating capacities and government-imposed periodic maintenance requirements;
3. numerous crews capable of operating only particular aircraft types, and limited by a litany of rules and regulations specifying how and when they can work; and
4. a large number of potential routes and origin–destination (O-D) markets to serve, each different in their volume of demand and customer demographics.

Although these considerations alone pose a tremendous challenge for airline planners, the task of airline schedule planning is made even more difficult by two major

The Global Airline Industry   P. Belobaba, C. Barnhart and A. Odoni
© 2009 John Wiley & Sons, Ltd

factors: (1) the sheer size of the problem (large airlines operate thousands of flights with hundreds of crews and aircraft daily); and (2) the fact that the system is subject to uncertainty affecting passenger demand, pricing and airline operations, among other things. Operations researchers have worked for decades to model and generate optimal aircraft and crew scheduling solutions. These tools have proven to be valuable, with one major US airline claiming that each year over $500 million in incremental profits are attributable to their schedule optimization systems (Cook, 2000). This chapter provides a description of the optimization problems that are solved in building airline flight schedules and presents an overview of the successes and impacts of these optimization approaches, as well as the remaining challenges. Sections marked with an asterisk indicate advanced sections that can be skipped without loss of continuity.

7.1 Schedule Optimization Problems

Operations research has been instrumental in helping airlines meet the challenges of designing aircraft and crew schedules. The practice, designed out of necessity due to the very large-scale nature of airline scheduling problems, has been to break the schedule planning problem into several subproblems:

1. **The schedule design problem** is to design the airline's flight schedule, specifying the set of flight legs to be operated by the airline, given the outcomes of the route evaluation and frequency planning processes. Each leg has an associated origin airport, an arrival airport, a scheduled departure time, (roughly determining the scheduled arrival time) and a frequency plan designating the days on which the flight leg is operated.
2. **The fleet assignment problem** is to find a profit-maximizing assignment of aircraft types to flight legs in the airline's network. Generally speaking, this involves trying to match passenger demand with the number of assigned aircraft seats on each flight leg. Although conceptually simple, assigning aircraft types to flight legs is complicated by limiting factors, including the composition of the airline's fleet (i.e., the types of aircraft comprising the fleet and the number of aircraft of each type), and fleet balance requirements that ensure a flight schedule can be repeated periodically (typically daily) by matching the number of aircraft of a type arriving at an airport with the number departing it during that period.
3. **The maintenance routing problem** is to assign to each flight leg a specific aircraft (tail number), maintaining consistency with the results of the fleet assignment problem and ensuring that each individual aircraft can be assigned a sequence of flight legs (a routing) that allows the aircraft to undergo periodic maintenance checks. If maintenance requirements restricting an aircraft to fly a limited number of hours between maintenance checks are not satisfied, the airline must ground that aircraft until a maintenance check is performed. Grounded aircraft result in flight cancellations; aircraft, passenger and crew disruptions; and increased workloads for gate and reservation agents. Hence, ill-planned maintenance rotations can be extremely costly to the airline.
4. **The crew scheduling problem** is to determine cost-minimizing assignments of crews, namely pilots and cabin crew (flight attendants), to each flight leg in the airline's schedule. A crew schedule specifies the sequence of flight legs and other activities (like training, vacation leave, etc.) to be executed by a crew member over a period

of time, typically 1 month. Crew schedules are constructed to satisfy all work rules specified by regulatory agencies and collective bargaining agreements between the airline and its employees.

Each of the above subproblems is typically solved in order, with the decisions made in solving earlier subproblems used as input to subsequent subproblems. For example, the set of flight legs and associated schedules that solve the schedule design problem are inputs to all subsequent problems, including the fleet assignment problem, the maintenance routing problem and the crew scheduling problem. In the following sections, detailed descriptions are provided of these subproblems and the progress and associated impacts that optimizers have made in addressing them. For ease of exposition, the fleet assignment problem is first addressed, and then the schedule design and crew scheduling problems are presented. Aircraft routing problems are discussed within the context of integrated development of crew schedules and aircraft routes.

7.2 Fleet Assignment

Given the completion of the timetable development step in the schedule development process described in Chapter 6 (or, as described in Section 7.3, after completing the schedule design step), the next step is *fleet assignment*. The problem of fleet assignment can be illustrated with the following small example. Assume that an airline wishes to operate the set of flights listed in Table 7.1, and depicted in Figure 7.1. Associated with each flight leg are its flight number, departure airport, arrival airport, departure time, arrival time, average fare and total number of passengers wishing to travel on that leg. Note that all times are represented here as Eastern Standard Time (EST) to allow planners to compute block times (i.e., the total elapsed time from push-back of the aircraft from the gate until arrival at the destination gate) and to compute ground times between arrivals and departures. Depicted in Table 7.2 are the details of the airline's fleet, including the types of available aircraft; the number of aircraft of each type; the capacity, or number of seats, provided by each aircraft type; and the operating costs for each aircraft type and each flight leg operated by the airline.

Table 7.1 Example: flight schedule, fares and passenger demands

Flight no.	From	To	Dep. time (EST)	Arr. time (EST)	Fare ($)	Demand (passengers)
CL301	LGA	BOS	11:00	12:00	150	250
CL302	LGA	BOS	12:00	13:00	150	250
CL303	LGA	BOS	14:00	15:00	150	100
CL331	BOS	LGA	08:00	09:00	150	150
CL332	BOS	LGA	11:30	12:30	150	300
CL333	BOS	LGA	14:00	15:00	150	150
CL501	LGA	ORD	12:00	15:00	400	150
CL502	LGA	ORD	13:00	16:00	400	200
CL551	ORD	LGA	08:00	11:00	400	200
CL552	ORD	LGA	09:30	12:30	400	150

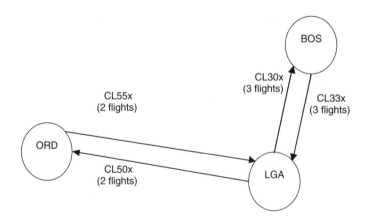

Figure 7.1   Example flight network

Table 7.2   Fleet information

Fleet type	No. of aircraft owned	Capacity (seats)	Per flight operating cost ($000)	
			LGA–BOS	LGA–ORD
DC9	1	120	10	15
B737	2	150	12	17
A300	2	250	15	20

Given this information, the daily fleet assignment problem is to find a profit-maximizing assignment of fleet types to flight legs in the airline's network, without using more than the available aircraft and ensuring balance by aircraft type at each airport location each day. The profitability of assigning a fleet type f to flight leg l can be computed as the product of its average fare and the expected number of passengers it carries, reduced by the associated operating cost. That is:

$$c_{l,f} = fare_l \times \min(D_l, Cap_f) - OpCost_{l,f}$$

with:

$c_{l,f}$: profitability of assigning fleet type f to flight leg l;
$fare_l$: fare of flight leg l;
D_l: demand of flight leg l;
Cap_f: capacity of fleet type f;
$OpCost_{l,f}$: operating cost of assigning fleet type f to flight leg l.

The profitability associated with assigning a fleet type to a flight leg, for each possible combination, is calculated and reported in Table 7.3.

A greedy solution, which assigns the aircraft type that maximizes profit for each flight leg, results in the assignment of A300s to flight legs CL301, CL302, CL332, CL502 and CL551; B737s to flight legs CL331, CL333, CL501 and CL552; and a DC9 to flight leg CL303. A closer look at the network of flight legs depicted in Figure 7.1, however, reveals

Table 7.3   Profit ($000 per day)

Flight no.	DC9	B737	A300
CL301	8	10.5	22.5
CL302	8	10.5	22.5
CL303	5	3	0
CL331	8	10.5	7.5
CL332	8	10.5	22.5
CL333	8	10.5	7.5
CL501	33	43	40
CL502	33	43	60
CL551	33	43	60
CL552	33	43	40

that this assignment is clearly infeasible: aircraft balance is not achieved for any of the fleet types. Moreover, it is difficult with the static network representation of Figure 7.1 to determine how many aircraft of a given type are needed each day to operate the assigned flights. For example, consider the pair of balanced flight legs, CL552 and CL501. Both are assigned the aircraft type B737. Although their total flying time is clearly achievable by a single aircraft in a day, the arrival of CL552 is later than the departure of CL501 and hence the flight legs would require the use of two, not one, B737 aircraft each day.

To capture the temporal nature of this problem, flight networks are modeled using *time–space networks* (an extension of the schedule map concepts introduced in Chapter 6). A time–space flight network is an *expansion* of the static flight network in which each node represents both a location and a point in time. Two types of arcs are present in the time–space network:

1. A *flight arc* representing a flight leg with departure location and time represented by the arc's origin node, and arrival location and arrival plus *turn time* represented by the arc's destination node. The minimum *turn time* is the minimum amount of time needed to service an aircraft after its arrival at a station and before its departure from that station. Minimum turn times, which involve refueling, cleaning and inspecting aircraft, can vary by fleet type, by airport, and by time of day. In this example, the minimum (and actual) turn times are assumed to be instantaneous.
2. A *ground arc* representing aircraft on the ground during the period spanned by the times associated with the arc's end nodes.

The time–space network corresponding to the airline's flight schedule is depicted in Figure 7.2.

Finding a feasible fleet assignment is analogous to selecting a *path* through the time–space network for each aircraft. In this example, no more than two paths can be selected for the B737 and A300 aircraft types, and at most one path can be selected for the DC9 aircraft. Because there is only one DC9 aircraft, its assigned path must begin and end at the same location to ensure that the schedule can be repeated each day. In the case of the B737 and the A300 paths, however, this condition is relaxed. For each aircraft type and each location in the network, the relaxed requirement is that the *number*

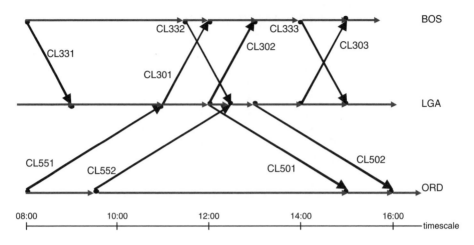

Figure 7.2 Time–space network

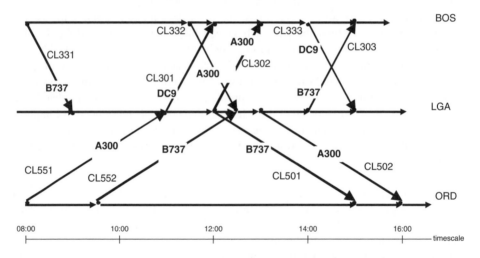

Figure 7.3 Optimal fleet assignment

of paths originating at a location must equal the number terminating there. The optimal solution, provided in Figure 7.3, selects the set of paths satisfying these conditions and maximizing total profit. The optimal fleet assignment uses all available aircraft, and, indeed, balances the A300 aircraft with one aircraft beginning the day in Boston and ending the day in Chicago and another beginning the day in Chicago and ending the day in Boston. The total profit associated with this optimal fleet assignment is $280 500. Note that a suboptimal solution with value $255 000, depicted in Figure 7.4, is achieved if each aircraft is required to begin and end the day at the same location.

7.2.1 The Fleet Assignment Model*

The example illustrating the selection of aircraft paths through the time–space network and the assignment of aircraft types to flight legs can be formalized and represented

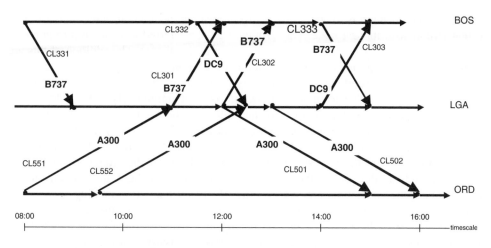

Figure 7.4   Suboptimal fleet assignment

as a mathematical optimization problem. Underlying this formulation is the time–space network presented above, with modifications, as depicted in Figure 7.5. One modification is to construct a time–space network for *each fleet type* to reflect fleet-specific differences in aircraft types. For example, different aircraft types operate at different speeds; different fleet types are subject to different noise and gate restrictions at certain airports; and different aircraft have different operating costs and characteristics (e.g., some aircraft cannot fly over water while others can).

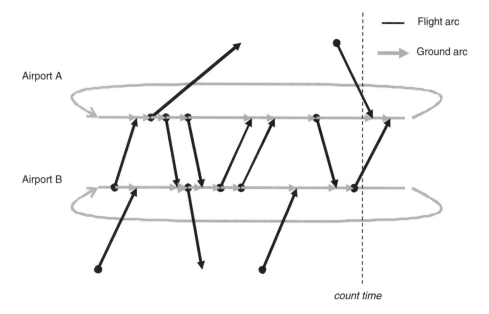

Figure 7.5   Fleet-specific time–space network with count time and wraparound ground arcs

Another modification is to add, at each airport location, an arc connecting the last node at that airport with its first node. This ground arc, also referred to as a *wraparound arc*, represents the option for aircraft to spend the night at the airport. The addition of this arc changes the fleet assignment problem from one of selecting aircraft *paths* in the network to one of selecting aircraft *cycles* (or *rotations*). Because aircraft rotations can contain wraparound arcs, and, hence, flight legs occurring over multiple days, the selection of a cycle corresponds to the assignment of multiple aircraft of a given type. To illustrate, consider a rotation with one wraparound arc and hence 2 days of flying. To cover this flying, the rotation must be assigned two aircraft, one each to operate each day of flying.

In general, to count the number of aircraft needed to cover the aircraft rotations in a solution, a point in time called the *count time* is selected. All flight arcs and all ground arcs spanning the count time are said to *cross the count time*. By counting the number of arcs assigned to each fleet type f that crosses the count time, the total number of assigned aircraft of fleet type f can be determined. This is true for any count time, because the cyclic structure of the time–space network ensures that the same number of aircraft is present in the network at every point in time.

The following is a fleet assignment problem formulation, often referred to as the *basic fleet assignment model* or FAM (Hane *et al.*, 1995). The set of flight legs to be operated is denoted by F, the set of fleet types is denoted by K and M^k represents the number of available aircraft of type k. A time–space network is constructed for each aircraft type, with N^k representing the set of nodes and G^k representing the set of ground arcs in the time–space network for fleet k. The set of flight legs originating at node n in fleet k's time–space network is denoted by $O(k, n)$ and, similarly, the set of flight legs terminating at node n in fleet k's time–space network is denoted by $I(k, n)$. The ground arc originating at any node $n \in N^k$ is denoted n^+ and the ground arc terminating at n is denoted n^-. The set of flight legs and the set of ground arcs in fleet k's network that cross the count time are denoted $CL(k)$ and $CG(k)$, respectively. The operating cost minus total revenue associated with assigning an aircraft of type k to flight leg i is represented by c_i^k. Each binary decision variable $f_i^k = 1$ if flight leg i is assigned to aircraft type k, and $f_i^k = 0$ otherwise. Ground variables y_a^k represent the number of aircraft of type k on the ground arc a:

$$\text{Minimize} \quad \sum_{i \in F} \sum_{k \in K} c_i^k f_i^k$$

subject to:

$$\sum_{k \in K} f_i^k = 1, \qquad \forall i \in F \tag{7.1}$$

$$y_{n^+}^k + \sum_{i \in O(k,n)} f_i^k - y_{n^-}^k - \sum_{i \in I(k,n)} f_i^k = 0, \qquad \forall n \in N^k, \forall k \in K \tag{7.2}$$

$$\sum_{a \in CG(k)} y_a^k + \sum_{i \in CL(k)} f_i^k \leq M^k, \qquad \forall k \in K \tag{7.3}$$

$$f_i^k \in \{0, 1\}, \qquad \forall i \in F, \forall k \in K \tag{7.4}$$

$$y_a^k \geq 0, \qquad \forall a \in G^k, \forall k \in K \tag{7.5}$$

The objective is to select the fleet assignment that maximizes profitability, or equivalently, minimizes operating costs less total revenue. The cover constraints (7.1), together with the binary constraints (7.4), ensure that each flight leg is assigned to exactly one fleet type. The balance constraints (7.2) guarantee that the total numbers of aircraft of each type arriving and departing each airport are equal, while the count constraints (7.3) limit the number of assigned aircraft of each fleet type to the number available. Finally, the non-negativity constraints (7.5), together with the balance and binary constraints, ensure that only non-negative integer numbers of aircraft are on the ground at all times.

It is important to note that this basic FAM captures some of the most important constraints of the problem, but by no means is comprehensive in its coverage. Some constraints that can be added to FAM relate to:

1. **Noise restrictions**: Some aircraft are noisier than others and some airports restrict noise levels, hence limiting the mix of fleet types that an airline can operate at that airport.
2. **Maintenance requirements:** The fleet assignment solution must enable each aircraft to visit one of its maintenance stations, located only at specific airports, at regular intervals.
3. **Gate restrictions:** Gate capacity might vary depending on the fleet assignments: for example, an airport might be able to accommodate more aircraft if narrow-body, rather than wide-body, aircraft are assigned; some aircraft types might not be able to be accommodated at certain airports due to gate limitations.
4. **Crew considerations:** Because crew costs are influenced greatly by the fleet assignment, there are various options for including *estimates* of the crew costs associated with various fleet assignment decisions (Clarke *et al.*, 1996). Resulting crew costs are estimates, rather than exact, because to compute exact crew costs requires enhancements to FAM that would render it mathematically intractable.

FAM is thus an integer, multi-commodity network flow problem with side constraints, which can be solved using off-the-shelf optimization software packages, such as CPLEX (2005) or Xpress-MP (2005). Airlines, even those with very large-scale flight networks, use this type of FAM to solve fleet assignment problems routinely. A representative example of a problem size and solution time needed is provided in Table 7.4.

Table 7.4 FAM: a representative problem size and solution time

Flight network	
No. of flight legs	2044
No. of fleet types	9
FAM problem size	
No. of variables	18 487
No. of constraints	7827
Solution time (seconds)	974

7.2.2 Fleet Assignment Solutions/Impacts

Fleet assignment optimization, which has been applied widely in practice, is attributed with generating solutions that lead to significant improvements in operating profit. Examples include the following:

1. FAMs at American Airlines have led to a 1.4% improvement in operating margins (Abara, 1989).
2. FAMs have achieved annual savings at Delta Airlines of $100 million per year (Wiper *et al.*, 1994).
3. An audit based on *actual* results at USAir indicates annual savings of $15 million attributable to the use of a fleet assignment optimizer (Rushmeier and Kontogiorgis, 1997).

These impressive results notwithstanding, there is still significant room for improvement in modeling and optimizing the fleet assignment problem.

7.2.3 Extending Basic Fleet Assignment Models

There are several shortcomings of basic FAMs, including:

1. In basic FAMs, *spill costs* and *recapture* are ignored or modeled only approximately. *Spill cost* on a flight is the revenue lost when the assigned aircraft for that flight cannot accommodate all passenger demand. *Recapture* occurs when the airline spills passengers from one flight leg and then books them on other flights in the airline's own network. A leg-based model, like FAM, requires the estimation of spill costs for each *flight leg*. Fares and passenger demand, however, are specified for *itineraries* in an O-D market . To compute leg-based spill costs, itinerary fares must be allocated to flight legs (see discussion of revenue allocation in Section 6.3). A standard approach is to prorate the total itinerary fare to the flight legs contained in the itinerary such that the sum equals the total fare. Such proration is typically performed on the basis of the relative distance of the legs traversed by the itinerary. Some airlines, however, assign the entire itinerary fare to each leg in the itinerary. The rationale behind this full-fare proration scheme is that the cost associated with spilling a passenger from a particular flight leg is the entire itinerary fare, as a passenger spilled from one flight leg will not travel on any of the legs in the itinerary. Whatever the approach, a leg-based FAM that optimizes based on estimates of spilled demand and revenue on each leg while ignoring the effects on other legs in the network will be inaccurate in its estimation of spill and recapture for the airline's overall network. These inaccuracies, caused by *flight leg interdependencies* or *network effects*, occur whenever some passengers travel on itineraries containing more than a single flight leg, as detailed later in this chapter.
2. Most FAMs consider only aggregate demand and average fares, either by flight leg or by passenger itinerary. This can significantly impact spill cost estimates, particularly because most airlines employ sophisticated revenue management systems to manage inventory and deliberately spill demand for certain fare classes in order to protect seats for higher-value passengers (as described in Chapter 4). Without a more detailed representation of demands and fares for different fare classes, the accuracy of estimated spills and spill costs in FAMs can be seriously compromised.

3. Most FAMs assume that demand is static over the schedule period being optimized, even though it varies by day of week and season.

To illustrate the effects of the proration issue above, consider the example detailed in Tables 7.5 and 7.6. Table 7.5 includes demand and fare data in three markets X–Y, Y–Z and X–Z (connecting through Y). Note that given these itinerary demands, the maximum possible revenue is 75($200) + 150($225) + 75($300) = $71 250. In Table 7.6, the seating capacity for each fleet type is reported, while in Table 7.7, operating costs for each possible fleet to flight assignment are depicted. In Table 7.8, each possible fleet assignment combination, or *fleeting*, for this small example are listed, along with the associated operating costs.

For the following analysis, assume that the airline has full discretion in determining which passengers it wishes to accommodate. If fleeting I is selected, then each flight leg has a capacity of 100 seats. The demand for flights 1 and 2 is 150 and 225 passengers, respectively. Therefore, 50 of the passengers who desire travel on flight 1 are spilled and 125 passengers who desire travel on flight 2 are spilled. Because the fare for the X–Z itinerary is less than the sum of the two local itineraries, the revenue-maximizing strategy is first to spill 50 passengers on the X–Z itinerary ($15 000). The remaining demand for flight 1 no longer exceeds capacity. Because the local fare for flight 2 is less than the fare for the X–Z itinerary, 75 passengers are spilled from the Y–Z itinerary ($16 875). Therefore, the minimum total spill cost for fleeting I is $15 000 + $16 875 = $31 875. The spill costs for all fleet assignments are shown in Table 7.9.

Table 7.5   Demand data

Market	Itinerary (sequence of flights)	Number of passengers	Average fare
X–Y	1	75	$200
Y–Z	2	150	$225
X–Z	1–2	75	$300

Table 7.6   Seating capacity

Fleet type	Number of seats
A	100
B	200

Table 7.7   Operating costs

Fleet type	Flight 1	Flight 2
A	$10 000	$20 000
B	$20 000	$39 500

Table 7.8 Possible fleeting configurations

Fleeting	Flight 1	Flight 2	Total operating cost
I	A	A	$30 000
II	A	B	$49 500
III	B	A	$40 000
IV	B	B	$59 500

Table 7.9 Minimum spill costs and resulting contributions for each fleeting combination

Fleeting	Operating costs	Spilled passengers	Spill costs	Contribution
I	$30 000	50 X–Z, 75 Y–Z	$31 875	$9375
II	$49 500	25 X–Z, 25 X–Y	$12 500	$9250
III	$40 000	125 Y–Z	$28 125	$3125
IV	$59 500	25 Y–Z	$5 625	$6125

Defining contribution as the maximum potential revenue less both the spill costs and operating costs, the contribution (profit) for fleeting I is $71 250 − ($30 000 + $31 875) = $9375. Analyzing all other fleeting solutions similarly, the optimal fleeting for this tiny example is fleeting I, as shown in Table 7.9.

Consider now the leg-based case in which a total itinerary fare valuation scheme is employed and the objective is to minimize the spill cost for each individual flight leg, independent of the effects on other flights in the network. The spill cost-minimizing strategy for each flight leg in this case is to spill passengers greedily in order of increasing fare until the number of passengers exactly equals the capacity for that flight. In this example, local passengers are always spilled in favor of keeping the connecting passengers with a higher total fare. For fleeting I, 50 X–Y passengers are spilled at a fare of $200 and 125 Y–Z passengers are spilled at a fare of $225. The resulting spill costs and contribution for each fleet assignment are in Table 7.10.

Using this "greedy" approach, fleet assignments II or IV are equally preferred. The reason for the difference in fleeting solutions between the greedy leg-based heuristic and the network itinerary-based approach is that the greedy model does not capture flight interdependencies or *network effects*. The best set of passengers to spill from one flight leg is a function of the demands and assigned capacity on other flight legs. The myopic, greedy

Table 7.10 The contribution using a greedy algorithm

Fleeting	Operating costs	Spilled passengers	Spill costs	Contribution
I	$30 000	50 X–Y, 125 Y–Z	$38 125	$3125
II	$49 500	50 X–Y, 25 Y–Z	$15 625	$6125
III	$40 000	125 Y–Z	$28 125	$3125
IV	$59 500	25 Y–Z	$5 625	$6125

solution can be improved by taking a network-wide view, with connecting passengers spilled and local passengers accommodated.

In this small example, it is possible to enumerate possible fleeting combinations and compute the minimum spill costs accordingly. However, with a network of hundreds or thousands of flight legs, enumeration is computationally expensive, if not entirely impossible. Consequently, researchers have developed mathematical models and optimization approaches for large-scale problems. They conclude that the benefits of modeling network effects can be significant. For example, the network-based fleet assignment approach at American Airlines has yielded annual improvements in revenue of 0.54% to 0.77% (Jacobs et al., 1999). Increased annual contributions from $30 to over $100 million have also been reported as achievable at United Airlines when fleeting decisions are made with a network-enhanced FAM instead of a leg-based FAM (Barnhart et al., 2002).

Modeling and Solving Extended Fleet Assignment Problems*

One approach to modeling network effects and flight leg interdependencies is to extend basic FAMs to include passenger spill decisions (Barnhart et al., 2002). The approach is to find the profit-maximizing fleet assignment, given the following inputs:

1. *An airline's flight schedule.* A schedule is a set of flight legs with specified O-D locations and fleet-specific times.
2. *Itinerary-based passenger demand.* An itinerary is a sequence of one or more flight legs. Associated with each itinerary are: (a) a fare value; and (b) an unconstrained mean demand value that represents the average number of passengers wishing to book the itinerary offered by the airline if seats are available.
3. *Aircraft operating cost data.* This includes fuel, crew, maintenance, ownership and various other costs that are independent of the number of passengers traveling. Operating costs are specified for each fleet type–flight leg combination.

To construct FAM with network effects, the basic FAM is expanded to include variables representing the mean number of passengers assigned to each itinerary in the airline's network. These variables, denoted t_p^r, represent the expected number of passengers desiring to travel on itinerary p that are spilled to a different itinerary r. The recapture rate, denoted b_p^r, is the estimated fraction of passengers spilled from itinerary p that is recaptured and travels on itinerary r. For the case of $r = p$, $b_p^p = 1$, indicating that all passengers desiring travel on itinerary p will accept that itinerary if it is offered. Hence, the number of passengers traveling on itinerary r that preferred itinerary p is $b_p^r t_p^r$.

Using this notation, FAM, constraints (7.1) through (7.5), is augmented to include: (1) constraints (7.6) that limit the total number of passengers assigned to itinerary p to the number of available seats; and (2) constraints (7.7) that limit the total number of passengers traveling on or spilled from itinerary p to the unconstrained demand of p, denoted D_p. The number of passengers assigned to any itinerary is restricted to be non-negative (constraints 7.8), rather than integer, because both recapture rates and unconstrained demand values can be fractional. The objective of FAM is expanded to minimize the sum of the following two components: (1) the operating cost of flying flight leg i with an aircraft of type k, denoted c_i^k, for all flight legs and fleet types; and (2) the negative of total revenues.

Let P denote the set of itineraries, and let $\delta_i^r = 1$ if itinerary $r \in P$ contains flight leg $i \in F$, and $\delta_i^r = 0$ otherwise. These additional factors are added to the basic FAM to create the *itinerary-based fleet assignment model*, or IFAM, as follows:

$$\text{Minimize} \quad \sum_{i \in F} \sum_{k \in K} c_i^k f_i^k - \sum_{p \in P} \sum_{r \in P} fare_r b_p^r t_p^r$$

subject to:

$$\sum_{k \in K} f_i^k = 1, \qquad \forall i \in F$$

$$y_{n+}^k + \sum_{i \in O(k,n)} f_i^k - y_{n-}^k - \sum_{i \in I(k,n)} f_i^k = 0, \qquad \forall n \in N^k, \forall k \in K$$

$$\sum_{a \in CG(k)} y_a^k + \sum_{i \in CL(k)} f_i^k \leq M^k, \qquad \forall k \in K$$

$$\sum_{p \in P} \sum_{r \in P} \delta_i^r b_p^r t_p^r \leq \sum_{k \in K} CAP^k f_i, \qquad \forall i \in F \qquad (7.6)$$

$$\sum_{r \in P} t_p^r \leq D_p, \qquad \forall p \in P \qquad (7.7)$$

$$f_i^k \in \{0, 1\}, \qquad \forall i \in F, \forall k \in K$$

$$y_a^k \geq 0, \qquad \forall a \in G^k, \forall k \in K$$

$$t_p^r \geq 0, \qquad \forall p \in P, \forall r \in P \qquad (7.8)$$

The cover, balance, count, integrality and non-negativity constraints are the same as those in the basic FAM . They ensure that the solution to IFAM represents a feasible fleeting. The remaining additional *passenger assignment or flow* variables t_p^r and the *capacity* (7.6), *demand* (7.7) and *non-negativity constraints* (7.8) in IFAM are used to model spill and recapture and estimate revenues more accurately, thereby allowing improved fleeting solutions.

IFAM, like FAM, is a planning model designed to identify optimal fleeting decisions and not to determine actual passenger flows over the network. (Hence the use of unconstrained demand values representing an average – typically not an integer value – of demands over time.) IFAM's passenger flows are used to guide the fleeting process, recognizing that capturing network effects when making fleeting decisions can provide a significant economic benefit. These economic benefits, however, come at the expense of increased problem size and, sometimes, tractability issues, primarily caused by the addition of the passenger flow variables. An example, comparing the solution times of FAM and IFAM, using actual data provided by a large US airline, is presented in Table 7.11. Long solution times and tractability issues for IFAM, however, are diminished by the rapid pace of computing and hardware advances, coupled with improvements in optimization modeling and theory.

Table 7.11 FAM vs. IFAM: problem size and solution times

Flight schedule: 2044 flight legs and 9 fleet types		
	FAM	IFAM
Problem size		
No. of columns	18 487	77 284
No. of rows	7 827	10 905
No. of non-zero entries	50 034	128 864
Solution time (seconds)	974	>100 000

7.3 Schedule Design Optimization

In theory, optimization models and methods can be applied to the integrated problems of network structure, route selection and scheduling, as detailed in Chapter 6. In practice, however, this is impossible. The host of challenges associated with this task includes, but is not limited to, the following:

1. Optimizing an entirely new schedule requires data which might or might not be available to the airline.
2. Building an improved new schedule from scratch is operationally impractical and computationally difficult, if not intractable.
3. Optimized schedules can result in dramatic changes to the schedule, yet airlines prefer a degree of consistency from one planning period to the next, especially in business markets in which reliability and consistency are highly valued by travelers.

Even if these obstacles were to be overcome, the strategic and competitive importance of airline schedules would lead to difficulties in mathematically capturing the many objectives and considerations underlying schedule design. Hence, it is not likely that the task of schedule design will be relegated to an optimization program anytime in the near, or perhaps even extended, future. Operations researchers have therefore focused their attentions on *incremental schedule design*: that is, optimizing schedule and fleeting decisions while allowing only limited changes to a given, often current, schedule. The advantages of incremental optimization include the following:

1. Airlines are able to use historical booking data and other traffic forecasts to forecast future demands for their own routes.
2. Required planning efforts and time are manageable.
3. Fixed investments at stations can be utilized efficiently.
4. Consistency can be maintained by introducing only a limited number of changes to the schedule.

With incremental optimization, the decisions to be modeled can be restricted to tactical and operational issues, like retiming of flight legs or replacement of a small set of unprofitable flight legs. Focusing optimization efforts on decisions that are well suited to mathematical optimization provides analytical support for airline decision makers when strategic issues require a myriad of objectives to be balanced.

An example of incremental schedule design optimization is illustrated in the case of redesigning airline hub connections. With service between a hub and several other airports, a hub airport allows an airline to serve many markets by providing connecting opportunities for passengers and by consolidating demand from several markets onto each flight leg (see Section 6.3). In designing these hub-and-spoke networks, carriers, in an effort to provide short connection times for passengers, implemented banks, or complexes, of flight legs at their hubs.

As introduced in Section 6.3, hub de-banking allows for significant improvements in resource productivity, and can yield significant operating cost savings. These benefits, however, must be weighed against the revenue losses resulting from degraded passenger service, with some passengers experiencing increased connection times, increased travel times and possibly reduced frequency of service. No single optimization model exists that can capture the many industry forces and competitive pressures facing an airline and recommend to it that some of its hub operations be de-banked. Models do exist, however, that can help to determine *how* to de-bank once the decision is made to do so. In the simplest case, one in which the airline decides to operate the same set of flights before and after de-banking, the problem is to determine when to reschedule current flights to achieve both a smoother arrival and departure pattern at the hubs to be de-banked, and a profit-maximizing flight schedule. More complex versions of the problem would include the ability to add new flight legs or remove existing flight legs from the current flight network. Even the simplest representation of the problem is difficult, however, involving several complexities including:

1. **Scheduling decisions must be made for *all* flight legs**, not just those at the hub, because a single aircraft operates several flight legs in sequence. Adjusting the timing of one leg will necessitate the rescheduling of other flight legs to be covered by the aircraft, and perhaps will require a change in the set of legs operated by the aircraft.
2. **Fleeting decisions must be renewed**. Flight retimings both eliminate existing opportunities and provide new opportunities for an aircraft to operate consecutive flight legs, resulting in the possibility that a fleeting change can provide economic gain. For example, extending the time between the arrival of a flight leg f and the departure of another flight leg g can allow the consecutive operation of the flight legs by a greater number of aircraft types, in particular by larger aircraft with longer turn times. That is, before the retiming, the assignment of larger aircraft to *both* flight legs f and g requires two aircraft rather than one. If the airline does not have two larger aircraft, or if the large aircraft achieve greater profit by being assigned to other flight legs, then smaller aircraft are assigned to flight legs f and g, from which potentially large numbers of passengers are spilled. In the retimed schedule, however, covering flight legs f and g requires only a single larger aircraft. In this case, re-fleeting flight legs f and g can increase the profitability of the flight schedule. In general, whenever flight schedules are rescheduled, re-fleeting can lead to improved solutions.
3. **Fleeting and scheduling decisions must be determined simultaneously**. Simultaneous scheduling and fleeting decisions are required because the value of a schedule cannot be computed without knowing the optimal fleet assignment that corresponds

to it, and because the number of possible flight schedules is, practically speaking, unlimited. A sequential approach, in which schedules are repeatedly generated and the fleet assignment problem is then solved for each schedule, is both impractical and non-optimal in the sense that it cannot guarantee that optimal flight schedules will be found.

The observation above, that flight schedule changes accompanied by new fleeting decisions can result in more productive use of the airline's fleet, is true even when the airline's bank structure at its hubs is maintained. An incremental schedule optimization model was solved for a large US airline to determine revised flight schedules and fleet assignments, allowing small shifts from the current schedule in the departure times of flights, while maintaining the airline's connecting banks at its hubs. The result was a reduction of 0.55% (two aircraft) in the number of aircraft needed to operate the flight schedule, and additional expected annual operating cost savings and increased revenue capture of $50 million (Rexing *et al.*, 2000).

Increased aircraft productivity results in less aircraft idle time, allowing the same set of flight legs to be operated with fewer aircraft, or, conversely, providing the airline with the opportunity to operate more flight legs. An incremental schedule optimization model, therefore, should allow for the inclusion of additional flight legs, and, hence, should also allow for the removal of existing flight legs. Associated with retiming, adding and eliminating flight legs in the network, however, is an important dynamic: that of shifting passenger demands. It is well documented that: (1) unconstrained market demand for a carrier is a function of its flight schedule (with frequency of service being one critical element); and (2) *total* market demand can change as a result of changes in the flight schedule (Simpson and Belobaba, 1992). As an illustration of this, consider the removal of a flight leg from a connecting bank. Its removal can impact passengers from *many* markets because in addition to carrying local passengers from the leg's origin to its destination, the removed leg carries a significant number of connecting passengers from other markets making use of that leg. From the viewpoint of the passengers in those markets, the quality of service is deteriorated because the frequency of (connecting) service is decreased. The result then can be that the carrier experiences a decrease in its unconstrained market demands in the affected markets. The situation is the opposite when a flight leg is added to the bank.

Because flight frequencies and departure times are among the most important factors affecting passengers' choices of an air carrier when there is a high level of competition, approaches have been developed to capture these supply and demand interactions and integrate them into a single optimization model including incremental schedule design and fleet assignment decisions. Using such an approach, the schedules that optimized the integrated problems of a large US airline were estimated to provide the airline with annual potential savings in excess of $200 million (Lohatepanont and Barnhart, 2004).

7.3.1 *Modeling the Optimization of Flight Retiming and Fleet Assignment Problems**

A special case of the more general integrated schedule design and fleet assignment problem is the problem of flight retiming and fleet assignment in which schedule and fleeting decisions are made but the set of flight legs to be operated is specified. By altering the time–space network to include one flight arc copy for each possible departure time of each flight leg, the problem is then to select the flight copy to be operated (i.e., to select the scheduled departure, and, hence, scheduled arrival, time of the flight leg) and to assign an aircraft type to each selected flight leg.

This problem is modeled as a simple variant of FAM. Specifically, another subscript is added to each fleet assignment variable in FAM to identify the particular flight arc copy; for example, $f_{i,b}^k = 1$ implies that fleet type k is assigned to operate flight leg i *and* the departure time of flight leg i is that corresponding to the time of flight arc copy b. The set of copies for flight leg i is denoted B^i. The resulting model is as follows (Rexing *et al.*, 2000):

$$\text{Minimize} \quad \sum_{i\in F}\sum_{k\in K}\sum_{b\in B^i} c_{i,b}^k f_{i,b}^k$$

subject to:

$$\sum_{k\in K}\sum_{b\in B^i} f_{i,b}^k = 1, \qquad \forall i \in F \tag{7.9}$$

$$y_{n+}^k + \sum_{(i,b)\in O(k,n)} f_{i,b}^k - y_{n-}^k - \sum_{(i,b)\in I(k,n)} f_{i,b}^k = 0, \qquad \forall n \in N^k, \forall k \in K \tag{7.10}$$

$$\sum_{a\in CG(k)} y_a^k + \sum_{(i,b)\in CL(k)} f_{i,b}^k \le M^k, \qquad \forall k \in K \tag{7.11}$$

$$f_{i,b}^k \in \{0, 1\}, \qquad \forall i \in F, \forall b \in B^i, \forall k \in K \tag{7.12}$$

$$y_a^k \ge 0, \qquad \forall a \in G^k, \forall k \in K \tag{7.13}$$

The assignment variables and the constraints corresponding to cover (7.9), balance (7.10), count (7.11) and binary variables (7.12) are modified slightly from the FAM formulation to account for the multiple flight arc copies.

To adapt the Rexing *et al.* model to the de-banking problem with a fixed set of flight legs to be operated in the planning period T, let $z(i, b)$ represent the time associated with the bth copy of flight leg i; H be the set of hubs; α_i^h equal 1 if flight leg i originates at hub h and equal 0 otherwise; and δ_i^h equal 1 if flight leg i is destined to hub h and equal 0 otherwise. Then, the following set of constraints must be added to limit the number of flight departures and arrivals at hub h within each time interval t to some threshold value e_t^h:

$$\sum_{k\in K}\sum_{i\in F}\sum_{b\in B^i}\sum_{(i,b):z(i,b)\in t} f_{i,b}^k(\alpha_i^h + \delta_i^h) \le e_t^h, \qquad \forall h \in H, \forall t \in T$$

The Rexing *et al.* model can be adapted further to allow new flight legs to be added and existing flight legs to be eliminated from the schedule. The set of flight legs is partitioned into two categories, *mandatory* and *optional*. Mandatory flight legs are those that must be operated, while optional flight legs are current legs that can be omitted or new legs that can be added to the new schedule (Lohatepanont and Barnhart, 2004). Let L be the set of optional flight legs and D the set of mandatory flight legs, with $L \cup D = F$. Then the only additional change needed to the Rexing *et al.* model is to replace the cover constraints with:

$$\sum_{k \in K} \sum_{b \in B^i} f_{i,b}^k \leq 1, \qquad \forall i \in L$$

$$\sum_{k \in K} \sum_{b \in B^i} f_{i,b}^k = 1, \qquad \forall i \in D$$

7.4   Crew Scheduling

After the flight schedule is developed and aircraft are assigned to cover all the flight legs in the schedule, crew work schedules are constructed, with the help of optimization techniques. Typically, due to tractability issues, the optimization of crew schedules is broken into two stages. First the *crew pairing problem* is solved, resulting in the generation of *mini-schedules*, called *pairings*, typically spanning 1–5 days; and then the *crew rostering problem* is solved in which the pairings are assembled into longer crew schedules, in the form of either *rosters* or *bidlines*, typically spanning about 30 days. A roster is a work schedule generated for each individual crew member, taking into consideration his or her needs (such as training etc.) and preferences (such as requested holidays or days off, or desired destinations of assigned flight legs, etc.). A bidline, however, is a generic schedule that is (eventually) assigned to a crew member through a bidding and allocation process based on seniority. Crew members express their preferences for different bidlines, and crews are then processed in decreasing order of seniority, with each member receiving the schedule that is still unassigned and ranks highest on his or her list of preferences. The use of bidlines is common in the USA, but rosters are more commonly used in the rest of the world and are increasingly accepted in the USA.

The objective of the crew pairing problem is to minimize the crew costs associated with covering all flight legs in the flight schedule, while the objective of the crew rostering problem is typically to assemble pairings into schedules that maximize the satisfaction levels of crews. Planners try to maximize crew satisfaction by developing schedules to meet various criteria, including: (1) maximizing the total number of individual crew requests that are satisfied by the schedules, or maximizing the minimum number of requests satisfied for any one individual; (2) maximizing the minimum number of hours of flying time in any one individual's schedule to achieve equity in the workloads of individuals; and (3) optimizing the spacing between periods of work to keep diurnal clocks consistent and to allow crew members to take advantage of extended time-off periods.

There are two types of crew that fly each flight leg: a cockpit crew, charged with flying the aircraft; and a cabin crew, responsible for in-flight passenger safety and service. Although both types of crews are required to operate each of the legs in the flight schedule, the two crew types are scheduled differently. Because each cockpit crew is qualified to fly a specific fleet type or set of closely related fleet types known as a *fleet family*, much of the schedule of individuals in these crews tends to be the same, with the same set of crew members working together throughout the day, or longer. Cabin crew members, however, have more flexibility in the aircraft types to which they can be assigned. Moreover, the size of the cabin crew for a given flight leg can vary with the number of passengers on board: the greater the number, the larger the crew. For these reasons, cabin crews tend to be scheduled as individuals, rather than as a crew cohort, with one crew member working with potentially many different crews on any single day.

Another important difference between these crew types is their costs. Because cockpit crew members are paid substantially more than cabin crew members, most of the efforts to optimize airline crew scheduling models have focused on cockpit crews. In this section, the presentation is similarly limited, referring to *cockpit crews* simply as *crews*. Moreover, the discussion is focused on the crew pairing problem because the pairing, bidline and rostering problems are similar in structure, and similar models and solution approaches have been developed to solve them. Although the focus is on the crew pairing problem, similarities and differences of the crew pairing and crew rostering and bidline problems are highlighted.

7.4.1 The Crew Pairing Problem

The crew pairing problem has provided operations researchers a popular challenge for some time. In fact, a survey of research activities on the topic was published as early as 1969 (Arabeyre *et al.*, 1969). The problem's popularity stems from two sources: first, pairing optimization is an important problem to the airlines because crew costs represent a significant portion of the airline's operating costs, second only to fuel costs; and second, the crew pairing problem, characterized by its myriad of work rules governing feasibility, its nonlinear functions that represent the costs of each solution, and its very large-scale nature with millions or billions of possible pairings, is very complex and difficult to solve. Constructing even feasible, let alone optimal, solutions is difficult if not impossible to do without the aid of automated decision support.

Because crews are qualified only to fly aircraft in a particular fleet family, one crew pairing problem is defined for each fleet family. Hence, the crew pairing problems for a particular fleet family are concerned only with the flight legs assigned to that family. To facilitate the presentation in the remainder of this section, fleet-specific crew scheduling problems are referred to simply as crew pairing problems, without reference to the particular fleet family.

A crew pairing is composed of a sequence of flight legs, with the flight legs comprising a set of daily work activities, called *duty periods* or *duties*, separated by overnight rest periods. The first and last legs of the pairing must begin and end at the same *crew base*, i.e., the same domicile location for crews, with crews staying together throughout the duration of the pairing. Collective bargaining agreements typically limit both the total number of duties in a pairing and its total duration, called the *time-away-from-base*. Other collective bargaining agreements and rules mandated by regulatory agencies provide a multitude of additional restrictions that define whether or not a pairing is *legal*: that is, whether or not it can be operated by a crew. The overall set of rules include:

1. Flights in a pairing must be sequential in time and space; that is, the destination of an arriving flight must be the same as the departure location of the next flight in the pairing.
2. The elapsed time between the arrival of a flight leg and the departure of the subsequent flight leg in the pairing is bounded by a *maximum* and a *minimum* threshold. For two successive flight legs, whether within a duty or separated by overnight rest, a maximum

allowable connection time (referred to as maximum rest time in the case of two flight legs separated by overnight rest) is designed to ensure that crews are not idle on the ground for long periods of time. Idle crews are undesirable both from the viewpoint of both the airline (crews are not operating aircraft and providing revenue) and the crews (who are typically paid less for idle time than for flying time).

Within a duty, the minimum allowable crew connection time is the minimum time required for the crew to connect between arriving and departing flight legs. If two flight legs are assigned to the same crew but are operated by different aircraft, the amount of required connection time is greater than that if the same aircraft is scheduled to operate both flights. This reduced connection time results because the crew does not have to *check out* from its current flight, walk to the departure gate of its next flight, and *check in* at that flight leg. In addition to crew transfer between aircraft, minimum crew connection times are affected by factors such as aircraft size, with larger aircraft requiring more time for passengers to disembark (crew are the last to leave the aircraft), and airport configurations that determine the expected travel time to the departure gate of the next flight leg.

The crew connection time *between* duties is required to exceed the minimum allowable rest time, a quantity that varies depending on the amount of flying performed by a crew. If a crew has a *heavy* flying load, then additional rest beyond the minimum number of hours might be required.

To protect further against crew fatigue and unsafe operating conditions, duty period durations are limited to some *maximum* value, and the amount of flying time, or *block time*, in a duty period is limited to some *maximum* number of hours.

These multifaceted work rules are combined with often complex cost functions that define crew costs. How crew costs are calculated can vary widely by airline, with significant differences existing between airlines in different countries or regions. The majority of European, Asian and South American airlines pay their crews a fixed monthly salary, with additional per diem or bonus compensation provided, depending on the assigned routing of the crew. Because accommodation costs and per diem can be as much as 30% of the total cost for cabin crew, crew optimization models can play a critical role in minimizing these expenditures.

In other countries, such as Australia, Canada, the UK and the USA, the crew cost function is more complex, often defined as a function of the amount of work performed. For US airlines, crews are paid at one rate for flying time and at another, lower rate for time spent on duty but not flying. Typically, for these airlines, the crew cost associated with a duty period is the *maximum* of three costs: (1) the flying time cost; (2) the duty duration cost; and (3) the minimum guarantee cost. At a minimum, crews are paid for the amount of time spent flying during the duty. Duty duration and minimum guarantee costs, however, are established to protect crews from earning low wages when assigned to duties that have little flying time or have short durations. In the USA, pairing costs are similarly structured, with the crew cost of a pairing being the *maximum* of three costs, namely: (1) the sum of the costs of the duties contained in the pairing; (2) the time-away-from-base cost; and (3) the minimum guarantee cost. Again, this ensures that crews are paid at least for the amount of time spent flying, with additional compensation possible if the work schedule is very short or if it includes relatively large amounts of idle time. This additional compensation, referred to as *pay-and-credit*, is the variable portion

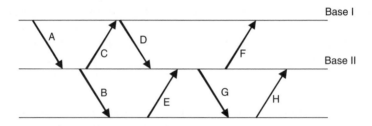

Figure 7.6   Sample crew pairing network

of crew costs, as flying costs are determined by the flight schedule and are independent of how crews are scheduled.

As an illustration of the crew pairing problem, consider the example flight network depicted in Figure 7.6. There are two crew bases, denoted *Base I* and *Base II*, and eight flight legs, each of which must be contained in exactly one crew pairing. Assume that associated with this network are four feasible pairings: two from Base I and two from Base II. The pairings are: A–C–D–F with cost of $1000, A–B–E–F with cost $2000, C–D–G–H with cost $4000, B–E with cost $3500, G–H with cost $3500 and B–E–G–H with cost $6000. Given these pairings, the set of feasible solutions and their costs are:

1. A–C–D–F, B–E–G–H with a cost of $7000;
2. A–B–E–F, C–D–G–H at a cost of $6000; and
3. A–C–D–F, B–E, G–H with a cost of $8000.

For this tiny example, the optimal solution is easily identified as option (2) above. The challenge for larger problems, however, is that it is not possible to enumerate in a timely manner all possible solutions to the problem. In fact, a fight network with several hundred flights per day has billions of pairings, and even enumerating all possible pairings is not practical. As a consequence, the crew pairing problem is typically solved using a combination of decomposition, optimization and heuristic methods.

Modeling and Solving the Crew Pairing Problem*

The crew pairing problem can be expressed quite simply as a set partitioning problem, with F the set of flight legs, J the set of feasible pairings, c_j the cost of pairing j, a_{ij} equal to 1 if flight leg i is covered by pairing j, and the decision variable x_j, for all all $j \in P$, equal to 1 if pairing j is included in the solution and equal to 0 otherwise.

The crew pairing problem is then modelled as:

$$\text{Minimize} \quad \sum_{j \in J} c_j x_j$$

subject to:

$$\sum_{j \in J} a_{ij} x_j = 1, \quad \forall i \in F \tag{7.14}$$

$$x_j \in \{0, 1\}, \quad \forall j \in P \tag{7.15}$$

The *cover constraints* (7.14), together with the binary constraints (7.15), ensure that each flight leg is included in exactly one pairing in the solution. The objective is to minimize the costs of the pairings in the solution. This basic crew pairing model is sometimes enhanced to include *crew base balance* constraints. These constraints typically limit, from below and above, the total number of flying hours in the selected pairings to ensure that the number of crews domiciled at a base roughly matches the total flying time of selected pairings originating from that base.

As stated earlier, the crew rostering problem is similarly formulated. Like the pairing problem, one rostering problem is solved for each fleet family. Unlike the pairing problem, however, separate rostering problems are solved for each *crew type*, i.e., for captains, first officers, flight engineers, etc. Like pairings, each roster is a sequence of activities separated by *off-duty* periods. In the case of rosters, these activities include pairings, flight training and vacation periods. One major difference between the pairing and rostering problems is the manner in which time is handled. Typically, for pairing problems, the *daily* pairing problem is solved under the assumption that the schedule is repeated daily. Hence, the set of selected pairings must cover each flight leg exactly once, and each pairing commences every day. The result is that multiple crews might be assigned to the same pairing, depending on the duration of the pairing.

Consider, for example, a three-day pairing p. Because each flight leg in pairing p must be flown every day, and because only pairing p in the crew pairing solution contains the flight legs in p, three crews, denoted A, B and C, must be assigned to pairing p. On day 1 of the schedule, crew A operates the flight legs comprising the work on the first day of pairing p. Similarly, crew B operates the flight legs making up pairing p's second day of work, and crew C operates the flight legs in p's third day. On day 2, each crew advances to the next day of work in pairing p, with crew A advancing to day 2, crew B advancing to day 3, and crew C cycling back to operate the flight legs comprising pairing p's first day of work. In rostering problems, a *dated* schedule is constructed for each crew member type, with each activity assigned both the requisite number of specific individuals of the type needed to operate the activity and an *activity start date*. Hence, the rostering model contains one constraint for each pairing commencing on a given day, for each day in the rostering period. Because individuals, rather than crews, are assigned to these rosters, an additional constraint is added to the rostering problem ensuring that exactly one schedule from each individual's set of alternative schedules is selected for that individual. If the airline is not required to use all crew members, a crew member might be assigned to an *empty* or *null* schedule – that is, a schedule containing no work.

Although relatively simple to formulate, the crew pairing and rostering problems are difficult to solve for all but extremely small problem formulations. The difficulty of identifying and even checking the feasibility of the decision variables, and the large number of non-continuous variables, poses tremendous tractability challenges. Binary constraints on the decision variables in the crew pairing problem require that an integer programming solution approach, such as *branch-and-bound*, be employed. Branch-and-bound is a smart enumeration technique in which the optimal integer solution is found by repeatedly solving the linear programming (LP) relaxation of the crew pairing problem with the binary constraints relaxed to allow variables to take on values greater than or equal to 0 and less than or equal to 1. With billions of possible pairings, however, it is impossible to solve directly even the linear programming relaxation of the crew pairing problem.

Instead, an approach called *column generation* is employed. In this case, *column* is synonymous with *variable*, and hence the approach is to *generate* variables on an as-needed basis. Because an optimal solution will contain only a small subset of all possible variables, the goal is to identify those variables that comprise an optimal solution, without explicitly considering *all* variables. In so doing, the tractability issues associated with crew pairing problems are overcome: an optimal solution is found by solving a problem with only a small subset of the original set of variables. The challenge, of course, is to find this subset of variables as efficiently as possible. To accomplish this, the following steps of the *Column Generation Algorithm* are repeatedly executed:

Step 1: *Solve the restricted master problem:* The restricted master problem is the crew pairing (or rostering) LP relaxation containing only a subset of all variables.

Step 2: *Solve the pricing problem:* The pricing problem is to generate one or more pairing (or roster) variables that might improve the current solution to the restricted master problem. If no variables are identified, STOP: the crew pairing (or rostering) problem LP relaxation is solved.

Step 3: *Construct a new restricted master problem:* Add the variables generated in solving the pricing problem to the restricted master problem and return to Step 1.

The Column Generation Algorithm for the crew pairing (or rostering) LP relaxation is thus embedded within a branch-and-bound algorithm to create a tailored *branch-and-price* algorithm for large-scale problems. Critical to the success of branch-and-price solution approaches are the following: (1) a tractable solution algorithm for the pricing problem; and (2) branching strategies in the branch-and-price algorithm that allow the *same* algorithm to be used to solve the pricing problem, while ensuring that all branching decisions are enforced in the solution. Details of challenges and successful strategies associated with solving large airline crew scheduling problems are provided in Barnhart *et al.* (1998b), Desrosiers *et al.* (1995) and Klabjan *et al.* (2001).

7.4.2 Crew Scheduling Problem Solutions and Impacts

Airline crew schedule optimization has been one of the great successes of operations research. Large airlines use optimization tools to generate partial or complete crew schedules, and attribute significant cost savings to their use. As early as the 1960s, airlines used decision support tools to automate the generation of solutions to the difficult crew pairing problem (Arabeyre *et al.*, 1969). And with more recent advances in operations research theory and computing software and hardware, airlines have continued to reap ever-increasing rewards from the application of optimization tools to the crew scheduling problem (Clarke and Smith, 2004). A decade ago, solutions to crew pairing problems had pay-and-credit costs exceeding flying costs by 10–15%, while crew pairing solutions today typically exceed flying costs by at most 1–2%. For large airlines, this improvement in solution quality translates to savings on the order of $50 million annually. Moreover, beyond these economic benefits, crew scheduling optimization tools can be used in contract negotiations to quantify the effects of proposed changes in work rules and compensation plans.

While significant, the impact of crew scheduling optimization at airlines is limited by the sequential nature of the planning process. With the flight schedule and aircraft assignment and routing decisions fixed, the range of crew scheduling possibilities is limited. To broaden the set of possible solutions, FAMs have been *expanded* to capture, at least approximately, the downstream effects of fleeting decisions on crews (Clarke *et al.*, 1996; Barnhart *et al.*, 1998a). Another approach has been to expand the set of feasible crew solutions by integrating aircraft routing and crew scheduling decisions (Cohn and Barnhart, 2003; Cordeau *et al.*, 2001; Klabjan *et al.*, 2002), This approach is discussed in further detail in the next section.

7.5   Aircraft Maintenance Routing and Crew Pairing Optimization

Even in countries that have implemented economic deregulation of airline markets, government regulations require that aircraft undergo periodic maintenance checks at regular intervals that satisfy limits on the amount of flying time between maintenance checks. These maintenance checks are performed at maintenance stations that in some cases are capable of servicing only specific aircraft types and which can be located at various airport locations in the flight network. In the USA, regulations require inspections of aircraft approximately every 3–5 days.[1] If violated, the uninspected aircraft is grounded and not allowed to fly without first undergoing an inspection. To avoid this costly outcome, airlines typically build aircraft routings that provide "more frequent" maintenance opportunities, thereby providing slack in the schedule and allowing schedule changes to delay maintenance checks without the consequence of grounded aircraft.

To address the problem of aircraft routing given maintenance requirements, a typical approach is to define one maintenance routing problem for each subnetwork of flight legs assigned to a common fleet. The goal then is to route individual aircraft so that each flight leg has exactly one aircraft assigned to it; each aircraft route begins and ends at the same location (thus forming a cycle or rotation) and each aircraft (or tail number) visits a maintenance station at regular intervals. When flight schedules, fleet assignments and crew pairings are built for daily operations, aircraft routing models are similarly constructed. As in the case of daily crew pairings, this implies that a rotation spanning *n* days requires *n* aircraft, each one performing the flight legs contained within a 24-hour period and each advancing day by day to the set of flight legs contained within the next 24-hour period in the aircraft rotation.

The aircraft maintenance routing problem is sometimes cast simply as a *feasibility problem*, and sometimes as the problem to maximize incremental *through-revenues* that can be obtained with *through-flights*. A through-flight is a pair of sequential flight legs that is flown by the same aircraft, providing the possibility to generate additional revenue from passengers who prefer not to transfer between aircraft when connecting. Of course, passengers prefer that there be no connection, but for some markets and certain time frames, non-stop flights are not a possibility. In these cases, through-flights can be desirable to passengers by obviating the need to walk between gates, and protecting passengers from misconnecting when disruptions occur.

In addition to providing potential through-revenues, aircraft routing decisions define the set of feasible crew pairings. If two flight legs have limited time (less than the minimum

---

[1] Inspection is of the aircraft's landing gear, control surfaces such as flaps and rudders, fluid levels, oxygen systems, lighting and auxiliary power systems (http://www.airlines.org/products/AirlineHandbookCh6.htm, December 19, 2008).)

allowable crew connection time) between the arrival of the first leg and departure of the second, but the same aircraft is assigned to operate both legs, then a single crew can be assigned to both flight legs. When no crew transfer between aircraft is needed, *short connects*, i.e., connections that are shorter in duration than the minimum, can be allowed. Hence, aircraft routing decisions define the set of potential crew short connects.

Because crew costs represent the second-largest operating cost to the airline, solving the aircraft maintenance routing problem first, without considering the impacts on crews, can limit the set of feasible solutions to the crew pairing problem and result in increased crew costs. One resolution of this issue is to change the order in which the two problems are solved, first solving the crew pairing problem (allowing any short connect to be used by the crews) and then solving a constrained aircraft maintenance routing problem in which pairs of flight legs are forced to be assigned to the same aircraft if they form a short connect used in the crew pairing solution (Klabjan *et al.*, 2002). This is a computationally attractive approach to the problem, but has the downside that feasible solutions to the resulting constrained routing problem might not exist. Another approach is instead to address the problem by integrating the crew pairing and aircraft routing problems and solving the integrated problem (Cordeau *et al.*, 2001). This has the advantage that feasible, and even optimal, solutions are theoretically guaranteed, but has the disadvantage that intractability can result for problems faced by large airlines.

Yet another approach is to reverse the solution order, first solving an *enhanced* crew pairing problem and then solving the aircraft maintenance routing problem (Cohn and Barnhart, 2003). The enhanced crew pairing problem has additional constraints and variables ensuring that each short connect used in the pairing solution can be included in a feasible solution to the aircraft routing problem. Solutions to the enhanced crew pairing problem thus represent both a cost-minimizing crew pairing solution and a feasible maintenance routing solution. The optimal maintenance routing solution is determined after solving the enhanced crew pairing problem by solving a variant of the aircraft routing problem, one requiring each short connect in the crew solution to be covered by the same aircraft, with varied objectives and additional constraints capturing preferred qualities of a routing solution.

7.5.1   Modeling and Solving the Enhanced Crew Pairing Problem*

The enhanced crew pairing formulation is most easily described by considering the following scenario in which crews of some type k must be assigned to a set of flight legs, and every feasible aircraft maintenance routing solution can be identified for the given set of flight legs. For simplicity, and without loss of generality, assume that the flight legs to be covered by type k crews correspond to the flight legs to be covered by aircraft of type k. Each maintenance-feasible routing solution is represented solely by the list of short connects it contains. The only solutions of interest are those that are *maximal*; that is, those for which no other solution exists containing *all* of the short connects it contains. In the enhanced crew pairing problem formulation, one binary variable is included for each maximal solution and one constraint is added to ensure that exactly one maximal aircraft routing solution is selected in the solution to the enhanced crew pairing problem. Additionally, one constraint is added for each possible short connect to allow it to be contained in the crew pairing solution only if it is included in the selected routing solution. With these modifications, the crew costs associated with the enhanced crew pairing solution are guaranteed to be at least as good as those produced by the basic crew pairing model. Moreover, the approach is guaranteed to yield at least one maintenance-feasible aircraft routing solution that includes the short connects contained in the pairing solution.

To formulate the enhanced crew pairing problem, let F^k represent the set of flight legs that must be covered by crews of type k and let T^k represent the set of short connection possibilities in the network defined by F^k. As in the basic crew pairing problem, $y_p = 1$ if pairing p in the set of pairings P is included in the crew pairing solution, and $y_p = 0$ otherwise. The new variables, x_s, one for each solution s in the set of routing solutions S^k, are defined such that $x_s = 1$ if maintenance routing solution s is selected in the enhanced crew pairing problem, and $x_s = 0$ otherwise. The parameter $\alpha_{ts} = 1$ if short connect t is included in routing solution s, and $\alpha_{ts} = 0$ otherwise. Similarly, parameter $\beta_{tp} = 1$ if short connect t is used in pairing p, and $\beta_{tp} = 0$ otherwise; and let $\delta_{fp} = 1$ if pairing $p \in P^k$ contains flight leg $f \in F^k$, and $\delta_{fp} = 0$ otherwise. The enhanced crew pairing problem for fleet type k is then formulated as:

$$\text{Minimize} \quad \sum_{p \in P^k} c_p y_p$$

subject to:

$$\sum_{p \in P^k} \delta_{fp} y_p = 1, \qquad \forall f \in F^k \tag{7.16}$$

$$\sum_{s \in S^k} \alpha_{ts} x_s - \sum_{p \in P^k} \beta_{tp} y_p \geq 0, \qquad \forall t \in T^k \tag{7.17}$$

$$\sum_{s \in S^k} x_s = 1, \tag{7.18}$$

$$x_s, y_p \in \{0, 1\} \qquad \forall s, p \tag{7.19}$$

Cover constraints (7.16) and binary constraints (7.19) ensure that each flight leg is assigned to exactly one pairing, while *short-connect constraints* (7.17) allow selected pairings to include a short connect only if the selected aircraft maintenance routing solution also contains that short connect. *Routing selection* constraints (7.18), together with binary constraints (7.19), require exactly one routing solution to be selected. The objective is to minimize total crew costs; however, this objective can be modified. For example, each maintenance routing solution variable can have an assigned cost representing the negative of its associated through-revenues. In this case, the enhanced crew pairing objective is to optimize the integrated aircraft maintenance routing and crew pairing problems.

One challenge in solving the enhanced crew pairing problem is to generate the set of maximal maintenance routing solutions. Just as in the case of pairings, it is not practical to generate *all* maximal routing solutions, and hence column generation is used to solve the enhanced crew pairing problem. Experience with this approach indicates that only a small fraction of all maintenance routing solutions must be generated to solve the enhanced crew pairing problem (Cohn and Barnhart, 2003). For example, in a small instance involving 61 flights, only 35 maximal routing solutions existed, although the number of possible maintenance routing solutions far exceeded 25 000. Moreover, after generating just 4 of these maximal solutions, the provably optimal solution to the enhanced crew pairing problem was identified.

7.6 Future Directions for Schedule Optimization

Substantial progress in optimization techniques, computing power and scheduling theory has allowed significant progress to be made in the optimization of many aircraft and crew scheduling problems. Nonetheless, significant challenges and opportunities remain. Two important areas of ongoing and future research are:

1. **Integrated schedule planning:** Although practical, the sequential nature of aircraft and crew schedule optimization leads to suboptimal plans, with potentially significant economic losses. As discussed in this chapter, improved plans can be generated by building and solving integrated models of some (and eventually all) of these schedule design, fleet assignment, maintenance routing and crew scheduling subproblems. Another area of integration is to expand schedule planning models to include pricing, revenue management decisions and competition. By integrating scheduling models (which are effective at minimizing operating costs) with pricing, revenue management and competition models (which are effective at maximizing revenues), schedules with improved overall profitability should result.

2. **Operations recovery and robustness:** The schedule planning and optimization processes at airlines produce plans that are rarely, if ever, executed. During operation of the schedule on any given day, disruptions caused by mechanical failures, crew illness and adverse weather conditions result in the need to replan and create recovery plans in response to these disruptions. Creating these responses is particularly challenging, with the need to develop feasible and cost-effective plans in a short period of time, often within minutes. As a complement to research on recovery planning, some researchers have focused on developing *robust* plans that are both more resilient to disruptions (allowing less frequent replanning) and easier to repair when replanning is necessary. Current research on robust schedule optimization attempts to achieve robustness by isolating causes of disruption and the resulting downstream effects, or by reallocating slack time in the schedule to where it is most needed. Much of the work to date, however, focuses on the recovery or robustness of a particular resource, namely aircraft or crews. Disruptions, however, typically affect multiple resources. Hence, the challenge moving forward is to enhance current capabilities to generate *integrated* passenger, crew and aircraft recovery decisions and robust plans, and to evaluate these new plans and recovery policies using simulation and optimization tools. Additional details of the challenges of robust optimization and irregular operations, representations of the robust optimization and schedule recovery problems and solutions to the associated optimization models are provided in Chapter 9.

References

Abara, J. (1989) "Applying Integer Linear Programming to the Fleet Assignment Problem," *Interfaces*, Vol. 19, pp. 20–28.

Arabeyre, J.P., Fearnley, J., Steiger, F.C., and Teather, W. (1969) "The Airline Crew Scheduling Problem: A Survey," *Transportation Science*, Vol. 3, pp. 140–163.

Barnhart, C., Boland, N., Clarke, L., Johnson, E., Nemhauser, G., and Shenoi, R. (1998a) "Flight String Models for Aircraft Fleeting and Routing," *Transportation Science*, Vol. 32, pp. 208–220.

Barnhart, C., Johnson, E., Nemhauser, G., Savelsbergh, M., and Vance, P. (1998b) "Branch-and-Price: Column Generation for Solving Huge Integer Programs," *Operations Research*, Vol. 46, pp. 316–329.

Barnhart C., Kniker, T., and Lohatepanont, M. (2002) "Itinerary-Based Airline Fleet Assignment," *Transportation Science*, Vol. 36, pp. 199–217.

Clarke, L., Hane, C., Johnson, E., and Nemhauser, G. (1996) "Maintenance and Crew Considerations in Fleet Assignment," *Transportation Science*, Vol. 30, pp. 249–260.

Clarke, M. and Smith, B. (2004) "The Impact of Operations Research on the Evolution of the Airline Industry," *AIAA Journal of Aircraft*, Vol. 41, pp. 62–72.

Cohn, A. and Barnhart, C. (2003) "Improving Crew Scheduling by Incorporating Key Maintenance Routing Decisions," *Operations Research*, Vol. 51, pp. 387–396.

Cook, T. (2000) *Creating Competitive Advantage Using Model-Driven Support Systems*, MIT Global Airline Industry Study Distinguished Speaker Seminar Series, Cambridge, MA.

Cordeau, J., Stojkociv, G., Soumis, F., and Desrosiers, J. (2001) "Benders Decomposition for Simultaneous Aircraft Routing and Crew Scheduling," *Transportation Science*, Vol. 35, pp. 375–388.

CPLEX, ILOG (2005) http://www.ilog.com/products/cplex/.

Desrosiers, J., Dumas, Y., Solomon, M.M., and Soumis, F. (1995) "Time Constrained Routing and Scheduling," *Handbook in Operations Research/Management Science, Network Routing*, ed. M. Ball, North-Holland, Amsterdam, pp. 35–139.

Hane, C.A., Barnhart, C., Johnson, L.E., Marsten, R.E., Nemhauser, G.L., and Sigismondi, G. (1995) "The Fleet Assignment Problem: Solving a Large-Scale Integer Program," *Mathematical Programming*, Vol. 70, pp. 211–232.

Jacobs, T.L., Johnson, E.L., and Smith, B.C. (1999) "O&D FAM: Incorporating Passenger Flows into the Fleeting Process," Presentation to AGIFORS Symposium, New Orleans, LA.

Klabjan, D., Johnson, E.L., and Nemhauser, G.L. (2001) "Solving Large Airline Crew Scheduling Problems: Random Pairing Generation and Strong Branching," *Computational Optimization and Applications*, Vol. 20, pp. 73–91.

Klabjan, D., Johnson, E.L., Nemhauser, G.L., Gelman, E., and Ramaswamy, S. (2002) "Airline Crew Scheduling with Time Windows and Plane Count Constraints," *Transportation Science*, Vol. 36, pp. 337–348.

Lohatepanont, M. and Barnhart, C. (2004) "Airline Schedule Planning: Integrated Models and Algorithms for Schedule Design and Fleet Assignment," *Transportation Science*, Vol. 38, pp. 19–32.

Rexing, B., Barnhart, C., Kniker, T., Jarrah, A., and Krishnamurthy, N. (2000) "Airline Fleet Assignment with Time Windows," *Transportation Science*, Vol. 34, pp. 1–20.

Rushmeier, R. and Kontogiorgis, S. (1997) "Advances in the Optimization of Airline Fleet Assignment," *Transportation Science*, Vol. 31, pp. 159–169.

Simpson, R.W. and Belobaba, P.P. (1992) "The Demand for Air Transportation Services," Lecture Note for 16.74 Air Transportation Economics, MIT, Cambridge, MA.

Wiper, D.S., Quillinan, J.D., Subramanian, R., Scheff, R.P. Jr., and Marsten, R.E. (1994) "Coldstart: Fleet Assignment at Delta Air Lines," *Interfaces*, Vol. 24, pp. 104–120.

Xpress-MP Suite, Dash Optimization (2005) http://www.dashoptimization.com/.

8

Airline Flight Operations

Alan H. Midkiff, R. John Hansman and Tom G. Reynolds

8.1 Introduction

Most airlines and other types of air carriers manage their flight operations under a system of prioritized goals including safety, economics and customer service (e.g., on-time departures and arrivals). The airline flight operations department is responsible for the safe and efficient movement of passengers and/or cargo which ultimately generate the revenue for the airline. The major components needing to be coordinated for any given flight include the aircraft and support equipment, cockpit and cabin crews, maintenance and ground service personnel. Although the maintenance and ground crew activities are critical to support flight operations, the emphasis in this chapter is on the regulation and scheduling of the flight crews to conduct a given flight, followed by a detailed discussion of the activities during the phases of a typical revenue passenger flight.

Note that this chapter provides a snapshot of how flight operations are conducted in the current air traffic management (ATM) system (Nolan, 2004). Future ATM evolutions, for example, the US Next Generation Air Transportation System (NextGen) (National Research Council, 2005) and the Single European Sky ATM Research (SESAR) programme (Cook, 2007) are likely to involve new technologies and procedures, such as the introduction of four-dimensional trajectory management, that should permit more efficient handling of traffic, but the overall flight operating principles are likely to be broadly similar. This chapter does not attempt to address detailed aviation skills and flight maneuvering topics and only includes such information in the context of the overall flight operation. However, specific flight procedures that may have a direct impact on the operational goals are included to aid in understanding the nature and complexity of the factors involved.

The Global Airline Industry   P. Belobaba, C. Barnhart and A. Odoni
© 2009 John Wiley & Sons, Ltd

8.2   Regulation and Scheduling

Before any flight operations can occur, regulatory requirements must be met for the aircraft and flight crew, who also need to be scheduled to meet the demands of the overall network of the airline. A summary of these aspects is given in this section.

8.2.1   General Regulatory Requirements

Air carrier operations are generally regulated by the country of registration and the sovereignty in which the operation is conducted, including the Federal Aviation Administration (FAA) in the USA and the Joint Aviation Authority (JAA) in Europe. International flights may also fall under the jurisdiction of ICAO (International Civil Aviation Organization) when operating abroad.[1] In the USA, the Code of Federal Regulations (CFR Section 14)[2] Parts 119, 121 and 135 cover commercial operations,[3] while Part 91 addresses general operating and flight rules. Parts 61, 65 and 67 govern air crew certification[4] (Spence, 2007). Many larger airlines also utilize fleet-specific flight operation guidelines under agreements with the local governing authorities.

8.2.2   Flight Crew Regulation

The principal human component of flight operations is the flight crew, comprising both cabin and cockpit crews. At large carriers, personnel rosters for cabin and cockpit crews may exceed 20 000 and 10 000 employees respectively. In many cases, crew members may have never worked together prior to a particular flight. In order to maintain a safe, efficient and smoothly functioning operation, airlines and regulators have developed very detailed procedures to be executed by the crew members with little room for improvisation. These procedures, including normal, abnormal and emergency conditions, are detailed in the crew members' operating manuals[5] and backed up through a system of checklists which are cross-checked between flight crew members. It is the responsibility of the training or flight standards department to establish crew member proficiency and currency. The captain, however, is always ultimately responsible for the safe and efficient conduct of the flight and in extraordinary circumstances may deviate from a procedure or regulation under his or her command authority (i.e., Captain's Emergency Authority).

The cabin crew is primarily responsible for passenger safety during the flight. Other duties include providing customer service products (meals, entertainment, etc.) and assistance with boarding. Flight attendants receive specialized training in aircraft emergencies, evacuation procedures, medical issues and health hazards, care of special

---

[1] US-based carriers are governed by both when conducting international operations.

[2] Previously known as Federal Aviation Regulations (FARs).

[3] Part 119 governs the certification for air carriers and commercial operators; Part 121 governs operating requirements for domestic, flag and supplemental operations (generally commercial operations with more than 30 seats), while Part 135 governs the operating requirements for commuter and "on-demand" air-taxi operation.

[4] Part 61 governs certification of pilots, flight instructors and ground instructors; Part 65 governs certification of persons other than flight crew members and Part 67 governs medical standards.

[5] The Standard Operating Procedures (SOPs) of an air carrier are detailed in a number of documents including the generalized flight manual (non-fleet-specific operating rules), flight procedures (Jeppesen or other approach and navigation publications), as well as aircraft-specific operating manuals, performance manuals and Minimum Equipment List (MEL, described later in this document).

needs passengers, flight regulations and meal service. Most flight attendants' pre-airline experience includes some type of customer service, medical care or marketing background. Post high school education is preferred but not required. A flight attendant's initial training regime usually lasts 4 to 7 weeks including final in-flight experience. Cabin crews are required to undergo annual recurrent training for each of the aircraft on which they are qualified. The number of flight attendants assigned to a given flight varies with both the seats available on the aircraft and their working agreement. Minimum FAA staffing requirements include one flight attendant for a seating capacity between 19 and 50; two flight attendants for capacities of 51–100 seats; and two flight attendants plus one additional flight attendant for every 50 seats (or fraction of 50) above 100.

Most airlines today operate equipment requiring a cockpit crew of two pilots: the captain in the left cockpit seat and first officer (FO) in the right cockpit seat. Older aircraft (such as B747-100/200 "Classics," DC10, B727) may also require a flight engineer to manage the systems which are now automated in newer aircraft. In addition, an augmented crew of relief pilots may be required to staff long-haul flights due to duty time restrictions imposed by working agreements and/or regulations. Although the captain is always Pilot in Command (PIC), augmented crew members must also hold an Airline Transport Pilot (ATP) type-rating certification[6] in the aircraft to meet the requirement that a licensed type-rated pilot is always at the controls during the captain's rest period.

In the USA, many flight deck crews at Part 121 carriers have experienced some level of military flight training through regular, reserve or air guard duty. Many maintain their reserve or guard status while employed by the airline and must accommodate their military duty requirements while flying full time. Civilian pilots come from a variety of backgrounds including commuter/regional, corporate and general aviation. A pilot's flight experience is primarily measured in hours of flying time, which is further detailed by type of aircraft and conditions of flight. Typical "new hire" flight experience ranges from 1500 to 5000 hours, including 1000 hours of multiengine and/or jet time. Military pilots tend to be in the lower range because of the type of specialized flying they perform and the means by which they log their flight hours. Complete (ab initio) training programs are offered by some educational institutions (including airline training departments) which train pilots with limited or no flight experience to a level of proficiency with which they can operate as part of a flight crew, usually at a regional airline. Many non-US carriers use this type of training for their flight crews and often send them to the USA to offset the higher costs associated with operating in their respective airspace. Cockpit crews typically have at least a two-year college education with most having a four-year degree.

Cockpit crews require licensing by their respective national authorities, including some level of commercial/transport certification as well as individual qualifications in specific aircraft for larger types (type ratings). Crews must complete the training regimen established by their airline before meeting the qualification requirements for operating that carrier's aircraft. Initial qualification training durations vary between aircraft but usually require 4 to 6 weeks, including ground training and simulator sessions. Improved simulator technology precludes the utilization of actual aircraft during flight training in all

---

[6] Large aircraft require certification in the specific type, e.g., B777, as opposed to smaller aircraft which are covered under a broader system of categories and classes, e.g., multiengine land, or single engine sea. These ratings are added to pilots' licenses which comprise different grades including private, commercial or ATP. ATP grade certification or equivalent is required to act as PIC of air carrier aircraft.

but the most unusual circumstances. The simulators can safely replicate a wide variety of environmental, flight and mechanical conditions in order to achieve flight crew proficiency in both normal and abnormal procedures. In almost all cases, the first instance a pilot operates the controls of an airliner is on a revenue flight with paying passengers aboard.[7] It should be noted, however, that as advanced as simulator technology has become, not all tactile and visual stimuli can be entirely replicated, resulting in the aircraft being easier to operate than the simulator. Recurrent training cycles also contain both ground and simulator components and, depending on the carrier's self-administered FAA-approved program (in the USA), usually occur on a basis of 6, 9, 12 or 24 months.

In most airlines, flight crews also participate in crew resource management (CRM) training as part of their recurrent itinerary. CRM is the effective use by the flight crews of all resources available to them, including people (flight attendants, dispatchers, maintenance, ground/gate personnel and air traffic control (ATC)) and other aviation systems. CRM training emphasizes the skills necessary to optimize the human/machine interface including situation awareness, use of automation systems, team building, task delegation, information transfer, problem solving and decision making.

Pilots are also required to maintain a minimum health standard which is validated by a licensed medical examiner. Crew members are subject to periodic flight physicals on a recurring basis (6 or 12 months in the USA), which is determined by the certification requirements commensurate with their flight duties. However, pilots' "fit-to-fly" decision for a given flight is based on a self-assessment of their current physical and mental condition and may result in the crew members temporarily disqualifying themselves for flight activities. In addition, specified guidelines concerning other activities including alcohol consumption, prescription drug use, blood donation and scuba diving must be adhered to and may also result in restrictions from flight duties.

At many air carriers, flight crews are represented by labor unions that negotiate collective bargaining agreements. The work rules contained in these contracts are usually more restrictive than those imposed by federal regulations and are driven by both safety and labor considerations. The collective bargaining process always has the potential of disrupting the carrier's operations and consideration must be given with respect to the effect of a work stoppage by one group on the others.

8.2.3 Flight Crew Scheduling

Cockpit and cabin crews are assigned duty to a given flight by a variety of means (discussed in greater detail elsewhere in this book). The majority of crew members are assigned to a flight as part of their normal schedules or "rosters." A roster describes a crew member's flight activity for some period (typically a month), and consists of sequences of flight duty and days off (e.g. three days on, two days off). In some airlines, "bidlines" are made available based on the flying requirements for a given crew base, then awarded in order of seniority through a bidding process. Each package of bidlines targets

---

[7] Before a crew member can conduct unsupervised flight duties, he or she must complete the final phase of flight training, which is usually referred to as Initial Operating Experience (IOE) or Line Operating Experience (LOE). This phase of a pilot's qualification consists of the first 15 to 25 hours of actual flight time, and is conducted under the supervision of a check pilot who acts as the legal PIC (regardless of whether he or she occupies the left or right seat).

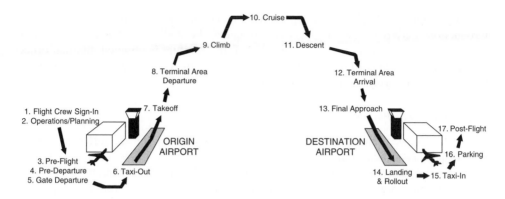

Figure 8.1 Typical flight phases

a specific flying "job" which may include crew base, equipment type, seat (captain, first officer, etc.) and division (e.g. LaGuardia – B767 – Captain – International). Rosters and bidlines may range from 65 to 85 hours spread over 10 to 18 duty days, depending on the carrier's working agreement. A typical duty day consists of one to four flight legs, but on some days the crew may operate none at all (sit at hotel), or fly more than five legs.

Many departures require filling an "open seat" because of a sick call, vacation, misconnect (including "commuting" crew members who reside out of base), legality (duty time) issues, or the job was never assigned as part of the normal bid process (open time). It is the responsibility of the crew scheduling department to fill these open seats and satisfy the crewing requirements for each flight. Open time can be proffered or traded with other lineholders,[8] or assigned to reserve crews. Each roster spot or bidline has a contingency of reserve crew members who are "on call" and available for the company to use at short notice (in some cases ready reserves are standing by at the airport). Once the first flight of a crew sequence departs, crewing responsibility shifts from crew schedule to crew tracking who intercede in cases of mid-sequence disruptions including illness, misconnects, equipment problems and cancellations.

8.3   Flight Crew Activities During a Typical Flight

A typical revenue flight involves the flight phases illustrated in Figure 8.1. The flight crew activities involved in each of these phases are described in detail in the following sections, numbered according to the phases outlined in the figure. The primary focus is on cockpit crew activities, but cabin crew activities are also discussed where relevant.

8.3.1   Flight Crew Sign-in

Once assigned to a flight sequence, crew members are required to sign in at the airport flight operations office (nominally) 1 hour prior to the departure of the first leg. Crews normally arrive earlier than 1 hour in order to accommodate international flight

---

[8] Lineholders are crew members whose monthly schedule consists of a bidline, as opposed to reserve crew members whose monthly schedule consists of days of availability and days off.

planning, publication/flight manual updating, or other administrative responsibilities. Once introductions between crew members are complete, the flight crew begins the planning tasks. In situations where the time available before departure is minimal, the first officer may proceed to the aircraft to begin the pre-flight duties there, while the captain completes the requisite paperwork in the flight operations office.

8.3.2 Operations/Planning

Most airlines have a central Airline Operations Control Center (AOCC) staffed by certified flight dispatchers. Their duties include the planning and "flight following" for as many as 20 (or more) concurrent flights. The flight planning task involves selecting the best routing (in terms of time, fuel burn, ride conditions, etc.) given the available information, and generating a "flight plan" that can be programmed into the aircraft automation. This accounts for aircraft type, forecast weather conditions, aircraft performance, loads and operating weights, aircraft mechanical condition, marketing constraints, airport limitations/curfews and company priorities (e.g., minimum fuel versus minimum time trajectory). This last issue is quantified via the "cost index" parameter which is the ratio of time-related costs to fuel-related costs and is a major driver of flight plan optimization given that minimum time and minimum fuel trajectories can be quite different (Airbus, 1998). The resulting flight plan is the means by which the dispatcher communicates the details of a flight to the cockpit crew and is usually available for retrieval via computer terminal approximately 1 hour before departure time. The flight plan details various aspects of the flight, such as routing, weather, alternate airport options, fuel requirements, takeoff performance and loads (which are subject to last-minute changes).

The flight plan is printed out and its details are examined by the incoming cockpit crew. Agreement is typically indicated by the captain's signature on the paper or electronic "station copy" of the flight plan. In many cases, no direct communication with the dispatcher is necessary unless changes in fuel load or routing, weather, aircraft operational status/mechanical discrepancies and/or any anticipated delays need to be discussed.

Excerpts from a sample flight plan for a flight from Newark, NJ (EWR/KEWR) to San Juan, Puerto Rico (SJU/TJSJ) are illustrated in Figure 8.2. Contents of the major sections of this sample flight plan (header, waypoints, fuel and takeoff planning) are described in detail next.

Header

The flight plan header contains the flight plan summary and information concerning aircraft type and registration, the filed routing, planned cruise Mach/altitude and the en route ATC sectors to be traversed. In the sample flight plan, this information is as follows:

1. Flight plan summary: IFR flight plan for flight OAL1234, registration NXYZ from Newark (EWR) to San Juan (SJU) with takeoff alternate Washington Dulles (IAD) and destination alternate St. Croix (STX) – see later discussion.
2. Fuel summary: minimum required; release amount; planned fuel burn and reserves.
3. Alternate destination routing: SJU direct to STX via waypoint PALCO.
4. Aircraft type: B757-200 followed by codes defining aircraft equipage.

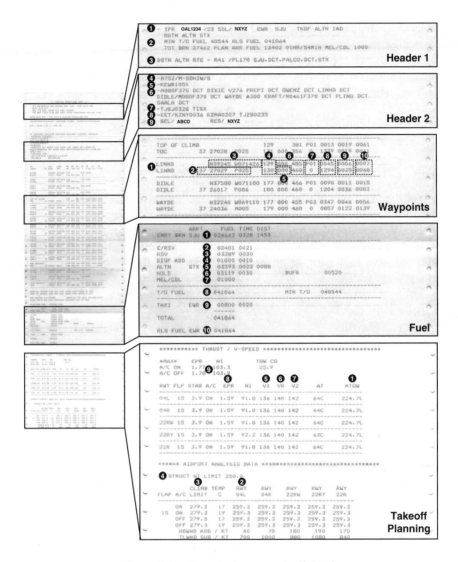

Figure 8.2 Excerpts from sample flight plan

5. Departure details: from Newark (ICAO code KEWR) planned at 10:51 Universal Time (UT).
6. Cruise Mach: 0.80.

   Cruise altitude: flight level (FL) 370 (37 000 feet altitude at standard pressure: see discussion later).

   Filed routing: direct to waypoint DIXIE; then via Victor airway V276 to waypoint PREPI; then direct to waypoints OWENZ, LINND, DIDLE (the oceanic entry point, after which maintain Mach 0.8 and FL370) and WAYDE; then via oceanic airway A300 to KRAFT (after which maintain 461 knots); then direct to waypoints PLING and SAALR.

7. Destination: San Juan (ICAO code TJSJ).
   Estimated en route flight time: 3 hours 28 minutes.
   Destination alternate: St. Croix (ICAO code TISX).
8. Estimated flight times to ATC sector boundary crossings: 36 minutes to NY sector (KZNY); 2 hours 7 minutes to Miami sector (KZMA); 2 hours 35 minutes to San Juan sector (TJZS).
9. Aircraft registration: NXYZ; SELCAL designation: ABCD (see later).

Waypoints

The body of the flight plan contains specific information corresponding to each waypoint in the filed route, including the flight level/altitude, winds aloft forecast, course/heading, Mach/airspeed/groundspeed, terrain elevation data, forecast turbulence, temperature deviation from standard, segment/cumulative distance, segment/cumulative flight time and segment/cumulative fuel burn. In the sample flight plan, the following information is given at waypoint LINND:

1. Waypoint ID: LINND.
2. Flight level: FL370.
   Forecast wind: 270° at 29 knots.
   Wind component relative to course: 25 knots tailwind.
3. Latitude/longitude of waypoint: N39°24.5′/W071°42.6′.
4. Magnetic course to LINND: 129°.
   Magnetic heading (course adjusted for wind): 130°.
5. Terrain height: 000 (sea level).
6. Mach: 0.80.
   True airspeed (TAS): 460 knots.
   Groundspeed (GS = TAS adjusted for wind component): 485 knots.
7. Temperature deviation from International Standard Atmosphere (ISA): +1°.
   Forecast turbulence index: 0 (mostly smooth or better).
8. Distance from previous waypoint: 45 nautical miles.
   Total distance remaining: 1294 nautical miles.
9. Segment time from previous waypoint: 6 minutes.
   Total flight time from departure point to LINND: 25 minutes.
10. Segment fuel burn: 700 lb.
    Total fuel used from departure to LINND: 6800 lb.

Fuel

One of the key factors in flight planning is determining the fuel load. It is also critical to finalize the fuel quantity required as early as possible because of the inherent delay in the ordering and appropriation of fuel. Considerations in determining the fuel load include: fuel requirements to destination including reserves (which vary depending on the type of flight, e.g., over water), destination weather and alternates, off-optimum speed or altitude requirements (which may be driven by marketing or ride conditions), ferrying fuel to destinations where it is cost effective and mechanical discrepancies of the aircraft. The fuel load will also affect takeoff and landing performance and may influence the payload the aircraft can carry. Information regarding fuel load in the sample flight plan is as follows:

1. Expected en route fuel/time/distance from EWR to SJU: 26 662 lb/3 hours 28 minutes/1453 nm.
2. En route reserves required for overwater operations: 2401 lb/21 minutes (10% of the flight time).
3. Regular reserves: 3289 lb/30 minutes.
4. Dispatch addition: 1000 lb/10 minutes (to account for possible en route chop as indicated in remarks).
5. Fuel to get to destination alternate (St. Croix (STX)): 3593 lb/23 minutes/88 nm.
6. Holding fuel: 3119 lb/30 minutes.
   Buffer fuel: 520 lb (amount of holding fuel that is allowed to be used before takeoff to account for unexpectedly long taxi times).
7. MEL/CDL fuel (757/767 fuel pump airworthiness directive adjustment): 1000 lb.
8. Takeoff fuel: 41 064 lb.
   Minimum takeoff fuel: 40 544 lb (takeoff fuel minus buffer fuel).
9. Taxi from gate to runway at EWR (including anticipated delays): 800 lb/20 minutes.
10. Total fuel load at release from EWR: 41 864 lb.

Takeoff Planning

The ability of a flight to generate revenue is driven by how much payload can be carried, which, in turn, is limited by aircraft performance. Takeoff performance must be evaluated once the load is determined and available runways considered. Planned takeoff performance data is normally calculated by the dispatcher and/or load manager during the preparation of the flight plan using computer-based tools.[9] The cockpit crew generally only have access to the results, although different runway/flap combinations from the baseline are available as needed (after gate departure). Adjustments to takeoff plans may also have to be made to deal with weather changes, and weight penalties associated with allowable equipment downgrades (discussed later) or runway contamination (e.g. snow, slush, standing water, rubber deposits, etc.). Other considerations include maximum runway and climb limited weights, and takeoff engine-out performance, specifically takeoff aborts and obstacle clearance.

The amount of payload that may be accommodated on a given flight is ultimately limited by the maximum takeoff weight given in (1) in the appropriate section of Figure 8.2. The maximum takeoff weight can be restricted by a number of factors including runway-limited takeoff weight (2), climb-limited takeoff weight (3), structural weight limit of the airframe (4) and the maximum landing weight. The runway-limited weight is derived from the most restrictive performance based on runway length, slope, obstacle clearance, brake energy and tire speed. The climb-limited weight is independent

---

[9] Some airlines provide computer-based tools to the flight crew and make takeoff, landing and load planning a flight crew responsibility.

of the runway used, and is based on the ability of the aircraft to climb at minimum angles with and without all engines operating normally. The structural weight is the maximum certified weight based on structural limitations, regardless of phase of flight or ground operations. The maximum landing weight is limited by landing runway length available and/or the ability of the aircraft to execute an aborted landing while still meeting minimum climb gradient requirements. The planned takeoff weight is often greater than the maximum landing weight and consideration must be given to the en route fuel burn (or fuel jettison, when available) to ensure that the aircraft weight is less than the maximum landing limited weight on touchdown.

Takeoff performance data also includes significant reference airspeeds, or "V-speeds," that are used by the cockpit crew for critical maneuvering and decision making. The takeoff decision speed, V_1 (5), is the maximum speed at which an abort can be initiated with adequate runway remaining for stopping the aircraft. Once the aircraft reaches V_1, there is sufficient speed that the aircraft can take off with one engine failed and the takeoff must continue. Any problems encountered after V_1 are resolved in the air or upon landing. The rotation speed, V_R (6), is the airspeed at which the nose of the aircraft is raised for the purpose of lifting off the runway. The takeoff safety speed, V_2 (7), is the target airspeed that ensures obstacle clearance if an engine fails between V_1 and V_2. All three V-speeds are provided in the flight plan performance section; however, after the performance data is updated with the final load values, they are reviewed by the captain and first officer and recalculated if needed.

In many cases, the actual planned takeoff weight is well below any of the previously mentioned limitations, or those imposed by aborted takeoff considerations. In these cases it is often beneficial to take off at reduced or "de-rated" thrust to minimize engine wear and noise impacts immediately surrounding the airport. The balanced field length for a given takeoff weight is defined as the distance required to accelerate to V_1 and safely stop the aircraft on the remaining runway or continue the takeoff so as to reach V_2 by 35 feet above the takeoff surface at the end of the runway. If the balanced field length is less than the actual runway available, the engine thrust setting (as measured by fan rotation speed (N1) or engine pressure ratio (EPR)) used for a departure (8) may be de-rated by a calculated amount up to 25% from the maximum available (9) while still meeting takeoff safety limits. The reduced takeoff thrust parameters are included in the performance data. It is, however, not always appropriate to perform a de-rated takeoff. It is precluded when there are reports of wind shear, tailwind, anti-icing fluid applied, runway contamination, equipment failures or for certain noise abatement purposes (e.g., when population distributions around an airport mean it is more important to climb as quickly as possible with maximum thrust followed by a thrust reduction when the higher population densities are being overflown further out). Often the decision to use maximum or de-rated engine power is not finalized until reaching the departure runway after taxi-out.

Alternate Airports

Although the intent of every departure is to land at the published destination airport, contingencies such as weather or traffic may require an alternate destination airport. In the sample flight plan, alternates and routes to get there are listed in the header section. Takeoff, en route and destination alternate airports are stipulated to satisfy certain weather requirements or routing limitations. Takeoff alternates are required whenever the option to return to the departure airfield is in question. The criteria for determining the necessity for a takeoff alternate is often driven by the possibility of downgraded operational status of the aircraft (engine-out approach minimums), or the fact that takeoff weather requirements are often less restrictive than landing weather requirements. Additionally, takeoff alternate options are normally limited to within a certain distance of the departure airfield (i.e., 360 nm). En route alternates are required when operational considerations dictate specialized contingency diversion procedures such as an engine-out, loss of cabin pressure over mountainous terrain, or diversions while operating over water. En route alternates and diversion decision making are discussed in further detail later in the chapter.

Destination alternate requirements are driven by forecast weather conditions at the airport of intended landing, and in cases where weather conditions are good and not expected to be a factor, no alternate may be required. The requirements to include destination alternate(s) in the flight plan are determined by the forecast weather and approach navigational facilities at the destination airport. If the weather is forecast to go below certain minimums, an alternate is stipulated in the flight planning process, including the extra fuel to fly from the destination to the alternate, plus 30 or 45 minutes of reserve fuel depending on the type of operation. The suitability of an alternate airport is also limited by the forecast weather and approach procedures available.

8.3.3 Pre-flight

The crew must determine the airworthiness of the aircraft and address any open issues before departure. The term "pre-flight" is typically used to describe the interior and exterior inspections of the aircraft, but in a general sense can be used to describe any activity involved with preparing the aircraft for departure. The aircraft inspection is usually divided among the cockpit crew and includes an exterior walkaround examination, interior cockpit setup and systems checks. These pre-flight inspections are outlined in a checklist[10] format (an example of which is given in Figure 8.3) which is used by the crew to aid in ensuring completeness and maintaining an acceptable level of standardization.

The exterior walkaround pre-flight consists of a visual inspection in which the crew member checks for obvious damage to the fuselage, engines, wings and flight control surfaces. Other important items include tire wear and pressure, brake wear indicators, absence of leaks or fluid on the ramp, condition of antennas, probes and lights, necessity for deicing and any other factors which may affect the safe conduct of the flight. A typical walkaround inspection can take anywhere from 8 to 20 minutes depending on the size and condition of the aircraft and the number of wheels/tires.

The ground servicing equipment available at a typical airport gate is shown in Figure 8.4. If the aircraft was completely powered down after the previous leg, a source of electric power and air-conditioning will need to be utilized to enable operation of aircraft systems during pre-flight. In most cases the options available include: the auxiliary power unit (APU), external power and external air. The APU is a small turbine engine which is usually located in the aircraft's tail cone section and is designed to

---

[10] Nearly all phases of flight are accompanied by a checklist of some type, which may be read aloud or completed silently by the crew.

757 / 767
Pre-Flight Inspection
12-17-03

The entire Pre-Flight Inspection is accomplished prior to every flight except for the items in Additional Checks. When maintenance has been performed, the pre-flight item(s) associated with the system, component, etc., should be checked. Also check that any associated c-b's that may have been pulled have been reset.

NOTES
- On all flights without Flight Attendants on board, determine that all galley drawers and doors are closed and latched prior to taxi.
- A 🟤 identifies a 757 only item and a ⑥ identifies a 767 only item. Items without an identifier apply to both airplanes.

Exterior Safety Inspection

Wheel Chocks	INSTALLED
Flight Control Surface Areas	CLEAR
APU Exhaust Area	CLEAR

Cockpit Preparation

Cockpit Safety Inspection

Battery Switch	ON (Guarded)
Standby Power Selector	AUTO
Electrical Panel	CHECK
Emergency Lights Switch	ARMED
Hydraulic Panel	SET
Gear Handle	DOWN and IN
Altn Flaps Selector / Switches	CHECK
Flap Handle / Flap Position	AGREE
Circuit Breakers	CHECK

Cockpit Initial Preparation

Starting the APU is at the Captain's discretion. Normally it will be started about ten minutes before departure. Earlier starts should be considered:
- When necessary to satisfy air conditioning or special electrical requirements
- On critical flights and international flights to avoid delays resulting from APU start problems

APU Power / External Power	ESTABLISH

-1-

IRU Mode Selectors	NAV
FMS Active Nav Data Base	CHECK
IRUs	INITIALIZE
Pneumatics and Air Conditioning	ESTABLISH
Parking Brake	SET
🟤 Power Transfer Unit (PTU) Switch	OFF (Guarded)
⑥ Bulk Cargo Heat Selector	NORM / VENT
Flight Recorder	NORM
Service Interphone Switch	OFF
⑥ Reserve Brakes and Steering Reset Panel	CHECK
Observer's Audio Selector Panels	CHECK
Log Book and Flight Forms	REVIEW
Shoulder Harnesses	CHECK

Cockpit Area Inspection

Crew Life Vests	CHECK ABOARD
Cockpit Emergency Equipment	CHECK ABOARD

Walk-Around Inspection

Walk-Around Inspection	ACCOMPLISH

Cockpit Final Preparation

Yaw Damper Switches	CHECK ON
Electronic Engine Control Switches (if equipped)	ON (Guarded)
Overhead Annunciator Panel	CHECK
Evacuation Command Switch	OFF (Guarded)
Passenger Oxygen Switch	BLANK (Guarded)
Voice Recorder	TEST
Ram Air Turbine Switch	BLANK (Guarded)
🟤 Engine Limiter Control Switches (if equipped)	ON
Engine Start Panel	SET / CHECK
⑥ Fuel Jettison Panel (if equipped)	SET
Fuel Panel	SET
Wing Anti-Ice Switch	BLANK
Engine Anti-Ice Switches	BLANK
⑥ Cargo Heat Switches	OFF
Window Heat Switches	ON
No Smoking Signs	ON
Fasten Seat Belt Signs	OFF
Pressurization System	SET
🟤 Equipment Cooling Switch	BLANK
⑥ Equipment Cooling Switch	AUTO
Magnetic Standby Compass	CHECK

-2-

🟤 Reserve Brakes Switch	BLANK
⑥ Reserve Brakes / Steering Switch	BLANK
Standby Flight Instruments	CHECK
Caution and Warning System	CHECK
Standby Engine Indicator	AUTO
Auto Brakes Selector	OFF
EICAS Display Switch	ENGINE
EICAS Display Switch	STATUS
EICAS Computer Selector	AUTO
Thrust Reference Selector	BOTH / IN
HSI Hdg Ref Switch	NORM
Altn Gear Extend Switch	OFF (Guarded)
Gnd Prox / Flap Ovrd Switch	BLANK (Guarded)
Gnd Prox / Config Gear Ovrd Switch	BLANK (Guarded)
Spoiler Handle	DOWN
Reverse Levers and Throttles	DOWN and CLOSED
Stabilizer Trim Cutout Switches	NORM (Guarded)
Fuel Control Switches	CUTOFF
Fire Handles	IN
Fire Bottle Discharge Lights	EXTINGUISHED
Transponder / TCAS	TEST / SET
Aileron and Rudder Trim	SET
Fuel Panel, Quantity and Distribution	SET and CHECK
ACARS	DATA / SET

Additional Checks

Accomplish the following on the first flight of the day.

Position & Anti-Collision Lights	CHECK
Standby Power	TEST
Indicator Lights Test Switch	PRESS
Fire and Overheat Detection System	TEST
Fire Extinguisher and Overwing Slide Squibs	TEST

-3-

Figure 8.3   Sample pre-flight checklist

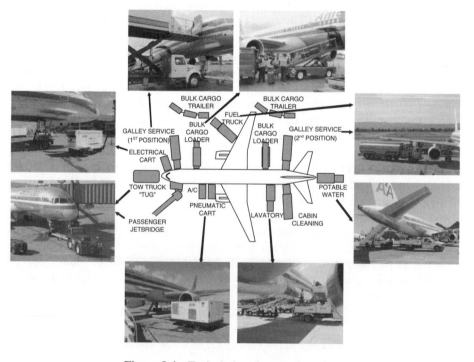

Figure 8.4   Typical aircraft ground services

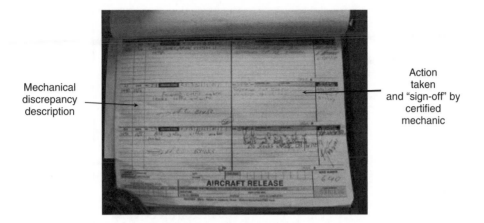

Mechanical
discrepancy
description

Action
taken
and "sign-off" by
certified
mechanic

Figure 8.5   Example mechanical logbook page

meet the electric and pneumatic demands of the aircraft when the main engines are shut down.[11] However, if external air and power are available for the aircraft, they will often be used in lieu of the APU to save on fuel, maintenance costs and to reduce emissions. External electric power may be provided to the aircraft either by a cable from the jetbridge, or from an electrical cart. External low-pressure conditioned air can be provided through a flexible duct to the belly of the aircraft or from a dedicated unit mounted on the jetbridge. External high-pressure air may be provided to the aircraft by one or more "air start" carts for the purpose of starting an engine. Optimally, the station will provide external power and conditioned air from the jetbridge units until 5–10 minutes prior to departure when the crew will switch over to internal (APU) power.

Once electric power and air are available on the aircraft, interior pre-flight and cockpit "cleanup" checklists are conducted to confirm that each system is operational. Some of these systems employ self-tests (e.g., fire warnings), while others only require verification that acceptable parameters are satisfied (e.g., oil quantity). Some systems may require an initialization procedure, such as the inertial reference unit (IRU) alignment from which on-board navigation can be based. The cabin crew also has certain pre-flight duties which include checking the status of catering and cabin emergency equipment, as well as general cabin condition.

The pre-flight also includes verification that all required manuals and paperwork are on board and complete. The aircraft mechanical logbook (Figure 8.5) serves as a means for flight and cabin crews to convey mechanical discrepancies to station maintenance personnel and subsequent flight crews. Any discrepancy entered into the logbook must be balanced with an entry by a certified aircraft mechanic[12] who either resolves the problem or defers it according to specified guidelines. Some items can be deferred based on time (hours of flight, or days/weeks), type of maintenance available, or whether they are listed

---

[11] The APU may also provide supplementary air/electric power during abnormal situations such as engine/generator failure or high-altitude operations.

[12] If company maintenance is unavailable at an airport station, an outside contractor is used. Maintenance personnel handle discrepancies and perform required inspections and logbook endorsements which may be mandated for certain international flights.

Section 6
Page 32-3
7-28-89

Minimum Equipment List Manual

B-757/767

SYSTEM 32 LANDING GEAR	NORMAL COMPLEMENT OF EQUIPMENT	MINIMUM EQUIPMENT REQUIRED FOR DEPARTURE	ADVANCE NOTIFICATION TO DISPATCH REQUIRED	FLIGHT CREW PLACARDING PERMITTED	
-5 Alternate Anti-skid Valves	4	*	YES	NO	* (M)(O) Any or all may be inoperative provided: -Manual braking capability on alternate system is confirmed per MM 32-42-99 prior to first flight with this condition, and -If failure affects normal anti-skid system, anti-skid is placarded per MEL item 32-4. FLIGHT CREW NOTE: Refer to OM Section 16, Abnormal Procedure titled ANTI-SKID. MAINTENANCE NOTE: Braking of one wheel may be deactivated by placarding MEL item 32-2 or 32-4.
-6 ANTI-SKID Light	1	*	-	NO	* (M) May be inoperative and anti-skid braking used provided anti-skid system is determined to be operative per MM 32-42-99 prior to first flight with this condition. Not required for inoperative anti-skid when placarded per MEL item 32-4.
-7 Autobrake System	1	*0	-	YES	* May be inoperative.
				NO	(M) If AUTO BRAKES light illuminates with AUTO BRAKE selector switch OFF, the autobrakes valve must be verified closed (prior to first flight with this condition) or module removed per MM 32-42-99.
-8 Parking Brakes System ALWAYS REQUIRED	1	1	-	-	

Figure 8.6 Minimum Equipment List (MEL) sample page

in the Minimum Equipment List (MEL). The MEL identifies the components which may be inoperative on a given aircraft while still maintaining legality for dispatch as well as the deferral rules. An example from an MEL is presented in Figure 8.6.

Crew responses associated with MEL items range from simple awareness to complex critical procedural changes. An example of a trivial MEL item would be a failed navigation light (two are available), which involves no performance penalty or crew procedures (except to remember to select the operative one). However, the deactivation of a wheel brake would be considered a critical MEL item. In this situation only one deactivation is allowed and dispatch must include a major weight penalty in the takeoff planning. It is the cockpit crew's responsibility to familiarize itself with any procedural changes that result from an MEL item. The Configuration Deviation List (CDL) is similar to the MEL, but references airframe components that are more structural in nature (e.g., a missing flap track fairing).

Once the exterior/interior inspections and system checks are complete, the crew undertakes the Flight Management System (FMS) and autoflight initialization programming to allow their use during the flight. Modern aircraft have extensive autoflight capabilities that allow many of the navigation[13] and performance optimization tasks to be handled automatically if desired. Autoflight initialization and FMS programming are conducted during the pre-flight phase.

[13] Lateral navigation over the ground is provided by LNAV (Boeing) or NAV (Airbus) automation modes, while vertical navigation is provided by VNAV (Boeing) or PROFILE (Airbus) automation modes.

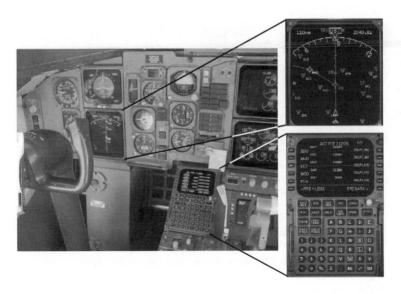

Figure 8.7   FMS Control Display Unit (bottom right) and Navigation Map Display (top right)

Use of automation has been the focus of recent human factors research which has resulted in air carriers adopting "automation policies." These policies attempt to address workload and situation awareness issues which have proven to be contributing factors in some aircraft accidents and incidents. In general, operating philosophies are being included in flight training that emphasize use of appropriate levels of automation and confirmation of inputs. Basically there are three automation levels: manual control (hand flying), tactical modes (directly dialing in flight parameters such as heading, altitude, airspeed) and strategic lateral and vertical navigation modes (in which the aircraft is flown automatically by the autopilot along a flight path programmed into the FMS). Pilots must understand the functionality of each level, the importance of correctly setting and confirming flight parameters, and when the use of a given level is most appropriate. Some procedures mandate high levels of automation (e.g., autoland), while at other times it is beneficial to manually control the aircraft in order to maintain proficiency in basic flying skills. Although the required trajectory parameters are usually programmed into the automation, use of the autopilot is often discretionary and most pilots will make a manual takeoff and engage the autopilot sometime during climb-out or just after level-off. The autopilot is typically disengaged during descent anywhere from terminal area arrival to final approach depending on the procedure being conducted.

The appropriate flight plan information can be entered manually into the FMS via the Control Display Unit (CDU)[14] illustrated in Figure 8.7. Various menu interfaces and databases of standard procedures and navigation aids (NAVAIDs) are available via the CDU to assist with the programming tasks. Some airlines have information systems which allow information required to initialize the autoflight systems to be uploaded automatically via the Aircraft Communication and Reporting System (ACARS) datalink unit. When programming is complete, the crew performs a route check, where one crew member

---

[14] The CDU is both a keypad and multi-line text display that the flight crew uses for input/output to the FMS.

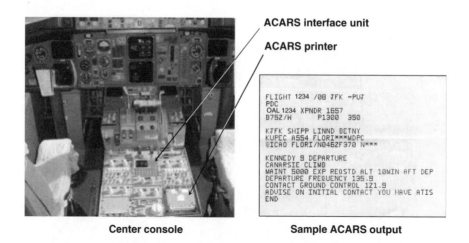

| **ACARS interface unit** |
| **ACARS printer** |

```
FLIGHT 1234 /08 JFK -PUJ
PDC
OAL 1234 XPNDR 1657
B752/H        P1300   350

KJFK SHIPP LINND BETNY
KUPEC A554 FLORI***MDPC
@ICAO FLORI/N046ZF370 N***

KENNEDY 9 DEPARTURE
CANARSIE CLIMB
MAINT 5000 EXP REQSTD ALT 10MIN AFT DEP
DEPARTURE FREQUENCY 135.9
CONTACT GROUND CONTROL 121.9
ADVISE ON INITIAL CONTACT YOU HAVE ATIS
END
```

Center console   Sample ACARS output

Figure 8.8   ACARS unit and sample printout

reads the FMS waypoints from the CDU and steps through the map depiction on the navigation display (also shown in Figure 8.7), while the other compares the waypoints read to the paper copy of the flight plan.

In addition to the trajectory-based flight plan routing and waypoint data, planned aircraft performance and operating weights are entered and checked against the load information. The V-speeds and configuration cleanup speeds are often depicted directly on the flight instruments via movable indicators or "bugs" (visible as white markers on the airspeed indicator at the top left of the panel in Figure 8.7), or by using icons on digital flight displays.

The cockpit setup also includes initialization of the on-board data communications system. Many airlines use the ACARS system which may be co-located with the FMS as an ACARS page on the CDU, or as a standalone terminal: see Figure 8.8.

ACARS typically utilizes a very high-frequency (VHF) datalink and alphanumeric interface to facilitate company-specific communications between the aircraft and AOCC. When used, there are a minimum set of ACARS messages that are downlinked for every flight including the OUT time (brakes released, cabin doors closed), OFF time (weight off landing gear after takeoff), ON time (weight on landing gear after landing) and IN time (cabin door opened). These time events are automatically sent and are used in determining on-time performance, arrival estimates, crew member compensation, and a number of other statistics. The engine monitor log (giving engine parameters) is also automatically downlinked by many carriers and is useful in determining engine performance and maintenance requirements. The crew can also downlink weather and position reports, estimated arrival times, holding and diversion notification, delay categories and times, aircraft maintenance requests and virtually any free text message. Uplinked communications include FMS flight plan routing and performance data during pre-flight, updated takeoff performance information, flight closeout data (final actual payload, fuel and takeoff data), messages from dispatch including weather updates, pilot reports (PIREPs),[15] and arrival

---

[15] Pilot Weather Reports (PIREPs) are reports of inflight weather conditions issued by pilots to ATC or the company, which are passed along to other flight crews or facilities.

gate information including availability of ground power. Recently, more ATC datalink functions have been utilizing the ACARS interface. Examples include the Pre-Departure Clearance (PDC, as seen in the sample output in Figure 8.8), international overwater and North Atlantic Track (NAT) clearances, Automatic Terminal Information Service (ATIS)[16] and Controller–Pilot Data Link Communications (CPDLC). In general, the use of ACARS by air carriers satisfies the requirement that their aircraft are continuously able to be contacted by dispatch during the entire flight. For overwater segments which are beyond VHF radio and radar coverage, utilization of high-frequency (HF) communications and selective calling (SELCAL)[17] are standard procedure, although Automatic Dependent Surveillance – Broadcast (ADS-B) systems where aircraft transmit information regarding their current state (e.g., position, speed) and intentions (e.g., programmed trajectory) are becoming available.

As the departure time approaches, a fuel slip is provided to the crew by the fueler to corroborate the fuel quantity and distribution (between different fuel tanks) with the flight plan and on-board sensors. The fuel slip also indicates which type of jet fuel was loaded, which may be a factor in cold weather operations due to minimum fuel temperature and fuel freezing point. Other required flight documents may include: water and lavatory servicing verification, security inspection confirmation, any paperwork accompanying restricted cargo (e.g., hazardous material (HAZMAT) slip with required signatures), armed passengers (including prisoner and deportee escorts, law enforcement personnel and air marshals) and customs declarations for international flights. In addition, jumpseat[18] riders require identification confirmation from their designated agencies.

Communication between the cockpit and cabin crew members is critical to the safety and efficiency of the flight. At some point during or before pre-flight activities and passenger boarding, the captain conducts a briefing with the purser or lead flight attendant. This includes standard information covering en route flight time and destination weather, as well as taxi-out time (in the case of a short taxi, the flight attendants must start the safety video/demonstration as early as practicable), security issues and alerts, ride conditions and turbulence, inoperative cabin components, requirement of overwater flight passenger life vest demonstrations, augmented crew, crew meal service and any other relevant safety or operational issues. The captain may also discuss adherence to the sterile cockpit period in which access to the flight deck is limited to reduce distractions during critical flight phases, nominally anytime the aircraft is below 10 000 feet above mean sea level (MSL). During the sterile period the cockpit crew is restricted from performing any duties or activities that are not directly required for the safe operation of the aircraft (e.g., eating meals, non-essential conversation, etc.). Exceptions to the sterile period include (as

---

[16] ATIS is a recorded message available via VHF or datalink detailing airfield weather conditions, runway/approach operations and other pertinent safety-related information in high-activity terminal areas. ATIS is usually updated every hour or when a change in the weather conditions or operational status of the airport dictates otherwise.

[17] The SELCAL function of HF radios allows ATC to initiate discrete contact (typically via third-party routers such as ARINC, Gander Radio, Shanwick Radio, etc.) with individual aircraft and must be checked upon initial contact with each facility. When communication with the aircraft is required, a distinctive audio tone is transmitted which activates an alerting chime in the cockpit and signals the crew to respond on the assigned HF channel. With the availability of SELCAL, the crews are not required to continuously monitor the frequency, which in the case of the continuous static of HF would be both distracting and fatiguing.

[18] Air transport cockpits normally contain one or two extra seats or "jumpseats" that are used primarily by augmented crew members, check pilots, FAA inspectors and other officials. In addition, other pilots may occupy the jumpseat with the permission of the captain, which enables them to commute when the cabin is full.

briefed) cabin emergencies requiring cockpit notification and short flight legs where the cruise altitude is below 10 000 feet MSL. In some terrain-critical regions such as South America, the sterile period may continue to as high as 25 000 feet MSL.

8.3.4 Pre-departure

As the scheduled departure time approaches, the captain, lead gate agent and ground crew chief coordinate their efforts to see that all pre-departure requirements are met. The pilots finalize the FMS and autoflight parameters by obtaining an update on weather conditions and runway utilization through ATIS. In addition, the crew must receive confirmation of the flight's routing from ATC. Prior to the scheduled departure (usually at least a few hours before), the airline's dispatch office files a requested routing based on the flight plan optimization with ATC. Approximately 20 minutes prior to departure, the ATC route clearance[19] is requested, preferably through the ACARS PDC function previously discussed. The ATC route clearance received by the crew may differ from the filed routing and the changes must be addressed (fuel/performance/dispatch considerations) and reprogrammed.

Since the takeoff performance parameters are calculated an hour or more prior to takeoff, any changes in the weather, runway usage and loads must be closely monitored and accounted for right up to the takeoff roll. Once the clearance is received, the crew can perform the "before starting engines" checklist. At approximately 10 minutes prior to departure, the captain turns on the "Fasten Seat Belt" sign which signals the flight attendants to ready the cabin for departure and deliver the requisite public address (PA) announcements.

In addition to possible routing changes, ATC may also adjust the planned departure time as a result of current airspace demands. The ATC clearance may include a "gate hold" or expected "wheels-up" time due to traffic congestion, routing conflicts or adverse weather conditions. Gate hold and departure delay procedures vary from airport to airport and may actually be issued to station operations personnel before the crew arrives, or by ATC directly to the crew approaching the takeoff runway during taxi-out. A PDC gate hold is most common at busy US airports and may include instructions for the crew to contact a separate gate hold frequency for further information and to monitor any changes. Once a hold is issued, the crew must decide on the appropriate action depending on the anticipated length of the delay and specific station requirements. If the delay is substantial (and not likely to decrease), the agent or captain may elect to postpone boarding or offload boarded passengers. Ramp and gate capacity requirements may dictate, however, that the hold be absorbed off the gate. In those cases, passenger boarding and aircraft servicing must be completed before the aircraft is relocated from the ramp area to another location on the airfield to wait out the delay. The pilots must account for any potential additional fuel consumption due to extended APU usage, or long taxi routings with one or more engines shut down.

It is desirable for the agent to complete the passenger boarding process as early as possible in case special situations or security issues need to be addressed, including positive bag matching. If any passengers require removal because of illness or misconduct,

---

[19] Airlines typically operate their aircraft under Instrument Flight Rules (IFR) as opposed to Visual Flight Rules (VFR). When operating under IFR, an ATC clearance is required along with special aircraft and air crew certification. Operating under an IFR flight plan provides positive ATC for the entire flight which enables flight operations into weather conditions which would preclude a VFR-only flight.

or are not on board as the flight nears departure time, all of their checked bags may have to be removed. This can be a time-consuming process if the location of the bag is unknown and/or "buried deep" in the cargo compartment. Airline performance criteria tend to put pressure on gate agents to achieve an "on schedule" OUT time; however, in some cases, the captain can authorize a late departure if he or she determines that the lost time can be made up en route and a scheduled arrival can still be achieved or to satisfy customer service requirements. If conditions permit and expected passengers are on board, an early departure (within certain limits) may be initiated.

Once all passengers are on board and seated, the agent coordinates closing the aircraft doors with the captain and lead flight attendant. In the USA, the seated passenger requirement exists only at the gate. If a passenger leaves his or her seat during taxi-out the aircraft is usually not stopped as this would disrupt the ground traffic flow (although the flight attendant is still required to inform the cockpit).

In order to prepare the aircraft for movement, the ground crew completes the baggage and cargo loading, including late bags, and closes the cargo doors. If necessary, any required external power or air is removed from the aircraft, unless required for engine start. The tug is connected to the aircraft via a towbar unless a "powerback"[20] is planned. The flight deck crew performs the "just prior to push-back" portion of the checklist which includes, among other things, confirmation that all the doors are closed and that the anti-collision (red flashing) beacon is operating. At this time the flight attendants arm the escape slide mechanism of the entry doors in case a ground evacuation becomes necessary.[21] When the checks are complete and the aircraft is ready for gate departure, the ground crew becomes the push-back crew. Typically, the captain communicates to the tug driver (or other ground crew member) through an "interphone" link, while the first officer communicates to ramp control and/or ATC via the VHF radio.

8.3.5 Gate Departure

Once the agent moves the jetbridge out of the way, the push-back crew advises the cockpit that the wheel chocks are removed and that it is safe to release the parking brake. The captain acknowledges release of the parking brake and signals the first officer to call ramp control (or ATC, depending on local requirements) for push-back clearance. When received, the aircraft is pushed out of the gate area (Figure 8.9) and engines are started when the cockpit crew is advised by the push-back crew that the area is clear. At this stage, any late bags may be loaded only after contacting the flight deck.

Under certain weather conditions, ice or frost may be present on the airframe or airfoil surfaces which require removal before takeoff. In situations where deicing or anti-icing is required, the captain delays the engine start while the push crew positions the aircraft in a designated deice location which may be just off the gate (see Figure 8.10). In some cases only the engine inlets are deiced and the aircraft taxis to another location for completion.

---

[20] During a powerback, the aircraft backs out of the gate using reverse thrust. Powerback operations are normally planned if there is a logistical advantage in doing so (such as if lengthy delays would be incurred waiting for a tug to become available), and must be balanced against the adverse effects of engine noise and wear, and, to a lesser degree, fuel consumption.

[21] Once armed, opening the door activates the auto-inflate device in the slide. The cabin crew must be very diligent about the arming status of the door since inadvertent slide deployment can result in major ground delays and possible injuries to passengers, agents or flight crew.

Figure 8.9 Push-back

Figure 8.10 Aircraft deicing
Source: Photo courtesy Den Pascoe, used with permission, www.airliners.net

At many airports, secondary deicing locations are established nearer to the departure runway in order to keep the time to takeoff below the holdover time (the length of time (in minutes) that the anti-icing fluid is effective and is determined by the flight crew from tables in their flight manuals). The time may vary according to temperature, type and intensity of precipitation, and type and concentration of fluid used. Deicing fluid is used to remove snow and ice from the aircraft and is normally a mixture of glycol and hot water. Anti-ice fluid is applied after all contamination has been removed and is used to prevent further build-up of snow or ice during taxi. This fluid is specially designed to shear off under aerodynamic forces during the takeoff roll. Once the anti-ice application is complete, the icing coordinator advises the cockpit crew when the holdover time begins. It is then the responsibility of the captain to monitor the holdover time against takeoff time, and/or accomplish visual inspections to ensure no further contamination is taking place. If the delay before takeoff is too long, the aircraft may have to return to a deicing location to be retreated.

After the engines are started and the towbar is disconnected, the captain gives the guide crew member permission to disconnect the interphone headset. The guide crew member then steps into a position that is visible from the flight deck, presents the nulling pin (used to disable the aircraft's nosewheel steering system during push-back) and gives a salute which confirms the ramp area is clear to taxi. The captain acknowledges the salute and the first officer calls for taxi clearance. Once clearance is received, the captain begins the taxi-out only after both pilots have visually checked outside and verbally announced "clear left" and "clear right."

8.3.6 Taxi-out

As in the case of push-back, anytime ground movement is initiated, permission must be received from the controlling authority. At some point before the aircraft leaves the ramp area, the first officer contacts ground control to get taxi clearance and routing to the active runway. Operational considerations (such as high takeoff weight) may dictate the request for a special runway which can result in a taxi and/or takeoff delay while ATC works out a modified route and sequence. During the taxi, the load closeout is received via ACARS or by VHF radio. The load closeout typically includes finalized aircraft and fuel weights, stabilizer trim settings, center of gravity data, passenger count, cargo loading, live animal and security information. The FO uses the updated information to calculate finalized takeoff performance data, either in the CDU or by reference to flight manuals. The FO will also reset the stabilizer trim and set takeoff reference speeds through the bugs on the airspeed indicators. In cases where the closeout weights are greater than planned, adjustments may have to be made to the flap and/or power settings, or an alternate runway may be required. Similarly, any wind or temperature changes may also need to be addressed.

Once the closeout information is processed, the crew completes the "taxi" and "before takeoff" checklists. At some point, the captain conducts a takeoff briefing which includes which pilot will be making the takeoff,[22] initial heading, altitude and departure procedure requirements, obstacle clearance and noise abatement issues, airport elevation and the normal cleanup altitude.[23] In addition, the briefing must address runway abort considerations, engine-out procedures and associated cleanup altitudes, and emergency contingencies requiring return to the departure point or other proximate landing options.

In situations where there will be a long taxi from the gate to the departure runway, single engine taxi procedures may be utilized in the interest of fuel conservation and emissions reduction. If there are numerous departures ahead in sequence for takeoff, it is desirable for the captain to make a PA announcement informing the passengers and cabin crew of his or her best estimate of the length of the delay. This is typically done by counting the number of aircraft ahead in the takeoff queue. If the delay is significant, the company may have to be updated via ACARS or VHF radio with a new estimated time "OFF" (ETO). As the aircraft approaches the departure end of the runway, the captain makes a departure PA announcement to inform the flight attendants that the takeoff is imminent and they should secure themselves at their stations. The captain must ensure that the passenger briefing (video) has been completed, which may be a factor in short taxi-out situations. In addition, the crew needs to verify that minimum fuel requirements remain satisfied and that FAA crew duty time limitations have not been exceeded.

8.3.7 Takeoff

In order to make most efficient use of runway resources, the local tower controller often issues a "position and hold"[24] clearance to an aircraft in preparation for final takeoff clearance. This allows the aircraft to taxi into position and hold on the departure runway

---

[22] Pilots normally alternate flight legs unless other requirements dictate.

[23] The cleanup altitude is the height where climb pitch is reduced and the aircraft is accelerated so as to enable flap/slat retraction and reduce the overall drag.

[24] "Position and hold" in some countries, "line up and wait" in others.

while waiting for other traffic to vacate or cross, runway restrictions or an ATC-issued departure time. If this hold time is not required or a departure needs to be expedited, the tower may clear the flight for takeoff without holding in position. At this time the crew makes final checks of the wind/weather and the presence of runway contamination. If the flight is following the departure of a large aircraft, adequate wake separation requirements must be assured by confirming that an acceptable interval of time has elapsed before commencing the takeoff roll (this requirement is normally handled by ATC as part of the takeoff clearance, but usually corroborated by flight crew timings).

Once the takeoff clearance is received, the pilots' roles of captain/FO change to "Pilot Flying"/"Pilot Monitoring"[25] (PF/PM) in order to accomplish the procedures commensurate with which pilot is flying the leg. Even if the captain assumes the PM role, at all times he or she is still PIC responsible for the flight and may choose to take over the PF role at his or her discretion. Certain weather conditions (low visibility) or crew experience levels may dictate that the captain remain PF during some or all of the flight.

The engine throttle levers are pushed forward and during the resulting takeoff roll the crew monitors the aircraft centerline tracking, engine parameters and conditions both inside and outside of the aircraft. The PM calls out each V-speed as part of the normal procedure. Should a critical problem occur before the abort decision speed, V_1, the takeoff is rejected and the aircraft is stopped on the runway. Once the aircraft is brought to a standstill, the crew performs the "abnormal" checklist including notification of the tower, which is especially important in low-visibility conditions. Depending on the problem, the crew may require fire and rescue emergency equipment, maintenance/tow-in or passenger ground transportation (in the case of evacuation or deplanement by airstairs). In any event, the tower needs to be appraised of the situation so that it can accommodate the runway closure. Often high-speed aborts are accompanied by multiple tire failures (due to heavy braking) resulting in extended periods of runway unavailability while any contamination is removed.

An uneventful takeoff is followed by a normal initial climb-out which includes cleaning up the aircraft (gear raised, flaps/slats retracted) while conforming to any noise abatement departure procedure and/or obstacle requirements. It is also important (and in accordance with procedure) that, regardless of the operational status of the Traffic Alert and Collision Avoidance System (TCAS),[26] at least one crew member accomplishes a "traffic watch" (heads up looking outside the aircraft).

8.3.8 Terminal Area Departure

The climb flight profile is determined by both ATC/airspace requirements and performance characteristics which may be aircraft specific. When clear of the immediate airport traffic area, the aircraft is accelerated to maximum low-altitude climb speed (nominally 250 knots below 10 000 feet MSL in the USA) unless a restriction has been issued by ATC. Terminal area airspace may be very complex and certain standard procedures have been developed for both departing and arriving flights at high-density locations. During climb-out the flight typically conforms to a standard Departure Procedure (DP) (also

---

[25] "Pilot Monitoring" is sometimes known as "Pilot Not Flying" (PNF).

[26] TCAS provides alerting and proximity information to assist the crew in visual acquisition of other aircraft. The resolution function of TCAS provides the pilots with vertical maneuvering guidance to avoid conflicting traffic.

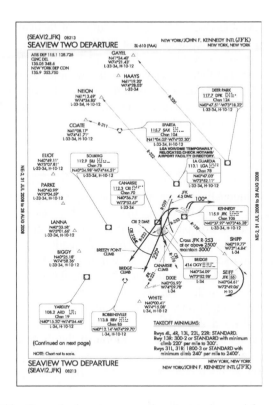

Figure 8.11 Example departure procedure (not to be used for navigation)

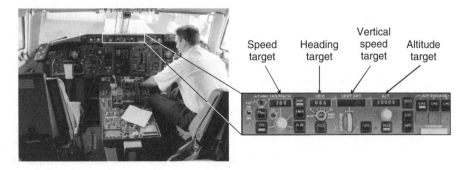

Figure 8.12 Autopilot Mode Control Panel

commonly called a Standard Instrument Departure (SID)), an example of which is given in Figure 8.11.

ATC commonly provides radar heading assignments ("vectors") to shorten the ground track or to provide traffic spacing, which are often implemented through an autopilot Mode Control Panel (MCP) when available, as shown in Figure 8.12. This interface is used in many flight phases when specific altitude, heading and/or speed target values are required as communicated to the cockpit crew from ATC in order to achieve separation, sequencing or efficiency objectives. This is an example of tactical automation described earlier, as compared to strategic control via the FMS.

The flight attendants typically commence their in-flight service duties during climb-out, unless the captain has briefed them otherwise, or the weather and/or ride conditions are prohibitive. When 10 000 feet MSL is reached, the aircraft is accelerated from 250 knots to the optimal climb speed, which can range anywhere from 270 to 350 knots. In addition, the flight attendants are chimed to indicate the end of the sterile cockpit period.

8.3.9 Climb

The dynamics of the flight environment, including accommodation of ATC directives, require the crew to continuously monitor aircraft performance in order to realize the best possible flight profile. At some point during the climb, the cockpit crew checks the FMS and/or performance charts to compare the optimal and maximum cruise altitudes with the planned data and desired cruise Mach. This information is used to coordinate an optimal cruise altitude and speed with ATC accounting for the fuel consumption versus time priorities (which were factored into the FMS-calculated optimal trajectory based on the cost index from the flight plan described earlier). Other factors include wind data and ride (turbulence) conditions, en route convective weather, MEL contingencies and traffic-induced speed restrictions. Winds aloft notwithstanding, in most cases higher altitudes provide for more efficient engine operation. If the flight is restricted to a lower altitude due to weather or traffic, the crew must consider the effects on total fuel burn and reserves. In addition, some aircraft types are more fuel sensitive to off-optimal cruise Mach than others, which may also limit the cruise altitude options.

As the aircraft climbs through the transition altitude[27] (18 000 feet MSL in the USA but can be as low as 4000 feet in other parts of the world), the crew resets the altimeter referencing from a local barometric pressure setting to the standard atmospheric pressure reference (1013 mbar or 29.92 in Hg) so that all aircraft are using a common pressure reference. The altitudes are then measured in flight levels (FLs) of hundreds of feet. Depending on the outside air temperature and indicated airspeed, the aircraft typically changes over to Mach reference[28] between FL270 and FL330. Above FL290, eastbound and westbound cruise levels providing for 1000 feet vertical separation are available for those aircraft that meet the equipment requirements of RVSM (Reduced Vertical Separation Minimums), otherwise 2000 feet vertical separations are required.

Passenger-related activities during the climb include beginning the meal and/or beverage service, delivering any marketing PA announcements and activating any entertainment systems. In addition, the captain usually makes a PA describing en route flight time

---

[27] Transition altitude and transition level are determined by the local governing authority and may vary by location and/or time of day based on terrain considerations or other airspace limitations. Below these heights, barometric altitude is referenced to the local pressure setting (or "QNH") which is issued to the crew via ATIS or ATC. Above these heights, the mean standard barometric pressure is used and the aircraft height above sea level is depicted in flight levels (FL180 vs. 18 000 feet MSL). The transition altitude is used during the climb (altitudes to flight levels) while the transition level is used during descent (flight levels to altitudes), although they are often identical such that the distinction is not always observed.

[28] The airspeed/Mach transition occurs to account for the difference in the nature of the critical limits at low and high altitudes. Most cruise performance data is specified in terms of Mach number because the critical limits during this phase are due to controllability issues associated with transonic flows over portions of the wings. At lower altitudes, the critical limits are associated with pressure loads on the airframe governed by the true airspeed.

and weather conditions, points of interest, arrival estimate, destination weather and, if applicable, any information concerning the presence of an augmented crew. Seat belt sign usage is at the captain's discretion and is typically activated in the presence of adverse ride conditions, or at the flight attendants' request such as during the meal service. The segment of the captain's PA which informs the passengers that "while in their seats they are to keep their seat belts fastened" is included by many airlines as a standard procedure and a mandatory disclaimer.

8.3.10 Cruise

As cruise altitude is reached, the power settings/Mach target are established. The crew also performs various administrative duties, including downlinking any departure delay ACARS codes and recording the engine monitor log (if not automated). The aircraft is usually equipped with at least two VHF transceivers and, if overwater certified, HF radios. VHF radio management usually requires one tuner to be set to the current ATC frequency, while the other is utilized for company communications or to maintain a listening watch on the universal emergency channel (121.5 MHz). If the flight extends beyond line-of-sight VHF range (oceanic and low population regions such as parts of South America), the HF units and SELCAL are used for required communications, including position reporting. Satellite communications (SATCOM) are also used where available for both ATC and company communications. SATCOM systems offer the benefit of worldwide communication coverage without the signal degradation, time-of-day variability, and other deficiencies associated with HF, but are often more expensive. In addition, when out of VHF contact with ground facilities, the crew typically maintains a listening watch on the air-to-air frequency of 123.45 MHz. This channel is used to pass along operational information such as ride reports and en route weather directly between aircraft.

During cruise, the crew must maintain a time/fuel log in order to compare planned time and fuel burn performance with the Actual Time of Arrival (ATA) and Fuel On Board (FOB) over each flight plan waypoint. The baseline departure (OFF) time and takeoff FOB are used to generate the Estimated Time of Arrival/Estimated Fuel On Board (ETA/EFOB) log which is usually very accurate. Consideration must be given by the cockpit crew to the possible causes of any deviations from the waypoint ETAs/EFOBs (including fuel imbalance) and the effect on the destination arrival time and fuel. Potential sources of time/burn variation include winds aloft greater or less than forecast, cruise speed or altitude being different than planned or mechanical problems such as a fuel leak. The cockpit crew also continuously evaluates altitude options. As the aircraft weight decreases due to fuel burn, the optimum cruise altitude typically increases due to better engine efficiency at higher altitudes. Available altitude options may be limited by ATC, or the 2000/4000-foot vertical increments available for a given direction of flight associated with the nominal vertical separation requirements, resulting in step–climb options. The decision to climb must include the effects of head/tailwinds and ride conditions (icing may also be a consideration for lower cruise altitudes). Most often, the sources of information available for cruise-level decision making include PIREPs from other flights, ATC, the crew's own experience, dispatch and the flight plan. In some instances, descent to a lower cruise altitude may be preferable to take advantage of more favorable winds, better ride conditions, or in the case of a traffic issue, a more desirable cruise Mach.

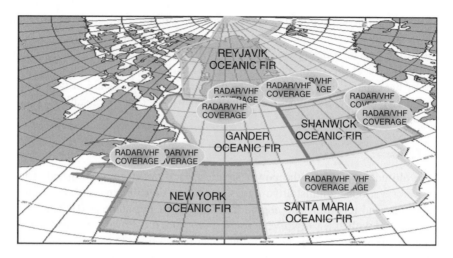

Figure 8.13   North Atlantic airspace and radar/VHF radio coverage

On international flights, transitioning through airspace boundaries under the jurisdiction of other national sovereignties may require supplementary procedures to address local restrictions. These Flight Information Region (FIR) boundaries normally require advance notification via the flight planning process (filed flight plan), and preliminary contact by the aircraft as the flight approaches the boundary. Generally, separate ATC clearances must be issued at each boundary crossing, including entering the oceanic airspace. Before entering such airspace, it is the responsibility of the crew members to familiarize themselves with any specific procedural requirements including position reporting, use of datalink, VHF or HF communications, and any other airspeed or operational limitations (holding speeds, speed limit below a given altitude, etc.). For example, large areas of North Atlantic airspace (see Figure 8.13) are beyond VHF radio communications and radar surveillance, and therefore well-defined procedural separation and communications (e.g., reporting position by HF radio every 10° of longitude) protocols are in place.

The need to deviate from the desired track due to adverse weather is always a possibility. The nature of hazardous weather en route varies with the geographical region (e.g., transcontinental, Caribbean, North Atlantic, etc.) as well as the type of aircraft and the equipment on board. The procedures and available options for coping with adverse weather are also airspace dependent. In the CONUS (continental USA), convective weather and thunderstorms often require deviations from planned routings, but this is facilitated through coordination with ATC in this VHF/radar environment. Along the North Atlantic tracks, thunderstorms are very rare, but clear air turbulence is often present and typically requires a change of altitude to find smoother air. However, given that communication and surveillance in oceanic airspace can be much more limited (see Figure 8.13) and the traffic separation requirements are consequently much larger, deviation requests are rarely granted. In the Caribbean and Western Atlantic Route System (WATRS), the presence of convective weather is made more difficult by the non-radar environment, but special deviation procedures have been instituted to allow more flexibility.

As in other phases of flight, the crew must be constantly prepared for the possibility of contingencies requiring diversion of the aircraft to an en route alternate airport. In

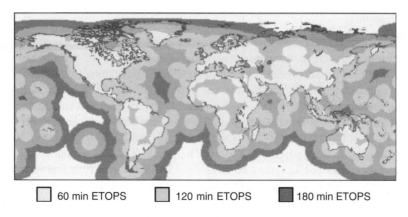

| 60 min ETOPS | 120 min ETOPS | 180 min ETOPS |

Figure 8.14   ETOPS airspace

addition to the possible closure of the destination airport (due to weather, power outages, or other field situations), reasons for diverting include medical emergencies (sick passengers/crew), aircraft equipment problems, terrorist activities in flight, unacceptable holding times and fuel diversion due to wind or traffic delays. The decision to divert usually includes input from dispatch and must include a clearance from the controller – unless the captain declares an emergency. If the situation warrants the declaration of an emergency, the flight is given priority handling en route, and the necessary ground and rescue services are assembled to meet the aircraft upon arrival.

Other, more routine duties that the crew performs during cruise include monitoring the aircraft flight path and systems, maintaining lateral fuel balance within limits (if not automated), cabin temperature control and ATC/AOCC communications requirements. Operations such as flight over mountainous terrain or extended range over water (including extended-range twin-engine operations, ETOPS) require special procedures to handle contingencies and emergencies in these cases due to the limited availability of alternate diversion airports. When flying over mountainous terrain, an emergency descent may be required due to a cabin pressurization problem, or inability of the aircraft to maintain altitude due to an engine malfunction. In either case, escape routes must be available for all terrain-critical segments of the flight and the crew must continuously update the navigation system with decision points and acceptable alternate routes. In the case of extended overwater operations with twin-engine aircraft types, limits are placed on maximum flying time to suitable alternate airports that can be used in the event of an emergency (such as loss of an engine). For example, 180 min ETOPS require twin-engine aircraft to remain within 3 hours' flying time of a suitable alternate (with one engine inoperative) at all times. This affects the airspace available to ETOPS aircraft (see Figure 8.14) and can therefore significantly alter the route flown on transoceanic flights compared to the more direct routings available to aircraft with three or four engines.

In addition, ETOPS-qualified twin-engined jets are subject to more stringent MEL requirements compared to an equivalent non-ETOPS model. During ETOPS, alternate airports must be constantly evaluated for acceptability and, as in the case of terrain-critical situations, decision points must be established for each potential alternate.

8.3.11 Descent

In future air traffic management environments (e.g., US NextGen and European SESAR), four-dimensional trajectory management should enable all phases of flight to be optimized for each aircraft given the overall network demand and constraints. In theory, the descent could then be initiated to accommodate an optimal profile all the way to landing to minimize fuel burn and emissions. In the present system, the descent profile is determined by both ATC limitations and optimal aircraft performance. An aircraft operating at typical cruise altitudes (FL310–410) will nominally initiate the descent at 100 to 130 nautical miles from the destination airport. The distance varies primarily due to ATC restrictions/procedures but also may be influenced by equipment type and environmental conditions such as winds aloft and turbulence. The initial descent takes place with about 30 to 40 minutes remaining in the flight, at which time the crew begins the approach and landing preparations. An "In Range" message is often transmitted to the destination airport station either through ACARS or by VHF radio. This message includes the latest touchdown estimate, special passenger requests (wheelchairs/connections) and, if not already transmitted, any maintenance discrepancies. The station transmits or uplinks the arrival gate assignment, ground power unit status and any other relevant status message such as a "tow-in only" requirement for the assigned gate.

During the descent, ATC may issue crossing restrictions which can be part of a published standard arrival procedure (such as a Standard Terminal Arrival Route (STAR), an example of which is shown in Figure 8.15) or as a response to a traffic sequencing requirement.

Crossing restrictions are generally issued to the cockpit crew in terms of altitude over a fix, e.g., "cross STW at FL240," or may include a speed restriction as well, e.g., "cross CAMRN at 11 000 feet at 250 knots." If the clearance is not issued as an immediate descent, it is the responsibility of the cockpit crew to determine a Top of Descent (TOD) point which satisfies the crossing restriction. The VNAV function of the FMS can be used to determine an "optimum" TOD point. Ride conditions notwithstanding, it is desirable from a fuel burn and environmental performance perspective to delay the descent as long as possible, then descend at or near engine idle at an optimum speed and meet the restriction within a few miles *before* the fix to allow a margin to ensure the restriction is met. Such engine-idle descents help to reduce fuel burn, emissions and noise, improving overall operating economics and environmental performance. Factors which must be taken into consideration during descent planning include wind direction and intensity for the relevant altitudes, possibility of speed restrictions if not already assigned, high barometric local pressure at the transition level and turbulence.[29] In addition, the descent may be shallowed by the use of anti-ice, which requires higher engine idle speed, and in some older aircraft partial power descents are required to meet cabin pressurization demands. The FMS is the primary resource available to the crew for descent planning as restrictions can be programmed directly via the CDU and a profile calculated. There are other ad hoc methods for determining the distance required to lose a given amount of altitude. The "3 to 1" rule is still used by most pilots to back up the FMS solution in which 3 miles are required for every 1000 feet of altitude loss, e.g., 30 000 feet would require 90 miles.

---

[29] Depending on the intensity of the chop or turbulence, the aircraft may have to be slowed to its "turbulence penetration speed." In most cases, this speed is less than the optimum cruise or descent speed and results in a shallower descent profile but reduces the impact of the turbulence on the airframe.

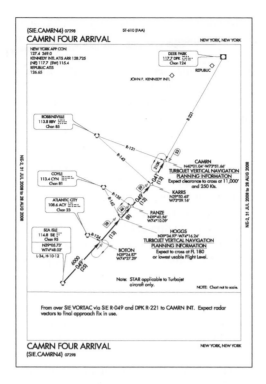

Figure 8.15   Example Standard Terminal Arrival Route (not to be used for navigation)

Adjustments are then made to accommodate headwinds/tailwinds and anticipated speed restrictions (such as the 250 knots speed restriction below 10 000 feet).

Destination weather and the expected approach/runway procedures are major considerations in planning the arrival. The primary source of this information is the ATIS previously described, although holding delays, weather conditions and runway operations may be passed along via ATC and/or dispatch. ATIS provides the current weather, instrument approach procedures in use and active runways, as well as details concerning runway and taxiway closures, windshear reports, precise visibility values for individual runways, braking capability, bird activity, temporary obstructions (e.g., construction), Land and Hold Short Operations (LAHSO)[30] utilization and any other relevant safety-related information.

Once the crew has received the destination weather and approach information, it begins setting up the navigation equipment for the expected arrival procedure. Of primary concern are the current weather conditions compared to the available approach procedures: low ceilings and visibility mandate specialized procedures which in turn require specific navigation equipment for executing the approaches. Approach procedures are segregated into categories of operation (CAT)[31] depending on the Runway Visual Range (RVR, a

---

[30] During LAHSO, landing aircraft are expected to stop or exit the runway before reaching a designated crossing runway or taxiway.

[31] CAT I: RVR ≥ 1800 feet or $1/2$ mile, DH at 200 feet; CAT II: RVR ≥ 1200 feet (≥1000 feet if autoland is used), DH at 150/100 feet; CAT IIIA: RVR ≥ 700 feet, DH at 50 feet; CAT IIIB: RVR ≥ 150 feet, DH zero; CAT IIIC: RVR and DH both zero. Conditions below CAT II require an autoland or head-up display (HUD) guidance.

precise measurement of the horizontal visibility) and the Decision Height (DH, the height from which the runway environment must be visually acquired in order to proceed with the landing, otherwise a missed approach must be performed). The baseline precision approach procedures (i.e., those that require lateral and vertical path information to be continuously available to the flight crew) for most well-equipped runways are the Instrument Landing System (ILS) CAT I or satellite-based Global Positioning System (GPS) procedures, both of which allow landings at visibilities as low as $1/2$ mile and ceilings as low as 200 feet above the ground. Non-precision approaches (i.e., where lateral track information is provided by a local navigation aid or GPS, but vertical guidance is received through barometric referencing or other means not directly associated with the specific runway ground-based NAVAIDs) have higher minimum values. Higher-precision systems and procedures can result in lower minimum CAT II or even "zero/zero" conditions of CAT IIIC approaches. If the current weather is below the minimums available for the procedure in use, or the necessary equipment is unavailable or inoperative, the crew must consider other options which include holding (if weather improvement is anticipated) or diverting to an alternate airport. Either course of action requires coordination with ATC and the airline's dispatch office.

After the crew has programmed the navigation systems and FMS for the anticipated procedure, the PF goes over the approach with the PM, using the published approach procedure as a reference. Some low-visibility weather situations may mandate that the captain always perform the PF duties to meet standardization requirements during critical low-visibility operations. In addition to designating the PF, the approach briefing includes information about the required NAVAIDs, key segment and crossing altitudes, approach minimums vs. current weather, obstacles and terrain awareness, and the missed approach procedure.

The cabin crew activities during the descent include preparing the cabin and galleys for landing, forwarding connecting gate information to the passengers, completing customs-related documents, forwarding any cabin-related discrepancies to the cockpit and verifying that seat belt compliance requirements are satisfied. The captain's descent PA announcement usually includes updates of arrival estimates and weather conditions. Any anticipated adverse weather or delays are usually briefed to both cabin crew and passengers. Delays due to limited runway capacity could require the aircraft to enter a holding pattern until such time as a landing slot can be guaranteed, at which point the descent and approach can continue as described below.

As the aircraft descends below the transition level (FL180 in the USA) the altimeters are reset to the local barometric setting. The PM works on completing the "descent" checklist which includes monitoring the pressurization, correcting any accumulated fuel imbalance, and calculating and/or reviewing landing data (approach speeds, runway limits). While passing through 10 000 feet, the captain alerts the cabin crew (by chime or PA) that the sterile cockpit period is in effect and that the final cabin preparations for landing should be completed.

8.3.12   Terminal Area Arrival

Terminal area maneuvering generally begins when the aircraft descends below 10 000 feet about 30 to 40 miles from the destination airport. At this point the flight path is usually defined by the STAR and/or radar vectors from ATC. Radar vectors consist of heading directives issued to the pilots and are used by ATC for the sequencing and/or spacing of air traffic. Increasingly, noise abatement arrival procedures are also being used when traffic conditions permit. The most common procedure involves variants of a Continuous Descent Approach (CDA), designed to keep aircraft higher and at lower thrust (and hence lower fuel burn and emissions rates) for longer by eliminating the level segments in conventional step-down approach procedures. These can either be accomplished by ATC giving "track distance to go" estimates to the flight crew, who then determine appropriate descent rates needed to eliminate (or minimize) periods of level flight, or be flown by the FMS when specific Precision Area Navigation (P-RNAV) CDA procedures are available (Reynolds *et al.*, 2007). In non-radar environments, the flight is operated along established airways or feeder routes to an initial approach point defined by the approach procedure in use. In any case, the crew must keep a vigilant traffic watch and maintain terrain awareness (using electronic aids, charts, visual, etc.) especially in mountainous areas and/or areas of high traffic congestion.

As the flight nears the position where it will commence the approach, the crew may be issued additional real-time landing information or instructions. Braking action reports are given by previous arrivals and include a qualitative ranking of the braking effectiveness during rollout after touchdown (good, fair, poor, nil in the USA). Certain braking action conditions may require the utilization of specified on-board systems such as autobrakes and/or autospoilers, or may dictate that the flight enter holding until the runway condition can be improved through plowing or chemical treatment. In low-visibility conditions, real-time RVR reports are issued to the arriving flights for the purpose of determining approach legality, or applying other operating restrictions to the flight (e.g., crosswind limitations, autoland requirements, etc.). ATC may also issue LAHSO instructions to arriving aircraft. The crew may refuse a LAHSO restriction (and possibly be subject to additional vectoring/sequencing), but, once accepted, the restrictions must be adhered to. Microburst alerts and airspeed loss/gain reports from prior arrivals are also passed on to the crew. Adjustments may have to be made to reference landing speeds to operate under such conditions. In many cases, the flight will have to enter holding to wait out low-visibility, poor braking, or windshear/microburst conditions.

8.3.13   Final Approach

The aircraft operated by most air carriers are usually equipped to satisfy the navigation requirements of a variety of approach procedures. Precision approaches include GPS autoland, GPS LNAV/VNAV and CAT I, II and III ILS approaches as previously described. Many runways at larger airports utilize ILS to provide guidance to aircraft during instrument conditions along a well-defined path made up of lateral and vertical elements called the localizer and glide slope respectively, as illustrated in Figure 8.16.

As expected, precision approaches provide for operations in much lower ceiling and visibility conditions than non-precision approaches. When very low-visibility conditions exist, CAT III approaches are mandated which require autoland or head-up display (HUD) guidance. Arrival delays should be anticipated during these operations due to reduced runway capacity. When VFR conditions exist, pilots are encouraged to use all available NAVAIDs as a backup even during visual approaches. In addition, many airfields employ instrument procedures during VFR conditions in order to manage aircraft sequencing or noise restrictions.

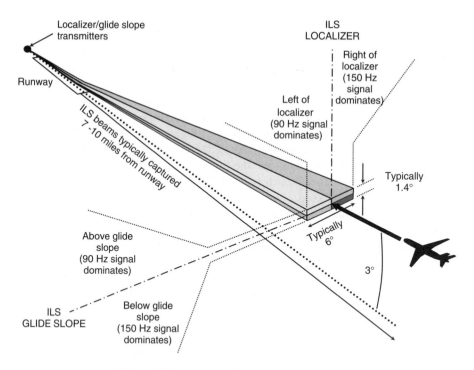

Figure 8.16 Instrument Landing System (ILS)

At some point during the vectoring or feeder segment, the flight will be "cleared for the approach." An approach clearance by ATC authorizes the crew to start execution of the approach procedures as published. As mentioned earlier, it is the cockpit crew's responsibility to determine approach legality. The current weather conditions must be compared to the procedures and equipment available, both ground based and airborne. Downgrades in on-board automation or displays may dictate higher landing minimums and/or unavailability of certain procedures. Likewise, the inoperative status of any ground components may result in additional landing restrictions. An adverse condition of any required approach facilities is usually reported by ATC or ATIS, but also may be detected by on-board alerting systems.

Most authorities designate a specific location in the procedure where the current weather must be at or above weather minimums in order for the aircraft to continue the approach. In the USA, the final approach fix is designated for non-precision approaches, while glide slope intercept (at a "normal" intercept altitude) is the requirement for precision approaches. Other countries may utilize the outer marker or specific altitudes, but generally, if the flight passes the designated position with reported weather at or above minimums, it may continue to the Missed Approach Point or Decision Height (MAP/DH), as applicable. If acceptable conditions exist at the MAP/DH[32] and the aircraft is stabilized

---

[32] In some cases, visual acquisition of the approach lighting system allows for further descent to a lower height; however, visual contact with the runway environment must still be achieved before landing. The runway environment usually includes the runway surface and any lighting aids at or beyond the threshold.

or "in the slot,"[33] the approach may be continued to landing. If weather conditions are below minimums at the designated position, the procedure must be aborted and other alternatives considered, e.g., diversion, holding, etc.

Runway wind conditions must be addressed by the crew during the final approach and landing. Depending on the wind direction, intensity and presence of gusts, adjustments may have to be made resulting in a higher planned approach speed. Maximum crosswind limitations vary among equipment type and weather conditions. The demonstrated crosswind limitations of modern transport aircraft typically range from 25 to 35 knots. This value may be further reduced by runway conditions (wet vs. dry), low visibility or inoperative equipment (e.g., rudder malfunctions). In any event, during high-crosswind situations the crew must maintain awareness of the aircraft's limitations, as well changes in surface winds and gusts.

Unless weather requirements dictate, use of the autopilot during the approach (or for any phase of the flight) is at the pilot's discretion. Many carriers encourage their crews to manually operate the flight controls, when appropriate, for the purpose of maintaining proficiency. Cockpit crews are expected, however, to utilize all available information (such as the autopilot flight director guidance cues presented on the primary flight display) during all phases of flight including approaches and landings, whether under visual or instrument conditions. As mentioned earlier, some procedures mandate use of the autoflight system to touchdown and rollout (CAT III), while others require autopilot disconnection before landing (non-autoland operations). It is the responsibility of the cockpit crew to determine the appropriate level of automation for a given phase of flight or procedure and to keep the resultant workload of managing the automation to a reasonable level.

Unless the flight is cleared for a visual approach, ATC is responsible for aircraft and wake vortex separation (i.e., keeping enough distance between consecutive aircraft so the wake vortex from one does not adversely affect those following and depends on relative weight categories). The crew, however, must be diligent in high-density arrival operations involving minimum spacing between landing aircraft and can utilize on-board resources such as TCAS to supplement a visual traffic watch.

In the event that the requirements for completing the approach and landing are not satisfied, a "go-around" is executed and the standardized missed approach procedure and/or ATC instructions must be followed. Options available following a missed approach include entering holding to wait out whatever unacceptable condition resulted in the aborted landing, diverting to an alternate airport or, most commonly, accepting ATC vectors to initiate another approach. Many aborted landings are initiated by ATC or the cockpit crew due to traffic on the runway. In most cases a prior arrival failed to clear the runway in a timely manner, but a delayed takeoff by an aircraft sitting in position at the threshold can also result in an aborted landing.

---

[33] When an aircraft is "in the slot," it is in a normal position for landing in stable flight. This requirement may vary between carriers, but is typically specified as within vertical and lateral deviation limits from the target track and airspeed within limits. In a more general sense, the "stabilized approach" concept adopted by most carriers requires the aircraft to be in landing configuration (on speed, gear down, flaps at landing setting) when passing through 1000 feet above ground level (AGL) in Instrument Meteorological Conditions (IMC) or 500 feet AGL in Visual Meteorological Conditions (VMC). Aircraft exceeding any of these parameters during the approach are expected to abort the landing and execute the published missed approach procedure.

Figure 8.17   Thrust reverser and ground spoiler deployment
Source: Photo courtesy Marinus Bergsma, used with permission, www.airliners.net

   If the runway is in sight and clear when at the decision point, the cockpit crew continues the descent until initiating the landing "flare" maneuver where the descent rate is reduced just before touchdown.

8.3.14   Landing and Rollout

After touching down on the runway, the PF uses a combination of reverse thrust, ground spoilers and wheel braking to decelerate to taxi speed and vacate the runway (Figure 8.17). Reverse thrust may not be allowed at some airports and at some times of the day for noise abatement reasons. Spoilers can be automatically activated upon touchdown and/or wheel spin-up in order to increase drag and reduce wing lift (which increases the weight on the wheels and improves braking effectiveness) to lower the landing rollout distance. In order to reduce wear on the brakes and engines, reverse thrust is often utilized at higher speeds (above 60 knots), while wheel brakes are employed at low speeds (below 100 knots). For safety and maximization of runway capacity, it is desirable to minimize runway occupancy during rollout and the controller expects the crews to vacate at the earliest safe opportunity.[34] Many airports have "high-speed turnoffs" to allow pilots to leave a runway at higher speeds than perpendicular turnoffs would allow. As the aircraft slows to turnoff speed, the captain and FO assume the taxi and communications tasks as per normal ground operations. Once clear of the runway, the crew reports any adverse wind or braking conditions to the tower (in low-visibility conditions the crew may also be required to report clear of the runway). After the aircraft has exited the runway, the FO contacts ground control for taxi-in instructions, completes the "after landing taxi" checklist and calls the local ramp control to confirm the arrival gate assignment and occupancy status.

8.3.15   Taxi-in

The pilots typically use taxiway diagrams of the destination airport to assist in the execution of taxi clearances given to them by ATC. Pilots must be diligent during ground operations at airports with which they are unfamiliar or that are undergoing construction.

---

[34] Some high-density airport operations mandate specific runway exits as part of the published arrival procedure.

Figure 8.18   Taxi-in guidance

Airport markings and lighting, although standardized in the USA, can be variable in other locations. Operations during nighttime or heavy precipitation also require special consideration and may substantially impede the overall traffic movement on the airfield. At some point during taxi-in, the captain determines the necessity of starting the APU. In the interest of fuel conservation, an engine may be shut down which may require utilizing the APU, depending on the aircraft type. Normally the APU is started while the aircraft is a few minutes from the gate area, unless it has been determined that ground power will be used. In that case, the APU is not started and an engine is left running after gate arrival until the ground electric power is connected by the ground crew.

If the arrival gate is occupied, the aircraft may be required to wait out the delay at a remote location. Occupied gates are often the result of a delayed departure or other operational issues with the aircraft currently positioned at the gate, and the anticipated delay to the arriving flight should be passed on to ATC and the passengers. Once clearance to the gate is received, the captain taxis to the ramp area and visually acquires the marshallers. After the crew confirms the gate area is unobstructed, the marshallers utilize lighted wands to signal clearance to taxi to the stop point adjacent to the jetbridge (Figure 8.18). Often delays are encountered at this point due to the unavailability of the ground crew, carts or vehicular traffic in the gate area, or a tow-in requirement for the assigned gate (in which the aircraft engines are shut down and a tug is used to tow the aircraft to the gate position). Some stations utilize automatic parking systems which employ an arrangement of lights and/or signs that the captain uses for lead-in line and stopping position guidance. In the absence of self-guidance, the marshaller uses wand signals to direct and stop the aircraft at the desired location. Once the brakes are parked, the agent moves the jetbridge into position at the entry door, or in the case of airstair disembarkation, positions the truck(s) under the appropriate exit door(s).

8.3.16   Parking

In most cases, setting the parking brake and opening a cabin door trigger the "IN" event. The "IN" time is used to determine a number of metrics including the length of the flight (which is used to calculate flight crew compensation and legality for subsequent trips), FAA on-time arrival report card, customs and immigration data, and company-specific performance monitoring and scheduling adjustments.

Once a source of ground power is connected (APU or external power cable), the engines are shut down and the crew completes the "engine shutdown" checklist. The agent verifies the disarming of the doors with the flight attendants and opens the designated exit door(s) to commence passenger disembarkation. Usually any wheelchair passengers or unaccompanied minors are accommodated last. After engine shutdown, the ground crew begins unloading and processing the baggage and freight. The flight crew secures the cockpit and cabin before departing the aircraft. At specified outstations, an exterior post-flight walkaround inspection may be required which is completed by the cockpit crew before leaving the gate area. Post-flight inspections are not as thorough as pre-flight ones and are usually mandated after the last flight of the day at stations with limited or contract maintenance. Finally, for international operations, any customs/immigration requirements may need to be addressed.

8.3.17 Post-flight

Upon completion of post-flight duties at the aircraft, the cockpit crew accomplishes any required debriefing reports while the flight attendants make liquor and duty-free deposits, usually at the station operations. Debriefs/reports are required by the cockpit crew in instances of a declared emergency or ATC violation, significant mechanical failures (e.g., engine shutdown), fuel dumping, illness, injury or death of a passenger or crew member, passenger misconduct/smoking, overweight landing, HAZMAT issues, diversions, high-speed aborts, lightning strikes, near midair collisions and a number of other situations involving non-standard operations or issues. Once all cockpit and cabin obligations are fulfilled, the crew begins preparations for the next flight leg. In situations where the same aircraft is to be used, a new flight plan is pulled up and the pre-flight sequence starts again. In many cases, however, the crew must change aircraft and relocate to the new departure gate where the planning/pre-flight duties are repeated. If possible, the inbound flight crew should attempt to meet the outbound flight crew and discuss any aircraft discrepancies or operational issues. If this is the last flight of the day, the crew is released from duty and typically proceeds to the crew hotel in cases of out-of-base layovers, or the bus to the employee parking lot if the inbound flight was the last leg of a sequence.

If the aircraft is to be "turned around" for use in a subsequent leg, maintenance personnel will attempt to meet the flight upon gate arrival. Any discrepancies are discussed with the inbound flight crew and the necessary repairs are begun as soon as possible. If the discrepancy has been reported in flight, the mechanics often meet the aircraft with replacement parts such as Line Replaceable Units (LRUs) which can often allow repair in the normally scheduled turnaround window. If the aircraft is not to be used right away, required maintenance may be performed during periods of less demand. In situations where the aircraft is finished for the day, it may be towed or taxied to a remote location, or the hangar, where the requisite maintenance and/or inspections are completed. In addition to maintenance requirements, other post-flight activities conducted by ground personnel include aircraft cleaning and de-catering, security checks and any required customs inspections. When customs or security inspections are required, delays are often incurred since the outbound crew cannot access the aircraft until the inspection is complete. In any event, it is desirable from an efficiency standpoint that ground service activities associated with the inbound flight such as catering, cleaning and baggage handling should dovetail with the departure cycles of subsequent flight legs.

8.4 Summary

- Airlines operate under a system of prioritized goals including safety, economics and customer service.
- Flight operations comprise a carefully choreographed sequence of tasks which must be performed in accordance with a fine-tuned schedule. These tasks, occurring in both series and parallel, demand temporal and spatial coordination and success depends on the synchronization of many different responsible parties. Flight and ground crews including cockpit and cabin personnel, dispatch, maintenance, ticketing/gate agents, guide/push crews, baggage handlers, caterers and fuelers each must abide by specific duties and procedures to keep the operation flowing smoothly.
- Cockpit and cabin crews come from a variety of backgrounds and work experiences. An airline may employ thousands of flight crew members who must function together at a high level of safety and efficiency. This can only be achieved through extensive training and adherence to well-established procedures and protocols.
- Activities before takeoff include flight planning, passenger processing, aircraft pre-flight, fueling, and other required ground processes. The pre-flight activities are orchestrated so as to achieve an "on-schedule" push-back from the departure gate, although any maintenance and gate hold issues must also be considered.
- During taxiing to the departure runway, the cockpit crew prepares the aircraft for takeoff and updates the takeoff performance parameters with actual load data. In addition, any departure delays, aircraft icing and environmental conditions such as runway contamination may need to be addressed.
- The takeoff is a highly critical maneuver where a number of factors must be taken into consideration including airport/runway, environmental and emergency/abnormal contingencies. Departure from the terminal area, which may include route and speed restrictions, is followed by the climb to cruise altitude.
- The optimum cruise altitude is determined by a number of factors including company priorities of fuel burn and time, efficiency, ride comfort, other traffic and/or airspace limitations. During cruise flight, the cockpit crew must continually evaluate contingency options in the event of a passenger or mechanical disruption. In addition, some flight situations require adherence to specialized procedures including international routing, mountainous terrain and extended overwater operations.
- During descent, the crew begins preparing the cockpit and cabin for landing. In addition to conforming to ATC restrictions, the cockpit crew plans the approach to landing, which includes consideration of the meteorological conditions, the approach procedures available, environmental performance, aircraft mechanical condition and, if necessary, available fuel for holding. In the event of an unsuccessful approach, diversion to a more suitable landing point may have to be considered.
- After landing, the aircraft is taxied to the arrival gate while the crew readies the cockpit and cabin for parking. Once the aircraft is parked, passenger disembarkation

is completed as well as baggage/cargo unloading. The flight crews proceed to the next departure gate, or to the ground transportation area if their duty day is complete.

8.5   Appendix: List of Acronyms

ACARS	Aircraft Communication and Reporting System
ADS-B	Automatic Dependent Surveillance – Broadcast
AGL	Above Ground Level
AOCC	Airline Operations Control Center
APU	Auxiliary Power Unit
ARINC	Aeronautical Radio, Inc.
ATA	Actual Time of Arrival
ATC	Air Traffic Control
ATIS	Automatic Terminal Information Service
ATM	Air Traffic Management
ATP	Airline Transport Pilot
CAT	Category
CDA	Continuous Descent Approach
CDL	Configuration Deviation List
CDU	Control Display Unit
CFR	Code of Federal Regulations
CONUS	Continental United States
CPDLC	Controller–Pilot Data Link Communications
CRM	Crew Resource Management
DH	Decision Height
DP	Departure Procedure
EFOB	Estimated Fuel On Board
EPR	Engine Pressure Ratio
ETA	Estimated Time of Arrival
ETO	Estimated Time Off
ETOPS	Extended-range Twin-engine Operations
FAA	Federal Aviation Administration (USA)
FIR	Flight Information Region
FL	Flight Level
FMS	Flight Management System
FO	First Officer
FOB	Fuel On Board
GPS	Global Positioning System
HAZMAT	Hazardous Material
HF	High Frequency
HUD	Head-Up Display
ICAO	International Civil Aviation Organization
IFR	Instrument Flight Rules
ILS	Instrument Landing System
IMC	Instrument Meteorological Conditions
IOE	Initial Operating Experience

IRU	Inertial Reference Unit
ISA	International Standard Atmosphere
JAA	Joint Aviation Authority (Europe)
LAHSO	Land And Hold Short Operation
LNAV	Lateral Navigation
LOE	Line Operating Experience
LRU	Line Replaceable Unit
MAP	Missed Approach Point
MCP	Mode Control Panel
MEL	Minimum Equipment List
MSL	Mean Sea Level
N1	Engine Fan Rotation Speed
NAT	North Atlantic Track
NAVAID	Navigation Aid
NGATS/NextGen	Next Generation Air Transportation System
PA	Public Address
PDC	Pre-Departure Clearance
PF	Pilot Flying
PIC	Pilot In Command
PIREP	Pilot Report
PM	Pilot Monitoring
PNF	Pilot Not Flying
P-RNAV	Precision Area Navigation
QNH	Local barometric altitude above sea level
RVR	Runway Visual Range
RVSM	Reduced Vertical Separation Minima
SATCOM	Satellite Communication
SELCAL	Selective Calling
SESAR	Single European Sky ATM Research
SID	Standard Instrument Departure
SOP	Standard Operating Procedure
STAR	Standard Terminal Arrival Route
TCAS	Traffic Alert and Collision Avoidance System
TOD	Top Of Descent
VFR	Visual Flight Rules
VHF	Very High Frequency
VMC	Visual Meteorological Conditions
VNAV	Vertical Navigation
WATRS	Western Atlantic Route System

References

Airbus Flight Operations Support & Line Assistance (1998) *Getting to Grips with the Cost Index*, Airbus Customer Services, France, Issue 2, www.iata.org/NR/ContentConnector/CS2000/ Siteinterface/sites/whatwedo/file/Airbus_Cost_Index_Material.pdf (December 22, 2008).

Cook, A. (ed.) (2007) *European Air Traffic Management: Principles, Practice and Research*, Ashgate, Aldershot, Hants.

National Research Council (2005) *Technology Pathways: Assessing the Integrated Plan for a Next Generation Air Transportation System*, National Academies Press, Washington, DC.

Nolan, M.S. (2004) *Fundamentals of Air Traffic Control*, 4th edition, Thomson Brooks/Cole, Belmont, CA.

Reynolds, T.G., Ren, L., and Clarke, J.-P.B. (2007) "Advanced Noise Abatement Approach Activities at a Regional UK Airport", *Air Traffic Control Quarterly*, Vol. 15, No. 4, pp. 275–298, www.atca.org.

Spence, C.F. (ed.) (2007) *AIM/FAR 2008: Aeronautical Information Manual/Federal Aviation Regulations*, McGraw-Hill Professional, New York.

9

Irregular Operations: Schedule Recovery and Robustness

Cynthia Barnhart

9.1 Introduction

The optimized aircraft and crew schedules generated using approaches detailed in Chapter 7 are rarely, if ever, implemented exactly as planned on a daily basis. Impeding their implementation are two major sources of disruption:

1. *Airline resource shortages* caused by aircraft mechanical problems, crew unavailability due to illness or upstream missed connections or delays, aircraft delays caused by lack of gates or other ground resources, and longer than expected passenger boarding and disembarking times.
2. *Airport and airspace capacity shortages* caused by factors such as airport security delays and inclement weather conditions that reduce throughput at airports.

These disruptions often result in *irregular operations* in which the needed resources, such as crews, aircraft, gates and landing slots, become unavailable and the planned schedule becomes inoperable. To respond to these disruptions as they occur, airlines operate control centers at which controllers, who are provided with up-to-the-minute information about the network-wide status of the airline's operations, make decisions regarding how to reassign resources and adjust the flight schedule in a manner that *best* repairs the disrupted schedules and allows the airline to resume planned operations.

This chapter focuses on the *schedule recovery* process involving the replanning and recovering of aircraft, crews and passengers from irregular operations. In the initial sections of this chapter, delays and economic impacts associated with irregular operations are described, and details are provided of schedule recovery approaches that can be executed during operations to make effective resource reassignment decisions when irregular operations arise. In the latter part of the chapter, another approach to irregular

The Global Airline Industry   P. Belobaba, C. Barnhart and A. Odoni
© 2009 John Wiley & Sons, Ltd

Table 9.1 On-time performance statistics, January through April

Year	On-time arrivals	On time (%)	Arrival delayed	Delayed (%)	Flights cancelled	Cancelled (%)	Diverted	Flight operations
1999	1 341 961	75.04	379 012	21.19	62 408	3.49	4947	1 788 328
2000	1 400 240	75.23	394 954	21.22	61 582	3.31	4407	1 861 183
2001	1 554 426	75.71	427 757	20.83	66 629	3.25	4258	2 053 070
2002	1 405 705	81.64	291 654	16.94	22 170	1.29	2379	1 721 908
2003	1 762 713	82.41	328 431	15.35	44 230	2.07	3586	2 138 960
2004	1 839 241	79.20	440 538	18.97	39 180	1.69	3286	2 322 245
2005	1 818 924	77.33	476 734	20.27	52 290	2.22	4339	2 352 287
2006	1 777 550	77.18	485 521	21.08	35 270	1.53	4761	2 303 102
2007	1 768 575	72.45	598 259	24.51	69 257	2.84	4929	2 441 020
2008	1 734 538	72.60	584 583	24.47	64 442	2.70	5654	2 389 217

Source: http://www.transtats.bts.gov/HomeDrillChart.asp

operations is presented. This approach, referred to as *robust planning*, is applied in the planning stage[1] detailed in Chapter 7, and is aimed at *reducing* the complexity of or the need for schedule recovery. Its goal is to generate *robust schedules* that are more resilient to disruptions, allowing planned schedules to be operated more consistently, or requiring less extensive schedule adjustments to maintain operability. An overview of schedule recovery and robust planning approaches is presented and successful applications, impacts and remaining challenges are discussed.

9.2   Irregular Operations

The impacts of airport and airspace capacity and airline resource shortages are significant, as depicted in Table 9.1. On-time performance, defined by the US Department of Transportation as the percentage of flights arriving no later than 15 minutes after their scheduled arrival time, reached record lows in 2007 for the period January–April, 1995–2008 (Table 9.1). More than 24% of all flights in the first four months of 2007 were delayed (by more than 15 minutes), 2.84% were cancelled, and almost 5000 flights were diverted. These statistics are particularly noteworthy in light of an earlier finding (Bratu and Barnhart, 2005) that the US Department of Transportation's on-time performance metrics can underestimate passenger delays by a significant margin and mask true schedule reliability (or lack thereof).

Because aircraft, crews and passengers all flow in an interconnected manner over a network, and all weave their way through the airline network in different manners, a disturbance to one flight leg can have widespread disruptive effects that are seemingly unconnected to the original disturbance. For example, a delayed aircraft with its correspondingly delayed crew and passengers can:

1. cause delays to *other aircraft* to which delayed passengers or crew members are connecting;

---

[1] An overview of the steps involved in schedule planning is provided in Barnhart *et al*. (2003).

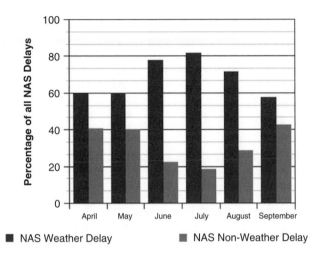

Figure 9.1 Weather's share of national aviation system (NAS) delays (November 2007 to April 2008)

Source: http://www.transtats.bts.gov/OT_Delay/ot_delaycause1.asp?type=4&pn=1

2. cause passenger misconnections;
3. cause *reserve crews*[2] to be called to work to replace delayed crews; or
4. cause delays to the crew and passengers scheduled on subsequent flight legs of the delayed aircraft.

Even a single disruption can wreak havoc on the operations plan, leading to severed connections for passengers and crews and significant levels of disruption *throughout* the network. These same network propagation phenomena enable *local* weather delays in one geographical area to affect other locations far removed from the inclement weather, thereby impacting network operations *globally*. With network propagation magnifying its effects, inclement weather accounted for between 60% and 75% of the national aviation system delays during the period from November 2007 through April 2008, as depicted in Figure 9.1.

These network effects are exacerbated and disruption levels are elevated by two major factors:

1. Airline schedules are *optimized* under an assumption of normal weather and other operating conditions, with non-productive or *slack* time largely stripped from the schedule to maximize resource productivity, thus providing little opportunity to absorb unexpected delays, which, thus, propagate throughout the network (Lan *et al.*, 2006).
2. Unlike air traffic demand growth, capacity growth at many of the world's congested airports is stagnant.

---

[2] Reserve crews are backup crews, not originally assigned to the flight schedules, but prepositioned at certain locations and available to report to duty, if needed. They are typically guaranteed a minimum monthly salary, whether or not they are called into work, and they are limited to a maximum number of flying hours per month.

Table 9.2 Air Transport Association direct operating cost estimates, July 2007

Direct (aircraft) operating costs Calendar year 2007	$ per block minute	Annual delay costs ($ millions)
Fuel	27.86	3 727
Crew – pilots/flight attendants	12.71	1 700
Maintenance	9.57	1 281
Aircraft ownership	7.70	1 031
Other	2.62	350
Total DOCs	**60.46**	**8 089**

Notes:

1. Costs based on calendar year 2007 US DOT Form 41 data as reported by US scheduled passenger airlines with annual revenues of at least $100 million.

2. Arrival delay minutes reflect operations at 77 US airports and are taken from the FAA Aviation System Performance Metrics (ASPM) database.

Source: http://www.airlines.org/economics/specialtopics/ATC+Delay+Cost.htm

The direct (aircraft) operating costs of schedule delays in 2007 were estimated at about $8 billion (Table 9.2), with additional estimated passenger delay costs of about $4 billion per year. With the airline industry suffering tremendous losses in recent years, as discussed in Chapter 1, coupled with the current threat to profitability brought about by soaring fuel costs, it is imperative that airlines manage these disruptions effectively to control the costs of irregular operations.

9.2.1 Managing Irregular Operations: Airline Operations Control Centers

Airlines operate *Airline Operations Control Centers* (AOCCs) to: (1) ensure safety of operations; (2) exchange information with Air Navigation Service providers and other airlines; (3) manage aircraft, crew and passenger operations; and (4) recover from irregular operations. The AOCC comprises:

- **Airline operations controllers** who, at the helm of the AOCC, are responsible for aircraft rerouting, and for decisions on flight cancellations, ground delays and diversions for various types of aircraft.
- **Crew planners** who find efficient recovery solutions for crews and coordinate with airline operations controllers to ensure that considered operational decisions are feasible with respect to crews.
- **Customer service coordinators** who find efficient recovery solutions for passengers and coordinate with airline operations controllers to provide an assessment of the impact on passengers of possible operational decisions.
- **Dispatchers** who provide flight plans and relevant information to pilots.
- **An air traffic control group** that collects and provides information, such as the likelihood of future ground delay programs, to airline operations controllers.

The AOCC is complemented by *Station Operations Control Units*, located at airport stations, whose staff is responsible for *local* decisions, such as the assignment of flights to gates, ground workforce to aircraft and personnel for passenger services.

9.2.2   Options for Schedule Recovery from Irregular Operations

When disruptions occur, airline controllers adjust scheduled operations by:

1. delaying flight departures until aircraft and/or crews are ready;
2. cancelling flight legs;
3. rerouting or *swapping aircraft* (i.e., reassigning aircraft among a subset of flight legs);
4. calling in new crews or reassigning existing crews;
5. postponing the departure times of flight legs to prevent connecting passengers from missing their connections; and
6. reaccommodating *disrupted passengers*, i.e., passengers who must be reaccommodated on itineraries other than planned, due to flight leg cancellations or missed connections caused by flight leg delays.

All schedule adjustments must satisfy crew work rules, aircraft maintenance and safety regulations, and passenger and aircraft positioning requirements. A major potential stumbling block to achieving this is the need for up-to-the-minute information detailing the recent flying history of aircraft, crews and passengers and estimates of future aircraft and crew *ready times*. These ready times are estimated using forecasted *gate-to-gate times* for each flight leg. In practice, of course, gate-to-gate time, which is the sum of taxi-out time, en-route time and taxi-in time, varies with weather conditions and congestion levels.

Additional details of the schedule recovery strategies involving aircraft swapping, and flight leg cancellations and departure postponements are provided in the following sections.

Aircraft Swapping

To illustrate the idea of aircraft swapping, consider Figure 9.2 containing a flight schedule as planned and as operated. The schedule contains flight legs *f1*, *f2*, *f3* and *f4*, with aircraft *a1* assigned in the flight plan to operate *f1* and *f3*, and aircraft *a2* to operate *f2* and *f4*. Assume, for the sake of illustration, that *f1* is delayed and *f2* arrives as planned. Without reassigning aircraft, *f3* will be delayed and passengers from *f3* will potentially miss their connections to downstream flight legs. Consider, however, the aircraft swapping strategy in which *a1* is reassigned from *f3* to *f4* and *a2* is reassigned from *f4* to *f3*; that is, *a1* and *a2* swap flight leg assignments. With these swaps, the revised schedule absorbs the arrival delay of aircraft *a1* without delaying departing flight legs and without disrupting downstream passenger connections.

Operations controllers can also swap aircraft of different types, although several conditions must be satisfied before making such swaps. First, to leave crew schedules intact, pilots must be able to operate the aircraft types to which they are reassigned. Second, because different aircraft types are likely to have different numbers of seats, resulting passenger disruption costs must be outweighed by the benefits of swapping. Third, all swaps must allow aircraft maintenance requirements to be satisfied. Notwithstanding these restrictions, numerous swap opportunities often exist, especially for large airlines at a hub airport, where typically many aircraft will be on the ground at the same time.

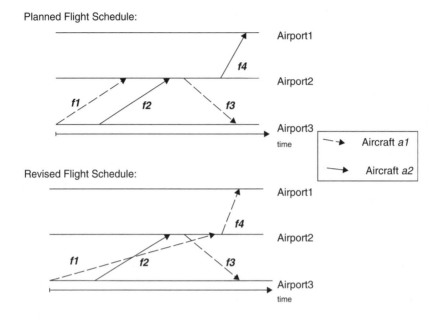

Figure 9.2 Aircraft swapping example

Flight Leg Cancellation and Departure Postponement

Reasons for flight leg cancellations include: mechanical problems; crew absence, delays and disruptions; upstream flight leg delays; and reduced airport departure and landing capacity due to weather conditions and resulting Ground Delay Programs (see Chapter 14 for details). Flight leg cancellations, like aircraft swapping, are invoked to stem the effects of delay propagation and decrease system-wide delays.

So that aircraft are positioned where needed to fly downstream flight legs, controllers must ensure that cancellation decisions maintain *aircraft flow conservation* (often referred to as *balance*) at each airport location. Normally, this requires that cancellations involve cycles of two or more flight legs. To cancel only a single flight leg and still be able to execute the remaining schedule, it is necessary to deploy to the destination of the cancelled flight leg a spare aircraft of the type originally assigned. Because spare aircraft are typically in very limited supply, cancelling only a single flight leg is not usually an option. Fortunately, hub-and-spoke flight networks often include a number of potential cancellation cycles, each with a small number of flight legs. Specifically, for large hub-and-spoke airline networks with a high degree of network connectivity, many cancellation cycles contain only two flight legs.

Additional cancellation options are provided by aircraft swaps, particularly at hub airports. Consider for example, aircraft $a1$ assigned to a late evening flight leg $f1$ and aircraft $a2$ assigned to an early morning flight leg $f2$. If $a1$ is unavailable to operate $f1$ but $a2$ is available, operations controllers have the option of reassigning $a2$ to $f1$ and cancelling $f2$ later on, if $a1$ is still unavailable for $f2$. By cancelling a morning flight leg rather than a late evening leg, passengers are not disrupted overnight and resulting delays can be significantly shortened.

To avoid crew disruptions, operations controllers typically try to cancel flight legs that are all assigned to the same crew. If this is not feasible, they limit the extent of the crew disruption, if possible, by cancelling flight legs with crews ending their workday so that reserve crews are not needed. In the worst case, when cancellations require reserve crews, operations controllers prefer to cancel flight legs that allow the disrupted crews to be repositioned to recommence their assigned future work as soon, and as inexpensively, as possible.

Operations managers often have the option not to cancel flight legs but, instead, to wait for the missing resources to be available or replaced. In some cases, controllers choose to postpone the departure of a flight leg even beyond its *ready time*, i.e., the earliest time at which the departure is possible, so that late connecting passengers are not left behind. This obviates the need to reaccommodate disrupted passengers and avoids the associated passenger disruption costs.

9.2.3   Schedule Recovery from Irregular Operations: Objectives and Process

Schedule recovery plans try to achieve one or more of a set of possible objectives, such as: minimizing the cost of reserve crews and spare aircraft used; minimizing passenger recovery costs; minimizing loss of passenger goodwill; and minimizing the amount of time until it is possible to resume the original plan. Whatever the objective, the recovery problem must often be solved within a limited number of minutes, otherwise the recovery solution can become infeasible. This time limitation renders it impractical to solve large, detailed optimization models and hence, to meet these objectives, most airline recovery processes are sequential. The first step is to recover aircraft, making decisions involving flight leg cancellations, delays and aircraft reroutings. The second step is to determine crew recovery plans, assigning crews to uncovered flight legs by reassigning them or utilizing reserve crews. Finally, the third step is to develop passenger reaccommodation plans for disrupted passengers. Although the AOCC decision process is hierarchical in nature, airline operations controllers, crew planners and passenger service coordinators consult with one another during the process to assess the feasibility and impact of possible decisions.

The sequential decision process, first aircraft, then crew, and finally passenger recovery, is reflected in the decision support tools that have been developed for airline recovery. Decision models typically focus on recovery of one type of resource, namely aircraft, or crews, or passengers, with the aircraft recovery decisions taken first, followed by crew recovery and then passenger recovery. While this approach greatly simplifies and speeds up the generation of recovery solutions, the decisions themselves, especially those made early in the process, must be based on approximate and incomplete information about the impact of these decisions on the other resources for which recovery decisions will follow.

To understand the sequential recovery process, rich in complexity and necessary trade-offs, consider the small example in Figure 9.3. The example, a subnetwork of the example presented in Figure 7.3, contains a set of four flight legs (CL302, CL332, CL502, CL551) operated by two identical A300 aircraft (A300-1 and A300-2). The schedule and assigned aircraft for each of the flight legs is shown in Table 9.3. As indicated, the plan is for A300-1 to fly CL551 followed by CL302, and A300-2 to fly CL332 followed by CL502.

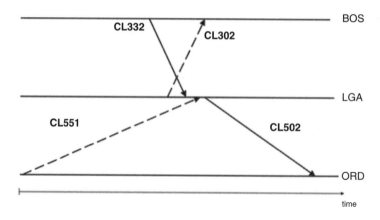

Figure 9.3 A300 flight schedule

Table 9.3 Example of flight schedule and aircraft assignments

Flight no.	From	To	Dep. time (EST)	Arr. time (EST)	Assigned aircraft
CL551	ORD	LGA	07:00	10:00	A300-1
CL332	BOS	LGA	10:30	11:30	A300-2
CL302	LGA	BOS	11:00	12:00	A300-1
CL502	LGA	ORD	15:00	18:00	A300-2

To demonstrate the recovery process, assume that, before the departure of flight leg CL551, it is known that the arrival of CL551 into LGA will be delayed until 15:00, exactly the planned departure time of flight leg CL502. (To simplify this discussion, let the minimum turn times be zero, thereby allowing aircraft, crew and passengers to depart immediately upon arrival.) When a delay occurs, the first step is to determine if recovery is needed. In some cases, a delay can be absorbed by slack in the schedule, either immediately or during downstream operations, with little to no disruption or cost to the system of aircraft, crews and passengers. In other cases, when the delay is more extensive or the schedule is tight with quick aircraft turns, or short crew and passenger connections, the do-nothing option can be very costly and intervention is warranted.

From Figure 9.3, it is clear that there is insufficient slack in the schedule of Table 9.3 to absorb the arrival delay of CL551 and, hence, some delay will have to propagate to downstream operations. The first action that the airline recovery team must take, then, is to identify the options for recovery. Following the sequential process, the options for recovery are generated first for the aircraft, as described in the next section.

Aircraft Recovery

When schedule disruptions occur, the aircraft recovery problem is to determine flight departure times and cancellations, and revised routings for affected aircraft. Rerouting options include: ferrying (repositioning an aircraft containing no passengers to another location, where it can be utilized); diverting (flying to an airport other than the scheduled destination); and most commonly, swapping. Each modification must satisfy maintenance requirements, station departure curfew restrictions and aircraft balance requirements, especially at the start and end of the recovery period. At the end of the period, aircraft types must be positioned at their originally scheduled locations in order for operations to resume as planned.

In our example, the aircraft recovery options available to the airline, include:

1. Delay propagation (no aircraft rerouting): This option, illustrated in Figure 9.4, involves operating aircraft A300-1 along its planned route, i.e., CL551 followed by CL302, with the departure time of CL302 delayed until 15:00. The result is an arrival delay of 5 hours to CL551 and an arrival delay of 4 hours to CL302. Aircraft A300-2 is unaffected and operates its planned route of CL332 followed by CL502 without delay.

2. Aircraft swapping (aircraft rerouting): This option, illustrated in Figure 9.5, requires the swapping of aircraft assigned to CL302 and CL502, creating a new route for aircraft

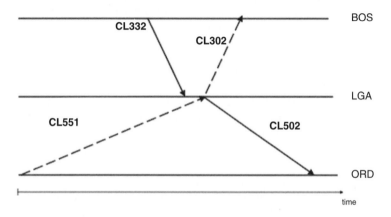

Figure 9.4   Delay propagation recovery action

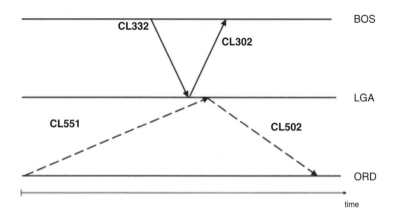

Figure 9.5   Aircraft swapping recovery action

A300-1 of CL551 followed by CL502, and a new route for aircraft A300-2 of CL332 followed by CL302. The result for A300-1 is an arrival delay of 5 hours for CL551 but no delay to CL502. For A300-2, because the planned (and actual) arrival time of CL332 is 11:30 and the planned departure time of CL302 is 11:00, the departure of CL302 must be delayed by 30 minutes to allow a feasible connection.

3. Flight cancellations: Because flight cancellations must leave the flight schedule balanced, with the number of cancelled arrivals equal to the number of cancelled departures at *each* station, it is not possible to cancel only the flights legs along the route of A300-1. CL551 begins in ORD and CL302 ends in BOS, leaving an unbalanced schedule. In this case, if the flight legs of A300-1 are to be cancelled, all the flight legs along the route of A300-2 must also be cancelled. Alternatively, if the aircraft swapping option is employed with A300-1 assigned to CL551 and CL502, and A300-2 assigned to CL332 and CL302, then fewer flight cancellations are needed to maintain aircraft balance over the network. Specifically, aircraft swapping enables two additional cancellation options, namely: cancellation of CL551 and CL502, or cancellation of CL332 and CL302. Cancellation options, while halting the propagation of aircraft delays, must be exercised carefully as aircraft maintenance plans, crew schedules and passenger itineraries can all be disrupted, with considerable repair time and costs involved.

Even for this very simple network of four flights and two identical aircraft, the recovery options become quite involved, especially when the myriad of operational constraints and preferences are considered. Take, for example, the requirement that aircraft fly no more than a specified number of hours between maintenance checks. Consider the aircraft swapping option above, and assume that: (1) A300-1 is due for maintenance after operating CL302; (2) BOS is a station at which maintenance checks can be performed; and (3) ORD is not a station at which maintenance checks are performed. In this scenario, the aircraft swapping option, and the swapping–cancellation options, are at best unattractive and, perhaps, impossible due to maintenance requirements. If a maintenance check must be performed immediately after CL302, a swap would result in the grounding of A300-1. Grounding of aircraft is typically prohibitively expensive, and hence this outcome will be avoided, if at all possible.

To further illustrate the complexity associated with even this very simple network, consider the task of identifying which recovery option is the best. If minimizing delay is the objective, the aircraft swapping option above seems to dominate the delay propagation option. The total delay for the delay propagation option is 9 hours, while the total delay for the aircraft swapping option is only 5.5 hours. Moreover, the delay to each flight leg in the delay propagation option is at least as great as in the aircraft swapping option. The aircraft swapping option, however, requires potentially very costly changes to the planned crew schedules, passenger itineraries and, perhaps, the maintenance plan.

Another fundamental issue that arises in comparing these options is the very definition of the metric by which they are compared. In the discussion above, delay is calculated relative to the flight schedule. If, however, delay is calculated for each passenger, the total delay associated with each option, especially the flight cancellation options, is significantly different and the relative ranking of the options could change dramatically. An analysis of the operations of a major US airline to compare flight and passenger delays (Bratu and Barnhart, 2005) has shown that:

1. Flight delays are not necessarily indicative of the magnitude of delays experienced by disrupted passengers. For example, on the same day that disrupted passengers experienced average delays of 419 minutes, the average flight delay was only 14 minutes.
2. Delays suffered by disrupted passengers and the associated costs account for a very significant part of total passenger delays, even when the number of disrupted passengers is relatively small. In the Bratu and Barnhart study, disrupted passengers represented just about 4% of all passengers but accounted for more than 50% of the total passenger delay. Moreover, associated with these disrupted passengers are significant recovery costs, including lodging, meals, rebooking (possibly on other airlines) and loss of passenger goodwill.

The study concluded that estimates of delay costs that consider only aircraft delays and ignore passenger disruptions can be inaccurate and misleading. Hence, a sequential process for developing recovery strategies must, at the least, consider downstream effects.

Aircraft Delays and Cancellations Model[3]

We now present, for purposes of illustration, a typical optimization model used to investigate aircraft recovery options. The purpose of this *integrated aircraft delays and cancellations model* is to generate sequentially, for each fleet type, a set of aircraft routes that minimize delays, cancellations and rerouting costs (Arguello *et al.*, 1997). Aircraft balance is ensured by matching aircraft assignments with the actual aircraft locations at the beginning of the recovery period and with the planned aircraft locations at the end of the period. In the example above, the recovery period begins at the planned departure time of flight leg CL551 and ends with the arrival of CL502.

The model includes two types of binary decision variables: maintenance-feasible aircraft routes and schedules, and flight cancellation decisions. An aircraft route is a sequence of flight legs spanning the recovery period, with the origin of a flight leg the same as the destination of its predecessor in the sequence, and the elapsed time between two successive legs at least as great as the minimum aircraft turn time. Routes for aircraft with planned maintenance within the recovery period are not altered to ensure that the modified routes satisfy maintenance requirements.

Let P be the set of aircraft routes, Q the set of aircraft, F the set of flight legs and T the set of airport locations. In our example, the set of aircraft routes includes:

1. routes corresponding to the delay propagation recovery option, namely A300-1 on CL551 followed by CL302 and A300-2 on CL332 followed by CL502; and
2. routes corresponding to the aircraft swapping recovery option, namely A300-1 assigned to CL551 followed by CL502, and A300-2 assigned to CL332 followed by CL302 with a 30-minute delayed departure.

The set of aircraft includes two identical aircraft, denoted A300-1 and A300-2, and the set of flight legs is CL551, CL332, CL302 and CL502.

---

[3] This advanced section can be skipped without loss of continuity.

In the solution to the optimization model, aircraft route variable x_j^k will be set equal to 1 if the recovery decision is to assign aircraft k to route j, and 0 otherwise. Its cost, denoted d_j^k, equals the sum of the delay costs associated with delays implied by assigning aircraft k to route j. As discussed above, quantifying these costs requires estimation of the downstream delay costs of aircraft, crew and passengers impacted by this recovery decision.

A flight cancellation variable, denoted y_i, is set to 1 if flight leg i is cancelled and 0 otherwise. In our example, if CL551 and CL502 are cancelled, this corresponds to A300-1 staying on the ground in ORD during the recovery period and A300-2 operating CL332 and CL302. Similarly, cancelling CL332 and CL302 corresponds to A300-1 operating CL551 and CL502 and A300-2 staying on the ground in BOS for the duration of the recovery period. The cancellation of all flight legs implies that both aircraft are on the ground during the recovery period. The cost associated with the cancellation of each flight leg i is c_i. As stated earlier, this is an approximation of the direct and indirect costs associated with recovering from the cancellation; that is, the costs of reassigning crews to the revised flight schedule and reaccommodating disrupted passengers.

Other parameters of the optimization model are: h_t equal to the number of aircraft needed at airport location t at the end of the recovery period to ensure that the plan can be resumed; δ_{ij} equal to 1 if flight leg i is covered by route j; and b_{tj} equal to 1 if route j ends at the airport t.

The aircraft recovery model is:

$$\text{Minimize} \quad \sum_{k \in Q} \sum_{j \in P} d_j^k x_j^k + \sum_{i \in F} c_i y_i$$

subject to:

$$\sum_{k \in Q} \sum_{j \in P} \delta_{ij} x_j^k + y_i = 1 \text{ for all flight legs } i \in F \tag{9.1}$$

$$\sum_{k \in Q} \sum_{j \in P} b_{tj} x_j^k = h_t \text{ for all airports } t \in T \text{ at the end of the recovery period} \tag{9.2}$$

$$\sum_{j \in P} x_j^k = 1 \text{ for all aircraft } k \in Q \tag{9.3}$$

$$x_j^k \in \{0, 1\} \text{ for all routes } j \in P \text{ and aircraft } k \in Q \tag{9.4}$$

$$y_i \in \{0, 1\} \text{ for all flights } i \in F \tag{9.5}$$

Constraints (9.1), together with constraints (9.4) and (9.5), require each flight leg either to be included in an assigned route or to be cancelled. Constraints (9.2) ensure that at the end of the recovery period, aircraft are repositioned so that the plan can be resumed. Finally, constraints (9.3) enforce the requirement that each aircraft be assigned to exactly one route, with the route commencing at the location of the aircraft at the start of the recovery period. The objective is to minimize flight cancellation and delay costs.

This aircraft recovery model was applied to a relatively small Boeing 757 flight schedule provided by Continental Airlines. For over 90% of the instances tested, with 42 flights,

16 aircraft and 13 airport locations, the approach produced a solution within 10% of the lower bound within 10 CPU seconds.

Crew Recovery

Although aircraft recovery decisions repair disrupted aircraft schedules, they often result in the disruption of crews. Flight cancellations, delays, diversion and swap decisions, together with crew illness, all result in the unavailability of crews at the locations needed.

To recover these disrupted crew schedules, one recovery option is *deadheading* crew members (i.e., repositioning crew members by flying them as passengers) from their point of disruption to the location of a later flight leg to which they are assigned. Once repositioned, a crew can then resume its original work schedule. Another option is to assign a reserve crew to cover the flight legs left *unassigned*, or *open*, by the crew disruption. Using reserve crews can be quite costly, however, because it can entail payment to *both* the reserve crew *and* the replaced crew. Airlines often must pay the *replaced* crew the *entire* amount originally planned, even if the work is performed by a different crew. A third recovery option is to reassign a crew from its original schedule to an alternative schedule. In this case, the new assignment must satisfy all collective bargaining agreements and work rule regulations, including maximum crew work time, minimum rest time, maximum flying time, maximum time away from home, etc. When reassigned, crews are typically paid the *maximum* of the pay associated with the original schedule and that of the new schedule to which they are assigned.

The crew recovery problem, then, is to construct new schedules for disrupted and reserve crews to achieve coverage of all flights at minimum cost. Crew recovery, however, is rife with challenges. First, the crew recovery solutions must be generated quickly, often in minutes to allow for real-time recovery. Second, information pertaining to the location and recent flying history of each crew member must be known at all times in order to generate recovery plans that satisfy the myriad of crew rules and collective bargaining agreements. Finally, the objective in solving the crew recovery problem is multi-dimensional; it is often cast as a blend of minimizing the incremental crew costs to operate the modified schedule, while returning to the plan as quickly as possible and minimizing the number of crew schedule changes made to do so. By limiting the number of crews affected, the quality of the original crew plans will be preserved to the greatest extent possible. Moreover, returning to plan as quickly as possible helps to avoid further downstream disruptions to aircraft, crew and passengers.

Compounding these complexities is the fact that aircraft and crew schedules are tightly linked. To understand how these linkages further complicate the recovery problem, consider again the example in Figure 9.3. For simplicity, assume that the original plan is for the crews to stay with the aircraft. Hence, assume that crew 1 is assigned to A300-1 and crew 2 is assigned to A300-2.

The aircraft recovery option involving delay propagation (Figure 9.4) necessitates a crew change if the delayed departure of CL302 results in crew 1 having to work beyond the maximum allowable hours. In this case, a reserve crew has to be called in to operate CL302, and crew 1 is forced to rest at LGA for the night, out of position and unable to fly the next assigned flight legs. Reserve crews, or other available crews, then have to be assigned to cover newly "open" flight legs, or the legs have to be cancelled. Moreover, the displaced crew either has to be deadheaded to a location where it can resume its work, or remains idle (but earning pay) in LGA until it can continue its scheduled flying. All of these options can lead to significant costs to the airline.

Aircraft swapping, like delay propagation, can also result in crew disruptions. The aircraft swapping option above, with A300-1 flying CL551–CL502 and A300-2 flying CL332–CL302, requires that the crews similarly swap flights. Crew 1 must follow CL551 with CL502 and crew 2 must follow CL332 with CL302. Keeping the original crew assignments is infeasible as CL302 is (re)scheduled to depart before the (delayed) arrival of CL551. With the required change in crew assignments, crew 1 is out of position in BOS instead of ORD, and crew 2 is in BOS instead of ORD, at the end of the recovery period. While this does not create an imbalance with respect to the system-wide need for crews to operate the schedule (one A300 crew ends its day in BOS and one ends its day in ORD), it does point to the many issues that can arise due to the non-interchangeability of crews. For example, if crew 1 is domiciled in BOS, then it is likely that the crew members are scheduled to return home for rest and time off after flying CL302. If, instead, the crew members complete their flying with CL502, they will have to be deadheaded to BOS after arriving in ORD, at significant expense to the airline. Moreover, this option might not even be feasible if the earliest arrival of crew 1 in BOS results in the crew being on duty longer than the maximum allowed time. (A deadheading crew, while not flying the aircraft, is considered to be on duty.)

The non-interchangeability of crews is also evident when flying time is considered. Again, consider the swapping option above and observe that CL502 involves more flying time than CL302, and CL551 requires more flying time than CL332. It is possible, then, that the reassignment of crew 1 from CL551 to CL502 instead of CL302 could cause crew 1 to exceed the limit on the maximum daily flying time allowed. In this case, the assignment of crew 1 to CL502 would not be allowed, and a reserve crew would have to be called in to fly CL502, or CL502 would have to be cancelled.

From the above discussion, it is clear that evaluating the aircraft recovery options without considering the impacts on crews can lead to non-optimal, and indeed very costly, and even infeasible (non-implementable) recovery strategies. Beyond the inherent complexities of rerouting aircraft, scheduling delayed flight departures and making cancellation decisions, an effective aircraft recovery solution approach must account for the downstream costs and impacts on crew.

Passenger Recovery

Just as aircraft recovery decisions can result in crew disruptions, aircraft and crew recovery decisions can lead to passenger disruptions. (Recall that a disrupted passenger is a passenger who, due to flight cancellations or missed connections caused by flight delays, must be reaccommodated on a different itinerary than planned.) The extent of the resulting passenger delay is a function of the number of alternative itineraries between the origins and destinations of the cancelled flights, the number of passengers disrupted by the cancellations and the number of available seats on these alternative itineraries. In a study of passenger delays for a major US airline (Bratu and Barnhart, 2005), it was found that passengers who misconnect due to flight delays were often reaccommodated on their *best* alternative itineraries; that is, on itineraries that arrived at their destinations at the earliest possible time, given the timing of the disruptions. In comparison, only about one-half of the passengers disrupted by flight leg cancellations were reaccommodated on their best itineraries. This occurs because the number of misconnecting passengers per flight leg is small compared to the number of passengers disrupted by a cancellation.

In our example involving aircraft swapping (Figure 9.5), all passengers connecting from CL551 to CL302 miss their connections, For the sake of illustration, suppose that the four flights in Figure 9.5 represent the entire daily flight schedule of the airline. Then, these disrupted passengers will have to remain in LGA overnight and return home on CL302 the next day, *if* seats are available. With high load factors, it is plausible that some of the disrupted passengers might not be accommodated on the first, or even second, third, etc., day of recovery.

Passengers, who cannot be accommodated the next day can be delayed to subsequent days, until the inventory of empty seats meets the backlog of disrupted passengers. In this example, it is clear that the cancellation option has large associated passenger costs, and, hence, it would be implemented only if the constraints and costs associated with other considerations, such as aircraft or crew, dominate.

Flight cancellation options pose significant challenges when it comes to comparing them to other options. While cancellations reduce delays to aircraft flights, they clearly create major passenger delays. In the *best* scenario above, the passengers connecting from CL551 to CL302 are delayed by 24 hours, while the delay to CL302 is only 30 minutes. From the perspective of passenger delay, the delay propagation option, even with its total associated aircraft delay of 9 hours, might be preferable. With this option, all possible passenger connections are preserved and the maximum passenger delay is (only) 5 hours. This serves to illustrate the point that minimizing aircraft delays does not always minimize passenger delays. Moreover, it reinforces the point made earlier relative to crews: if aircraft recovery decisions do not consider the associated passenger disruption costs, the true cost of the recovery plan will be underestimated, potentially by a large margin.

A complicating factor in estimating the cost associated with passenger recovery is that the number of seats available at the time of recovery and the passenger demand for these seats are functions of the disruptions, delays and recovery actions taken earlier. While seemingly straightforward, quantifying precisely, or even accurately, the number of available seats and the disrupted passenger demand for each flight leg is difficult, and even impossible, for dynamic, stochastic operations like those of the airlines.

To illustrate, first consider the task of computing the number of disrupted passengers for whom a particular flight leg is an option for recovery. Given known flight cancellations, delays and aircraft swaps, this number is straightforward to calculate. However, when facing extensive delays, some unknown number of these passengers will choose to cancel their trip or travel with a different air carrier or on a different mode of transportation, thereby reducing the pool of passengers. Moreover, future aircraft delays and recovery actions can result in the generation of an additional, unknown number of passengers for whom the best recovery option might be this same flight leg.

Unfortunately, the task of computing the number of available seats on flight legs yet to be operated is equally challenging. In a static operation, this number is calculated simply as the total number of seats minus the number of seats with assigned passengers. In a dynamic environment, however, after a recovery plan is developed and implemented, the number of available seats can change dramatically from what was planned if, for example, an aircraft swap or flight cancellation occurs. And if these changes do occur, the number of disrupted passengers can also change dramatically.

The principal message to be drawn from this discussion, as from that on the topic of crew recovery, is that expected passenger costs must be considered when deciding which aircraft recovery or crew recovery actions to take. Clearly, actions that are "optimal" from the aircraft or from the crew perspective can be very costly (and even infeasible) when aircraft, crew and passengers are jointly considered.

9.2.4 Evaluating the Costs of Recovery Options: The Challenges Imparted by Uncertainty and Downstream Effects

To decide whether to implement aircraft swapping, delay propagation, cancellation or some other recovery option, the airline recovery team must determine the costs associated with each. This is a non-trivial task in itself, with considerations including scheduled maintenance requirements, crew work rules and reserve crews, and passenger reaccommodations and load factors, to name a few. Nonetheless, estimation of these costs is a worthwhile and even necessary task, as the examples above serve to illustrate. Decisions made earlier in the sequential recovery decision process can have a tremendous impact on the recovery options (and associated costs) available later in the process.

At each stage of the recovery process, it is necessary to consider the downstream costs associated with all possible recovery actions. In the aircraft recovery model, terms are included in the objective function to capture the costs associated with flight delays and the costs associated with cancellations. These terms should reflect both direct and indirect costs to aircraft, crews and passengers that result from the recovery action under consideration. The delay cost of a recovery action, such as delay propagation, aircraft swapping or cancellation, must be an inclusive estimate of:

1. the aircraft-related costs imparted, including those involving maintenance and aircraft repositioning if necessary, to *all* aircraft as a result of the recovery action;
2. the crew-related delay costs associated with the delays, crew reassignments and repositioning, and the use of reserve crew members; and
3. the passenger-related costs associated with delays and resulting rebookings associated with missed connections and cancellations, and loss of passenger goodwill.

Even with sophisticated methods to estimate the costs of recovery options, uncertainty about future airline operations ensures that cost-minimizing recovery decisions can sometimes be far from optimal. Suppose, for example, that the recovery decision for delayed flight leg CL551 is to swap aircraft. The recovery plan is implemented and A300-2 flies CL332 and, upon arrival, departs on CL302. After the departure of CL302, the airline learns that CL502 will be delayed due to unexpected storms over ORD. Suppose that, with this delay, the crew assigned to CL551–CL502 will exceed the maximum workday limit, thereby rendering the revised plan infeasible. Further suppose that there are no reserve crews available, and hence flight CL502 must be cancelled. The end result is one aircraft out of position (in LGA instead of ORD) and two crews out of position (one in BOS instead of ORD, and one in LGA instead of BOS), all resulting in downstream (costly) impacts to operations. If the delay to CL502 had been known earlier, the airline might have decided not to swap aircraft (and crews) and the outcome might have been that all flight and crew schedules were operated as planned, albeit delayed.

Because recovery decisions must be made with incomplete and imprecise information about future operations, recovery plans are themselves subject to disruption, potentially resulting in extensive delays and additional recovery costs. To a large extent, then, the need for replanning is unavoidable. Nonetheless, including estimates (even inexact) of downstream costs in the aircraft recovery decision process can lead to improved decisions, as indicated by the study of Bratu and Barnhart (2006). Using actual airline operational data, they developed and solved an aircraft recovery optimization model to determine flight departure times and cancellations to minimize recovery costs, including the costs of reaccommodating disrupted passengers and crews, rerouting aircraft and cancelling flight legs. For problem instances containing 303 aircraft, 74 airport locations (3 of which are hubs), 1088 flight legs per day on average and 307675 passenger itineraries, they generated solutions achieving significant reductions in the expected delay to passengers. The expected benefits included reductions of more than 40% in the number of disrupted passengers, more than 45% in the number of passengers required to overnight at a destination other than that planned, and more than 33% in the total delay minutes of disrupted passengers.

9.3   Robust Airline Scheduling

Robust airline scheduling is an extension of schedule planning models, such as those presented in Chapter 7, that tries to account for the myriad of uncertainties that arise in the execution of a schedule. The idea is to extend planning models to include both the costs associated with executing the schedule as planned and the expected costs of recovering the plan from disruptions. By ignoring recovery costs, as has been done historically in the generation of "optimal" schedules, plans are generated that maximize resource utilization and minimize non-productive time on the ground, i.e., *slack time*. Lack of slack time, however, makes it difficult for disruptions to be absorbed in the schedule and limits the number of options for recovery. While this issue can be overcome by including recovery costs in optimization models, the associated modeling and computational challenges of doing so are great.

A number of researchers have begun to consider the challenge of including expected recovery costs in planning models, recognizing that robust planning is a problem rich in opportunity and potential impact. To facilitate the generation of robust plans, various proxies of *robustness* have been developed. These proxies typically reflect the objective of developing *flexible* plans that either provide a rich set of recovery options for passengers, crews and aircraft, or isolate the effects of disruptions, requiring only localized plan adjustments.

To highlight the linkages between robust and conventional optimization approaches, the next sections are organized to mirror the structure of Chapter 7. A typical approach to achieving robust plans is to identify a "robust" objective and to modify the optimization models presented in Chapter 7 to capture this new objective. In the next sections, robustness objectives are described for the schedule design, fleet assignment, aircraft maintenance routing and crew scheduling problems, and, for some, brief descriptions are provided of the associated impacts.

9.3.1 Robust Schedule Design

One approach to robust schedule design is to generate flight schedules that are likely to minimize passenger delay (Lan *et al.*, 2006). The idea is to select flight departure times to minimize the probability that passenger misconnections will occur during operations. This can be achieved by simply setting departure times "late enough" to effectively reduce to zero the probability of misconnecting. This can be done, in principle, by adding large amounts of slack time in the schedule. However, the total amount of slack time that can be added is limited by the number of aircraft available to operate the flight schedule. The challenge, then, is to find the most effective placement of slack, or, equivalently, to select flight leg departure times to minimize the number of disrupted passengers, while maintaining a high level of aircraft productivity.

Using this approach, the expected passenger delays and disruptions associated with the operation of a "robust" schedule with delay-minimizing departure times have been compared (Lan *et al.*, 2006) with those resulting from the actual operations of a major US airline. Even with flight leg departures in the robust schedule restricted to be at most 15 minutes earlier or later than in the original schedule, simulation results showed that the robust schedule could yield reductions in passenger delays of 20% and reductions in the number of passenger misconnections of 40%. Interestingly, these reductions are achieved with no increase in cost to the airline. This is possible because the robust optimization approach is restricted to select the solution that minimizes passenger delays and misconnections from among *only* the solutions with minimum total planned cost.

Another approach to achieving schedule robustness is to isolate the effects of disruptions (Kang, 2004): the legs of the flight schedule are partitioned into independent subnetworks, with aircraft and crew allowed to operate flight legs only within a single subnetwork. (Passengers, however, can travel within multiple subnetworks.) The subnetworks are prioritized based on the total revenue of the flights legs they contain, with the maximum revenue subnetwork having the highest priority. When disruptions occur, the top priority subnetworks are recovered first, shielding the associated crew, aircraft and passengers to the greatest extent possible from the resulting delays. This has the effect of relegating disruptions to the low-priority subnetworks, and minimizing the revenue associated with delayed and disrupted passengers.

An advantage of this approach is that it can simplify recovery operations. Because delays and propagation effects for aircraft and crew are contained within a single subnetwork, the recovery process needs only to take corrective action on the flights in the affected subnetwork and not on the entire airline network.

9.3.2 Robust Fleet Assignment

Because schedules are sensitive to disruptions at hubs, it follows that operational robustness is enhanced when hubs are "isolated" to the greatest extent possible. The degree to which a hub is isolated can be measured using a hub connectivity metric; the smaller the value of hub connectivity, the more isolated the hub. This concept, when applied to the fleet assignment problem, involves assigning aircraft types to flight legs to limit total hub connectivity and maximize the number of short cycles assigned to the same fleet type (Rosenberger *et al.*, 2004). A short cycle is a sequence of at least two connected flight legs that begins and ends at the same location and contains a small number of legs (two, for example). Short cycles allow an airline to limit the number of flights cancelled when a cancellation is necessary, thereby limiting the scope and impact of the disruption and facilitating recovery. When compared to solutions generated using a conventional model, "robust" fleet assignments generated in this manner have been shown to yield significant reductions in the number of cancelled flight legs, and improvements in on-time schedule performance, with only small increases in solution cost.

Another proxy for fleet assignment robustness is *purity* (Smith, 2004). Purity is achieved by limiting the number of different aircraft fleet types operating at spoke stations to at most one or two. While increasing solution cost, purity increases the number of possible crew swaps, thereby increasing the number of recovery options, and hence the recoverability of the solution. Purity also decreases maintenance cost because the need for spare parts is reduced.

9.3.3   Robust Aircraft Routing

Delay propagation occurs when an aircraft scheduled to operate a flight leg is not ready at the scheduled time due to an upstream disruption or a delay to one or more of its earlier assigned flight legs. Propagated delay, then, is a function of an aircraft's routing. This gives rise to a robust aircraft routing model aimed at minimizing delay propagation and, hence, passenger delays and disruptions, by optimizing aircraft routing decisions (Lan *et al.*, 2006). Optimizing routings, like optimizing departure times, can be cast as optimizing the placement of slack time in the schedule. Unlike the optimization of departure times, however, the allocation of slack time is achieved solely through routing decisions, not through schedule adjustments.

To illustrate the idea, consider the example in Figures 9.6, 9.7 and 9.8, where MTT is the minimum turn time. In the planned first-in, first-out aircraft routing, one aircraft operates flight leg *f1* followed by leg *f2*, and another operates leg *f3* followed by leg *f4*. Assume that *f3* is typically on schedule and *f1* is typically delayed, as depicted in Figure 9.6. Because insufficient turn time results from the delay, some of the delay to *f1* will propagate downstream to *f2*, as shown in Figure 9.7. Delay propagation can be avoided, however, by altering the aircraft routings to *f1* followed by *f4* for one aircraft, and to *f3* followed by *f2* for the other aircraft, as shown in Figure 9.8.

For a major US airline it has been shown that robust routings (when compared to conventional aircraft routings) can yield significant reductions in the total number of

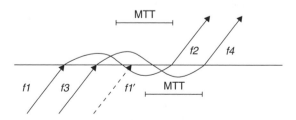

Figure 9.6   First-in, first-out routings and delayed flight leg *f1*

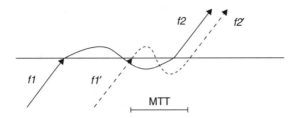

Figure 9.7 Delay propagation due to delay of flight leg *f1*

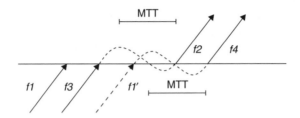

Figure 9.8 Revised routings minimizing delay propagation

passengers disrupted and in the total passenger delay. Specifically, in a Lan *et al.* experiment, the robust routings resulted in an average 11% reduction in the number of disrupted passengers and an average 44% reduction in the total expected number of propagated delay minutes. Moreover, the robust routings improved by 1.6% the airline's Department of Transportation (DOT) on-time arrival rate. This allowed the airline to improve its position in the DOT ranking; this is of significance because on-time performance is a widely publicized and often cited indicator of airline performance.

Another robustness approach is to build *flexible* aircraft routings that maximize the number of aircraft swap opportunities during recovery (Ageeva, 2000). In this case, a swap opportunity exists whenever the planned routings of two aircraft of the same type intersect at two or more locations. With this structure, aircraft routes can be operated by aircraft as planned up to the first intersection location, then the portion of the route between intersection locations can be operated by either aircraft, and finally the original assignments can be resumed after the second intersection location. As discussed in Section 9.2.2, aircraft swapping is an effective and often used recovery tool to reduce, and even eliminate, delays and disruptions. Flexible aircraft routings have been shown by Ageeva to improve schedule performance, with no increase in the cost of the routing solutions relative to those generated using conventional approaches.

9.3.4 Robust Crew Scheduling

To minimize both planned and unplanned costs for crews, it is necessary to balance the added costs of robust solutions with the reductions in recovery costs enabled by the solution's robustness. One approach for doing this is to compute, for each crew pairing, a value of *non-robustness* (Ehrgott and Ryan, 2002). The value of non-robustness is zero if crews do not change planes, but equals the potential disruptive effects of delays if the plan requires crews to connect between different aircraft. The objective of the crew scheduling problem, then, is to minimize the value of non-robustness, while maintaining the cost of the corresponding crew solution "close" to that of the minimum cost crew solution.

The experience of Ehrgott and Ryan with this approach suggests that small increases in cost allow considerable robustness gains. In one instance, by increasing costs by less than 1%, the non-robustness metric was increased by more than two orders of magnitude. The more robust solutions are characterized by longer ground times for the crews when they are required to connect between flight legs operated by different aircraft; fewer crew connections requiring the crew to transfer between two different aircraft; slightly longer crew duty times; and a slight decrease in the number of flight legs included in a crew's schedule.

Analogous to the idea of providing recovery flexibility through aircraft swapping is the concept of increasing flexibility in crew recovery using *move-up crews* (Shebalov and Klabjan, 2006). A move-up crew for a flight is a crew that is not actually assigned to that flight but can be feasibly assigned to it, if necessary. For this potential reassignment to be feasible it is required that: (1) the move-up crew has the same domicile as the assigned crew; (2) the move-up crew is available to operate the flight at its planned departure time; and (3) the move-up crew is available to operate the rest of the crew schedule that includes the flight. Using this concept, the objective of the robust crew optimization model is to find the set of crew schedules that maximizes the number of possible move-up crews. A large number of move-up crews translate to a large number of opportunities for crew swaps during recovery. Using this approach, flexible crew solutions are generated with only slightly higher planned costs and a 5- to 10-fold increase in the number of move-up crews, as compared to the conventional, cost-minimizing crew optimization solutions.

9.4 Directions for Ongoing and Future Work on Schedule Recovery from Irregular Operations

With direct and indirect economic costs amounting to several billion dollars annually, the airline industry has a vital stake in mitigating the effects of disruptive events and expediting recovery from "irregular" operations. There is growing recognition in the airline industry that planning for schedule robustness and reliability in the complex, highly stochastic and dynamic environment of air transportation might be just as important as planning for minimizing costs. Specific achievements that have been described herein include: development and implementation of approaches for efficient "recovery" of aircraft, crews and passengers following schedule disruptions; and the nascent appearance of increasingly viable models for introducing robustness in the scheduling of aircraft, crews and passengers.

At the same time, it is fair to describe all this work as still being in its early stages in many respects. On the side of the airlines, decision support software for recovery is perhaps at the stage where planning software was 15 years ago. While research is active and hardware and data support have improved substantially, optimization-based decision support tools for rapid recovery are still at an early stage of implementation at the major airlines. This represents a difficult, but crucial future challenge.

References

Ageeva, Y. (2000) "Approaches to Incorporating Robustness into Airline Scheduling," Master of Engineering thesis, Department of Aeronautics and Astronautics, MIT.

Arguello, M.F., Bard, J.F., and Yu, G. (1997) "Models and Methods for Managing Airline Irregular Operations," in G. Yu (ed.), *Operations Research in Airline Industry*, Kluwer Academic, Boston, MA, pp. 1–45.

Barnhart, C., Belobaba, P., and Odoni, A.R. (2003) "Applications of Operations Research in the Air Transport Industry," *Transportation Science*, Vol. 37, No. 4, pp. 368–391.

Bratu, S. and Barnhart, C. (2005) "An Analysis of Passenger Delays Using Flight Operations and Passenger Booking Data," *Air Traffic Control Quarterly*, Vol. 13, pp. 1–27.

Bratu, S. and Barnhart, C. (2006) "Flight Operations Recovery: New Approaches Considering Passenger Recovery," *Journal of Scheduling*, Vol. 9, pp. 279–298.

Ehrgott, M. and Ryan, D.M. (2002) "Constructing Robust Crew Schedules with Bicriteria Optimization," *Journal for Multi-Criteria Decision Analysis*, Vol. 11, pp. 139–150.

Kang, L.S. (2004) "Degradable Airline Scheduling: An Approach to Improve Operational Robustness and Differentiate Service Quality," Doctoral thesis, Operations Research Center, MIT, http://dspace.mit.edu/handle/1721.1/17659.

Lan, S., Clarke, J.P., and Barnhart, C. (2006) "Planning for Robust Airline Operations: Optimizing Aircraft Routings and Flight Departure Times to Minimize Passenger Disruptions," *Transportation Science*, Vol. 40, pp. 15–28.

Rosenberger, J., Johnson, E.L., and Nemhauser, G.L. (2004) "A Robust Fleet-Assignment Model with Hub Isolation and Short Cycles," *Transportation Science*, Vol. 38, pp. 357–368.

Shebalov, S. and Klabjan, D. (2006) "Robust Airline Crew Scheduling: Move-up Crews," *Transportation Science*, Vol. 40, pp. 300–312.

Smith, B.C. (2004) "Robust Airline Fleet Assignment," Ph.D. dissertation, The Logistics Institute, Georgia Institute of Technology.

10

Labor Relations and Human Resource Management in the Airline Industry[1]

Jody Hoffer Gittell, Andrew von Nordenflycht, Thomas A. Kochan, Robert McKersie and Greg J. Bamber

Labor relations (LR) and human resource (HR) management are two aspects of the employment relationship, reflecting the ways that firms relate to their employees. The employment relationship is especially important in the airline industry, given its service-intensive nature, its high ratio of labor costs to total costs, and the high level of union representation in the industry. Employees have the ability to affect airline performance in significant ways, both positively and negatively. Through collective bargaining, employees can achieve higher wages and employment security but this can leave firms with higher costs and less operating flexibility. Furthermore, employees can impose additional costs in the process of setting those wages and employment conditions, through strikes or other service disruptions. At the same time, employees can also add significant value, with the potential to play a major role in lowering unit costs through their coordinated and committed efforts and through their productivity-enhancing ideas, and a major role in delivering high-quality service through their friendly, responsive interactions with customers.

Evidence shows that front-line employees play a critical role in achieving quality and productivity outcomes in the airline industry (Gittell, 2003). As a result, airlines around

[1] For much more information and analysis of the issues in this chapter, see Bamber *et al.* (2009). The authors of this chapter participate in the Airline Industry Council's international research network. They acknowledge that this chapter draws on the work conducted by many others in the network. A series of articles from this network will be published in 2009 in the *International Journal of Human Resource Management*.

the world have looked to the management of their workforce for potential sources of advantage, striving to achieve either lower unit costs or superior customer service, or both. However, there are a variety of approaches for achieving such advantages and some are more effective than others.

In addition to firm-level strategies, the industry-level structures and processes that shape the employment relationship are also important. That is, airlines are part of an air transport system that involves competitors, partners in code sharing and other relationships, unions that represent employees at multiple airlines, and government regulatory agencies (such as the National Mediation Board (NMB) in the USA) that oversee labor relations in the airline industry and participate actively in the resolution of collective bargaining disputes. Effective structuring and management of the industry-level aspects of labor relations, including the regulatory framework, can yield outcomes that represent mutual gains for airlines, employees and the flying public.

Before discussing the lessons of what works in airline human resources and labor relations (HR/LR), it is useful to examine the basic structures and patterns of HR/LR that have long characterized the airline industry. Thus, this chapter addresses the following topics. First, we outline a general framework for identifying the key aspects of an employment relationship. Second, we describe the regulatory framework and history for LR in the USA. We then provide a discussion of the regulatory frameworks and history for LR in other parts of the world, including Europe, Scandinavia and Asia. Finally, we explore "what works?" in both airline LR and HR management to build high levels of productivity and service quality.

10.1   Alternative Strategies for the Employment Relationship

The employment relationship encompasses the broad range of ways in which firms interact with employees to set the terms of employment (e.g., wages) and to communicate and coordinate the work to be performed. Identifying a few major dimensions of this relationship and some basic alternative approaches to addressing those dimensions will help set the context for the chapter. At a conceptual level, the employment relationship can be segmented into three tiers of interaction: workplace; collective bargaining; and strategic decision making (Kochan *et al.*, 1986). The first tier encompasses the way in which employees and managers interact on a day-to-day basis and is the primary sphere of many HR approaches, policies and programs.

Within HR there are, broadly, two alternative models or approaches to the exercise of authority, which have been labeled control and commitment (Walton, 1985). In the control model, the interaction between managers and employees is hierarchical and management acts unilaterally. Employees are expected to comply with management's orders, but are not expected to go above and beyond in exercising initiative on behalf of the company. In the commitment model, managers consult more with employees on decisions and allow employees more discretion in the execution of their jobs. In return, employees are expected to demonstrate greater commitment to the firm and its customers, and to exert higher levels of discretionary effort.

Collective bargaining encompasses the structures and processes of how unions and firms negotiate and implement the labor contracts that set many of the basic employment terms, such as wages, hours, working conditions and disciplinary procedures. Within the collective bargaining realm, the relationship between a firm and a union can fall into roughly three categories: avoidance; accommodation; and partnership (Walton *et al.*, 1994). While the choice of how to approach the relationship depends partly on union representatives and rank and file employees, it has long been recognized that management tends to be more influential than labor in shaping the overall LR climate of an employment relationship (Kochan, 1980). In avoidance, the firm seeks a non-union environment either by preventing a union from forming or by bypassing or removing an existing union. In accommodation, the firm accepts the basic existence of the union, but maintains an adversarial stance and keeps interaction largely within the bargaining process. In partnership, the firm and union have a more cooperative, less adversarial stance, looking for opportunities for mutual gains arrangements and interacting more frequently and consultatively and beyond the bargaining table.

Finally, strategic decision making refers to the extent to which employees or their representatives participate in the governance and management of the firm. Clearly there are interactions between the different levels. For example, collective bargaining can shape workplace interactions, and the nature of workplace interactions can influence the tone and culture of collective bargaining.

Firms differ in their HR/LR strategies, but these strategies also differ from industry to industry and from country to country due in part to differences in regulatory frameworks. These differences can have an important impact on how effective specific HR/LR strategies will be for individual firms. Which choices offer the best potential for achieving high levels of firm performance in the airline industry? We discuss research findings on this question later in this chapter. But first we provide an overview of distinctive features of LR in the airline industry, focusing first on the USA and then on other parts of the world, especially Europe, Scandinavia, Asia and the Middle East.

10.2   Labor Relations in the US Airline Industry

The airline industry is one of the most highly unionized industries in the USA, with over 70% of eligible employees of the major airlines belonging to unions. Figure 10.1 shows the percentage of employees who are unionized at each of the major US airlines. Accordingly, the collective bargaining dimension plays a very central role in defining the employment relationships at most airlines.

A noteworthy characteristic of the unionization of airlines is the presence of a large number of different unions in the industry and even within a given firm. See Table 10.1 for a summary of the unions that represent employees at major US airlines. Unlike in other industries, airlines typically bargain with multiple unions representing distinct occupations. Even when the same union represents a particular occupational group across several airlines, collective bargaining and contract negotiations occur at the firm level, not at an industry-wide level. This decentralized bargaining structure has pros and cons, but, in any case, shapes the nature of the collective bargaining process. Collective bargaining is also shaped by the regulatory framework – that is, the laws governing how unions and firms interact. A large part of the airline industry's distinctiveness stems from its unique regulatory framework.

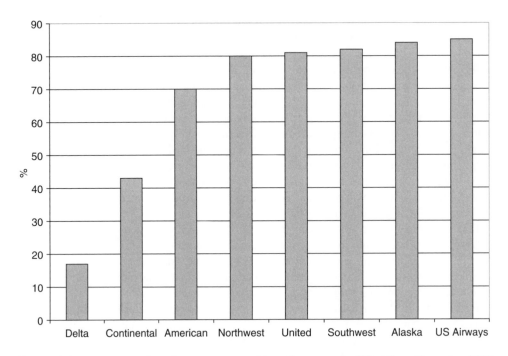

Figure 10.1   Percentage of employees represented by unions at major US airlines, December 2007
Sources: Airline Industrial Relations Conference and Airline Annual Reports

10.2.1   Regulatory Framework for US Labor Relations

In contrast to LR in almost all other private-sector industries, which are governed by the National Labor Relations Act (NLRA), LR for the US airline industry are governed by the Railway Labor Act (RLA), a statute first enacted to cover railroads in 1926 and then extended to air transportation in 1934. The RLA has a number of provisions designed to protect the public from work stoppages that are part of collective bargaining in this industry. For this reason, and because proposals for reform of the RLA have surfaced from time to time, we outline the process in some detail here, based on the descriptions of von Nordenflycht and Kochan (2003).

One notable difference between the NLRA and the RLA is that union representation under the RLA is craft based rather than workplace based. In other words, a given union will represent the employees in a specific occupation (e.g., pilots, mechanics) across all the sites of a single airline – in contrast to representing employees across all occupations, but only at one specific site (e.g., a given airport). Some observers attribute the high degree of unionization at airlines to this craft-based framework, since the craft model may make it easier to unionize; that is, a union can appeal to the particular needs of only one occupation, rather than trying to organize employees of very disparate skills, wage levels and backgrounds. But in other ways, it might make it harder, as unions must organize employees across a whole airline, rather than at a single site. Observers also link the craft-based framework to the variety of unions and the decentralized bargaining structure. However, it is unclear that the RLA itself has produced these structures. For

Table 10.1 Unions representing employees of the major US airlines (mid-2008)

Airline	Pilots	Flight attendants	Mechanics and related		Clerical/ agent
			Mechanics	Ramp/fleet service	
Alaska	ALPA	AFA-CWA	AMFA	IAM	IAM
American	APA	APFA	TWU	TWU	None
Continental	ALPA	IAM	IBT	None	None
Delta	ALPA	None	None	None	None
Northwest	ALPA	AFA-CWA	AMFA	IAM	IAM
Southwest	SWAPA	TWU	AMFA	TWU	IAM
United	ALPA	AFA-CWA	IBT	IAM	IAM
US Airways	USAPA	AFA-CWA	IAM	IAM	IBT/CWA

Note: AFA-CWA = Association of Flight Attendants (affiliated with Communications Workers of America); ALPA = Air Line Pilots Association; AMFA = Aircraft Mechanics Fraternal Association; APA = Allied Pilots Association; APFA = Association of Professional Flight Attendants; CWA = Communication Workers of America; IAM = International Association of Machinists; IBT = International Brotherhood of Teamsters; SWAPA = Southwest Airlines' Pilots Association; TWU = Transport Workers Union; USAPA = US Airline Pilots Association.
Source: Airline Industrial Relations Conference.

example, in the railroad industry, also governed by the RLA, bargaining has sometimes occurred at the industry level.

Another key difference from the NLRA is that under the RLA, contracts do not have fixed expiration dates. Instead, they have "amendable" dates. After the amendable date, the provisions of the existing contract remain in effect until the parties reach a new agreement. New contract terms cannot be imposed unilaterally and strikes or lockouts cannot be initiated until the parties have progressed through several steps that are regulated by the National Mediation Board (NMB).

If the parties cannot reach a contract agreement on their own, either side may then apply for mediation services from the NMB. Once in mediation, negotiations continue until an agreement is reached or until the NMB declares an impasse. At that point, the NMB offers the option of voluntary binding arbitration. If either party rejects the offer of binding arbitration, the NMB "releases" the parties. Once released, the parties then enter a 30-day "cooling-off period," during which time the existing contract provisions remain in effect. At the end of the cooling-off period, if the parties have still not reached an agreement, the NMB chooses whether to let the parties engage in "self-help" (i.e., a strike by workers, or a lockout or unilateral imposition of new contract terms by management) or refer the case to a Presidential Emergency Board (PEB). The PEB, composed of three neutral experts appointed by the President, is allowed 30 days to deliberate and formulate a recommended settlement. After the PEB issues its recommendations, another 30-day cooling-off period begins. Finally, at the end of the second cooling-off period, the parties are free to engage in self-help. As a final recourse, after the expiration of the second cooling-off period, the President can refer the case to Congress, requesting that Congress legislate a settlement.

In other words, once a contract becomes amendable, the parties are legally barred from self-help until the NMB releases them and the cooling-off period expires. Theoretically,

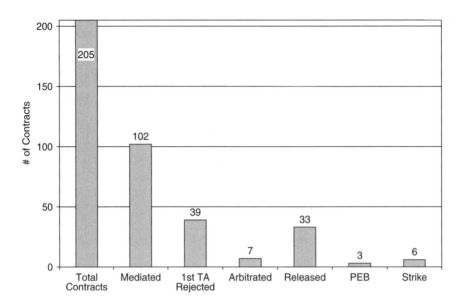

Figure 10.2   Airline dispute resolution under the Railway Labor Act, 1984–2001
Source: Airline Industrial Relations Conference

the parties could be prevented from self-help indefinitely, as the decision to release the parties while in mediation is at the discretion of the NMB. Even once the NMB releases the parties, it is then a minimum of 30 days and a maximum of 90 days (first cooling-off period, PEB, second cooling-off period) before the parties can engage in a strike or lockout.

Not surprisingly, strikes have been relatively rare in the airline industry (notwithstanding their high visibility when they do occur). Moreover, strikes have also become less frequent over time. The frequency of strikes since deregulation is much lower than that prior to deregulation and has been decreasing during the deregulated era, despite the occurrence of three major industry contractions (1981–1982; 1990–1992; 2001–2005) which has led to demands for major wage concessions. From 1982 to 2002, there were six strikes (out of 199 negotiations, or 3%) and there has been only one (Northwest's mechanics in 2005) since 2002 (see Figure 10.2). The increasing rarity of strikes is attributable not only to the procedures of the RLA, but also to the vastly increased costs of a strike since deregulation, when the traditional practice of running a joint strike insurance fund in which other airlines compensated the struck airline was disallowed and when airlines could freely invade each other's routes.

However, on the other side of the coin, the time required to reach agreement in the airline industry has been long relative to other industries. Some have argued that these lengthy contract negotiations are caused by the RLA procedures. For industries covered by the NLRA, 74% of contracts are settled before or within one month after the contract expiration date. By contrast, only 11% of airline contracts are settled within one month after the expiration date, as shown in Figure 10.3.

But long contract negotiations in airlines are not preordained. There is wide variation across collective bargaining relationships in the time required to reach

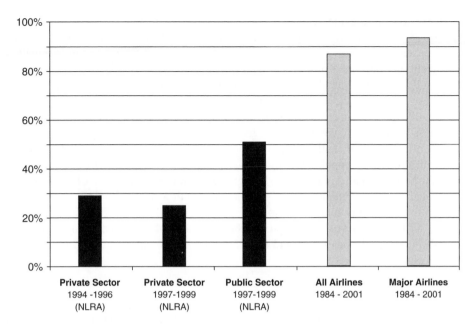

Figure 10.3 Percentage of contracts ratified more than one month past expiration date: NLRA vs. airlines industry

Sources: Airline Industrial Relations Conference (1984–2001) and Federal Mediation and Conciliation Service National Performance Review Surveys (1994–1996 and 1997–1999)

agreement – Continental, Southwest and a number of the regional carriers in recent years have achieved agreements in half or less of the average time (16 months) taken by all major carriers in recent years. Furthermore, there is some evidence that these differences are not related to the economic conditions facing either the airline or the industry at the time of negotiations (von Nordenflycht and Kochan, 2003). This suggests that the length of negotiations is not entirely caused by the regulatory environment or by the economic environment. Indeed, it seems that the success of negotiations is influenced more by the quality of the labor–management relationship than by labor law.

Overall, while the unique provisions of regulatory structures in each country are important to keep in mind and may well shape the outcomes of collective bargaining in the airline industry to some degree, they do not seem to determine the nature of union representation, the structure of bargaining, or the length of contract negotiations. Therefore efforts to improve the industry's LR *need not focus exclusively* on the regulatory framework. The history of the industry suggests that there are many opportunities for innovations by individual firms and unions, working alone or in partnership with each other, as we shall discuss later in this chapter.

10.2.2 US Airline Responses to Cost Pressures Post-deregulation

US airline employees have long earned a wage premium relative to comparably skilled workers in other industries, often attributed to the high unionization rates in the airline industry (Cremieux, 1996; Card, 1998; Hirsch and Macpherson, 2000). This was particularly true during the regulated era when government-regulated cost-plus pricing gave airlines little incentive to fight wage increases. When the industry was deregulated, many expected that wages might fall as legacy airlines faced a vastly more competitive environment and the threat of entrants with much lower labor costs.

As introduced in Chapter 5, new entrant airlines typically enjoy significant labor cost advantages relative to the incumbent or "legacy" airlines. A new entrant's workforce (a) is usually not unionized at the start, and hence is likely to have lower wage rates; (b) is most likely much younger or at least has lower tenure, leading to lower wage rates and offering significantly lower healthcare and pension costs, especially given the absence of any retiree population; (c) may also feel a higher degree of commitment to the airline, being a smaller, more tight-knit group and due to an absence of any history of labor-related conflicts. Finally, a new entrant starts with a relatively clean slate in terms of how to organize work and design jobs to achieve flexibility and efficiency, in contrast to legacy airlines whose half-century-old labor contracts often include dense thickets of work rules that strictly determine and limit how the airline can organize production and utilize its employees.

Concession Bargaining

After deregulation, the US legacy airlines sought ways to reduce their labor costs, citing threats from not only these low-cost new entrants, but also increased competition with other legacy airlines as well as periodic industry-wide downturns. Legacy airlines have attempted to reduce their wage rates in two basic ways. First, they have sought direct wage and work rule concessions. They have tried to persuade unions and employees that such concessions would be necessary to compete against old and new rivals with lower costs, backing up their position with threats to shrink the airline (and hence eliminate jobs) or even to enter bankruptcy, if the concessions were not forthcoming. In the first decade of deregulation, airlines pursued concessions very aggressively, buoyed by a perception that labor's position had been weakened both by the onset of competition in the industry and by the national climate that had emboldened employers to act aggressively in the aftermath of the air traffic controllers' strike in 1981.

However, despite setbacks soon after deregulation, labor unions regrouped thereafter and airlines have actually been generally unsuccessful in their efforts to achieve concessions. Research indicates that concessions were typically achieved only by airlines in great financial distress during industry downturns (Capelli, 1985; Nay, 1991). And even financial distress did not guarantee wage concessions, as a number of airlines were unable to reduce costs enough to survive and therefore ended up being acquired or ceasing operations entirely.

Two-Tier Wage Systems

The second approach was an attempt to create a two-tier wage system in some fashion, leaving the wages of existing employees alone while creating lower wage rates for new employees hired to staff new or growing operations. This attempt has come in a variety of forms. In one early attempt, a legacy airline, Texas International, formed a separate non-union subsidiary, New York Air. American Airlines pioneered a two-tier or "B-scale" wage system whereby wage rates for newly hired employees were significantly reduced, and this approach was rapidly adopted by rival airlines (Walsh, 1988). In the 1990s, two other forms arose. Legacy airlines developed alliances with smaller, regional airlines to fly shorter-haul flights that would connect with the legacy airlines' flights at their hub airports – in effect they were outsourcing short-haul flights to airlines with lower wage rates (even if unionized). In addition, a number of legacy airlines also experimented with forming separate short-haul operations within their own organization, built around a low-cost model including lower wages and fewer work rules.

However, unions were fairly quick to limit the ability of airlines to exploit these systems. For example, while airlines won major battles in the mid-1980s to establish the B-scale system, the B-scales were whittled away in subsequent negotiations, which closed the gap between existing and new employees by the early 1990s (Johnson, 1995). Furthermore, significant portions of labor contracts in the 1990s were devoted to "scope clauses" which specify a maximum number of aircraft that the airline can utilize in its low-cost operation or in its regional alliances. However, while unions have been able to negotiate limits on two-tier wage systems, wage differentials that exist across different routes (main line versus regional) within a legacy airline's network (as opposed to varying across old vs. new employees) have persisted.

In the end, legacy airlines experienced very limited success in reducing the wages from the time of deregulation through 9/11, with concessions occurring only at airlines facing financial crises. Various economic studies have found that wages in the industry declined slightly since deregulation but less than in the economy as a whole, and that declines were most pronounced during industry downturns but rose again afterwards (Johnson, 2001; Card, 1998; Hirsch and Macpherson, 2000). Legacy airlines' labor cost disadvantage vis-à-vis low-cost new entrants persisted for quite some time. Because the legacy airlines were able to develop significant advantages in other dimensions – including frequent-flyer programs, dominant hubs, ubiquitous networks (Levine, 1987) – several of them remained profitable and thus did not possess the leverage to force a reduction in wage rates and labor costs to match their low-cost rivals. Yet their attempts to do so have left no small amount of ill will among the industry's senior employees (Rosen, 1995).

Employee Stock Ownership Plans

The third response to these competitive pressures has been the employee-owned airline, in which employees collectively own a major share of their airline's equity and have the collective right to nominate several members of the board of directors. This participation by employees (and/or their unions) in corporate strategy and decision making has occurred through the use of Employee Stock Ownership Plans (ESOPs). Such ESOPs were in place at several major airlines for finite periods of time, including Western (1984–1986), Eastern (1984–1986), Northwest (since 1993), TWA (1992–2001) and United (1994–2002). One might be tempted to view these ESOPs as opportunities or attempts to improve the labor–management relationship and build employee motivation. But in actuality, these instances of "employee ownership" were all (essentially) the outcome of concessionary negotiations, in which employees granted concessions in order to stave off financial disaster, and demanded equity and board seats in return.

Not surprisingly, given the circumstances in which they were conceived, these ESOPs failed to deliver much advantage beyond the avoidance of bankruptcy. At each of the airlines that adopted an ESOP, there was an initial burst of cooperation and improved

LR, but it was short-lived. Within a few years or even months, new rounds of negotiations reignited mistrust and highlighted divergent expectations regarding wages and other issues (von Nordenflycht, 2002). Nor have the arrangements been long-lived. Three of the five airlines with ESOPs (Western, Eastern, TWA) were acquired by other airlines, two of which were acquired only two years after implementation of the ESOP. United's ESOP was terminated when the airline entered bankruptcy in 2002. Our quantitative analysis of LR at the major airlines indicates that ESOPs were weakly associated with productivity and profitability improvements at the airlines that adopted them, but not associated with higher or lower levels of profitability relative to airlines without ESOPs (Gittell *et al.*, 2004). The key lesson from a closer look at airline ESOPs supports the view expressed by those who have examined these arrangements in other settings: that is, a one-time change in the formal governance structure cannot, by itself, sustain a long-term improvement in firm performance (Levine and Tyson, 1990, Blasi *et al.*, 2003).

Union Avoidance

The fourth response has been to avoid unions altogether – either battling them outright, as in the case of Frank Lorenzo's union containment or suppression strategy at Continental Airlines, or attempting to offer employees a deal as attractive as they could hope to earn through negotiation, as in the case of Delta Air Lines' union substitution strategy. Lorenzo's tenure in the deregulated industry is the most noteworthy example of a union containment or union suppression strategy. Head of Texas International when the industry deregulated, he first founded the aforementioned non-union New York Air subsidiary. More notably, he acquired Continental Airlines in a hostile takeover in 1981 and, after failing to win major wage concessions, put the airline into bankruptcy in order to abrogate the labor contracts and reset wages at half of their existing levels. When the unions went on strike, Lorenzo assembled a new non-union workforce. He then acquired Eastern Airlines in 1986 and, in addition to demanding large concessions, also began transferring assets from Eastern to his other non-union airlines. However, while labor costs were drastically reduced in his airlines, service quality also declined and profitability was infrequent. By 1991 all of his airlines were in bankruptcy and he was in fact banned from the industry by a court ruling, suggesting that this approach was not only bitterly opposed by employees – Lorenzo's airlines were the target of seven of the industry's thirteen strikes between 1982 and 1991 – but also not rewarded in the marketplace.

Delta provides an example of union substitution. With the exception of its pilots and flight dispatchers, Delta has remained non-union by following a union substitution strategy. Delta's approach to LR as initially formulated involved a promise of "industry-leading" wages and lifetime employment. In return it gained strong employee loyalty (as well as votes against unionization), to the point where employees jointly purchased a Boeing 767 to demonstrate their appreciation to the company for avoiding layoffs in the recession of the early 1980s. In addition, Delta was traditionally one of the more profitable major airlines, with a reputation for superior customer service. This strategy was severely tested in the mid-1990s, when sustained losses led to layoffs and wage cuts, which were followed by a decline in service quality and unionization drives. But Delta reestablished enough loyalty to defeat the unionization drives. The price of Delta's union avoidance has been high labor costs, whose sustainability is very much in doubt in the newly emerging airline environment. It remains to be seen whether Delta employees will

Table 10.2 Unionization at selected start-ups as of December 31, 2007

Carrier	Pilots	Flight attendants	Mechanics	Ramp/fleet service	Customer service/res. agents	Carrier age (years)
Southwest	SWAPA	TWU	AMFA	TWU	IAM	36
America West	ALPA	AFA-CWA	IBT	TWU	IBT	22[a]
Midwest Express	ALPA	AFA-CWA				22
AirTran	NPA	AFA-CWA	IBT	IBT		14
Frontier	FAPA		IBT			12
JetBlue						7

[a] Now US Airways.
Source: Airline 10 K reports.

vote to unionize if the proposed merger of Delta and Northwest (under review as of this writing) is approved by the government.

Many start-up airlines adopt this approach including, e.g., JetBlue Airways, which has thus far avoided unionization by adopting a range of HR practices (related to the best practices in HR discussed later in this chapter) that foster employee commitment to the airline and weaken employee interest in union representation (Gittell and Reilly, 2001). Whether JetBlue will remain non-union is not an easy question to answer. On the one hand, a review of industry history suggests that even in non-union start-ups, it has only been a matter of time until at least one craft group votes for union representation, typically starting with the pilots (see Table 10.2).

At the opposite end of the LR strategic spectrum, Southwest Airlines exemplifies a highly unionized and profitable airline. In fact, Southwest is one of the most unionized airlines in the US airline industry (Figure 10.1) and its employees are represented by several traditional unions, including the Transport Workers Union and the International Association of Machinists. Not only has Southwest been the most profitable major airline for decades, but it has also achieved the lowest rates of labor conflict (see Figure 10.4) and one of the shortest times required to reach new labor agreements (Figure 10.5). Southwest's performance demonstrates that union representation by itself is not an impediment to strong relationships or high performance (Gittell *et al.*, 2004). In fact, union leaders can be highly supportive of an organization's performance goals, as they have tended to be at Southwest Airlines. One factor in Southwest's achievement of relatively high-quality LR is the decision of its leaders early in the airline's life to welcome union representation and to initiate a partnership approach.

10.2.3 US Labor Relations Post-9/11

The airline industry experienced a major shock from the terrorist attacks of September 11, 2001. Demand fell sharply and revenues dropped by nearly 20% in the months after the attack. By 2005, demand, measured in terms of passenger miles flown, had returned to pre-9/11 levels, although revenue per seat mile still remained below prior levels. The industry lost approximately $8 billion in 2001, and jaw-dropping losses continued through 2005. During this period the legacy airlines' higher cost position relative to low-cost

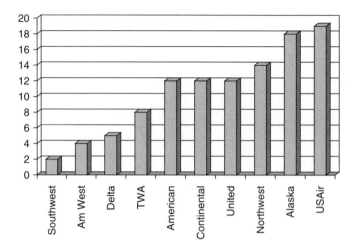

Figure 10.4 Labor conflict at the major US airlines: number of strikes and arbitrations, mediations and releases since 1985

Source: National Mediation Board

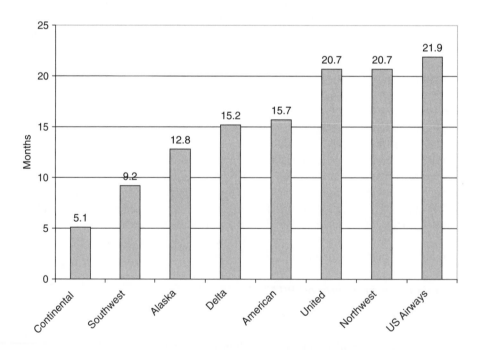

Figure 10.5 Average months required to reach contract agreement at major airlines, 1985–2002

Source: National Mediation Board

airlines was no longer viable, as they became less able to capture a revenue premium, for two reasons. First, customers became more price sensitive and less willing to pay premium fares. Second, and likely the cause of the first reason, the surge in the market presence of lower-cost airlines vastly reduced the number of markets in which legacy airlines were able to charge higher prices.

These conditions, and the corresponding billions in financial losses, led the legacy airlines to launch massive restructuring programs to reduce costs drastically. Labor costs were a major (though by no means exclusive) focus of these programs. Several legacy airlines negotiated enormous wage concessions with their employees (both inside and outside of bankruptcy protection). Thus, overall wage levels in the industry actually decreased between 2001 and 2006.

This restructuring movement may also lead to more fundamental changes to the industry's LR paradigms. On the one hand, the growing presence of low-cost airlines, some using approaches to the employment relationship that are quite different from the legacy airlines, will continue to exert pressure on the legacy airlines to modify their traditional practices. A few legacy airlines have also negotiated major changes to their work rules as well, opening up the possibility for innovations in the design of jobs and work at the workplace level. In addition, attempts are being launched to improve the interactions between unions and management, both at and beyond the bargaining table, perhaps moving toward the partnership approach found at Southwest Airlines.

10.3   Labor Relations in the Airline Industry in Other Countries

Historically, the USA could claim to have the world's largest civil aviation market. By 2010, if trends continue, Asia will be the world's largest aviation market for the first time in history. By 2025, it is predicted that both Europe and Asia will have larger aviation markets than North America. These estimates suggest that in 2025 the three largest economic regions will have the following approximate shares of the world's aviation traffic: North America, 25% (it was 31% in 2005); Europe, 27% (29% in 2005); and Asia, 32% (26% in 2005) (Herdman, 2007).

Airlines in other countries are typically different from their counterparts in the USA in three main ways. First, in the post-World War II period most of the legacy airlines in Europe and Asia were at least partly owned by governments, including such leading examples as Aer Lingus, Air France, Alitalia, British Airways, KLM, Lufthansa, Qantas, SAS and Singapore Airlines. Many of these airlines were launched after World War I by entrepreneurs who had been pilots in the war. Many of these new enterprises were financially unstable, and they were taken over by national governments. Government ownership was also encouraged by the bilateral forms of international regulation whereby each country was required to designate its national "flag carrier" (see Chapter 2). Governments wanted airlines that would be financially stable and that would project a good image of their country. Furthermore, after World War II, as in the USA, governments became increasingly aware of the strategic importance of airlines for defense purposes. State-owned airlines, in particular, were potential resources to be mobilized if necessary (Doganis, 2006).

A second difference between US airlines and their foreign counterparts is that outside the USA, most legacy airlines focus on international flights. This is because, apart from the

USA, most other countries have domestic markets that are relatively small. Moreover, in Europe, short-haul airlines face serious competition from surface transport, especially the growing high-speed rail network, while US airlines do not generally face such competition. Third, the US airline industry was the first to experience deregulation. As a result, it was US entrepreneurs who pioneered the development of low-cost, new-entrant airlines. Most other countries have been slower to deregulate, but the pace of liberalization continues to increase worldwide.

10.3.1   The International Regulatory Framework for Labor Relations

Like airlines in the USA, airlines throughout the world have historically been subject to high levels of regulation, as described in Chapter 2. The International Civil Aviation Organization (ICAO) developed regulations that cover almost every aspect of airline operations, including HR. For instance, in the past ICAO regulations have covered the "numbers of flight and cabin crew, their training and licensing, their duties and functions on board and their work loads and schedules" (Doganis, 2002). The International Transport Workers' Federation (ITF) organizes 654 unions in 148 countries representing almost 4.5 million transport workers including many airline work groups (as well as workers from other transport industries). As in the USA, pilots in most other countries are organized separately from the unions that organize other occupational groups. The International Federation of Air Line Pilots' Associations (IFALPA) claims to represent more than 100 000 pilots in more than 95 countries. ITF and IFALPA voice employees' interests at ICAO and other international regulatory bodies, especially with regard to aviation safety and security.

Regulatory agencies shape the context for management choices about LR strategies, but they do not regulate these strategies directly. In most cases, LR are shaped through collective bargaining and/or works councils. But in some cases, where there are no unions, LR are determined by managerial prerogative, or even by governments when airline staff are in effect government employees. The following sections discuss examples of LR at key airlines in selected countries in Europe and Asia.

Europe

European airline deregulation was more gradual than in the USA, since the EU is a federation of Member States, each with its own government, regulatory preferences and stakeholder interests. The unions generally have more influence over the EU than unions have on US administrations. Further, in comparison to the USA, unions in most EU Member States have more influence on governments. Such influence is exerted both through direct lobbying and through political parties (e.g., Labour or Social Democratic Parties) in which unions have strong influence.

In this overview of Europe, we offer a little more detail on the UK than on the other countries, for in terms of airline deregulation, the UK provides a closer parallel to the USA than most other countries. Beyond the USA, moves toward deregulation and privatization were led by the UK in the 1980s. As a result the UK, along with Ireland, led the way in the emergence of low-cost, new-entrant airlines.

The UK

The UK has a tradition of adversarial employment relations (Marchington *et al.*, 2004). The main airlines accommodate unions, but generally do not partner with them. British Airways, the UK's largest and oldest airline, was nationalized by the government in 1939. While preparing for privatization in the 1980s, BA laid off large numbers of employees in an effort to cut its labor costs quickly.

BA's rhetoric has long included campaigns to foster employee commitment and engagement saying that "people are our most important asset" (Colling, 1995). Nevertheless, BA periodically launches campaigns to cut labor costs. For instance, while BA was advertising its excellent customer service, its attempt to restructure allowances and pay scales for flight attendants prompted a strike in the summer of 1997. BA adopted a tough stance, threatening to fire strikers and to sue them for breach of contract. BA's stance was counterproductive, however, as it turned moderate staff opinion against BA:

> Although only 300 cabin crew joined the three-day strike in July, more than 2,000 went on sick leave which resulted in longer-term disruptions through August. The cost of the strike was estimated at $245 million. The effects on staff morale, service and company reputation (further damaged by the simultaneous introduction of a new baggage-handling system which led to record levels of lost baggage) could not be quantified.
>
> (Arrowsmith *et al.*, 2000)

In 2005, there was another serious dispute involving BA's dismissal of two ground support staff. Subsequently BA announced plans to restructure its business with a 35% reduction of its 1715 managers by 2008, to contribute towards a further £300 million employee cost reduction program (British Airways, 2006).

Before deregulation, BA dominated UK domestic and international scheduled aviation. From time to time others challenged BA's dominance of scheduled services, but most of the new scheduled airlines did not survive. Although BA is a key player in the OneWorld global airline alliance and although BA has continued to be profitable in most years, the airline now faces considerable competition from successful low-cost airlines based in the UK and elsewhere in Europe.

An English entrepreneur, Richard Branson, diversified from the record industry to establish Virgin Atlantic as a long-haul new entrant. But unlike many other new entrants, Virgin Atlantic's competitive strategy did not focus on few or no "frills." Rather it promised extra service. Branson started this airline on the basis that it would have to succeed within a year or he would exit the market. He had a one-year limit on everything associated with starting up to minimize costs if the venture failed. This included limiting employment contracts to only one year (Branson, 1999). Despite a strong BA competitive response against this new rival, Virgin Atlantic survived and continues to thrive as a transcontinental airline known for its high level of on-board "frills."

For its first 15 years, Virgin Atlantic adopted a union-avoidance strategy. It recognized unions only after it had to do so under the new Labour government, which reformed labor law in 1999. A majority of pilots and flight attendants voted for union recognition. Within a decade, there was a high degree of union representation among the pilots, though it was lower among the flight attendants. Like BA, Virgin has had an adversarial relationship with the unions that represent its pilots and flight attendants. It accommodated the unions rather than partnering with them. Further, its competitive strategy included trying to minimize labor costs.

In 1995 a Greek entrepreneur, Stelios Haji-Ioannou, launched easyJet. In contrast with the two full-service airlines, BA and Virgin Atlantic, easyJet would, first, be a low-cost airline and, second, would focus on only short- or medium-haul routes in the UK and Europe. It was founded on the principles of maximizing aircraft utilization and no-frills passenger service. Although easyJet is based in the UK, it has also developed operational focus cities in other parts of Europe. By 2007, easyJet claimed to be operating more flights per day than any other European airline. It was highly profitable and had plans to grow by 15% a year for the next few years. In its early years, easyJet pursued a union-avoidance strategy, but by 2007 the airline appeared to have adopted more of a union-accommodation strategy while attempting to move from a traditional control style of management toward a commitment approach.

The Republic of Ireland

Similar to British Airways, Ireland's national flag carrier Aer Lingus has not always experienced peaceful LR. When Aer Lingus launched productivity initiatives to cut costs in the 1970s and 1980s, workers responded by striking. In 2002 there was a strike and a lockout involving Aer Lingus and its 530 pilots (*Birmingham Post*, 2002). Aer Lingus also initiated pay freezes, voluntary layoffs and reorganizations. Against the background of attempts to introduce more social partnership in Ireland, however, Aer Lingus also introduced employee profit sharing and stock ownership. In each instance of change, Aer Lingus had to accommodate the unions. In early 2007, when Aer Lingus tried to introduce change unilaterally, its efforts were denounced by the Irish Labour Court. However, the Labour Court also recognized the need for Aer Lingus to change (Wallace *et al*., 2007). As such, Aer Lingus fits into the "accommodate" category of union–management relations.

Many of these change efforts at Aer Lingus were taken in response to competition from Ryanair. In 1985, before deregulation, Tony Ryan co-founded Ryanair. New entrant Ryanair was based in Dublin and its first international routes were to the UK. After a few setbacks for the struggling airline, Michael O'Leary took over its leadership in 1993. He transformed it into a fast-growing, highly profitable, low-cost airline. Ryanair's low fares helped to break customers' loyalty to Aer Lingus, even though Aer Lingus was a full-service airline. With increased competition, however, Aer Lingus found it increasingly difficult to operate as a legacy airline, so it reinvented itself in 2002 as a low-cost airline. Nonetheless, after Aer Lingus shares were available for trade on the stock market, Ryanair attempted to take over Aer Lingus.

Ryanair's employment strategy is to focus on low costs via wage minimization, union avoidance and employee control rather than commitment. According to an ITF survey, Ryanair is one of only a few airlines in Europe that does not recognize a union for collective bargaining. It aggressively avoids unions. This strategy has induced the ITF to launch a web-based campaign (Ryan-Be-Fair) attempting to mobilize Ryanair workers across Europe to take action against Ryanair. Nevertheless, Ryanair has become the low-cost, new-entrant airline leading the growth of the European market for cheap, no-frills flights.

Germany

Lufthansa is the national flag airline that has long dominated German aviation. While the original Lufthansa was born in 1930, the current Lufthansa was reborn in 1955. The German Federal Republic, the North Rhine-Westphalia State and the public-sector railroad each held major stakes in Lufthansa. The reborn Lufthansa also had private shareholders. Lufthansa has also involved its employees in profit sharing and has given them the opportunity to choose between cash and preference shares since 1970.

Apart from a few crisis years, Lufthansa has usually made a profit. Even in 2001, the year of the 9/11 crisis, when many airlines lost money, Lufthansa earned a small profit even while refusing to lay off employees. Lufthansa has developed into a global aviation group with nearly 95,000 employees. In contrast with British Airways, where Lufthansa has outsourced functions, the airline has kept these functions under the same collective bargaining umbrella. This is true even in major functions like maintenance and cargo. Together with United Airlines and others, Lufthansa co-founded the Star Alliance. Although such alliances are primarily marketing networks, they have implications for employment relations. In 2001 Lufthansa pilots sought and received a pay increase of about 30% on the grounds that their pay should keep up with that of pilots at United, one of their Star Alliance partners (Doganis, 2006, p. 132).

Most German firms are obliged to treat unions as "social partners." Lufthansa and other companies in Germany have a two-tier board. The supervisory (upper-level) board appoints, supervises and advises the executive board. The executive board is responsible for the implementation of corporate strategies and managing the company. The two boards collaborate (Keller, 2004). The supervisory board has 20 voting members. The 10 share-holder representatives are elected by the annual general meeting and the 10 employee representatives are elected by Lufthansa employees. Multiple stakeholders (employees, executives, other shareholders), then, are embedded in the company's structure.

This institutional requirement has fostered a continuing labor–management partnership. Lufthansa's partnership approach facilitated its success in restructuring to become profitable again after its losses and cash flow problems in the recession of the early 1990s. Union and works council involvement in Lufthansa's restructuring ensured that there was no major deterioration in working conditions, nor were there mass layoffs. Although Lufthansa also suffered a significant loss of revenues following 9/11, it did not implement any layoffs (Hatty and Hollmeier, 2003). The obligation of managers at Lufthansa to consult with their employees precipitates a focus on longer-term restructuring options. In terms of the cost competitiveness strategy of this legacy airline, such rigidities of the German labor market encourage Lufthansa to choose productivity-enhancing strategies over wage minimization strategies.

Scandinavia[2]

As in other European countries, civil aviation in Scandinavia was developed by private enterprises and it began to grow in the 1920s. World War II stopped developments in Denmark and Norway, which were occupied by Germany, while it continued in a limited

---

[2] We acknowledge with many thanks that this section draws on insights from former SAS staff, including Bernhard Rikardsen, former Director of Human Resources, SAS.

way in Sweden, which was neutral. Emerging from the war with scarce resources, the major private airlines, visionary industrialists and the governments of the three countries decided that a joint effort would be the best way to develop Scandinavian aviation on an international scale. Between 1946 and 1950, they created a pan-Scandinavian airline, Scandinavian Airline System (SAS). SAS was structured as a consortium owned by the three national private airlines, which in turn were listed on the three national stock exchanges. The shares of these three national airlines were half owned by each of the three countries' governments and half by private investors. SAS is still half owned by the three governments, but a holding company owns the three founding airlines; its shares are now listed in parallel on the three stock exchanges. The governments have generally not intervened financially or politically in the governance or operations of SAS, except after SAS overinvested in jet aircraft in the early 1960s.

LR in Scandinavia are broadly based on a tripartite model between employers and unions, with the state shaping the context. In general, the state encourages employment relations arrangements based on relative moderation by unions, which helps to ensure that industries are profitable and that they provide good jobs. The model is built on a welfare state fully funded by taxes. This eases the direct burden of health care, for example, for employers and employees. However, the context is one of relatively high total labor costs. This induces businesses to improve productivity, to develop innovative products and to focus on premium consumer markets. Another important aspect of the Scandinavian model is the representation of workers' interests on companies' boards of directors. This gives employees a voice in their company.

Before deregulation, SAS employees enjoyed terms and conditions of employment that were more generous than those which generally applied in most other industries. As the Scandinavian countries deregulated the airline industry in the mid-1990s, SAS's lack of international competitiveness became more apparent. SAS's Scandinavian employment relations model fosters relatively cooperative relations between the main parties. Nevertheless, it was more costly for SAS than the models used by most other airlines. This reflected SAS's tendency to offer its employees the best of the benefits from each of the three countries.

This tendency became a serious problem for SAS as competition in national and international markets increased. New-entrant airlines were attracted to Scandinavian markets because of the relative affluence of the population, frequency of air travel in Scandinavia, and the high fares charged by SAS. For a few years before 9/11, SAS focused on expansion. In this period, labor costs increased, but the underlying uncompetitive cost structure was hidden by growth during the economic boom of the late 1990s.

The increased level of competition became more obvious following the 9/11 attacks in 2001 and their immediate negative impact on world aviation. Management advisers strongly recommended radically reducing employment in the original legacy airline, SAS, while simultaneously growing the new-entrant airlines that it had acquired and which had become SAS subsidiaries or associates: Spanair in Spain, Braathens and Widerøe in Norway, Blue1 in Finland, Air Baltic in Latvia and Estonian in Estonia. Despite such recommendations, SAS chose to give unions a chance to work with the managers to help turn around the airline, in line with the cooperative tripartite norms of Scandinavian society and its political economy. Unlike several airlines in the USA, SAS did not make any major unilateral cuts in its pension plans. In short, SAS's strategy included cost reduction, the avoidance of non-essential costs, and increasing productivity.

The main stakeholders in Scandinavia generally see their employment relations model as effective, durable and flexible; this was the case even during the turmoil following 9/11. This perception is shared by most managers and unions at SAS, even though SAS's structural complexity blunted the effectiveness of the Scandinavian employment relations model. Further, there is usually more of a genuine partnership with unions in the Scandinavian countries when they have Social Democratic governments. Governments formed by right-wing parties tend to be less supportive of such partnerships.

Asia

Asia is a much more diverse region than North America or Europe. There are multiple governments and regulators involved in Asian aviation. In comparison with the USA, Canada and Europe, unions are weaker and more fragmented in Asia and are not well positioned to press for the inclusion of EU-style social partnership arrangements.

Since the 1970s, airlines have grown even more rapidly in Asia than in North America and Europe. This reflects the rapid growth of Japan, the East Asian "Tigers" and other economies. National airlines from Asia such as Malaysia Airlines, Singapore Airlines, Thai Airways and Garuda of Indonesia are mainly government owned. But some, like Cathay Pacific of Hong Kong, Eva Airways of Taiwan and Asiana Airlines of South Korea, are private companies.

In the early twenty-first century, on average, Asian airlines were more profitable than those in Europe, which have in turn tended to be more profitable than those based in the USA, largely because of the aforementioned demand growth as well as a slower pace of deregulation (Herdman, 2007).

However, markets are being deregulated, and consequently new-entrant airlines are growing in Asia. Notable new entrants in the region include: Virgin Blue (Australia, 2000), Lion Air (Indonesia, 2002), One-Two-Go (Thailand, 2003), Air Deccan (India, 2003), Thai AirAsia and Nok Air (Thailand, 2004), Valuair and Tiger Airways (Singapore, 2004), SpiceJet and Kingfisher Airlines (India, 2005), AWAir (Indonesia, 2005), Okay Airways (China, 2005), Oasis (Hong Kong, 2006) (Baker *et al*., 2005). Faced with this increased competition, many of the Asian legacy airlines have also been restructuring their operations.

Australia

Although Australia has less than 10% of the US population, Australia like the USA has an extensive domestic aviation market. For most of the post-World War II period, the Australian government had a "two-airline" policy that created, in effect, a duopoly. The domestic mainline routes were shared between Australian Airlines (later merged with Qantas) and Ansett. Qantas was founded in 1920 and is Australia's dominant legacy airline. Since it was privatized in the 1990s it has operated profitably in international and domestic markets. Ansett was mainly a domestic airline. The strategic position of both of these legacies was to offer full service with relatively high operating costs and fares.

Following several short-lived attempts since the 1980s to start a third domestic airline, Impulse and Virgin Blue both launched airlines in 2000. In the 1990s, Richard Branson supported a business plan to develop a Virgin-branded new entrant in Australia. Branson provided a one-off US $10 million equity investment. Brett Godfrey became CEO of Virgin Blue Airlines. His team saw Southwest Airlines as a role model (Godfrey, 2002).

Despite a significant recent decline in the number of labor disputes, in contrast with Germany or Scandinavia, Australia still has a tradition of adversarial LR (on the Australian context, see Lansbury and Wailes, 2004). Both of Australia's legacy airlines were highly unionized among all occupational categories. Both had a general LR strategy to accommodate unions. Nevertheless, in 1989–1990, the two legacy domestic airlines, with strong support from the federal government, fought a major dispute for more than six months with the Australian Federation of Air Pilots. This involved a lockout and mass resignations of most of the legacies' domestic pilots (Bray and Wailes, 1999).

In the early twenty-first century, in the face of new-entrant airlines enjoying a 30 to 40% cost advantage, Qantas was still accommodating the unions, but its rhetoric was increasingly tough in relation to its unions. Qantas periodically contemplated outsourcing various maintenance operations to other countries with lower labor costs, but such proposals were opposed by politicians and unions.

In comparison to their approaches when they were state-owned enterprises in regulated markets, the three old legacy airlines discussed above – BA, Aer Lingus and Qantas – appear to have been adopting increasingly tough management tactics, in relation to their employees and unions. They continued to accommodate unions, but did not partner with them.

South Korea

Korean Air and Asiana Airlines, still protected by government regulation, dominate the airline industry in Korea. Korean Air was established in 1969 after a company purchased the former government airline. Korean Air monopolized Korean air travel until 1988, when the government permitted Asiana to become a second national airline. More recently, two new entrants (Hansung Airlines and Jeju Air) were allowed to enter the Korean market following an "open skies" policy established with China (Lee and Cho, 2007).

In relatively stable operating conditions, the two legacy airlines have developed their policies to manage HR/LR, providing relatively high wages and associated conditions such as fringe benefits, job security, career paths and extensive training. Both legacy airlines provide a Japanese-style policy of "lifetime employment," where employees are given firm-specific training, opportunities for career growth and little job mobility (largely due to the duopolistic nature of the Korean industry). This, in part, is due to regulations in Korea that require statutory licenses for jobs as pilots, mechanics and flight attendants. The lifetime employment policy and associated HR practices, e.g., promotion from within rather than filling vacancies from the external labor market, considerable investment in training, relatively few rules limiting flexibility in job assignments, reflect a commitment approach to employees. Such practices also support the service-enhancement strategies emphasized by the Korean airlines.

Many pilots are recruited from the Korean Air Force; however, a shortage of pilots prompted Korea Air to develop a "pilots' academy," where pilots are trained and must remain at the company for 10 years following training. Flight attendants are usually graduates from Korean colleges.

Both airlines have accommodated unions. Two unions cover most of their workforce: one for pilots and the other for all other employees. The Korean Air Labor Union, covering all employees except pilots, has usually cooperated with managers. It has maintained 100% union membership. The other unions at the two legacies have been critical of management's authoritarian approach to LR. The pilots unions have also been concerned about the long working hours of their members (Lee and Cho, 2007).

Again these airlines accommodate unions rather than suppressing them or partnering with them. However, their approaches to employee relations display some elements of a commitment approach (e.g., greater employment security and flexible job assignments).

Malaysia

Malaysia Airlines (MAS) was founded in 1937. MAS lost nearly half a billion dollars in 2005, while some of its regional competitors reported strong profits. In 2006 the government divided Malaysia's domestic air routes between MAS and new entrant AirAsia, which meant that these airlines would not have to compete on their home ground. In fiscal year 2007, AirAsia reported a pretax profit of $79 million on revenue of $458 million, a 17% margin that is third best in the industry, behind only Brazil's Gol and Ryanair. It became profitable by adopting a new- entrant, low-cost model. There have been several other such transformations in Asia, with national airlines reforming as low-cost carriers in attempts to remain profitable (Poon and Waring, 2009).

After years of protectionist government policy toward MAS, in 2007 the government also endorsed AirAsia's sister company, AirAsia X – a low-cost, long-haul airline. AirAsia X began flying to Australia and other longer-haul destinations in 2007 and plans to start flying to Europe and the USA. The Virgin Group bought a 20% stake in AirAsia X in 2007. These Asian new entrants sometimes claim to be implementing the Southwest model. However, AirAsia does not even accommodate unions, let alone partner with them. Rather, AirAsia has emulated Ryanair's employment relations model. AirAsia follows a control approach with its employees and avoids unions.

While AirAsia has continued to grow, MAS has had to restructure (Poon and Waring, 2009). MAS launched a business turnaround plan that focuses, among other things, on the employment relationship, emphasizing its commitment to unleashing talents and capabilities of its employees: "We will work together with our employees to ensure that they have a working environment in which their talents can thrive. . . . We are dedicated to the creation of a company that will be a source of pride and admiration for its employees and indeed all its stakeholders" (Malaysian Airlines, 2007). At the same time, MAS is seeking substantial staffing cuts along with increases in labor productivity. Despite this language in its recent business plan about movement toward a commitment approach, MAS's approach thus far is best characterized as an accommodative, arm's-length relationship with its unions (neither suppressing them nor partnering with them) and a control approach to its employees.

10.3.2   Summary of Airline Labor Relations Strategies: What Works?

The above examples from airlines around the world illustrate that variations in LR strategies can be observed even within similar regulatory frameworks. Airlines thus have a choice among alternative LR strategies. The dominant approach continues to be the accommodation of unions, maintaining an arm's-length relationship with them, sharing information only as required and interacting primarily at the bargaining table. An alternative approach is to actively suppress unions as exemplified historically by Lorenzo in the USA, and currently by Ryanair in Europe and Air Asia in Asia, or to attempt union substitution through progressive HR as exemplified by JetBlue and Delta Air Lines. The third approach is to partner with unions, as exemplified by Lufthansa and Southwest Airlines.

Which LR strategy is most effective for achieving firm-level performance objectives? Research does not support the proposition that non-union approaches are more effective than partnering with or accommodating unions. Findings from a longitudinal study of the major US airlines from the late 1980s to 2000 indicate that higher levels of unionization, while associated with higher wage levels, were nonetheless associated with higher productivity and higher profitability. See Table 10.3 for a summary of results. This table shows the impact of LR on airline performance for 10 major US airlines over 15 years from 1986 through 2000. LR are measured in five ways: (1) the percentage of employees represented by unions; (2) shared governance in the form of equity ownership by employees and employee representation on the board; (3) union–management conflict as captured by the number of releases and strikes; (4) the presence or absence of a positive workplace culture; and (5) average wage levels. Airline performance is measured in four ways: (1) service quality based on an index of late arrivals, lost bags, customer complaints and pilot safety deviations; (2) aircraft productivity measured as block hours per aircraft day; (3) labor productivity measured as outputs (varying by employee function) per airline employee; and (4) operating margins (Gittell *et al.*, 2004).

Thus, it appears that efforts to avoid unions are not likely to produce a sustained improvement in airline performance. This finding suggests a caution for managers who actively seek to maintain a non-union workplace. Managers who actively avoid union representation will face a major challenge, if and when their employees choose to seek representation, to develop a partnership with the newly elected union representatives.

Our research, based on in-depth studies of several of the most successful US airlines of recent years (Southwest, Continental and JetBlue), comparisons to their main rivals, and quantitative analyses of airline performance over a 15-year period, indicates that building and maintaining *high-quality relationships* – between management and unions and between managers and employees – is the most effective approach to achieving high levels of firm performance. High-quality relationships are more consistently associated with superior productivity and profitability than any specific structural feature such as unionization, employee ownership or wage levels. High-quality relationships enable more cooperative and productive collective bargaining, marked by efficient negotiations, no work stoppages and minimal lingering ill will. High-quality relationships also foster better communication and more trust between supervisors and supervisees, and between employees in different functions.

Table 10.3 Impact of labor relations on airline performance

	Airline performance			
	Service failure	Aircraft productivity	Labor productivity	Operating margins
Union representation	−0.11 (0.633)	0.70*** (0.000)	0.11 (0.388)	0.02* (0.050)
Shared governance	−0.40*** (0.000)	−0.53*** (0.000)	−0.07 (0.375)	−0.02** (0.001)
Negotiating relationships	0.25* (0.016)	−0.46** (0.002)	−0.21 (0.113)	−0.07*** (0.000)
Workplace relationships	−0.22* (0.014)	0.54*** (0.000)	0.35*** (0.000)	0.03*** (0.000)
Wage levels (log)	−2.08*** (0.000)	−3.55*** (0.000)	−1.30*** (0.000)	−0.14*** (0.000)
Capital intensity (000 000)	4.25*** (0.000)	−2.73*** (0.000)	0.98+ (0.095)	0.41*** (0.000)
Leg length (000)		−0.78** (0.0007)	−1.46*** (0.000)	−0.12*** (0.000)
Aircraft size (00)		2.09*** (0.000)	0.58* (0.040)	0.11*** (0.000)
R squared	0.36	0.56	0.39	0.65

Notes: All models are random effects regressions with firm/quarter as the unit of analysis ($n = 485$ for labor productivity model and $n = 489$ for all other models) and firm ($n = 10$) as the random effect. Statistical significance is denoted $^{+}p < 0.10$, $^{*}p < 0.05$, $^{**}p < 0.01$, $^{***}p < 0.001$, and suggests the certainty that a change in the variables in the left-hand column will produce a change in airline performance, where a smaller p-value suggests a higher certainty. R squared denotes the percentage of the variation in performance that is explained by the model. Each model includes quarterly dummies to capture changes in the industry environment.

10.4   Human Resource Management at Airlines

As noted above, HR/LR are two aspects of the employment relationship. LR include the ways that management relates to employees as they are organized to gain representation in decision making, while HR management includes the ways that management uses conditions of employment to influence the work-related behaviors of employees. In contrast to LR strategies, HR management strategies are less dependent on national context. Similar to LR strategies, however, the effectiveness of HR strategies often depends on their ability to build high-quality relationships. Our research suggests that a commitment-based approach to HR, rather than the control-based approach that has typified the industry, holds the highest potential for superior airline performance (von Nordenflycht, 2004). In this section, we discuss what we have learned about HR strategies that build high-quality

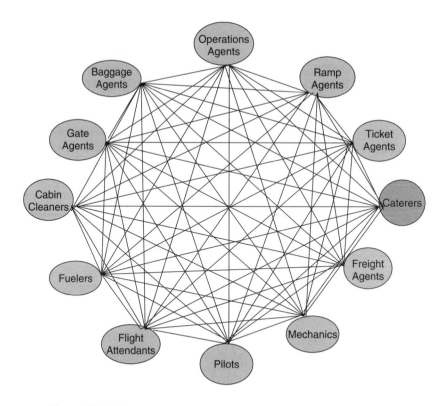

Figure 10.6   Functions involved in the coordination of flight departures

relationships and employee commitment, whether or not unions are "on the property," thus contributing to superior productivity and customer service outcomes.

A critical driver of performance in airlines is the ability to effectively coordinate the work of the multiple functional groups involved – pilots, flight attendants, mechanics, ramp workers, customer service agents, and so on. In airlines, employees do not affect service quality and productivity only as individuals, but also collectively as a group. For example, it is not just the skills and motivations of individual gate agents, ramp agents and flight crew members that determine whether customers will be kept well informed, whether their flight will depart on time and whether their baggage will arrive with them. Figure 10.6 illustrates the coordination challenge associated with flight departures, a critical work process for any airline, carried out dozens of times per day in each of hundreds of locations. For work processes like flight departures that are highly interdependent, with many separate tasks performed by individuals with distinct functional skills, coordination between the individuals performing these tasks plays a critical role in driving performance, over and above the individual efforts of employees.

Coordination is not just a technical process – it is also a highly relational process (Gittell, 2003). This reality is reflected in the concept of relational coordination – the coordination of work through relationships of shared goals, shared knowledge and mutual respect. When positive, these relationships enable airline employees to effectively coordinate their work by supporting frequent, timely, problem-solving communication. When negative, these same relationships serve as obstacles to coordination. Employees who feel disrespected by members of another function tend to avoid communication (and even eye contact) with members of that function. The absence of frequent dialogue in turn solidifies the existence of distinct "thought worlds" for each functional area, undermining relationships of shared knowledge. Without relationships of shared knowledge, employees are less able to engage in timely communication when circumstances change, not knowing with sufficient precision who needs to know what and with what urgency.

The lack of timely communication undermines relationships of shared goals, reinforcing the belief that each function is looking out for itself. Without shared goals, the easiest response to problems is to blame others for having caused the problem rather than to engage in problem-solving communication. The focus of communication around blaming rather than problem solving further undermines mutual respect. This negative cycle decreases the potential for effective coordination to occur. Figure 10.7 illustrates the

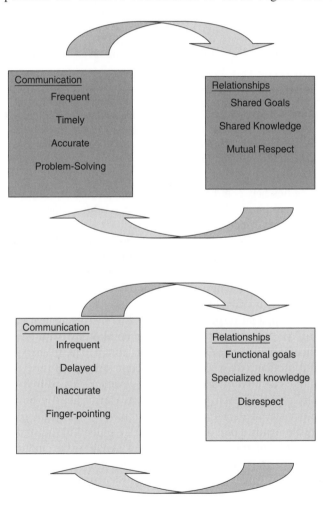

Figure 10.7   Alternative dynamics of relational coordination

Table 10.4 Impact of relational coordination on airline performance

	Airline performance				
	Turnaround time (aircraft productivity)	Staff time per passenger (labor productivity)	Customer complaints (service quality)	Lost bags (service quality)	Late arrivals (service quality)
Relational coordination	−0.21*** (0.000)	−0.42*** (0.000)	−0.64*** (0.000)	−0.31* (0.042)	−0.50** (0.001)
Flights/day	−0.19*** (0.000)	−0.37*** (0.000)	−0.30*** (0.000)	0.13 (0.287)	−0.22+ (0.065)
Flight length, passengers and cargo	0.79*** (0.000)	0.45*** (0.081)	0.13 (0.188)	0.12 (0.471)	0.54** (0.001)
Passenger connections	0.12** (0.004)	0.19** (0.008)	0.09 (0.329)	0.13 (0.287)	0.00 (0.987)
R squared	0.94	0.81	0.69	0.19	0.20

Notes: All models are random effects regressions with site/month as the unit of analysis ($n = 99$) and site ($n = 9$) as the random effect. Statistical significance is denoted $^+p < 0.10$, $^*p < 0.05$, $^{**}p < 0.01$, $^{***}p < 0.001$, and suggests the certainty that a change in relational coordination will produce a change in performance, where a smaller p-value suggests a higher certainty. R squared denotes the percentage of the variation in performance that is explained by the model.

mutually reinforcing ties between the communication and relationship ties that form relational coordination, and shows how these ties can reinforce one another in a positive direction, or in a negative direction.

Relational coordination helps airlines to achieve both improved productivity and service quality outcomes, in effect shifting out the quality/efficiency frontier by enabling airlines to achieve higher levels of both. These performance effects are summarized in Table 10.4 and Figure 10.8. Table 10.4 shows the impact of relational coordination on airline performance. Relational coordination, coordination carried out through relationships of shared goals, shared knowledge and mutual respect, is measured as the percentage of cross-functional ties that are "strong" or "very strong," based on an employee survey. Airline performance includes: (1) service quality captured by customer complaints, mishandled bags and late arrivals; (2) aircraft productivity, captured by aircraft turnaround time per departure; and (3) labor productivity, captured by staff time per passenger. Figure 10.8 summarizes these results in a scatterplot where each performance measure from Table 10.4 was adjusted for differences in product characteristics and combined into a single performance index, and where each circle denotes one of the nine sites included in the study (Gittell, 2003).

Given the importance of relational coordination in the airline industry, airlines benefit from adopting HR practices that focus on developing employees as team players rather than as individual functional experts. Because employees in this industry influence performance outcomes as team players, they should be hired, trained, monitored and rewarded not only for individual excellence but also for their ability to contribute to a larger coordinated effort.

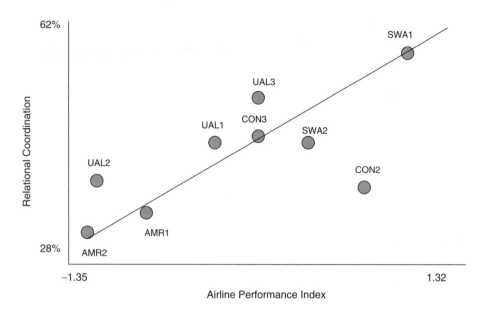

Figure 10.8   Impact of relational coordination on airline performance

The following sections describe several key HR management practices that are the building blocks for achieving relational coordination, which in turn drives high levels of productivity and service quality. These practices include hiring and training for relational competence, flexible boundaries between jobs, supervisory coaching and feedback, performance measurement at the process level, conflict resolution, and boundary spanning roles.

10.4.1   Hiring and Training for Relational Competence

Because different jobs require different abilities, one of the most important objectives of the hiring process is to find people who best fit the requirements of the job. But the critical skills to be identified in the hiring process go *beyond* the technical and cognitive realm to include personality traits (Day and Silverman, 1989). Service management experts Schlesinger and Heskett (1991) have made the case that service companies in particular should hire for "soft skills" such as customer orientation and teamwork ability.

Teamwork ability can be understood more specifically as relational competence – the ability to relate effectively with others. Particularly when hiring people for jobs that require high levels of expertise, organizations tend to underestimate the importance of relational competence. Yet even people who perform highly skilled jobs (e.g., pilots) need relational competence to effectively integrate their work with the work of their fellow employees.

Hiring for functional skills while ignoring relational competence is a common occurrence in airlines, particularly when hiring for positions like pilots and mechanics in which specific functional skills are highly valued. However, people who perform jobs that require high levels of functional expertise also tend to need high levels of relational competence to integrate their work with the work of fellow employees. Our study of flight departures suggests that airlines that recognize the importance of relational competence, look diligently for employees who have it, and then develop it to even higher levels through training, have a distinct performance advantage over airlines that do not.

10.4.2 Flexible Boundaries Between Jobs

In a highly interdependent and service-intensive industry such as airlines, flexibility among employees doing different jobs – pilots, flight attendants, passenger service agents, ramp workers, mechanics, etc. – is critical. In particular, flexible job boundaries between these employee groups were found to support high levels of relational coordination. Flexible job boundaries create more opportunities for communication across functional boundaries (Preuss, 1996), therefore enabling employees to develop a higher degree of shared goals, shared knowledge and mutual respect. Especially in settings that require a more spontaneous form of coordination, such as the flight departure process, it is important that jobs be defined flexibly enough that people can come to understand the jobs of those with whom they must coordinate. Flexible job boundaries enable people to directly experience the work of those with whom their work most closely relates, helping them to respond constructively when things go wrong.

In response to these benefits, there have been efforts by airlines like Southwest, easyJet and Virgin Blue to broaden job descriptions to allow employees to take the actions they see as necessary to accomplish the organization's goals. But broadening job descriptions is difficult to achieve because it threatens people's sense of security and introduces an element of the unknown into their jobs. These fears can prevent organizations from broadening job descriptions, particularly in unionized settings like the airline industry where job descriptions are contractually negotiated through collective bargaining. Airlines with high-quality LR are therefore more likely to sustain flexible job boundaries over time. For those airlines that do achieve flexible job boundaries, the benefits include higher levels of relational coordination across employee groups, leading to higher levels of aircraft productivity, staffing productivity and service quality.

10.4.3 Supervisory Coaching and Feedback

Many contemporary management thinkers have argued that the purpose of supervisors is to perpetuate bureaucracy by controlling and monitoring workers (Selznick, 1949). Today's competitive environment calls for teamwork among front-line employees, they argue, and supervisors tend to get in the way. Flat organizations with few supervisors should therefore perform better than more bureaucratic organizations (Piore and Sabel, 1984; Walton, 1985; Appelbaum and Batt, 1994). But in the airline industry, this argument was not supported by our data.

In the airline industry, the supervisory span was found to vary from one supervisor per 10 front-line employees at Southwest Airlines, to one supervisor per 35 front-line employees at American Airlines (Gittell, 2001). Likewise, the intensity of interaction between supervisors and front-line employees was found to range from infrequent and arm's-length at American, to frequent and intensive at Southwest. Higher levels of supervisory staffing at Southwest gave airline supervisors fewer direct reports, enabling them to engage in more frequent and intensive interaction with their direct reports, while supervisors with more direct reports engaged in less frequent and more arm's-length interaction with their direct reports.

With fewer direct reports, supervisors at Southwest had greater opportunities for work-ing side by side with the front-line employees they were responsible for supervising. Work-ing together reduced informational and social distance between Southwest supervisors and the workers they supervised, supporting the creation of shared goals. Shared goals made Southwest employees more receptive to supervisory coaching and feedback, reducing the role of supervisory monitoring due to mutual monitoring by employees of each other.

But at American Airlines, a different story emerged. At American, supervision was reduced drastically in the late 1980s to economize on staffing and to increase participation by front-line employees, as part of a program called "Committing to Leadership." The result was that supervisors had arm's-length relationships with their direct reports and played a largely bureaucratic role, relying on impersonal rules to allocate responsibility for late departures and other errors. Their role was to monitor compliance with performance targets set by headquarters, and compliance with basic rules of behavior such as being on duty at the scheduled times.

We found that high levels of supervisory staffing were correlated with high levels of relational coordination across functional groups, leading to higher levels of aircraft productivity, staffing productivity and service quality.

10.4.4 Performance Measurement at the Process Level

Like many organizations in other industries, airlines have traditionally relied on systems of functional accountability. Outcomes of the flight departure process are typically divided into departmental objectives, for which individual departments are held accountable. Each departure delay is traced to the department that is thought to have caused it. Then on a daily, weekly and monthly basis, the percentage of on-time departures is calculated for each department.

As total quality management expert J. Edward Deming would have predicted, this system of accountability tends to generate a search for departmental failure (Deming, 1986). Yet because of the task interdependencies in the flight departure process, it is often difficult to determine which department caused a particular delay. One rule of thumb often used is "whoever was off the plane last." If the gate agent who was boarding passengers was last off the plane, it is presumed to be a customer service delay. If the ramp agents loading baggage were last off the plane, it is presumed to be a ramp delay. If the fueler was the last one off, it is presumed to be a fueling delay. Therefore the common pattern is a race to finish one's own assigned task before the other groups finish their tasks, even when cooperation between the groups would improve the speed and quality of the process. Worse, participants tended to hide information to avoid blame, thus undermining the potential for learning.

Through these unintended dynamics, functional accountability undermines relationships of shared goals, shared knowledge and mutual respect among those who must coordinate closely in order to achieve high performance. To achieve quality outcomes in the face of weak coordination requires longer turnaround times and higher staffing levels, resulting in tremendous efficiency losses. However, there are constructive alternatives to these systems of functional accountability. Cross-functional performance measures encourage participants to focus on learning rather than blaming when things go wrong, thereby bolstering relationships of shared goals, shared knowledge and mutual respect, resulting in better performance (Gittell, 2000). Consistent with this argument, we found in our study of flight departures that allowing multiple functions to take responsibility for a given delay – like the "team delay" at Southwest Airlines – was associated with higher levels of relational coordination across functional groups, leading to higher levels of aircraft productivity, staffing productivity and service quality.

10.4.5   Conflict Resolution

In the flight departure process, conflict is a common occurrence. There is tremendous pressure to get the aircraft out on time, and at the same time there are multiple functional groups involved, each of which tends not to understand very well the perspective of the other. From pilots to cabin cleaners, the employee groups whose coordination is essential to achieving performance outcomes in the departure process tend to be divided by the lack of shared goals, shared knowledge and mutual respect. The resulting friction between these groups often contributes to process failures. This should not be a surprise. Conflicts are a fact of life in interdependent work processes that span multiple functions (Lawrence and Lorsch, 1967). Not only are conflicts more likely to occur in highly interdependent processes, but those conflicts are also more likely to have intensified effects (Gladstein, 1984).

To many people, conflicts appear to be destructive, and to be avoided at all costs. However, there are potentially constructive aspects to conflict as well. We argue that proactively identifying and resolving conflicts is a way to *strengthen* the relationships that underlie relational coordination. Conflict resolution is an often-overlooked opportunity to build a shared understanding of the work process among participants who do not fully understand each other's perspectives.

Certainly there are ways to *prevent* conflicts from occurring. Management theorist Louis Pondy noted in the late 1960s that one way to prevent conflict is to reduce interdependence by "1) reducing dependence on common resources; 2) loosening up schedules or introducing buffers, such as inventories or contingency funds; and 3) reducing pressures for consensus." He also noted, however, that "these techniques of preventing conflict may be costly in both direct and indirect costs," and that ultimately, "*interpersonal friction is one of the costs of 'running a tight ship'*" (Pondy, 1967; italics added).

In the flight departure process, the most common method for reducing interdependence and reducing conflict is to increase aircraft turnaround times – time that the aircraft is scheduled to be at the gate in between flights. But long turnaround times result in low levels of aircraft and employee productivity, creating a competitive disadvantage for airlines that use this strategy to reduce conflict. Rather than reducing cross-functional conflict by introducing costly buffers, and rather than simply accepting it as a cost of running a tight ship, airline managers can approach conflict as an occasion for learning (Van de Ven, 1976), and for developing a clear understanding about goals, expectations and behaviors (Van de Ven and Ferry, 1980). When airline managers *seek out* conflicts rather than allowing them to fester, and when the parties are brought together to better understand each others' perspectives, the problem-solving and learning processes are enhanced. When managers treat cross-functional conflict as an occasion for learning, they strengthen relationships between employees and boost performance of the work processes in which those employees are engaged. Accordingly, our study of flight departures showed that rewarding managers for being proactive about conflict resolution was associated with higher levels of relational coordination across functional groups, leading to higher levels of aircraft productivity, staffing productivity and service quality.

10.4.6 Boundary Spanning Roles

Many different employees play a critical role in coordinating flight departures – pilots, flight attendants, gate agents, ticket agents, baggage handlers, mechanics, and so forth. But there is one role that is particularly central for coordinating flight departures – the flight dispatcher, known also at some airlines as the "operations agent." In the airline industry the operations agent is at the center of communication among the various functional groups that are working to get the aircraft unloaded, serviced, reloaded and on its way. The tasks of the operations agent include collecting information about the passengers, bags, freight, mail and fuel going out on a particular flight, making calculations about how much of each could be loaded and where it should be loaded, consistent with weather and route information. Before the aircraft arrives, during its time at the gate, and after its departure, operations agents gather and process the needed information from each of the functional groups, make adjustments as needed and communicate those adjustments back to the other functional groups. In so doing, operations agents bring together and reconcile sometimes conflicting perspectives among the various departments regarding passenger needs, commitments to freight and mail customers, and the requirements of flight safety.

Operations agents in effect serve as "boundary spanners," managing the flow of information across functional lines. Organization design theorists tell us that boundary spanners are particularly important for coordinating work processes in which participants perform very different tasks and as a result have very different perspectives about what needs to be done (Galbraith, 1994). The boundary spanner has traditionally been seen as a mechanism for collecting, filtering, translating, interpreting and disseminating knowledge across organizational boundaries (Aldrich and Herker, 1977). However, an effective boundary spanner does *more* than process information. An effective boundary spanner is also engaged in relationship building, developing relationships of shared goals, shared knowledge and mutual respect among fellow workers to facilitate the coordination of work.

Boundary spanners are costly, however, because they are an employee group whose primary task is coordination (Galbraith, 1994). One way to reduce the cost of boundary spanners is to reduce their staffing levels – by increasing the number of flight departures they are assigned to coordinate. Since the mid-1980s, many airlines have been doing just this – attempting to make operations agents more efficient by relying more and more on computer interfaces to bring together the information required to dispatch a flight. And indeed, these new systems allow operations agents to be more efficient. With information technology, operations agents can be located centrally and can coordinate up to 15 departures at a time. Operations agents read a computer file into which each functional group has input the relevant information, make contacts when there is a discrepancy or need for further information, then make the necessary judgments and decisions before dispatching the flight. But the quality and nuance of communication is not very high in this arrangement. The operations agent is remotely located and is forced by workload to rely almost exclusively on the computer interface.

At some smaller airlines, operations agents had traditionally served as a source of social cohesion across functions in the stations. Operations agents used to be well-known personalities because they came into face-to-face contact with each functional group during the preplanning or implementation phase of each departure. The on-site operations center where they worked used to serve as "watering holes," as one of the few locations where members of diverse functional groups, like pilots, fuelers, baggage handlers, mechanics and customer service agents, could congregate comfortably. In their efforts to automate decision support for the flight dispatch process and improve productivity, airlines reduced the staffing levels of operations agents, losing many of the personal interactions that built strong relationships.

Boundary spanners play a critical role in coordinating work processes, but the boundary spanner is most effective when the position is conceived to be more than a conduit for information exchange. When the boundary spanner role is sufficiently staffed – with Southwest perhaps at the extreme, assigning each operations agent to coordinate just one flight at a time – the boundary spanner can help to create high levels of shared goals, shared knowledge and mutual respect across functional boundaries. Our study of flight departures found, indeed, that higher levels of staffing for operations agents were associated with higher levels of relational coordination across functional groups, and with higher levels of aircraft productivity, staffing productivity and service quality.

10.4.7 Partnering with Other Key Players

Coordination across internal functional boundaries does indeed appear to be an essential ingredient for high performance in the airline industry, and the HR practices described above combine into an HR system that supports high levels of relational coordination. However, airlines do not exist in a vacuum. Rather each airline belongs to a larger network of organizations. For the major airlines, this network typically includes a set of regional airlines that feed passengers from small markets into the major airlines' hubs, and a set of other majors with whom a marketing alliance has been formed based on code sharing (e.g. OneWorld, Star Alliance, SkyTeam).

Other organizations in an airline's network include suppliers such as travel agents, online reservations services, aircraft manufacturers, airports, heavy maintenance providers, fuelers, food service companies, airport security, and so on. The industry network further includes regulatory bodies such as the National Mediation Board, the Federal Aviation Administration and the Air Transport Safety Board, as well as industry associations such as the Air Transport Association and the Airline Industrial Relations Conference. Finally, as discussed earlier in this chapter, the industry network includes numerous employee unions, some of which are specific to the airline industry, such as the Air Line Pilots Association and the Association of Flight Attendants, and some of which represent workers from multiple industries, such as the International Association of Machinists, the Transport Workers Union and the International Brotherhood of Teamsters. Some unions are newcomers to the industry, including some that have entered after establishing a record in other industries (e.g., the Communication Workers of America), and some that have formed specifically to represent airline employees (e.g., the Aircraft Mechanics Fraternal Association).

One broad lesson from our research is that airlines can benefit from partnering with other key parties in their networks. Just as airlines can benefit from investing in their LR and in HR practices that strengthen relationships among front-line employees, airlines also benefit from partnering with key players in their networks like airports, air traffic controllers and aircraft manufacturers.

Efforts by airlines to reduce labor costs and reform work systems in recent years have highlighted another dimension of coordination and partnership of growing importance in the industry, namely relations between airlines and the various subcontractors that do work previously carried out by the airline's own employees. Maintenance of aircraft is one visible and highly contested example of work that is done partly by airline employees, partly by aircraft manufacturer representatives, and increasingly by specialized mechanical subcontracting firms. Indeed, the 2005 strike of mechanics at Northwest Airlines was largely over the company's decision to contract out much of its heavy (major overhauls of engines etc.) maintenance work to subcontractors and thereby cut its own mechanics' workforce in half. The advantage for firms of this arrangement is that it lowers their fixed labor costs. Moreover, often the outsourcing of work is accompanied by a reorganization of remaining maintenance work in an effort to increase the type of flexibility noted above that, if combined with other supporting HR practices, can enhance productivity.

There is great debate over whether the movement to contract out more airline functions leads to reductions in total costs (direct and indirect labor costs plus contracting fees plus associated contract management and coordination costs), and how this strategy affects the bottom-line safety, productivity, profitability and customer service outcomes. While the verdict on this is still out (there have been no studies to date comparing the bottom-line effects of these different strategies), evidence from other industries that make substantial use of contractors is that the results vary depending on the quality of the relationships among contractors and parent firms and the employee groups of the different enterprises that work together, often side by side (Kochan *et al.*, 1994). Ultimately, this ability to partner is an acquired skill like any other, and one with potentially significant effects on organizational success (Lorenzonni and Lipparini, 1999). By treating these external parties as partners, an airline can extend its sphere of influence beyond its employees to encompass its entire value chain.

10.5   Conclusions

While the airline industry has been characterized as a stronghold of traditional industrial relations because of its high levels of unionization, adversarial labor–management relationships and a control-based approach to HR, we are seeing innovations and new directions. Some airlines are striving to move toward a commitment approach to HR, with a focus on building high-quality relationships between employees. Some are seeking to move away from the adversarial approach toward labor unions to develop a partnership approach in its place. These changes are difficult to make and often there are setbacks. But as both legacy and new-entrant airlines face pressures to reduce their costs in the increasingly competitive environment, they are likely to continue experimenting with these innovative approaches, many of which have been pioneered by Southwest Airlines. Other airlines are responding to cost pressures in a very different way, by intensifying the traditional control approach toward employees and/or moving toward outright union avoidance, approaches that are associated instead with Ryanair.

In this chapter we have described the HR/LR system as composed of interrelated components found at multiple levels: the workplace, labor negotiations, interactions between labor and management at the level of business strategy and corporate governance, and in external relations among the firms, unions and government agencies that combine to form the overall industry. We have stressed that improvements are needed in all aspects of this system if the industry is to find a model for sustained profitability. Marginal improvements, or a hope that the industry can recover on its own and that relationships and practices can return at some point to past patterns, are not realistic. Yet this perspective is only slowly gaining acceptance among participants in the industry. The pace and scope of change will need to accelerate if the parties are to be successful in achieving the improvements necessary to cope with the new competitive environment.

Overall, two points stand out from our research. First, of crucial importance is an underlying philosophy of treating employees as valuable resources, rather than primarily as costs to be minimized, in order to build higher-quality relationships within the firm. Second, this philosophy should be supported by an integrated set of HR practices that fosters teamwork and coordination across the multiple functional groups involved in airline operations.

While HR/LR policies are often thought of as lying within the province of individual companies and unions, a strong rationale exists for developing an industry-wide approach to many of the airline industry's more difficult problems. Management, labor and government representatives are continuing to explore ways to improve labor and employee relations in the industry. Whether these discussions will produce the fundamental changes necessary to support the recovery of the overall industry remains to be seen. If they do, a new chapter in both airline HR/LR and industry performance will be written. If not, others – dissatisfied customers, investors or government officials – may write the next chapter for the industry.

References

Aldrich, H. and Herker, D. (1977) "Boundary Spanning Roles and Organization Structure." *Academy of Management Review*, Vol. 2, No. 2, pp. 217–230.

Appelbaum, E. and Batt, R. (1994) *The New American Workplace*, ILR Press, Ithaca, NY.

Arrowsmith, J., Edwards, T., and Sisson, K. (2000) "Industrial Relations at British Airways: Setting a New Course?" European Industrial Relations Observatory, http://www.eurofound.europa.eu/eiro/2000/04/feature/uk0004168f.htm.

Baker, C., Field, D., and Ionides, N. (2005) "Global Reach," *Airline Business*, Vol. 5, pp. 60–65.

Bamber, G., Gittell, J.H., Kochan, Y. and von Nordenflycht, A. (2009) *Up in the Air: How Airlines Can Improve Performance by Engaging their Employees*, Ithaca, NY: Cornell University Press.

Birmingham Post (2002) "Aer Lingus Grounds Fleet and Turns Back Pilots After Strike," June 1, http://www.highbeam.com/doc/1G1-86578704.html, June 17, 2007.

Blasi, J.D., Kruse, D., and Bernstein, A. (2003) *In the Company of Owners: The Truth About Stock Options and Why Every Employee Should Have Them*, Basic Books, New York.

Branson, R. (1999) *Losing My Virginity: The Autobiography*, Random House, Milsons Point, NSW.

Bray, M. and Wailes, N. (1999) "Reinterpreting the 199 Pilots' Dispute: The Role of Managerial Control and Labour Productivity," *Labor & Industry*, Vol. 10, pp. 79–105.

British Airways (2006) *British Airways Fact Book 2006*, http://media.corporate-ir.net/media_files/irol/ 69/69499/bafactbook/Fact_Book_2006.pdf, June 20, 2007.

Capelli, P. (1985) "Competitive Pressures and Labor Relations in the Airline Industry," *Industrial Relations*, Vol. 24, No. 3, pp. 316–338.

Card, D. (1998) "Deregulation and Labor Earnings in the Airline Industry," in J. Peoples (ed.), *Regulatory Reform and Labor Markets*, Kluwer, Boston, pp. 183–229.

Colling, T. (1995) "Experiencing Turbulence: Competition, Strategic Choice and the Management of Human Resources in British Airways," *Human Resource Management Journal*, Vol. 5, pp. 18–31.

Cremieux, P.-Y. (1996) "The Effect of Deregulation on Employee Earnings: Pilots, Flight Attendants, and Mechanics, 1959-1992," *Industrial and Labor Relations Review*, Vol. 49, No. 2, pp. 223–242.

Day, D. and Silverman, S. (1989) "Personality and Job Performance: Evidence of Incremental Validity," *Personnel Psychology*, Vol. 42, No. 1, pp. 25–36.

Deming, J.E. (1986) *Out of the Crisis*, MIT Press, Cambridge, MA.

Doganis, R. (2002) *Flying Off Course: The Economics of International Airlines*, Routledge, London, p. 26.

Doganis, R. (2006) *The Airline Business*, Routledge, London, pp. 132, 223.

Galbraith, J. (1994) *Competing with Flexible Lateral Organizations*, Addison-Wesley, Reading, MA.

Gittell, J.H. (2000) "Paradox of Coordination and Control," *California Management Review*, Vol. 42, No. 3, pp. 1–17.

Gittell, J.H. (2001) "Supervisory Span, Relational Coordination and Flight Departure Performance: A Reassessment of Post-Bureaucracy Theory," *Organization Science*, Vol. 12, No. 4, pp. 467–482.

Gittell, J.H. (2003) *The Southwest Airlines Way: Using the Power of Relationships to Achieve High Performance*, McGraw-Hill, New York.

Gittell, J.H. and Reilly, C. (2001) "JetBlue Airways: Starting from Scratch," *Harvard Business School Case*, Harvard Business School Publishing, Boston, MA.

Gittell, J.H., von Nordenflycht, A., and Kochan, T.A. (2004) "Mutual Gains or Zero Sum? Labor Relations and Firm Performance in the Airline Industry," *Industrial and Labor Relations Review*, Vol. 57, No. 2, pp. 163–179.

Gladstein, D. (1984) "A Model of Task Group Effectiveness," *Administrative Science Quarterly*, Vol. 29, pp. 499–517.

Godfrey, B. (2002) "Low Cost Airlines in the Asia Pacific Region," *Centre for Asia Pacific Aviation*, February, p. 32.

Hatty, H. and Hollmeier, S. (2003) "Airline Strategy in the 2001/2002 Crisis: The Lufthansa Example," *Journal of Air Transport Management*, Vol. 9, No. 1, pp. 51–55.

Herdman, A. (2007) "Full Service Airlines: Adopting New Business Strategies in the Crowded Sky," http://www.aapairlines.org/resource_centre/SP_AAPA-HerdmanLCACongressSingapore-24 Jan2007.pdf, June 16, 2007.

Hirsch, B.T. and Macpherson, D.A. (2000) "Earnings, Rents and Competition in the Airline Labor Market," *Journal of Labor Economics*, Vol. 18, No. 1, pp. 125–155.

Johnson, N.B. (1995) "Pay Levels in the Airlines Since Deregulation," in P. Capelli (ed.), *Airline Labor Relations in the Global Era*, ILR Press, Ithaca, NY.

Johnson, N.B. (2001) "Airlines," Working Paper, Gatton College of Business and Economics, University of Kentucky.

Keller, B. (2004) "Employment Relations in Germany," in G.J. Bamber, R.D. Lansbury, and N. Wailes (eds.), *International and Comparative Employment Relations: Globalisation and the Developed Market Economies*, Sage, London, pp. 211–253.

Kochan, T.A. (1980) *Collective Bargaining and Industrial Relations*, Richard D. Irwin, Homewood, IL.

Kochan, T.A., Katz, H., and McKersie, R. (1986) *The Transformation of American Industrial Relations*, Basic Books, New York.

Kochan, T.A., Smith, M., Wells, J., and Rebitzer, J. (1994) "Human Resource Strategies and Contingent Workers: The Case of Safety and Health in the Petrochemical Industry," *Human Resource Management*, Vol. 33, pp. 55–78.

Lansbury, R.D. and Wailes, N. (2004) "Employment Relations in Australia," in G.J. Bamber, R. D. Lansbury, and N. Wailes (eds.), *International and Comparative Employment Relations: Globalisation and the Developed Market Economies*, Sage, London, pp. 119–145.

Lawrence, P. and Lorsch, J. (1967) *Organization and Environment: Managing Differentiation and Integration*, Harvard Business School Press, Cambridge, MA.

Lee, B.-H. and Cho, S.-J. (2007) "Employment Relations of the Korean Airline Industry: Comparison of Koreana Air and Asiana Airlines," Unpublished case study.

Levine, D.I. and Tyson, L.D. (1990) "Participation, Productivity and the Firm's Environment," in A.S. Binder (ed.), *Paying for Productivity: A Look at the Evidence*, The Brookings Institution, Washington, DC.

Levine, M.E. (1987) "Airline Competition in Deregulated Markets: Theory, Firm Strategy, and Public Policy," *Yale Journal on Regulation*, Vol. 4, pp. 393–495.

Lorenzoni, G. and Lipparini, A. (1999) "The Leveraging of Interfirm Relationships as a Distinctive Organizational Capability: A Longitudinal Study," *Strategic Management Journal*, Vol. 20, pp. 317–338.

Malaysian Airlines (2007) "The MAS Way: Business Turnaround Plan," available at http://www. malaysiaairlines.com/uploads/en/downloads/common/BusinessTurnaroundPlan(BTP1).pdf (December 22, 2008).

Marchington, M., Goodman, J., and Beeridge, J. (2004) "Employment Relations in Britain," in G.J. Bamber, R.D. Lansbury and N. Wailes (eds.), *International and Comparative Employment Relations: Globalisation and the Developed Market Economies*, Sage, London, pp. 36–66.

Nay, L. (1991) "Determinants of Concession Bargaining in the Airlines," *Industrial and Labor Relations Review*, Vol. 44, No. 2, pp. 307–323.

Piore, M. and Sabel, C. (1984) *The Second Industrial Divide*, Basic Books, New York.

Pondy, L. (1967) "Organizational Conflict: Concepts and Models," *Administrative Science Quarterly*, Vol. 8, pp. 297–320.

Poon, T. and Waring, P. (2009) "Lean Production Aviation in Asia: The Case of AirAsia," *International Journal of Human Resource Management*, forthcoming.

Preuss, G. (1996) "The Structuring of Organizational Information Capacity: An Examination of Hospital Care," *Best Papers Proceedings of the 1996 Annual Meeting of the Academy of Management*.

Rosen, S.D. (1995) "Corporate Restructuring: A Labor Perspective," in P. Capelli (ed.), *Airline Labor Relations in the Global Era*, ILR Press, Ithaca, NY.

Ryan-Be-Fair, "Ryanair Workers Mobilised to Strike," http://www.itfglobal.org/campaigns/mobilised .cfm, July 30, 2007.

Schlesinger, L. and Heskett, J. (1991) "The Service-Driven Service Company," *Harvard Business Review*, Vol. 69, September/October, pp. 71–81.

Selznick, P. (1949) *TVA and the Grass Roots*, Berkeley University Press, Berkeley, CA.

Van de Ven, A. (1976) "A Framework for Organization Assessment," *Academy of Management Review*, Vol. 1, pp. 64–78.

Van de Ven, A. and Ferry, D. (1980) *Measuring and Assessing Organizations*, John Wiley & Sons, Inc., New York.

von Nordenflycht, A. (2002) "Alternation Approaches to Airline Labor Relations: Lessons for the Future," Presentation to the 2002 Annual Meeting of the IRRA.

von Nordenflycht, A. (2004) "The Transformation of Authority at Continental Airlines," *Best Papers Proceedings of the 2005 Annual Meeting of the Academy of Management*.

von Nordenflycht, A. and Kochan, T.A. (2003) "Labor Contract Negotiations in the Airline Industry," *Monthly Labor Review*, Vol. 7, pp. 18–28.

Wallace, J., Tiernan, S., and Whilte, L. (2007) "Industrial Adaptation in Aer Lingus: The Path from Legacy to Low Fares Airlines," Unpublished case study.

Walsh, D. (1988) "Accounting for the Proliferation of Two-Tier Wage Settlements in the US Airline Industry, 1983-1986," *Industrial and Labor Relations Review*, Vol. 42, No. 1, pp. 50–62.

Walton, R.E. (1985) "From Control to Commitment in the Workplace," *Harvard Business Review*, March 1.

Walton, R.E., Cutcher-Gershenfeld, J.E., and McKersie, R. (1994) *Strategic Negotiations: A Theory of Change in Labor–Management Relations*, ILR Press/Cornell University Press, Ithaca, NY.

11

Aviation Safety and Security

Arnold Barnett

In 1997, the US Air Transport Association held a conference titled "Safety: The Foundation of Our Business." That this title is more than a truism was made breathtakingly clear in the aftermath of the 9/11 calamity. US passenger traffic plunged 25% in the last quarter of 2001. And while traffic volumes did return in 2005 to 2000 levels, forecasters had predicted in 2000 that passenger enplanements would rise 20% by 2005. The 2005 shortfall is all the more striking because the projection did not anticipate the 21% drop in the average US domestic airfare that occurred between 2000 and 2005.

The fact is that, if a major air disaster strikes, the most meticulous aviation business plan can become moot in an afternoon. Beyond being important in its own right, therefore, aviation safety is essential to the industry's economic viability. A textbook about commercial aviation would be gravely deficient if it did not include a discussion of safety issues.

From the perspective of the air traveler, there are two components of safe travel. The first is avoiding aviation accidents, which – given that aircraft travel hundreds of miles per hour at thousands of feet above the ground – can easily prove fatal. The second is preventing acts by criminals or terrorists that can harm air passengers. In the first part of this chapter, we focus on the threat posed by accidents; in the second, on the threat posed by deliberate attacks.

Even when the focus is on accidents, aviation safety is a vast topic. Current practices in maintenance, pilot training, air traffic control, and airport and aircraft design reflect the accumulation and distillation of many decades of thought and experience. Many of these topics are discussed in Chapters 8, 12 and 13, and it would not be feasible to expand those discussions here. When a full treatment of the technical details of (say) deicing aircraft could require several hundred pages, we could not hope to do justice to any particular safety process.

Aviation security is likewise a large topic, and one could argue that it is inherently more difficult to discuss than safety. Once a physical hazard like windshear has been

The Global Airline Industry   P. Belobaba, C. Barnhart and A. Odoni
© 2009 John Wiley & Sons, Ltd

dealt with, it is essentially gone for good. Terrorists, by contrast, can continuously change their methods of attack, and it is hard to predict when, where and how they will try to strike next. It is equally hard to know whether the safeguards in place to prevent a terrorist attack have a real prospect of success.

Under the circumstances, we put some restrictions on the scope of this chapter. We limit ourselves to exploring aviation safety from a *statistical* perspective, with the bulk of the data analyses centered on the question "how safe is it to fly?" Yet data analysis can go beyond simply describing what has been achieved so far: it can also help identify the most prudent strategies to improve future air safety. We will illustrate this last point by excerpting two studies: one that influenced a major decision about air safety policy, and another concerning risks that may arise in the future.

For security, we also emphasize a statistical perspective. But we will see that what we can learn from data analysis is less clear than for safety. The discussion is sometimes tentative, and it serves to remind us of the difficulties that policy makers face every day. Yet neither they nor we can simply throw up our hands and declare security an intractable problem: decisions have to be made, and decisions based on imperfect but reasoned analyses would seem preferable to arbitrary choices. We will depict how some such analyses are being attempted today.

11.1   Safety

We begin our discussion of aviation safety in the next section, where we discuss a variety of ways to measure the mortality risk of passenger air travel. In Section 11.1.2, we settle on a "death risk per flight" metric (the Q-statistic) that, we will argue, has numerous features that make it a metric of choice. We then calculate Q-statistics in various settings in Sections 11.1.3, 11.1.4 and 11.1.5, and distinguish between comparative judgments about aviation safety that deserve to be taken seriously from others that should probably not be. In Section 11.1.6, we synthesize various statistical clues about the risk of airport runway collisions in the decades ahead; in Section 11.1.7, we move on to midair collisions.

11.1.1   Measuring Air Safety: Some Hazards

In this section, we assume that the air traveler's greatest fear is of being killed in a crash. It follows that statistics about the likelihood of that outcome are of inherent interest. But which particular statistics are most illuminating about passenger death risk is not obvious, as is made clear by a review of some of the indicators that are currently available.

The US National Transportation Safety Board (NTSB), for example, has advanced the statistic *fatal accidents per 100 000 hours flown*. In general terms, this indicator makes sense: it relates the number of adverse events to the amount of flying performed. It therefore provides something of a "cost–benefit" ratio.

Unfortunately, both the numerator and the denominator of the ratio are problematic. The term "fatal accident" obliterates the distinction between a crash that kills 1 person out of 300 on board and another that kills 300 out of 300. And the measure gives no weight to safety improvements (e.g., fire-retardant materials) that reduce fatalities but do not prevent them.

Moreover, the emphasis on hours flown misses the point that the overwhelming majority of fatal accidents occur in the takeoff/climb or descent/landing phases of flight. Boeing reports that, over the period 1998–2007, 13% of worldwide jet accidents with on-board fatalities occurred on takeoff, 22% during climb, 17% on descent/initial approach, 38% on final approach/landing, and only 10% during the cruise phase of flight, which is generally longer than the other four phases combined (Boeing, 2007). Larger proportions of passengers perish in fatal accidents at cruise altitude than in other fatal accidents: despite that circumstance, however, a minute at cruise altitude entails only one-sixth the death risk as a minute in another phase of flight.

Regardless of duration, every flight requires a takeoff/climb and descent/landing. Thus, a flight's fatal accident risk is nearly independent of duration: one flight of 4 hours does not entail the same risk as four 1-hour flights. Yet both these situations involve four flight hours, meaning that the denominator in the NTSB statistic treats them as equivalent. If the average duration per flight were to vary from one period to another, then the ratio "fatal accidents ÷ hours flown" could change for reasons having nothing to do with safety.

Another statistic that has long been used is *hull losses per 100 000 flight departures*. A hull loss is an accident that damages the aircraft beyond repair (the aerial equivalent of "totaling" a car). This index is of obvious interest to aircraft manufacturers. And in focusing on flight departures rather than flight hours, it recognizes that the distance or duration of the flight bears scant relationship to its accident risk.

If, however, the aim is to gauge passenger risk, then hull loss is a questionable proxy. Consider, for example, two passenger jet hull losses that occurred in the same month in 2005:

Location		Percentage of passengers killed	
Toronto	(August, 2005)	0%	(291 passengers on board)
Near Athens	(August, 2005)	100%	(115 passengers on board)

There have been many instances in which an aircraft landed with major damage but, because of well-executed emergency procedures, all passengers were evacuated before the aircraft was engulfed in flames and became a hull loss. Such a rescue is irrelevant to the hull loss statistic, but it hardly seems so to an assessment about the mortality risk of air travel.

Yet another statistic of interest is the ratio of *passengers killed to passengers carried*. This indicator has a plausible connection to the chance that a current passenger will be killed, for it literally reports what fraction of passengers were killed in a recent period. But weighting a crash by the number of fatalities can create difficulties. If a jetliner hits a mountain killing all passengers aboard, the implications about air safety are not three times as great if the aircraft is full rather than one-third full. And 32 deaths out of 32 aboard does not mean the same thing as 32 deaths out of 320 aboard. (In the latter case, an excellent crew response may have saved 90% of the passengers on board.) Risk statistics that use deaths in the numerator, in other words, are vulnerable to meaningless fluctuations in the proportion of seats occupied, yet insensitive to large differences in the fraction of passengers who survived the crash.

Beyond the three indicators just discussed, there are several others that also seem questionable as guides to aircraft mortality risk. See Barnett and Wang (2000), for example, for a discussion about such weird concoctions as airline "report cards." However, our point here is not to dwell on the shortcomings of individual metrics, but instead to stress that the very choice of a risk indicator is a major element of an air safety study. In the next section, we present the Q-statistic, an indicator developed to address many of these shortcomings.

11.1.2 The Q-statistic

The Q-statistic is the answer to the question: "Suppose that a person chose a flight *completely at random* from among the set of interest (e.g., scheduled UK domestic jet flights in the 1990s). What is the probability that he would not survive the flight?"

In this section, we are speaking of death in an aviation accident, and not from terrorism or natural causes. A flight is a non-stop trip; a journey that involves one intermediate stop would entail two flights.

The Q-statistic – which is essentially death risk per flight – assumes that there are N flights which can be indexed as $(1, 2, 3, \ldots, N)$. We define x_i as the fraction of passengers on flight i who do not survive it. If the flight lands safely, then $x_i = 0$; if it crashes and no one survives, then $x_i = 1$; if it crashes and 20% of the passengers perish, then $x_i = 0.2$.

To determine the Q-statistic, we note that each of the N flights has the same chance of $1/N$ of being selected at random. Given that flight i was selected, the traveler's conditional death risk is x_i. One way of dying because of the "flight lottery" would be to choose flight 1 at random and then to perish; the probability of this event is $(1/N)x_1$. Another way would be to select flight 2 and perish, an event of probability $(1/N)x_2$. The overall probability of dying would be the sum of the probabilities of all the N mutually exclusive ways that a fatal outcome can arise. In consequence, Q can be obtained under the formula:

$$Q = (1/N)x_1 + (1/N)x_2 + \cdots + (1/N)x_N = \sum_{i=1,\ldots,N} x_i/N \qquad (11.1)$$

The Q-statistic has several advantages compared to the metrics discussed in the previous section:

- The statistic weights each crash by the *fraction* of passengers killed. Thus, a crash into a mountain that killed everyone on the aircraft would be treated the same way regardless of how many people happened to be on board that day. And a high survival rate in a crash would be treated very differently from a low survival rate.
- In accordance with empirical evidence, the calculation gives no weight to the mileage or duration of the flight.
- The calculation of Q is surprisingly easy. N is generally known from publicly available data. The conditional probability x_i is almost always zero and, when it is not, official reports about the crash make clear the value of x_i.

- The intuitive interpretation of Q is straightforward. While Q can only be calculated for a period that has ended, a recent Q-value is a strong approximation of the chance of dying in an accident on a flight today.

In the next two sections, we present numerical values of Q.

11.1.3 Some Calculated Q-values

We limit ourselves in this section to scheduled passenger jet flights. Charter and propeller flights are not included. By any reasonable measure (e.g., travelers enplaned, passenger kilometers carried) scheduled passenger jet flights perform the overwhelming bulk of passenger air transportation. (Some Q-statistics about propeller flights appear in Barnett and Wang (2000) and Barnett (2006).)

Among scheduled passenger jet flights, the heavy majority are *domestic* services in First World countries like the USA, Canada, Germany, Japan and Australia. For example, Tokyo to Sapporo in northern Japan is the world's busiest air route. Thus, these domestic flights provide a natural venue for a first foray with Q-statistics.

In the 1990s, there were a total of approximately 70 million domestic jet flights in the developed world. All but 10 of those flights landed without any accidental deaths; on the remaining 10, the average fraction of passengers killed was 56%. Therefore, the sum of the x_i in the formula for Q (Equation (11.1)) was $10 \times 0.56 = 5.6$, and $N \approx 70$ million. Q is therefore estimated as $5.6 \div 70\,000\,000$, which is approximately 1 in 13 million.

This 1 in 13 million is obviously a low number, but how low? There are several ways of elaborating on this statistic. At a risk level of 1 in 13 million, a child taking off on a First World domestic jet would be roughly 10 times as likely to win a future Olympic gold medal as to fail to reach his or her destination today. In the Massachusetts lottery game called Megabucks, the chance of winning the jackpot is 1 in 5.2 million. Thus, a Massachusetts resident who buys a lottery ticket is 2.5 times as likely to win the jackpot as to "lose" disastrously on his next domestic jet flight. At a mortality risk of 1 in 13 million per flight, a person who took one flight at random per day could on average travel every day for *36 000 years* before succumbing to a fatal crash. This last outcome implies that even very frequent flyers have extremely low lifetime risk of dying in an air accident.

Moreover, the data for the first eight years of the new century are even more encouraging:

Period	Scheduled domestic jet flights performed in First World	Death risk per flight (Q-statistic)
1990–1999	70 million	1 in 13 million
2000–2007	80 million	1 in 80 million

Note that some approximations were used in estimating the number of flights performed.

The 1 in 80 million statistic reflects one fatal crash: a 2006 takeoff accident in Kentucky (USA) that killed all the passengers on board.[1] But might this improvement over the

---

[1] As this chapter was being finalized, a domestic jet crash occurred in August 2008 at Madrid, and it killed about 90% of passengers aboard.

1990s represent mere chance? Even in a statistical process that is stable over time, it is surprisingly common to get several events in quick succession, interspersed with long periods without any events.

Because the table shows nearly equal numbers of flights over 2000–2007 (i.e., 80 million) as over 1990–1999 (70 million), the answer to the question about "sheer luck" is quite straightforward. Based on the pattern of the 1990s, one would have projected 12 fatal accidents during the period 2000–2007 on First World domestic jets, with an average death rate around 56%. In fact, there was only one such event. Such a discrepancy would have an exceedingly small probability of arising by coincidence.

To amplify this last point, suppose that a six-sided die is tossed 72 times. If the die is fair, one would expect the number 1 to come up roughly $72 \times (1/6) = 12$ times. If it only came up once, one would strongly suspect that the die is not fair: the probability of a result at least 11 away from the expected outcome (i.e., the "p-value" of the result) is only 1 in 15 000. We are in essentially the same situation here: under any usual statistical standards, one would reject the hypothesis that mortality risk did not really improve between 1990–1999 and 2000–2007.

11.1.4 More Calculated Q-statistics

Here we expand on calculations in the last section in two ways:

- We consider scheduled jet flights from 1960 onwards, which is essentially the entire period during which passenger jet operations have taken place. We present the 1960–1999 data by decade.
- We consider all worldwide jet operations, breaking the flights into four non-overlapping categories, as follows:

First World Domestic:
Flights between two cities in the same First World country.
Example: Toronto–Vancouver.
First World International:
Flights that travel from a city in one First World country to a city in another one.
Example: Paris–Dublin.
Between First and Third World:
Flights from a Third World city to a First World one (or vice versa).
Example: Jakarta–Tokyo.
Within Third World:
Flights that originate and end in the Third World.
Example: Nairobi–Lagos.

As defined, these categories do not overlap, but they are not always conceptually distinct. If a First World airline suffers a takeoff crash at a First World airport, for example, the fact that it was headed for an airport in the Third World might be irrelevant. But this author believes that the advantages of working with a small number of well-defined categories outweigh the drawbacks.

The calculated Q-statistics for the four categories are as follows:

Period *First World Domestic*	*Q-statistic* *(death risk per flight)*
1960–1969	1 in 1 million
1970–1979	1 in 3 million
1980–1989	1 in 4 million
1990–1999	1 in 13 million
2000–2007	1 in 80 million
First World International	
1960–1969	1 in 400 000
1970–1979	1 in 1 million
1980–1989	1 in 4 million
1990–1999	1 in 6 million
2000–2007	1 in 9 million
Between First and Developing World	
1960–1969	1 in 200 000
1970–1979	1 in 300 000
1980–1989	1 in 600 000
1990–1999	1 in 1 million
2000–2007	1 in 1.5 million
Within Developing World	
1960–1969	1 in 100 000
1970–1979	1 in 200 000
1980–1989	1 in 400 000
1990–1999	1 in 500 000
2000–2007	1 in 2 million

These statistics do not include crashes caused by criminal or terrorist acts. The calculations entail some approximations about the numbers of flights performed; see Barnett and Higgins (1989) and Barnett and Wang (2000) for discussions of the methodology and data sources used.

The key patterns in the data are obvious. Throughout the world and without *any* exceptions, jet travel has consistently become safer decade by decade. Passenger mortality risk fell by over 90% between 1960–1969 and 2000–2007. The data offer no evidence that the percentage rate of improvement declined from decade to decade; this outcome is especially impressive because, as risk goes down, one might think that further improvement is harder to achieve.

It is also apparent, however, that death risk is far lower on jet flights in the First World than on those involving the Third World. In each of the five periods studied, the difference was a factor of at least 5, and the ratio has, if anything, grown in recent years. The statistical significance of the difference is beyond dispute.

11.1.5 Are Some Airlines Safer Than Others?

The numbers we have presented so far reflect the *average* performance of large numbers of airlines in the same category (e.g., First World International). But, if a person is flying non-stop from A to B, is there any reason related to safety to prefer one carrier traveling the route to another?

There are no statistically significant differences in mortality risk among airlines in the First World. But, given the (fortunate) scarcity of fatal accidents, this last statement is almost a foregone conclusion. Suppose, for example, that two First World airlines of equal size have a total of three fatal events over a period of two decades. If one of the carriers had two of the events and the other one, it would literally be true that the former had "twice the fatal accident rate" of the latter. But, to attribute great meaning to that difference might be as far-fetched as arguing that a coin is not fair because it came up heads two times out of three.

Some people have suggested that fatal accidents are merely the "tip of the iceberg," and that studying all accidents and incidents would provide useful comparisons across airlines based on large amounts of data. That approach is more problematic than it sounds: Barnett and Wang (2000) showed that, among established US carriers, there was a *negative* correlation between an airline's death risk per flight in 1990–1996 and its rate of non-fatal accidents and incidents. One possible explanation for this pattern is that non-fatal events raise awareness and caution, and thereby prevent some fatal accidents.

Pursuing the issue further, Czerwinski and Barnett (2006) studied not just tabulations about fatal jet accidents, but also records about all accidents and incidents in the USA in 1983–2002 on scheduled jet flights. They focused on events that had the potential to cause passenger deaths and had actually done so in some instances (e.g., loss of aircraft control). Applying a method that the Oakland Athletics professional baseball team had used to evaluate baseball players, they examined hundreds of untoward events among US airlines, weighting each type of event by the probability that it would lead to passenger deaths and also the average proportion killed given that fatalities occurred. For example, if a certain type of emergency on average led to the death of 2% of passengers, the baseball procedure would assign a penalty of 0.02 to an airline every time it encountered that emergency. Czerwinski and Barnett found no evidence that some US carriers were more "prone" to life-threatening emergencies than others. The authors also tested statistically whether some airlines might be better than others in recovering without fatalities from dire emergencies, given that such emergencies arose. To put it briefly, they found no evidence that some US carriers are systematically safer than others. All of the airlines have outstanding records.

But what about cases where one might expect a difference, such as between a First World airline and a carrier from the Third World? We know that, overall, these groups of carriers have markedly different mortality–risk records. But what does that difference mean about routes where First World and Third World airlines compete? If one is flying from (say) Tokyo to Jakarta, is it safer to fly Japan Airlines than Garuda Airlines of Indonesia?

To assert that the answer is "obviously yes" is to risk falling prey to what is known in statistics as the ecological fallacy. Suppose that one makes the three statements:

- John is better in mathematics than Bill.
- I have a problem in trigonometry.
- Therefore, I should contact John rather than Bill.

We can see the logical problem in this sequence of statements. Perhaps trigonometry is Bill's strong point and John's weak point. In that case, I would do well to contact Bill. More generally, a statement that might be true in the aggregate does not have to apply in each specific instance.

If we want to know whether First World carriers are safer on routes between the First and Third World than are Third World carriers, there is an obvious way to find out: we should compare fatal accident records on those specific routes. When Barnett and Wang (2000) did so, they found that the Q-statistic was the same for both groups of carriers over 1987–1996: approximately 1 in 600 000. When Barnett (2006) investigated the issue some years later, the result was the same: the Q-statistic was roughly 1 in 1.5 million on both First and Third World airlines. What may be happening is that the hazards in Third World aviation affect all airlines that fly in these regions, and not just the local carriers.

We therefore reach a surprisingly strong conclusion: when two airlines fly non-stop on the same route, very rarely is there an empirical basis for believing that one carrier is safer than the other. We cannot beat the overall odds; the good news, however, is that the odds are so excellent that we rarely have any need to beat them.

11.1.6 A Collision Risk Assessment

As we have seen, data analysis can offer an overview about the past and present safety of commercial air travel. But it can also help in making the resource allocation decisions that can affect the safety of future flights. Even though flying environments in the years ahead might differ markedly from any that we have seen, careful analysis of the "clues" available from historical data could prevent future fatal accidents.

As an example of such a future-oriented data analysis, we describe an actual study performed at the request of the US Federal Aviation Administration (FAA). The basic question asked by the FAA was: "How great is the potential danger posed by US airport runway collisions in the next 20 years?" The analysis posited the continued use of existing procedures and technologies (i.e., a status quo model). Indirectly, therefore, the issue was how large the benefits of investing resources would be to reduce future collision risk.

There are two distinct issues in any such risk projection. The first is the *frequency* with which runway collisions might occur over the period of interest. The second is the *consequences* of a given collision, especially with respect to human life. Our full analysis on these topics appears in Barnett *et al.* (2000); here we address a few of the issues.

It is widely recognized that airport operations (landings plus takeoffs) will increase in coming years. The question is what these increases will mean for the likelihood of runway collisions (which can involve not just two aircraft but also an aircraft and a land vehicle, or an aircraft and a physical obstacle). Will the number of collisions grow in proportion to operations (the linear model), meaning that a doubling of operations will cause a doubling of collisions? Will collisions grow with the square of operations, meaning a doubling of operations induces a quadrupling of collisions? Will they grow cubically, meaning a doubling of operations causes eight times as many collisions? Or might there be no increase in collisions at all: crowded highways – in which everyone drives slowly – do not have more deaths per mile driven than do free-flowing highways. (This last relationship is called the "zeroth-order" model.)

One might argue that, among the models just mentioned, the most plausible is the quadratic model. If M is an airport's total number of aircraft operations over a period such as a year, its probability of being the site of a runway collision could be roughly proportional to M^2. There are two general arguments for this viewpoint:

1. If there are M operations (numbered 1 through M), the total number of *possible* pairings of aircraft (e.g., (numbers 1 and 3), or (32 and 89)) that could theoretically collide is well known to follow the mathematical formula $M(M - 1)/2 = (M^2 - M)/2$. When M is large, $M^2 \gg M$, so the last expression approaches $M^2/2$ to within an exceedingly small percentage.

   This argument is a bit loose: air traffic delays have been growing, but it is hard to imagine that an aircraft that takes off on March 3 could collide with another that lands on August 9. But there is a more serious case for a quadratic model.
2. Consider a one-way suburban side street that has a stop sign at its intersection with a two-way main street. Suppose that vehicular traffic increases by 20% on both streets. How might we expect this increase to change the risk of an accident at the intersection, involving one car from each street?

Two effects are at work. Given the 20% growth in traffic on the side street, we might anticipate a 20% increase in the number of cars on that street that violate the stop sign and rush into the intersection. But, given the higher traffic density on the main street, there would be roughly a 20% greater chance than earlier that the car that violated the stop sign would actually hit another vehicle. In other words, we face a "double whammy": the overall chance of an accident per unit time would grow not by a factor of 1.2 but by a factor of $(1.2)^2 = 1.44$. This example supports a classic quadratic model.

Whenever possible, however, one should go beyond blackboard theorizing and test a hypothesis against available data. Fortunately, it was possible to do so for the quadratic model just described. For the year 1997, there were no fatal runway events at US airports. But there were 40 harrowing "close calls" that were both classified as having "extremely high accident potential" by a body of aviation experts (air traffic controllers, pilots) who reviewed what had happened, *and* occurred during conditions of reduced visibility (night, sunrise/sunset with attendant glare, or haze/fog). Observing how these 40 events were distributed across US airports allows a test of the quadratic hypothesis, assuming that the distribution of close calls among airports is a reasonable proxy for the corresponding distribution of fatal accident risk. A statistical test could explore whether, as the quadratic hypothesis predicts, airports with (say) 500 000 operations in 1997 averaged roughly four times as many close calls apiece as airports with 250 000 operations.

The testing regimen is described in Barnett *et al.* (2000). To put it briefly, the quadratic model passed with flying colors. Of very great interest is the point that the linear, cubic, and zeroth-power models did *not* pass the corresponding statistical tests. Collectively, the test outcomes meant that harrowing events did occur disproportionately at busier airports, but not to a wildly lopsided extent. For both conceptual and empirical reasons, therefore, we thought it reasonable to use the quadratic approximation in making projections about runway collision risk under higher traffic levels anticipated in the early twenty-first century. (See Barnett (2008) for further discussion of the quadratic model and its implications.)

Another key issue was how many passengers on a scheduled commercial flight would perish if it suffered a runway collision. The most frightening possibility would seem to be a collision between two passenger jets. In the USA, however, there had only been one such runway collision in the jet age, and 96% of the people on board the two aircraft – 190 out of 198 – had survived. While it was known that death rates had been higher in some foreign collisions, some analysts argued that the FAA should not consider such overseas events, because they would not have occurred under US air traffic control.

Barnett *et al.* (2000) contended, however, that an important distinction needed to be kept in mind. In estimating the *probability* of a US runway collision, restricting attention to US data could well make sense. But in assessing the consequences of such a collision, worldwide data might be more appropriate to use. For one thing, basing a mortality statistic on exactly one event would be statistically questionable. (One could not even compute a standard deviation.) For another, the heavy majority of deaths in runway collisions arise at once in conflagrations, meaning that differences in (say) the quality of local medical care are close to irrelevant. Under the circumstances, the authors studied all three runway collisions between passenger jets over the period 1970–1999 (each of which, incidentally, involved First World airlines). The results appear in Table 11.1.

In the table, the unweighted six-jet average is the simple average of the percentage killed for the various jets. The weighted six-jet average (70%) is the total killed out of the total aboard the six jets. Fractions in brackets present number killed over number on board. All the jets involved in these collisions were operated by First World airlines.

The authors believed that the simple "one-jet, one-vote" principle was the most defensible way of combining the data from the six jets involved in the runway collisions. Hence, they estimated as 60% the fatality rate per aircraft when two passenger jets collide (while recognizing the statistical uncertainty in such an estimate).

The quadratic risk model and the 60% statistic were part of a broader analysis that ultimately reached a set of collision risk forecasts. The *mid-range* projection in 2000 was that, assuming the continuation of the status quo, 15 passenger aircraft would be involved in fatal runway collisions at US airports with control towers over the period 2003–2022. These events would lead to 700–800 deaths and 200 serious injuries. Given the remarkable safety record of US commercial aviation, the authors argued, it was entirely possible that runway collisions would cause more accidental deaths among US air travelers than all other causes combined. A similar statement would apply for other First World travelers and, indeed, a runway collision in Milan in 2001 caused 110 deaths in the worst fatal

Table 11.1 Death rates in the three fatal runway collisions involving two passenger jets, 1970–1999

Location (year)	First jet	Second jet	Both jets combined
Tenerife (1977)	100% (248/248)	85% (335/396)	91% (583/644)
Madrid (1983)	100% (42/42)	55% (51/93)	69% (93/135)
Detroit (1990)	18% (8/44)	0% (0/154)	4% (8/198)
Six-jet average	60% (unweighted)		70% (weighted)

Source: Barnett *et al.* (2000)

accident on a First World airline in Western Europe over 2000–2007. Total European accident deaths over that period were approximately 160.

These projections were far higher than earlier ones circulating at the FAA, which had been based on linear models and exclusive reliance on US data. And the FAA did not ignore the new projections: the agency has publicly stated that the 2000 study contributed to a decision to deploy new collision-avoidance radar devices at 25 mid-sized US airports (34 larger airports were already slated to achieve such devices). The hope, therefore, is that the forecast of 700 deaths will be a "self-destroying prophecy" because it inspired new measures that saved lives. Instead, the forecast could help quantify how much was achieved by the aviation community, which could not fatalistically accept that "accidents will happen" and which took decisive steps to prevent them.

11.1.7 Midair Collision Risk

Collisions between aircraft can occur in the air as well as on the ground. Actually, midair collisions involving scheduled passenger flights have all but disappeared from First World skies. The last such event occurred in 1988, more than 200 million flights ago.[2] But the air traffic control arrangements in place now are likely to be changed soon. In Western Europe, there is strong pressure to replace the various national air traffic control systems with a harmonized one (a "single European sky"). In the USA, current airline itineraries – under which aircraft are confined to a network of prescribed flight paths – are slated to give way gradually to a set of "direct routings" from origin to destination. Such routings would lead to shorter flight times and, of growing importance, to lesser fuel consumption.

Such changes present safety challenges. Harmonizing dozens of air traffic systems is unlikely to be a straightforward process. And direct routings would certainly complicate any visual display of US flight paths. The moving dots that represent aircraft on controllers' screens – which line up today like points on a grid – could in the future resemble gas molecules in random scatter. Moreover, a fundamental notion of industrial sociology is the "learning curve," under which new procedures beget errors and difficulties that had not been anticipated. Thus, apart from specifics, any major changes in air traffic control could pose risk.

Mathematical models and data analysis help us to analyze the situation. They highlight an important point: direct routings can increase midair collision risk, but they can also do some things that should reduce risk. Such routings would change the geometry of flight paths in ways that, in themselves, could make collisions less likely. For example, consider Figure 11.1, which concerns one aircraft traveling from A to B and another from C to D. Under the present prescribed routes, the first aircraft might follow path A–E–F–B and the second, path C–E–F–D. These aircraft could therefore come in close proximity along their common segment EF. If each could take a direct routing, they would get nowhere near one another.

Moreover, there are reasons to believe that direct routings would reduce the angles at which flight paths would cross at a given altitude (see Barnett (2000), which examines data from the air traffic control sector over Albany, New York). Were such angles to drop, the characteristics of some collision warning systems would mean that pilots might get

---

[2] There was, however, a midair collision over Germany in 2002 involving a DHL cargo plane and a Russian charter flight.

Figure 11.1 Direct routings could reduce the overlap of different flight paths

more time to react to alerts about an impending collision. This extra time should increase
the likelihood of avoiding the midair crash.

To be sure, these advantages of direct routings must be weighed against their potential
drawbacks. With greater diversity of flight paths, the air traffic controller would no longer
be able to focus attention on a limited number of merging points. In principle, any point
on the screen could be the site of a midair collision. The key point is, however, that
new arrangements in air traffic control do not automatically bring greater dangers. With
sufficient care, midair collisions should continue to be exceedingly rare events.

How can we summarize this discussion about safety? In 1996, American Airlines CEO
Robert Crandall said that he hoped that, one day, air travelers could take their safety totally
for granted. We have not yet reached the point at which boarding an aircraft is considered
a safe as boarding an elevator; we have, however, made huge progress towards that goal,
progress that has accelerated since Mr. Crandall spoke. If the past is any prologue, his
hope is close to being realized.

11.2   Security

In this section, we discuss issues of aviation security, beginning with the proposition in
Section 11.2.1 that the attacks of 9/11 were not an isolated aberration but part of a far
broader pattern of aviation terrorism. Then in Section 11.2.2 we briefly review the quantifiable costs that security measures pose in the First World. We then turn to a description
of numerous specific security measures that are in use today or that are seriously contemplated (Section 11.2.3.). Thereafter, we broach the general issue of how much security is
warranted, and consider why cost–benefit analysis and traditional risk management are
hard to perform in the context of security (Sections 11.2.4 and 11.2.5). Then we illustrate
the dilemmas in setting security policy by considering an explosives-detection program
used in Europe, and also three particular security measures against on-board explosions
that are not now in use in the USA (Sections 11.2.6 and 11.2.7.).

11.2.1   September 11 in Context

Some public events are so terrible that they are etched forever in the minds of those who
lived through them. Among Europeans, September 1, 1939 – when Hitler's invasion of
Poland engulfed the continent in war – was one such event. In the USA, December 7,
1941 (the attack on Pearl Harbor) and November 22, 1963 (the assassination of President

Kennedy) fall into the category. There can be little doubt that September 11, 2001 – the worst day ever in the history of commercial aviation – has achieved such tragic status. On that day, four hijacked aircraft crashed, two of them destroying the World Trade Center in New York and a third causing grievous harm at the Pentagon in Washington.

But while 9/11 introduced many readers of this book to aviation terrorism, crimes against aircraft go back far longer. The first recorded hijacking took place in 1930, when several Peruvians took over a Pan American mail aircraft with the aim of dropping leaflets over Lima. Back in 1963, the Convention on Offenses and Certain Other Acts Committed on Board Aircraft (known as the Tokyo Convention) was developed by the International Civil Aviation Organization: it required the prompt return of hijacked aircraft and passengers. Martonosi and Barnett (2004) showed that, for a randomly chosen US citizen, the chance of being killed by terrorists on a *per hour* basis was about 600 times as high during an air journey as at other times. (They defined an air journey as the time between a passenger's first entry at the airport of origin and exit from the destination airport.) And that statistic was based on the *three decades prior to 9/11*, when Americans traveling by air died at terrorist hands on 23 separate occasions. Over that period, terrorists killed more Americans during air journeys than during any other form of activity.

Nor have attempts to harm air travelers been absent since 9/11. In late 2001, the "shoe bomber" tried to destroy a transatlantic jet. A 2002 shootout at Los Angeles International Airport left several dead and injured. A deliberate fire felled a Chinese jet in 2002, while suicide bombers simultaneously destroyed two Russian jets in 2004. In 2006, a plot was uncovered to destroy more than 10 transatlantic jets with liquid explosives, 2007 saw a firebombing at Glasgow Airport, while 2008 saw another attempt to crash a Chinese jet aircraft. And a post-9/11 terrorist plot was foiled which would have crashed a passenger jet into the tallest building on the US West Coast. In short, aviation terrorism neither began nor ended on 9/11, and the cataclysmic events of that date, though unforgettable, are far from the only reason to fear deliberate attacks on aircraft and their passengers.

11.2.2 Some Costs of Aviation Security

Protecting aviation security is an expensive undertaking. In the USA, in the years after 9/11, approximately $6 billion has been spent annually on security precautions for aviation, mostly under the auspices of the federal government's Transportation Security Administration (TSA). Most of the money is spent on airport security personnel, but some is spent on new equipment (e.g., CAT scan machines to search for explosives) and on research and development. In Western Europe, the comparable figure is $3 billion (approximately 2 billion euros), under a mixed system in which security is largely governmental in some countries and privatized in others (i.e., conducted by airports, airlines, and private contractors under government supervision) (Avia Solutions, 2004). Given roughly a billion enplaned passengers per year in the First World (the USA, Western and Southern Europe, Canada, Australia, New Zealand, Japan and Israel), the average security expenditures for these nations work out to approximately US $10 per passenger. Some of the cost is paid by air travelers in the form of user fees, but some is paid for from overall government revenues (and thus by taxpayers at large).

But security brings other costs as well, and they can be large. Precautions against terrorism mean that passengers spend more time in terms of "total trip time" for an air journey (see Chapter 3) than in years past. Under the principle that "time is money," the US Congress has estimated that each extra minute of trip time is equivalent to a monetary cost of 63 cents (Joint Economic Committee, 2008). It seems conservative to assume that current security measures cause First World air travelers to spend on average 20 minutes longer per one-way trip than they would have done otherwise. For example, they arrive earlier at the airport given security procedures, or they wait at their destinations for bags they had to check because the contents were forbidden in the passenger cabin. Twenty minutes has an FAA monetary equivalent of $12.60; given a billion First World air travelers per year, the total extra-time cost they incur can be estimated to be about $13 billion. This time cost to travelers is even larger than the direct cash expenditures on security.

Indeed, some substantial costs of security might not lend themselves to monetization. In the USA, screening devices are being tested that can look under the clothing of passengers in search of weapons or explosives. Not surprisingly, these devices can also see private body parts, which seems intrusive and distressing to many passengers. For security reasons, parking lots near airports have been closed, forcing passengers to walk further from their vehicles – luggage in hand – to air terminals. "No-fly" lists, which name individuals who are flatly not allowed to board aircraft, have caused great difficulty to many innocent individuals. (When US Senator Edward Kennedy was not allowed to board his flight from Boston to Washington because "Edward Kennedy" was on a terrorist watch list, he pointed out that he had been taking the flight for 42 years.) More generally, security measures can cause stress because they remind passengers about risks that they would rather not contemplate.

One can debate how to synthesize these various costs to estimate the total burden that security measures impose on air travelers. One cannot, however, debate the general premise that security has become a conspicuous and frequently exasperating aspect of air travel.

11.2.3 Some Security Procedures

The reader is likely aware of many security procedures from his or her most recent flight, but we now review security measures that are in use today or that might be soon. The procedures fall into three general categories:

1. measures to protect the passenger cabin and cockpit;
2. measures to prevent explosions in the baggage compartment; and
3. measures to prevent threats external to the aircraft.

We focus in this section on the situation in the First World, and, for brevity, use the word "terrorism" to refer to all criminal acts that can cause death among air travelers. Some of these acts are not political in nature: the first deliberate destruction of a US aircraft occurred in 1955, when a bomb exploded in the suitcase of a woman whose son had placed it there.

Passenger Cabin/Cockpit

In this category, we might start with *doors to the cockpit*, which have been reinforced since 9/11. A would-be hijacker would find it harder to burst in than before, although the door might still be opened by the cockpit crew for various reasons (e.g., a trip to the lavatory). Additional safety is achieved by having *double cockpit doors*, with which an outer door would remain locked when the cockpit door is opened, and the cockpit door would be locked before the outer door was opened. Strengthened cockpit doors, generally bulletproof, are close to universal on First World jets, and double doors have long been in use on the Israeli airline, El Al. Some European carriers also take this precaution and, in 2004, United Airlines became the first US carrier to install double doors in all its aircraft. United explained its decision by saying that it was making an investment for "the safety and security of passengers and crew."

At the same time, *armed personnel* have become increasingly common on aircraft. In the USA, thousands of aircraft captains and co-captains have brought guns into the cockpit after receiving specialized training. US airlines had opposed the measure, but the FAA overruled them after a bill allowing such guns passed by 87–6 in the US Senate. The idea is that, even if hijackers somehow make it into the cockpit, crew members will have a reasonable chance of using their weapons to prevent disaster. In addition, large numbers of armed air marshals ride in passenger cabins: in theory at least, they are indistinguishable from regular passengers. They can swing into action immediately if passengers behave in ways that could threaten the aircraft.

A great deal of emphasis is placed, however, on preventing dangerous individuals from entering the aircraft in the first place. The process begins when the individuals are asked to show photo identification before being allowed to board their aircraft. With more advanced technology, passengers can be asked to provide *biometric evidence* that they are, in fact, the person they are claiming to be. The indicators in use include fingerprints, digital photographs and retinal scans. Trying to board with a false identity is viewed as prima facie evidence that a passenger poses an unacceptable security risk.

Even if the person is indeed the person he purports to be – as was the case for the 9/11 hijackers – the question comes up: "is this person potentially dangerous?" In various countries, there are *terrorist watch lists* compiled about individuals whose activities or affiliations raise the possibility that their intentions are sinister. Especially in the USA, there has also been much discussion about Computer-Aided Passenger Prescreening Systems (CAPPS), which would seek to explore through publicly available information whether any aspects of the passenger's past life argue for especially stringent screening at the airport (see Barnett, 2004).

There are also attempts to monitor the passenger's behavior at the airport, to see whether there is any suggestion of nefarious intent. Some airlines (perhaps most conspicuously, El Al) interrogate some passengers at length, carefully assessing whether the answers given are incorrect, inconsistent or otherwise troubling. Such methodologies as SPOT (Screening of Passengers by Observation Techniques) look for physical and physiological reactions that might be expected in an individual planning an imminent attack on an aircraft.

The reader might well be familiar with another aspect of passenger screening: *physical inspections* of both the traveler and his or her carry-on possessions at airline security checkpoints. The aim is to preclude any traveler from bringing explosives or weapons aboard an aircraft. Passengers often pass through magnetometers, which determine whether they have an unusual amount of metal on their persons. These devices might signal that the traveler has a pistol strapped to his or her leg; the machines are unlikely, however, to uncover plastic explosives (except for those with detonators containing substantial amounts of metal). The magnetometer is sometimes supplemented by inspections conducted by security personnel, which might involve the use of wands in search of metal in particular places, or a "pat down" to discover dangerous objects close to the traveler's body. Machines that detect explosive traces by blowing air against the passenger have been tried, with limited success. At the leading edge of technology are machines mentioned earlier, which take fully revealing pictures of what is underneath the passenger's clothing. Passengers in the USA and elsewhere are routinely asked to remove their shoes and coats before entering the magnetometer, and pass them through X-ray machines.

Any items that the traveler plans to take into the passenger cabin are subject to inspection at the security checkpoint. Carry-on bags are passed through X-ray machines, which can detect metal and reveal unusually shaped objects, but which cannot detect explosives. Sometimes, bags are opened and searched physically by security personnel. Passengers are often asked to take laptop computers and other large electronic devices out of carry-on bags and pass them separately through the X-ray machine. Large amounts of liquid are rarely allowed through security checkpoints in single containers. Travelers are generally not permitted to bring on more than 100 milliliters of liquid (3.4 ounces) in individual containers, and all the containers the passenger is bringing on board must fit in a 1-quart clear plastic bag. That bag must be removed from the carry-on bag at the checkpoint. (Liquid items like beverages that are purchased beyond the checkpoint are allowed aboard aircraft in some First World countries.)

As for the agents now conducting the screening at passenger checkpoints, they are generally better paid, better trained and better scrutinized than their counterparts before 9/11. In some countries, only citizens are eligible to be part of the air security workforce. There is controversy about how much more effective today's screeners are in actually detecting weapons and explosives than are their predecessors, but the importance of their work is certainly more recognized.

Of course, one way to keep dangerous people off aircraft is to intercept plots against aviation before anyone reaches the airport. While the activities of First World intelligence services are necessarily confidential, there is every reason to believe that such activities have increased dramatically since 9/11. It is possible that these efforts have thwarted events that could have taken even more lives than 9/11, including the simultaneous destruction of 10 transatlantic jets in 2006.

Underlying these various methods to protect the aircraft is the philosophy of *layering*. Any given security measure is imperfect, the theory goes, but the combination of methods has a strong chance of preventing a passenger from destroying the aircraft. At the first layer, attempts are made to uncover anything suspicious in the passenger's identity, background or activities. At the next layer, the passengers and their carry-on luggage are physically inspected, so that even a dangerous person who passes the background/biometric tests cannot succeed in bringing dangerous objects onto the aircraft. In the last layer, even a passenger who manages to board the aircraft with weapons might not be able to use them, because of intervention by armed personnel or an inability to gain access to the cockpit. The cumulative effect of layering is believed to be psychological as well as practical: would-be terrorists might be deterred from even attempting to board the aircraft and destroy it because there are so many ways their plans might be foiled.

For that reason, the full benefit of security measures can be greater than the sum of its parts.

The Baggage Compartment

The fear is that an explosive device in the baggage compartment will destroy the aircraft. Such a device caused the crash of Pan American Flight 103 in 1988 at Lockerbie, Scotland, with the loss of 270 lives. As outlined earlier, United Airlines Flight 629 was destroyed near Denver in 1955 by a bomb placed in a suitcase by a passenger's son, who wanted to collect on her flight insurance policy that named him as the beneficiary. Numerous other First World aircraft have been destroyed by bombs, including a US domestic jet from Chicago to San Diego, a Swiss jet from Zurich to Tel Aviv, a British jet from Athens to Nicosia and a US jet from Athens to Rome.

Even though the events of 9/11 had nothing to do with the baggage compartment, efforts to prevent an explosion there increased drastically after that tragedy. The methods in place now in the First World include:

- Explosion detection equipment: Various technologies are employed to identify the elements of a bomb. Explosives detection is now routine for passenger luggage, and is to be extended to air freight in the near future (e.g., by 2009 in the USA). Some machines use CAT scan techniques that provide colored images of objects that could contain explosive compounds. Other devices seek traces of explosives, or evidence of wiring or other materials that could trigger an explosion. As we shall discuss, all technologies available now suffer both a *false positive* problem (indicating that a bag is dangerous when it is harmless) and a *false negative* problem (indicating that a bag is harmless when it is dangerous). Some explosives detectors have attracted attention because they raised false alarms because of chocolate or knitting wool.
- Canine teams: Dogs trained to react to the scents of various explosives have been used to assess passenger luggage and cargo. The dogs are quite adept when they are working, but the problem is that dogs, like people, sometimes get tired and do not perform. *The New York Times* once observed that bomb-sniffing dogs sometimes take work breaks without informing their employers.
- Hand searches of luggage: This method has its value, but bombs rarely take the form of bright orange devices that are ticking. The people performing the searches often do not know exactly which objects might pose dangers. In one instance, a bomb was built into the lining of a suitcase, with a thickness no greater than wax paper. Disaster was avoided because of the extraordinary security skills of El Al, but a similar success might not have occurred under different circumstances.
- Positive passenger bag match: Under this policy, a bag will not travel in the baggage compartment of an aircraft unless it is accompanied by a passenger who is known to have boarded the aircraft. In a full implementation, a bag is removed from the aircraft if the passenger who checked it does not show up to travel; furthermore, a bag not identified with any passenger on the aircraft is not loaded. (It was an unaccompanied bag of the latter kind that felled Pan Am 103.)
- Trusted shipper programs: These programs essentially exempt air cargo from physical screening for explosives if it is being sent by a known shipper who has been working with the airline for some time. Of course, even if a large organization that ships goods

by air has no links to terrorism, it is conceivable that some of its employees have such links. Dissatisfied by the dependence on "trust," the US Congress insisted that all cargo shipments be screened for bombs by explosives detection techniques by 2009, regardless of whether the shipper is known to the airline.

Security Threats External to the Aircraft

Passengers and shippers are not the only ones who can threaten aviation safety. Airport workers of various kinds (e.g., aircraft cleaning crews, food service personnel, airline employees and airport merchants) are in a position to help smuggle explosives aboard a aircraft. Individuals inside the airport can unleash terror against passengers, as can vehicles near the terminal laden with explosives. Surface-to-air missiles can shoot passenger aircraft out of the sky.

A potpourri of measures have been proposed or implemented against such threats. To guard against truck bombs (like the one employed in an earlier attack on the World Trade Center), physical barriers have been erected outside some airport terminals. Parking is sometimes prohibited within a certain distance from the terminal, the interiors of vehicles are sometimes inspected, and unaccompanied vehicles are almost never allowed to stand idle outside terminal buildings.

Inside the terminal, armed guards (sometimes carrying machine guns) are a common sight at large First World airports. So are canine units, and behavioral specialists who identify individuals who are acting suspiciously. Cameras continuously record what is going on throughout the airport, facilitating a fast response to anything that appears dangerous or unusual.

Since 9/11, there has been much more attention than before on the people who work at the airport. Background checks are sometimes mandatory and always encouraged. Individuals are given identification cards that (it is hoped) are difficult to replicate; biometric procedures to prevent impersonation are close to wide availability. At some airports, all employees who work beyond the security checkpoint must go through the same screening procedures as passengers. Even pilots are sometimes subject to these measures. Food and gift items that go through the security checkpoints often get scrutiny comparable to that of carry-on luggage.

There is also concern that individuals could furtively gain access to the airport to cause harm. Some airports (e.g., Los Angeles International) have installed electronic barriers at their perimeters, which could react when people try to sneak onto airport property. Many airlines have increased the observation of their airport hangars, and of aircraft that are parked overnight. The identification requirements for those entering sensitive areas of the airport have been increased. Getting beyond the passenger screening checkpoint without a boarding pass for an upcoming flight is in theory a thing of the past.

There is, in addition, the fear that terrorists will gain possession of surface-to-air missiles, and use them to destroy commercial aircraft. For this reason, suspicious behavior near the airport attracts more attention than otherwise: US authorities have distributed cell phones among fishermen in Boston Harbor, for example, and urged that any unusual activities be reported at once. Some of the most difficult discussions about this threat concern installing missile deflectors aboard commercial aircraft. (It is believed that the Israelis have already done so, or are planning to install such devices in the near future.) A program to equip entire fleets with deflectors would be very expensive, and the likelihood that a missile aimed at a aircraft will reach its target is well below 100%. Still, the issue is unresolved in the USA and elsewhere, and some passenger jets are flying with prototype deflectors in 2008.

11.2.4   Is It Worth It?

After reading such a long list of security procedures in aviation, the reader might be wondering, "is it all worth it?" Is it clear that aviation is at special risk today? Terrorists who wish to "top" the 9/11 achievements could turn to biological, chemical or nuclear attacks, or to strategies that cause convulsions in cyberspace or the economy. And even if terrorists use conventional weapons, why should they not turn from aircraft to subways, shopping malls or seaports, which are essentially unprotected? The US government's National Commission on Terrorist Attacks on the United States (2004) hinted at some sympathy for this viewpoint, warning that a preoccupation with aviation security because of the 9/11 events could be a case of "fighting the last war" (p. 351).

Perhaps this perspective makes some sense, but, as noted above, attempts to cause aviation disasters have not been absent since 9/11 in the First World; nor have actual successes been absent elsewhere, where three aircraft were destroyed by acts of sabotage. National security experts in the First World continue to rank aviation near the very top of sectors threatened by terrorism. And, paradoxically, the exceptional efforts to improve aviation security might themselves make aviation a more attractive target to terrorists. Public demoralization and anxiety might be maximized if the terrorists can find ways to circumvent even the most massive efforts to thwart them. Terrorism expert Bruce Hoffman (2007) thought it noteworthy that "instead of more accessible targets such as subways and commuter trains, hotels and tourist destinations, [the summer of 2006] plot to bomb more than ten US airliners was aimed at perhaps the most internationally hardened target since 9/11: commercial aviation."

But what of a systematic methodology to set security policy, taking into account the costs, drawbacks and advantages of individual methods and exploring the extent to which they cohere into a mutually reinforcing strategy? In principle, the techniques of cost–benefit analysis and risk management are ideally suited to such a task. In practice, however, they suffer real limitations in this context, as we discuss in the next section.

11.2.5   Two Economic Paradigms

Though an exact definition is elusive, we describe an *aviation security policy* as a set of measures taken primarily to reduce the risk of terrorism. Again, our concept of terrorism includes all criminal acts that threaten air travelers, whether or not they are politically motivated. Bulletproof cockpit doors might be part of such a strategy; so might be requiring passengers to show photo identification at the boarding gate (as was once part of US security policy, but is no longer).

As noted, two general economic paradigms about decision making might help in structuring an aviation security policy: (i) cost–benefit analysis; and (ii) risk management.

Cost–Benefit Analysis

This widely used technique can help an organization decide whether to adopt (or continue) a particular policy or procedure. The various costs of employing the procedure are tallied up. (Ideally, they are measured in common monetary units like dollars.) Then the benefits are summed. If the cost–benefit ratio exceeds 1, then the advantages of the procedure are deemed to outweigh its drawbacks; a ratio below 1 implies the opposite conclusion. Such analyses are simplest if both the costs and benefits accrue to the same parties.

Aviation security analysis presents formidable challenges to the cost–benefit framework. Certain costs associated with a new procedure are easy to estimate, such as the purchase price of new equipment or the labor costs of operating it. More elusive, however, are other costs, such as those borne by passengers. As we have noted, there are methods to "monetize" the cost of time delays, but they are not universally accepted. And there are intangible but real costs that arise if passengers feel a loss of privacy in submitting to the security measure, or if undergoing the scrutiny causes them greater anxiety about terrorism.

But if cost assessments raise difficulties, benefit assessments raise far more. These benefits depend on the answers to four questions:

1. Absent the measure under consideration, what is the probability per unit time that terrorists would attempt attacks on aviation?
2. What is the conditional probability that the measure would deter terrorists from their attempt?
3. Assuming that deterrence fails, what is the probability that the measure would foil the attack?
4. What would be the consequences of a successful attack on aviation (and thus the benefits of averting the attack)?

These are all vexatious questions. We understand too little about how terrorists think for us to predict what they are planning, and to gauge what probabilities of failure would deter them from proceeding. (See Martonosi and Barnett (2006) for some discussion of this last issue.) Estimating the price of successful terrorism seems particularly hapless: some people believe that the costs of 9/11 should include the trillion-dollar war in Iraq, which might never have occurred had the plot been thwarted. Attempts have been made, however, to place dollar values on terrorist successes, and we will encounter one of the major efforts later in this section.

Applying cost–benefit analysis to aviation security is obviously not simple. But, as we shall see, we can sometimes make progress with the approach. The question "is the ratio greater than 1?" is less demanding than its counterpart "what is the exact value of the ratio?" And cost–benefit analysis focuses on the easier of the two questions.

Risk Management

Risk management starts with the idea that there is a fixed budget for responding to a threat (whether of earthquakes or of terrorists), and that the aim is to identify those expenditures that yield greatest "bang for the buck." A familiar risk management procedure is to list the possible responses to the threat, rank them by cost-effectiveness, and spend available dollars so as to get the greatest overall reduction in risk.

There is a conceptual difference between risk management and cost–benefit analysis. In the latter, the cost–benefit ratio is compared to the value 1, and the decision is (in principle) based directly on the outcome. Under risk management, individual measures are added to the response strategy until the fixed budget has been spent. If money is tight, the last measure added might have a cost–benefit ratio that exceeds 1. In other situations, many if not most of the measures in the strategy have cost–benefit ratios less than 1.

The general idea of risk management makes eminent sense, but it engenders its own questions. First of all, what should the overall budget be? The magnitude of the threat obviously matters (and estimating the magnitude is often difficult). So does the likelihood that available measures would counteract the threat. As implied above, the need to rank potential responses by cost-effectiveness requires that a cost–benefit analysis be performed for each response. We have already seen that cost–benefit analysis is difficult in its own right.

Moreover, different responses to a threat cannot be viewed in isolation: the effectiveness of a given measure often depends on which other measures are in place. A simple example makes the point. Suppose that there are two adjacent doors to a house, and the threat to be managed is that of burglary. Installing massive triple locks on one of the doors yields little benefit if the neighboring one is left wide open. Likewise, measures to identify certain travelers as "high risk" do little good unless the extra security precautions applied to such travelers are effective in preventing terrorist acts.

11.2.6 A European Dilemma

It is helpful to move from general paradigms to examples of specific decisions, because it is at the level of individual decisions that a security policy takes shape. We start with a simplified version of a security arrangement widely used in Western Europe. When a checked bag is headed for an aircraft baggage compartment, it is passed through a system that has three distinct explosives detectors (named detectors 1, 2 and 3, in the order in which the detectors process the bags). Each of these detectors can declare the bag either "safe" or "suspicious."

None of the devices is perfectly accurate. To reflect their error rates, we require some definitions. If a bag is dangerous (i.e., contains a bomb), the conditional probability is P_1 that detector 1 will erroneously declare it safe (i.e., make a *false negative* error). If a bag is harmless, the conditional probability is Q_1 that the detector will classify the bag as suspicious (*false positive* error). We define two further conditional probabilities: P_2 is the conditional probability that detector 2 will declare a dangerous bag safe, given that detector 1 has (correctly) declared it suspicious, and P_3 is the conditional probability that the third detector declares a dangerous bag safe, given that detectors 1 and 2 have (correctly) declared it suspicious. Thus defined, P_2 and P_3 are probabilities that later detectors contradict earlier ones that had correctly refused to "vouch for" a dangerous bag. We further define Q_2 as the conditional probability that detector 2 declares a harmless bag suspicious, given that detector 1 has (erroneously) done so, and Q_3 the probability that detector 3 declares a harmless bag suspicious, given that detectors 1 and 2 have both done so.

The inspection procedure can play out in various ways under the European procedure. A bag first goes through detector 1. If the detector declares the luggage safe, the bag is immediately loaded onto the aircraft and does not go through the second detector. If the bag is deemed suspicious, however, it is sent on to detector 2. If detector 2 classifies the bag as safe, it is loaded onto the aircraft; if the second detector concurs with the first that the bag is suspicious, then the bag goes to detector 3. If detector 3 clears the bag, it is loaded, meaning that the bag is loaded if *any* of the three detectors deem it safe. If detector 3 does not approve the bag, then it is rejected and put through specialized procedures. These procedures might well cause the luggage to miss the flight even if it is eventually found to be safe.

Suppose that a bag is dangerous. Then, under the three-stage regimen, the chance that it will be loaded onto the aircraft is as follows:

$$P(\text{load given dangerous}) = P_1 + (1 - P_1)P_2 + (1 - P_1)(1 - P_2)P_3$$

To understand this expression, note that there are three mutually exclusive ways that the bag gets loaded: (i) it is cleared by the first detector; or (ii) it is declared suspicious by that detector, but cleared by the second one; or (iii) having been declared suspicious by the first two detectors, it is cleared by the third. Given the definitions of P_1, P_2 and P_3, the quantities added above are, respectively, the probabilities of these three paths to loading a dangerous bag.

We can also write:

$$P(\text{not loaded given harmless}) = Q_1 Q_2 Q_3$$

The probability of a triple "false alarm" is simply the product of the probabilities that each detector erroneously deems the bag suspicious, given that previous detectors did so. (The word "suspicious" might be a bit overstated: the bag is not cleared if anything about its content requires further explanation.)

For example, suppose $P_1 = 0.06$, $P_2 = 0.04$ and $P_3 = 0.02$. Moreover, suppose that $Q_1 = 0.24$, $Q_2 = 0.12$ and $Q_3 = 0.006$. (The last three statistics are based on the experience with three detectors at a major European airport in 2006, in the order in which the bags were processed; the first three are hypothetical numbers.) With these various numbers, the chance of loading a dangerous bag under the formula above would be $0.06 + 0.94(0.04) + (0.94)(0.96)(0.02) = 0.116.$, and the chance of rejecting a harmless one would be $0.24 \times 0.12 \times 0.006 = 0.00017$, which is about 1 in 6000.

Note that the expression:

$$P(\text{load given dangerous}) = P_1 + (1 - P_1)P_2 + (1 - P_1)(1 - P_2)P_3$$

can be rewritten as:

$$P(\text{load given dangerous}) = P_2 + (1 - P_2)P_1 + (1 - P_1)(1 - P_2)P_3$$

or:

$$P(\text{load given dangerous}) = P_3 + P_1(1 - P_3)(1 - P_2) + P_2(1 - P_3)$$

These equations mean that, when the P_i are all between 0 and 1, the probability of loading a dangerous bag under the three detector procedure exceeds P_1 and exceeds P_2

and exceeds P_3. (Note that the terms added to each P_i on the right-hand side must be positive). In other words, these three expressions show that the probability of loading a dangerous bag is *greater* using three detectors than for any single detector on its own! But then why use the detectors in this way?

We get some insight if we consider a different loading policy than the one just described. A simple alternative would eliminate the last two detectors altogether: the bag would be loaded if declared safe by the first and sent directly to special procedures if declared suspicious. That regime would yield a 6% chance of loading a dangerous bag, and a 24% chance of rejecting a harmless bag.

In summary, we have:

Loading policy	Conditional probability of loading a dangerous bag	Conditional probability of rejecting a harmless bag
Declared safe by any detector	11.6%	0.02%
Declared safe by first detector alone	6%	24%

The above numbers have been rounded off slightly for clarity. False rejection probabilities are based on statistics from a major European airport; false acceptance probabilities use illustrative numbers.

Note that it is not self-evident which of these policies is best. The "any detector" policy has the greater chance of loading a dangerous bag; on the other hand, it has a far smaller chance than the one-detector rule of rejecting an innocent one. In choosing between these policies, a trade-off is being made between two undesirable contingencies: a risk of the aircraft's destruction and the (far more common but also far less serious) risk of rejecting harmless bags.

How that trade-off should be made is not obvious, because several quantities we would need for a formal cost–benefit analysis are not available. What is the probability X per unit time that a dangerous bag will show up for loading? The chance of actually loading a bomb under the "any detector" rule is $0.116X$, but we do not know what X is. What is the likelihood that a loaded bomb will successfully explode, and that the explosion will destroy the aircraft? What price should be attached to the delays, anxieties and loss of goodwill that would arise for passengers who check harmless bags but are affected adversely by special security procedures?

In Europe, the authorities have decided to use the first of these rules (load as soon as any detector is satisfied). We see that this scheme – which gives a dangerous bag three separate opportunities to be loaded – is less effective against genuine threats than some other rules based on the detectors that would, however, cause far more disruption. Even with our hypothetical numbers about the error probabilities, the European decision can be defended. But, given uncertainties that cannot be dispelled, we cannot flatly assert that the decision is correct.

11.2.7 More Security Dilemmas

As we have seen, large numbers of security measures are undertaken now. But several others have been suggested, yet are not now performed. Here we discuss three such measures, all aimed at preventing explosions aboard aircraft. The discussions illustrate the kinds of difficult decisions that policy makers must face in connection with aviation security, many of which have no simple answers.

Match the Bags?

Under positive passenger bag match (PPBM), no checked luggage would travel unless accompanied by a passenger who has boarded the flight. PPBM is performed on all international flights, and also most European and other First World domestic flights outside the USA. But it is not required on US domestic services. Checked bags on US domestic flights are physically screened for explosives, and PPBM would be superfluous if these screening processes were perfect. But no one has suggested that this ideal situation prevails. Test results about the accuracy of explosives detection have not been heartening: in a widely publicized test outcome, a simulated bomb passed through an explosives detector at Newark and made its way to Amsterdam.

The conceptual argument for restoring PPBM within the USA is that, increasingly, the physical screening of checked luggage takes place in the absence of the passenger. A terrorist could therefore check luggage containing an explosive – perhaps using a fake identification document – and then flee the airport before this bag is examined. If the detectors failed to recognize the bomb, then – without PPBM – the luggage would travel despite the terrorist's absence, and could destroy the aircraft. If the detectors did identify the explosive, the catastrophe would be avoided, but the terrorist and those other terrorists in support would quite possibly escape capture. Without bag match, in other words, the explosive detector could become something of a roulette wheel, which a terrorist could play at almost no cost.

The situation changes when bag match at the passenger's point of origin becomes part of the security policy. To have any chance of destroying a aircraft, the terrorist would have to go to the boarding gate. Then, if an explosives detector uncovered the device, there is a real chance that the terrorist would be apprehended promptly. That prospect could be unnerving even to a suicidal terrorist: someone who would readily die in a successful explosion might nonetheless flinch at life imprisonment for a failed attempt. (Indeed, studies have shown that fear of ignominious failure does deter some terrorist acts.)

Why, then, not perform PPBM as an additional security measure? The arguments relate to its costs: PPBM can cause departure delays because passenger lists must be carefully checked for "no shows," and because the checked luggage of missing passengers must first be located in the baggage compartment and then removed from the aircraft. PPBM can cost money when an aircraft is delayed at the gate (e.g., because aircraft power units must run longer than usual, and costs associated with passenger delays and crews might accumulate). Extra expenses could grow rapidly if new employees had to be hired to perform bag match.

But we can go beyond framing the general cost–benefit issue because, in 2002, US carriers did perform bag match. Statistics arising from that experience – modified for various changes between 2002 and 2008 – imply that bag match at the passenger's point of origin could be expected to delay approximately 1 in 70 departures. Among the flights held back, the conditional mean departure delay would be about 13 minutes. There would be no need for additional employees given that none were needed in 2002; the mean dollar cost to passengers was estimated to be roughly 10 cents per flight (see Barnett, 2007).

As noted, it is not easy to estimate the economic cost of successful terrorism. But the RAND Corporation has tried, and has estimated that the cost of another successful terrorist attack against US aviation would be $15 billion (Chow *et al.*, 2005). Calculations using numbers in the last paragraph show that the annual cost of domestic bag match would be less than $100 million, even if we assign a monetary cost to passenger delay of $0.63 per minute (using the 2008 Congressional estimate). The comparison of $100 million and $15 billion implies that US domestic PPBM might be cost effective if it saved *one aircraft over 150 years*.

No Laptops on Board?

On September 10, 2006, *The New York Times* published an editorial titled "A Ban on Carry-On Luggage." It noted that "laptops, digital cameras, mobile phones and other electronic devices" could be used to trigger a bomb, and urged that they be prohibited from the passenger cabins of commercial aircraft. The newspaper acknowledged that "separating people from their laptops during flights would be painful," but added that "some people could surely use the time to go over reading material, or even revert to pen and paper." While expressing hope that the ban would eventually give way to technological advances, the editorial concluded that "for now, the surest way to keep dangerous materials out of the cabin is to keep virtually all materials out of the cabin."

The New York Times cannot be faulted for lack of audacity, or for the relentlessness of its logic. The main purpose of air transportation, after all, is to get people to their destinations swiftly and safely; providing a congenial environment en route is a desirable but secondary goal. If carry-on objects that could destroy the aircraft cannot be reliably detected, then maybe the objects should simply not be carried on.

But is the issue as straightforward as *The New York Times* suggests? The newspaper implies that taking away laptops could involve a form of rescheduling, in which passengers who work on the aircraft and read at home might simply do the reverse. But is that really the substitution that would occur? For many harried business travelers, the computer work that must wait until landing might replace time with the children, dinners with friends, or civic activities. And it might be facetious to suggest that modern business activity can be accomplished with "pen and paper." There is also the point that travelers might fear that their laptops – which they might have to place in checked baggage like large liquid containers – could be lost, stolen or damaged, which would cause sharp and enduring pain and inconvenience.

What is being proposed, in other words, is a major change for the worse in the lifestyles of business travelers. But might the horrific consequences of successful terrorism outweigh such considerations? As with PPBM, we can attempt to make some headway within the cost–benefit framework.

To perform a cost–benefit analysis, we would need to monetize the costs of a ban on laptops. The task is not straightforward, not least because passengers fall into two distinct categories: those who would not use laptops during the flight (and thus lose little or nothing because of a ban), and those others who would use them and would feel deprived if unable to do so.

In thinking about the issue, we might note that a typical First World jet flight is about 1.5 hours long (Boeing Company, 2007), during roughly 1.25 hours of which the use of laptops is now permitted. It seems clear that the total inconvenience to passengers caused by a laptop ban would be greater than that caused by a 1-minute arrival delay. It also seems clear that a 4-hour arrival delay would upset passengers more than 1.25 hours without laptops.

These last observations suggest an approximate way to monetize the cost of a laptop ban. There is presumably some number X such that an arrival delay of X minutes would cause the same total "unhappiness" to passengers on a 1.5-hour flight as a prohibition on laptop use. But, as noted, the monetized cost of a delay of X minutes is $0.63X$ under the latest US estimates. We thus reach the following chart, based on 1 billion First World passengers per year:

If the "delay equivalent" (X) of a laptop ban is:	Then the average cost per passenger would be	And the total First World cost would be roughly:[a]
10 minutes	$6.30	$6.3 billion per year
20 minutes	$12.60	$12.6 billion per year
30 minutes	$18.90	$18.9 billion per year
40 minutes	$25.20	$25.2 billion per year
50 minutes	$31.50	$31.5 billion per year

Based on 1 billion First World passengers per year.

Such numbers, although clearly approximate in nature, are of the same order of magnitude as RAND's $15 billion cost estimate for a successful terrorist act. It appears, therefore, that a laptop ban would pass the cost-effectiveness test if and only if it averted roughly one terrorist act *per year*. That is a different result, of course, than the corresponding one for PPBM.

Would a laptop ban meet this demanding criterion? It would not do so based on historical data, but that standard is problematic: in extreme form, it would mean that any means of attack that has not yet been used would be dismissed as a future threat. In this situation, however, there is an auxiliary point: especially in the USA, laptop computers already get more scrutiny at security checkpoints than practically any other carry-on items. Such scrutiny is presumably at least somewhat effective.

In summary, a ban on laptops might be so costly that it could be better to bear the risks of not imposing it than to accept the widespread disenchantment it would surely cause.

Out of Time?

One of the hallmarks of modern terrorism is that an initial act of violence often is followed almost immediately by other acts. The attacks of 9/11 clearly fit that pattern, as do more recent attacks on transportation systems in Spain, the UK, Russia and India. Other multiple near-simultaneous attacks have occurred in recent years in Japan, Israel, Morocco, Indonesia, Kenya, Saudi Arabia, Iraq, the Philippines, Jordan and Egypt. Generally, all the attacks within a given series were of the same form (e.g., time-actuated bombings).

Suppose, therefore, that an aircraft is suddenly destroyed by a terrorist bomb. Then there might be reason to fear that other aircraft aloft are in imminent danger of exploding. But what, if anything, can be done to save these aircraft? Because time is short, we might assume that any further bombs are highly unlikely to be located, let alone defused.

The first question to consider is how much time would elapse between the explosion and a warning to the pilots of other endangered aircraft? The fall to the earth after an explosion takes time: Korean flight 007 hit the sea 10 minutes after it was shot down, and Pan Am 103 crashed into Lockerbie 7 minutes after it exploded. Additional time would pass between the aircraft's hitting the ground and the confirmation of a crash, and further time would be required to notify aviation authorities, who could then contact air traffic controllers who in turn would notify pilots. In short, *10 minutes* is a very optimistic estimate of the time between the first airborne explosion and any warning to other aircraft.

But would the endangered aircraft actually have 10 minutes? The answer would depend on how simultaneous the bombings in the series would be. In this connection, it is useful to consider the *major bombings on air/rail systems since 9/11*, all of which claimed many lives:

Year	Place	Number of bombings
2003	Russia	One
2004	Madrid	Ten in 3 minutes
2004	Moscow	One
2004	Russia	Two in 2 minutes
2005	London	Three in 1 minute
2006	Mumbai	Seven in 11 minutes

The table shows that, in four of the six events, there were one or more follow-up bombings. The total number of such subsequent blasts was 18. With a 10-minute time lag, only one subsequent bombing – the last one in Mumbai – would not yet have occurred. Moreover, any sudden measures to depressurize an aircraft – probably the only feasible way to reduce the danger posed by a bomb explosion – are highly dangerous in their own right. (Loss of cabin pressure has caused nearly 200 deaths in two jet crashes since 2005.) Many lives could be lost in the attempts to mitigate a potential explosion, a good fraction of which would probably occur in response to false alarms.

Under these circumstances, we reach an agonizing conclusion: after an initial on-board explosion, the least dangerous response might be to *do nothing*. Most other aircraft in immediate danger could be beyond help, and rescue efforts could well cost more lives than they save. As optimization researchers know, even the "optimal" strategy against a terrible set of constraints can yield an extremely bad result.

How does it all add up for security? We might return to Mr. Crandall, who also expressed the hope in 1996 that air travelers would one day be able to take their security for granted. In the shadow of 9/11, few people would contend that we are approaching that point. Right now, airlines, airports and governments are making plans under the assumption that the threat of terrorism will persist unabated for many years to come. If air travelers have any reason to be fearful as they board their flights, it is because of the risk of terrorism.

11.2.8 Final Remarks

The aviation industry has many troubles – the joke is that, to make a small fortune in aviation, one should start with a large fortune. But in what is arguably the most important domain of all – preventing fatal accidents – the industry's record has been excellent. In virtually every period and every region of the world, passenger death risk has been dropping – and dropping sharply – since the start of the jet age.

Indeed, in the First World, commercial aviation accidents have been brought to the brink of extinction. Of course, many a near-extinct species has made a massive comeback, so there is no basis for complacency. The good news, however, is that there is no evidence of such complacency: with 10 million First World jet flights each year, any systematic lapses that could cost lives would be expected to show up sooner rather than later. And they have not done so.

The situation is less certain with respect to aviation security. Nothing can change the fact that, in 2001, aircraft were central to perhaps the worst peacetime terrorist act in world history. Nor can we ignore the terror scare in August 2006 that caused 75% of flights to be cancelled at the world's busiest international airport – London Heathrow. The threat of terrorism is one of the great sources of uncertainty about how commercial aviation will evolve in the twenty-first century.

While security activities have increased enormously since 9/11, we cannot be certain that we have reached an optimum. It does appear that, instead of trying to devise some grand overall security strategy in one fell swoop, it might be best to proceed one step at a time, considering specific policies against particular forms of aviation terrorism on an individual basis. We should not be paralyzed by the notion that, unless we can figure out everything about security policy, we can figure out nothing.

We might benefit from the example of the people of California in their dealing with the earthquake menace. They cannot reduce the likelihood of a gigantic earthquake, but they have worked relentlessly to mitigate its consequences. Having done that, they see no benefit to living in fear. The First World has done a great deal in response to the threat of terrorism, but there is an understanding that we cannot pay any price, however enormous, to achieve any improvement in security, however minuscule. We can debate further refinements in security policy but, at some point, we might take pride in our response and hope for the best. There is no benefit to living in fear.

References

Avia Solutions (Irish Airport Authority) (2004) "Study on Civil Aviation Security Financing," especially Section 8, http://www.ec.europa.eu/transport/air_portal/security/studies/doc/2004_aviation_security_s_0.pdf.

Barnett, A. (2000) "Free-Flight and En Route Air Safety: A First-Order Analysis," *Operations Research*, Vol. 48, No. 6, pp. 833–845.

Barnett, A. (2004) "CAPPS II: The Foundation of Aviation Security?" *Risk Analysis*, Vol. 24, pp. 909–916.

Barnett, A. (2006) "World Airline Safety: The Century So Far," *Flight Safety Digest*, Vol. 33, pp. 14–19.

Barnett, A. (2007) "Match the Bags Again," *Airport Innovation*, Vol. 1.

Barnett, A. (2008) "Is It Really Safe To Fly?" in *Tutorials in Operations Research 2008*, Institute for Operations Research and the Management Sciences, Hanover, MD, pp. 17–30.

Barnett, A. and Higgins, M.K. (1989) "Airline Safety: The Last Decade," *Management Science*, Vol. 35, No. 1, pp. 1–21.

Barnett, A. and Wang, A. (2000) "Passenger Mortality-Risk Estimates Provide Perspective on Aviation Safety," *Flight Safety Digest*, Vol. 27, pp. 1–12.

Barnett, A., Paull, G., and Iaedeluca, J. (2000) "Fatal US Runway Collisions Over the Next Two Decades," *Air Traffic Control Quarterly*, Vol. 8, No. 4, pp. 253–276.

Boeing Company (2007) "Statistical Summary of Commercial Jet Airplane Accidents 1959-2006," www.boeing.com/news/techissues.

Chow, J., Chisea, P., Dreyer, M., Eisman, T., Karasiki, J., Kvitky, S., Lingel, D., Ochmanek, D., and Shirley, C. (2005) "Protecting Commercial Aviation Against the Shoulder-Fired Missile Threat," RAND Corporation, available at www.rand.org/pubs/occasional_papers/2005/RAND_OP106.pdf.

Czerwinski, D. and Barnett, A. (2006) "Airlines as Baseball Players: Another Approach to Evaluating Air Carrier Safety Records," *Management Science*, Vol. 52, No. 9, pp. 1291–1300.

Hoffman, B. (2007) "Scarier than Bin Laden," *The Washington Post*, September 9, p. B1.

Joint Economic Committee of US Congress (2008) "Your Flight Has Been Delayed Again," May 22, p. 4, available at http//:www/jec.senate.gov.

Martonosi, S. and Barnett, A. (2004) "Terror is In the Air," Chance (American Statistical Association), Vol. 17, No. 2, pp. 25–27.

Martonosi, S. and Barnett, A. (2006) "How Effective Is Security Screening for Airline Passengers?" *Interfaces* (Special Issue on Homeland Security), Vol. 36, No. 6, pp. 545–552.

National Commission on Terrorist Attacks on the United States (2004) *The 9/11 Commission Report*, US Government Printing Office, Washington, DC.

12

Airports

Amedeo Odoni

12.1 Introduction

The airline industry depends vitally on a global infrastructure that consists of regional and national air traffic management systems (Chapter 13) and airports. This chapter provides a brief overview of the characteristics, operations and finances of large commercial airports, those that handle more than roughly 3 million passengers per year. These airports are truly central to the development, operation and economics of the airlines. Although in 2007 the total number of airports in this category was only about 280 – and the number of those with more than 15 million only 90 – they served an overwhelming percentage of the world's passengers and air freight. The top 30 passenger airports alone (Table 12.1) accounted for almost one-third of the 4.5 billion airport passengers[1] worldwide in 2007 and the top 100 for about two-thirds.

Airports consist of: (a) the "airside" (or "airfield") which includes the runways, taxiways, aprons, aircraft maintenance areas, and air traffic control facilities and equipment, as well as the land that surrounds all of these; and (b) the "landside", i.e., the complex of passenger buildings, cargo terminals and other supporting buildings (e.g., airport administration, utility plants, catering facilities, etc.), ground access facilities (curbside, access roads, automobile parking areas and buildings, railroad stations, etc.), and any additional non-aviation facilities (e.g., hotels, office buildings, shopping areas, etc.) that may be located within an airport's boundaries.

This chapter presents a summary of a number of important topics related to airports. The objective is to familiarize the reader with a broad range of issues and with the relevant terminology. References where one can find more detailed coverage are also given. Section 12.2 provides general background through an overview of some fundamental

---

[1] The total number of an airport's passengers is the sum of enplaning and of deplaning passengers; transit passengers are counted once, transfer passengers twice (when deplaning and when enplaning). The former are passengers who continue their trip on the same flight on which they arrived without a change of aircraft, while the latter connect to different flights.

The Global Airline Industry   P. Belobaba, C. Barnhart and A. Odoni
© 2009 John Wiley & Sons, Ltd

Table 12.1 The 30 busiest airports in the world in terms of annual number of passengers in 2007. Numbers in square brackets in column (2) indicate the rank of the airport, among the airports in the table, with regard to number of aircraft movements

		(1) Annual passengers (million)	(2) Annual aircraft movements (thousand)	(1)/(2) Passengers per movement
1.	Atlanta	89.4	994 [1]	90
2.	Chicago/O'Hare	76.2	928 [2]	82
3.	London/Heathrow	68.1	481 [13]	142
4.	Tokyo/Haneda	66.7	332 [24]	201
5.	Los Angeles	61.9	681 [4]	91
6.	Paris/CDG	59.9	553 [8]	108
7.	Dallas/Ft. Worth	59.8	685 [3]	87
8.	Frankfurt	54.2	493 [11]	110
9.	Beijing	53.7	400 [20]	134
10.	Madrid	52.1	483 [12]	108
11.	Denver	49.9	614 [5]	81
12.	New York/JFK	47.8	443 [18]	108
13.	Amsterdam	47.8	454 [15]	105
14.	Las Vegas	47.6	609 [6]	78
15.	Hong Kong	47.0	305 [25]	154
16.	Houston	43.0	604 [7]	71
17.	Phoenix	42.2	538 [9]	78
18.	Bangkok	41.2	266 [27]	155
19.	Singapore	36.7	223 [29]	165
20.	New York/Newark	36.4	444 [17]	82
21.	Orlando	36.4	360 [23]	101
22.	Detroit	36.1	467 [14]	77
23.	San Francisco	35.8	380 [22]	94
24.	Tokyo/Narita	35.5	195 [30]	182
25.	London/Gatwick	35.2	267 [26]	132
26.	Minneapolis/St. Paul	35.2	450 [16]	78
27.	Dubai	34.3	260 [28]	132
28.	Munich	34.0	432 [19]	79
29.	Miami	33.7	387 [21]	87
30.	Charlotte, NC	33.4	523 [10]	64
	Total	**1421.2**	**14 251**	**100**

Source: For columns (1) and (2), ACI (2008)

characteristics of airports in different regions of the world. Section 12.3 discusses the physical characteristics of airside and landside facilities, including international design standards and specifications. Section 12.4 addresses the critical topic of airport capacity, on airside and on landside, with an emphasis on clarifying several points which have often caused confusion even among air transport specialists. Airside delays and airport demand management are also reviewed briefly. Finally, the subject of Section 12.5 is airport user fees, including some regulatory approaches to the setting of charges for aeronautical facilities and services. Alternative methods for financing airport capital projects are also reviewed.

Many other airport-related topics are covered elsewhere in this book. The regulation of access to congested airports and the international trend toward airport privatization are discussed in Chapter 2; the interfacing of airline operations with airports in Chapter 8; airline and airport safety and security in Chapter 11; air traffic management at airports (and in en route airspace) in Chapter 13; the environmental impact of airports and relevant mitigation measures in Chapter 14; and some IT-related developments at airports in Chapter 15.

12.2   General Background

Commercial airports everywhere must satisfy a set of international technical standards intended to ensure aircraft safety and interoperability across national boundaries. Handling procedures for both passengers and cargo are also being increasingly standardized on a global scale as a result of airports adopting "best practices" developed elsewhere. Despite such commonalities, airports around the world still display a wide range of characteristics in such respects as land area, geometrical layout, equipment used, services offered, intensity of facility utilization, and economic performance. Denver International Airport (DEN), for example, occupies an enormous tract of 136 million m^2 (13 600 hectares or roughly 34 000 acres) while New York LaGuardia (LGA) occupies 2.6 million m^2 (260 hectares, 640 acres). In other words, more than 50 "copies" of LGA can fit into the land area of DEN![2] Even if one excludes such "outliers" as DEN and LGA, most of the busiest airports in the world occupy land areas of between 8 and 35 million m^2 (2000 and 9000 acres) – still quite a range. Figures 12.1–12.3, which show the layouts of three of the world's 50 busiest airports, make the same point: note the differences in the number and geometric configuration of the runways and of the passenger terminals, the land area occupied, etc. These will be discussed further later on.

Significant *regional* differences also exist in the *traffic* characteristics of airports around the world, as suggested by Table 12.1. Of the 30 busiest airports in the world in 2007, as measured by number of passengers, 16 were in the USA, 7 in Europe and 7 in Asia, of which 1 (Dubai) is in the Middle East. This underscores a point already made several times in this book: air traffic activity in the USA is still more intensive than anywhere else. On the other hand, growth rates at the busiest European and Asian airports have been significantly higher, on average, than those in the USA since the early 1990s. Thus, more European and, especially, Asian airports will probably be joining "busiest airport" lists over the next few years.

The fleet of aircraft (or "aircraft mix") using the busiest Asian airports is heavily tilted toward large, wide-body jets. By contrast, the aircraft mix at many North American airports has few wide-body aircraft and is dominated by a combination of narrow-body jets, regional jets and non-jets flown by major and regional airlines and of private, general aviation aircraft. A large percentage of the passengers on the regional jets and non-jets connect to/from flights operated with the larger aircraft. European airports fall somewhere in the middle between these two extremes. In fact, following the 1993 liberalization of the

---

[2] DEN served twice as many passengers as LGA in 2007 (50 million vs. 25 million).

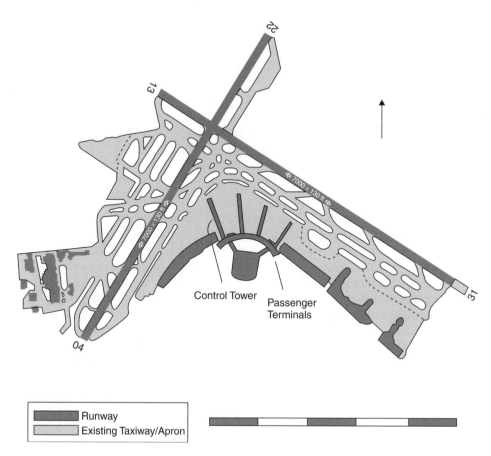

Control Tower

Passenger
Terminals

| | Runway |
| | Existing Taxiway/Apron |

Figure 12.1 New York/LaGuardia is an airport with two intersecting runways. It normally operates with two active runways, one used primarily or exclusively for arrivals and the other for departures. In the presence of strong winds or in poor weather conditions, all movements may have to be accommodated on a single runway. The terminal buildings have been constructed, rebuilt or expanded in several stages over the years. It is built on a small (for airports) tract of only 2.6 million m^2 (~650 acres)

Source: de Neufville and Odoni (2003)

airline industry (Chapter 2) in the European Union (EU), there has been a sharp increase in the number of regional flights and feeder flights with smaller aircraft at major European airports. Europe thus has moved closer to the "North American model" in recent years with respect to aircraft mix.

The overall effect is that, in order to handle the same number of annual passengers, North American airports must typically serve more annual aircraft movements than European airports and many more than their counterparts in Asia. This is shown in the third column of Table 12.1. Note that Asian airports rank much lower within the table (square brackets in column 4) in terms of aircraft movements than in terms of passengers. The net result is a much smaller number of passengers per movement at most of the North American airports in the table, as compared to those in Asia (right-most column). The number

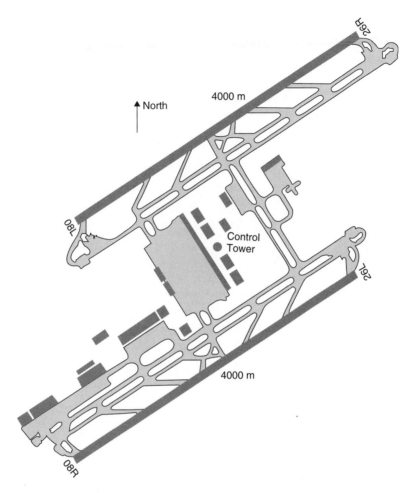

Figure 12.2 Munich International is a relatively new airport (opened in 1991) with two independent parallel runways. It usually operates with one runway used primarily or exclusively for arrivals and the other for departures. The original terminal building (Terminal 1, shown in the middle) is linear with decentralized processing. A new Terminal 2, to the east of Terminal 1, is now operating with a linear configuration but centralized processing. The airport occupies a land area of 15 million m^2 (∼3750 acres)

Source: de Neufville and Odoni (2003)

of passengers per movement is also strongly correlated with the presence of significant long-haul traffic at an airport – compare, in this respect, New York/Kennedy, Los Angeles International, San Francisco, London/Heathrow and London/Gatwick, all with strong intercontinental traffic, to other airports in North America and in Europe. This state of affairs means that North American and, to a lesser extent, European airports *need more runways* than Asian ones to accommodate the same number of passengers: practically all the airports that operate with three or more active runways much of the time are in North America (mostly) and in Europe.

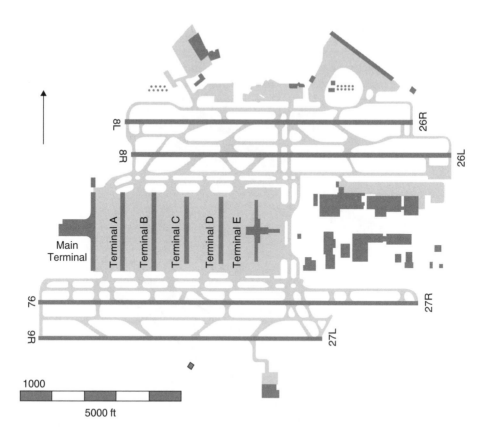

Figure 12.3 Atlanta International is the busiest airport in the world in terms of annual passengers. The main runway system shown here consists of two pairs of close parallel runways. Typically, runways 08L/26R and 09R/27L are used for arrivals and the other two for departures. The airport is well known for its passenger terminal that consists of a main terminal and five linear mid-field satellites (Terminals A–E), all connected via an underground passage and automated people mover system. It occupies a land area of 15 million m^2 (3750 acres). A fifth, shorter parallel runway to the south of the complex shown here went into operation in 2006. This runway, built on additional land, is used primarily for regional jet and general aviation operations

Source: de Neufville and Odoni (2003)

In general, runways are a scarce resource at major airports throughout the world and the main locus of airport capacity shortfalls and traffic congestion. It would then seem logical that the number of passengers per aircraft movement would increase over the years, as airlines are faced with runway capacity constraints. This, however, has not been the case. Since the early 1980s – and despite the economies of scale, in direct operating costs per seat, associated with larger aircraft – the airlines have chosen to compete in large part, especially in the USA, by increasing their frequencies, rather than increasing aircraft size (see also Chapter 3 – the "S-curve"). Moreover, this tendency is unlikely to be reversed significantly on a global scale in the near future, as the spread of market liberalization (Chapter 2) is leading to: (i) fast growth in domestic and regional short-haul traffic with narrow-body jets and smaller aircraft in much of the world; and (ii) increased airline competition and continued emphasis on flight frequency. Airlines in North America and in Europe have also aggressively purchased narrow-body jets and, during the late 1990s and early 2000s, regional jets with 100 or fewer seats.

Finally, the split between domestic and international passenger traffic[3] also differs greatly across regions. International traffic is secondary or non-existent at all but a few major airports in the USA, but is still dominant at many or most of the busiest airports in Europe, Asia and the rest of the world. This often has important implications for the size of the required airport terminal facilities and the complexity and cost of passenger processing.

12.3 Physical Characteristics

We turn next to a discussion of some of the main physical characteristics of airports. We shall first consider the airside (Sections 12.3.1 and 12.3.2) and then turn to passenger terminals (Sections 12.3.3 and 12.3.4) and ground access (Section 12.3.5). For much more detail on these and on other types of facilities (e.g., cargo terminals, baggage handling systems, etc.) the reader is referred to de Neufville and Odoni (2003).

12.3.1 Airside Design Standards

Because of the importance of safety for aviation operations, airfield design must comply with a detailed set of design standards and recommended practices adopted over the years by the International Civil Aviation Organization (ICAO) and published in Annex 14 to the International Convention on Civil Aviation (cf. Chapter 2). Annex 14's (ICAO, 2007) initial version appeared in 1951 and, by 2007, had been amended more than 40 times by the ICAO Council, usually following reports and studies by various committees and panels of experts. Annex 14 specifies, among other things: the dimensional requirements for runways, taxiways and apron areas; the limits on the heights of physical and human-made obstacles in the vicinity of runways and taxiways; and the placement and other physical characteristics of air navigation equipment, airport lighting, airport signage, etc. The Annex is supplemented by several other documents that provide more detailed guidance on some specific aspects of airport planning and design.[4] Any national recommended practices that differ from those described in Annex 14 must be reported to and are published by ICAO for the information of all its Member States.

In practice, the US Federal Aviation Administration (FAA) has also played an important role in this respect, as the design standards that it has adopted have often preceded the adoption of identical or very similar standards by the ICAO. The principal relevant document is Advisory Circular 150/5300–13 (FAA, 2007) on Airport Design, which is amended periodically by the FAA. The FAA also publishes numerous other related advisory circulars and federal aviation regulations (FARs).[5]

---

[3] The distinction between domestic and international passengers for purposes of airport processing is blurred in a number of cases. For example, EU Member States that have signed the Schengen Treaty process passengers to/from other signatory States as essentially domestic passengers.

[4] The site http://www.icao.org/cgi/goto.pl?icao/en/sales.htm provides the list of publications and online order forms.

[5] The list of FAA publications can be found at http://www.airweb.faa.gov/Regulatory_and_Guidance_Library/rgAdvisoryCircular.nsf/MainFrame?OpenFrameSet. Many of these are available online at no cost. FARs are published in the Code of Federal Regulations.

Table 12.2 ICAO airport reference codes; the upper and lower limits shown apply to the corresponding variable (e.g., for Code number 3, the RFL is greater than or equal to 1200 m and less than 1800 m)

ICAO Code element 1		ICAO Code element 2		
Code number	Aeroplane reference field length (RFL)	Code letter	Wing span (WS)	Outer main gear wheel span (OMG)
1	RFL < 800 m	A	WS < 15 m	OMG < 4.5 m
2	800 m ≤ RFL < 1200 m	B	15 m ≤ WS < 24 m	4.5 m ≤ OMG < 6 m
3	1200 m ≤ RFL < 1800 m	C	24 m ≤ WS < 36 m	4.5 m ≤ OMG < 9 m
4	1800 m ≤ RFL	D	36 m ≤ WS < 52 m	9 m ≤ OMG < 14 m
		E	52 m ≤ WS < 65 m	9 m ≤ OMG < 14 m
		F	65 m ≤ WS < 80 m	14m ≤ OMG < 16 m

Source: ICAO (2007)

Table 12.3 FAA airport reference codes; the upper and lower limits shown apply to the corresponding variable (e.g., for Aircraft approach category C, the AS is greater than or equal to 121 knots and less than 141 knots); 1 knot = 1 nautical mile per hour ≈ 1815 m per hour, 1 ft ≈0.3028 m

FAA Reference code element 1		FAA Reference code element 2		
Aircraft approach category	Aircraft approach spped (AS) in knots	Airplane design group	Aircraft wing span (WS) in feet (m)	Tail height (TH) in feet
A	AS < 91	I	WS < 49 ft (15 m)	WS < 20
B	91 ≤ AS < 121	II	49 ft (15 m) ≤ WS < 79 ft (24 m)	20 ≤ WS < 30
C	121 ≤ AS < 141	III	79 ft (24 m) ≤ WS < 118 ft (36 m)	30 ≤ WS < 45
D	141 ≤ AS < 166	IV	118 ft (36 m) ≤ WS < 171 ft (52 m)	45 ≤ WS < 60
E	166 ≤ AS	V	171 ft (52 m) ≤ WS < 214 ft (65 m)	60 ≤ WS < 66
		VI	214 ft (65 m) ≤ WS < 262 ft (80 m)	66 ≤ WS < 80

Source: FAA (2007)

Both the ICAO and the FAA use two-element *reference codes* to specify their design standards and recommended practices, as shown in Tables 12.2 and 12.3. For each type of aircraft, the first element of ICAO's code is determined by the *aeroplane reference field length*, i.e., the minimum field length required by that aircraft for takeoff at maximum certificated takeoff weight (MTOW), sea level, standard atmospheric conditions,[6] no wind and level runway. The second element is determined by the most restrictive of two physical characteristics of the aircraft: its wingspan and the distance between the outside edges

---

[6] The standard atmosphere is defined as a temperature of 15 °C and a pressure of 76 cm Hg at sea level, with a temperature gradient of −0.0065 °C/m from sea level to an altitude of 11 000 m.

of the wheels of the main gear.[7] The reference code that applies to an airport (e.g., "the airport is designed to comply with Code 4-E") thus corresponds to the code for the *most demanding* type of aircraft (the "critical aeroplane" in Annex 14 terminology) served by the airport. In an analogous, but not identical, way, the FAA uses aircraft approach speed to determine the first element of its reference code and wingspan along with tail height to determine the second. The aircraft approach speed is defined as 1.3 times the stall speed in the aircraft's landing configuration at maximum landing weight.

From the practical viewpoint, it can be noted that virtually all the major commercial airports have ICAO airport code number 4, as the most demanding aircraft using these airports almost always have a reference field length greater than 1800 m. For the same reason, the first FAA reference code element of all major commercial airports is C or higher.

Another observation is that the second element of the ICAO reference code at major airports is determined, for all practical purposes, by the wingspan of the most demanding aircraft. This is because, for the existing types of important commercial jets, the distance between the outside edges of the wheels never places an aircraft in a code letter category higher than the one it would be assigned to on the basis of its wingspan. For exactly the same reason, it is also the wingspan, not the tail height, that determines the second element of the FAA reference code.

Note, as well, that the wingspan limits in the FAA and the ICAO reference codes in Tables 12.2 and 12.3 are identical. Thus, an aircraft with a code letter A per the ICAO classification will be in Group I of the FAA, an aircraft with an ICAO code letter B will be in the FAA's Group II, and so on. This is important, because the second element of the ICAO and FAA reference codes (i.e., the wingspan of the most demanding aircraft) is also the element that largely determines the geometric design specifications to which an airfield is built. Since the dimensional specifications used by the FAA and ICAO for aircraft in each of the wingspan categories are, for the most part, either identical or very similar, *it makes little difference in the case of most of an airfield's geometric characteristics whether an airport is designed to FAA or ICAO standards*.[8]

It should be clear that, in planning for new airfields or for major improvements to existing ones, the choice of the most demanding type of aircraft is critical. Of the principal existing types of commercial aircraft, only the new, very large transport, the Airbus 380, belongs to the code letter F class (or FAA Group VI). Among other common aircraft types, the wide-body commercial jets have either code letter E or D (FAA Group V or IV): the Boeing 747 and 777 and Airbus 340 and 330 series, as well as the forthcoming Boeing 787 and Airbus 350 aircraft, have code letter E, while the Boeing 767, MD-11 and Airbus 300 and 310 have code letter D. Of the narrow-body commercial jets, the Boeing 757 also has code letter D (Group IV), but all the others (such as the Boeing 717, 727 and 737 families and the Airbus 318, 319, 320 and 321 families) have code letter C (Group III). Most regional jets currently in service have code letter B (Group II). Finally, most general aviation aircraft, including most of the "new-generation" light executive jets, have code letter A (Group I).

---

[7] Note that the outer main gear wheel span limits for ICAO code letters D and E are identical.

[8] But a few significant exceptions do exist (see, e.g., de Neufville and Odoni, 2003, Chapter 9).

12.3.2   Geometric Configuration on Airside

The airside typically occupies 80–95% of the land area of an airport. As suggested by Figures 12.1–12.3, the geometric configuration of airsides is largely determined by the number of runways and their location relative to one another and relative to the terminal complex (landside). Many major airports have only a single runway, with limited, if any, possibility of adding more, due to various constraints. At the opposite extreme, Dallas/Ft. Worth has seven, while Denver, which currently has six, has been designed to accommodate up to twelve runways.

The required length and other dimensions of a runway depend on the characteristics of the "most demanding" aircraft that will operate on the runway. The principal factors that determine the runway length required by any given aircraft operation on any given day are: the weight of the aircraft on takeoff or on landing; the stage length (non-stop distance) to be flown; weather conditions, particularly temperature and surface wind; airport location, notably airport elevation and the presence of any physical obstacles in the general vicinity of the runway; and such characteristics as runway slope, runway exit locations and runway surface conditions (e.g., wet or dry pavement). Roughly speaking, *at sea level*, a runway of about 3100 m (~10 200 ft) is needed to accommodate the great majority of long-range flights, while runways 3500–4000 m long (~11 500–13 000 ft) will accommodate flights of practically all stage lengths at full aircraft loads in other than extremely high temperatures. At the opposite end, a 2000 m (~6600 ft) runway is sufficient for practically all regional jets and for many short-range flights, while a 2700 m (~8900 ft) runway can serve most short- and medium-range flights of 4500 km (~2800 miles) or less. Required runway lengths increase with the elevation of an airport – by roughly 7% for every 300 m (~1000 ft) of elevation.

Runways are identified by a two-digit number, which indicates the magnetic azimuth (i.e., the orientation) of the runway in the direction of operations to the nearest 10°. For example, a runway with a magnetic azimuth of 224° is designated and marked as "Runway 22." Obviously, the identification numbers at the two ends of any given runway will differ by 18, e.g., the opposite end of Runway 22 is designated as Runway 04, and the runway will be referred to as "Runway 04/22." In the case of two parallel runways, the letters R, for "right," and L, for "left," are added (e.g., 22L and 22R) to distinguish between the runways (see the case of Munich in Figure 12.3). With three parallel runways, the letters R, C (for "center") and L are used. If four parallel runways are present, then one pair is marked to the nearest 10°, with the additional indications R and L, and the other to the next nearest 10°, with the additional indications R and L (e.g., Atlanta in Figure 12.3 has the pair "08L/26R and 08R/26L" and the parallel pair "09L/27R and 09R/27L").

Designers of airports with two or more runways prefer configurations involving combinations of parallel runways, as these facilitate air traffic operations and are helpful in achieving high airside capacities. Depending on the separation between their centerlines, parallel runways can be *close, medium-spaced* or *independent. Independent parallel runways* permit simultaneous parallel approaches under Instrument Flight Rules (IFR) and thus maximize the capacity that can be obtained from a pair of runways under all weather conditions. ICAO specifies that parallel runways may be approved for fully independent operations when the separation between their centerlines[9] is as little as 3000 ft (~915 m). However, most countries still require 5000 ft (~1525 m) or more.[10]

At the opposite end, full coordination is required between any pair of operations on *close parallel runways* under IFR. For example, when a takeoff on one of the runways will follow an arrival on the other, the takeoff roll cannot begin before the landing aircraft has reached its touchdown point on the other runway. In the US, close parallel runways are pairs with centerline separations of less than 2500 ft (762 m). *Medium-spaced parallel runways* make possible independent departures from the two runways or independent "segregated" parallel operations, meaning that one runway can be used for arrivals and the other, independently, for departures. However, arrivals on two different medium-spaced parallel runways are not independent.

It is important to realize that close and medium-spaced parallel runways can generate capacities approximately equal to those of an independent pair in good weather, when operated according to Visual Flight Rules (VFR), as they often are in the USA. Simultaneous parallel runway operations, with some restrictions, can be conducted under VFR on runway pairs separated by as little as 700 ft (214 m) according to the FAA and the ICAO. However, the FAA recommends at least 1200 ft (366 m) of separation between centerlines when the runways are used by aircraft in Group V or higher.

The space available between pairs of close or medium-spaced parallel runways is generally insufficient for the development of a landside complex with high capacity. Thus, at airports with such runway systems, landside facilities are typically located at one or both sides of the runway pair. An important disadvantage in such cases is that aircraft operating on the runway farthest from the passenger building must cross the other (active) runway or its extension. This, in turn, may decrease runway capacity and increase taxi times and delays on the airport's surface, as well as the workload of air traffic controllers.

Independent parallel runways usually provide sufficient space between them to accommodate the bulk of an airport's landside facilities, especially when the spacing is 5000 ft (1525 m) or more, as is most often the case. The main landside facilities are then built along the central axis of the airport either in parallel or perpendicularly to the parallel runways. Examples of airports with this type of layout include several of the world's busiest and many of the newest ones that started operations after 1990, such as Singapore, Beijing, Kuala Lumpur/International, Munich[11] (Figure 12.2), Hong Kong and Athens. The principal advantages of this layout are: efficient utilization of the vast area between the independent runways, which would otherwise have been largely underutilized; reasonable proximity of passenger and cargo buildings to both runways; better airfield traffic circulation as aircraft can taxi to/from either runway without having to cross another active runway; and ability to isolate the airport's landside from the surroundings of the airport and thus better control both the landside's development and ground access to the airport.

---

[9] The separation between centerlines is not the sole criterion for allowing independent operations on parallel runways. Additional criteria include airport instrumentation and the availability of diverging aircraft flight paths after takeoff. Diverging paths are also needed to ensure the safety of missed approaches.

[10] The FAA typically requires 4300 ft and has certified independent operations at some airports with parallel runways that are separated by as little as 3400 ft.

[11] Note that the runways of Munich are "staggered", i.e., their thresholds are offset. This offers further advantages for air traffic control and for aircraft circulation on the airport's surface.

Many airports have runways that intersect, either physically or along their projected centerlines.[12] New York/LaGuardia (Figure 12.1) is an example. Airfield geometries with intersecting runways may be necessary at sites that either have limited land area (usually older airports located near city centers) or often experience strong winds from several different directions. In the latter case, the different orientations of the runways make it possible to operate the airport under most weather conditions and should provide 95% or greater coverage for crosswinds,[13] as recommended by the ICAO and the FAA. Airports with intersecting runways are more difficult to operate from the air traffic management (ATM) point of view, as aircraft movements on each runway must be carefully coordinated with those on the other when both runways are active. Moreover, the capacity of the runway pair will vary depending on the direction in which the operations take place and the location of the intersection. Note, as well, that when strong winds in one direction force one of the two runways to close down, the airfield capacity is affected in a major way. Airports with intersecting runways often present difficult operational challenges.

Some major airports provide two independent parallel runways for operations in a primary direction, but also offer reduced capacity in a secondary direction through one or more "crosswind" runways. Brussels and Tampa are examples. When winds are calm, such airports occasionally operate with three active runways, if needed.

The runway configuration of Atlanta (Figure 12.3) is a natural extension of configurations with two independent parallel runways.[14] It provides two *pairs of close parallel runways*, each pair operating independently of the other. This configuration is well suited to airports with very large volumes of traffic: with the right fleet mix, up to 100 million passengers per year could eventually be served.[15] Los Angeles/International and Paris/de Gaulle also have four-runway layouts of this type.

The airside geometries already described do not exhaust all possibilities. Local factors (e.g., at the airports of Zurich, Amsterdam, Boston, Chicago/O'Hare, San Francisco) or the availability of sites with extraordinarily large land areas (e.g., at Dallas/Ft. Worth, Denver, Orlando/International) have led or may lead in the future to development of airfields that do not fit any of the more common layouts described. Among airports with such less common layouts, Boston has five runways in three different directions, Amsterdam six in three directions and Chicago/O'Hare seven in four directions (but will be simplified in the future to six runways in three directions).

Taxiway systems are the second major component of the airfield. At busy airports they can be extensive and complex in geometry. They are also very costly to construct and to maintain as they typically have a total length that may easily reach 15, 20 or more kilometers at the largest airports and far exceed the combined length of the runways. Taxiway systems may also include expensive "bridge taxiways" that provide overpasses

---

[12] Two runways that intersect only at a notional point along their projected centerlines are often called "converging" or "diverging," depending on the direction of operations.

[13] Crosswind refers to the component of the surface wind velocity vector, which is perpendicular to a runway's centerline. For larger aircraft, crosswinds should not exceed 20 knots (~37 km/h) when the runway surface is dry, according to the ICAO and the FAA. In general, such limits vary with aircraft type and with braking action on the runway.

[14] Figure 12.3 shows Atlanta's airport prior to the addition in 2006 of a fifth runway to the south. This new, shorter runway, parallel to the other four, and far removed from them, is used as a reliever of the four main runways and is used primarily by smaller general aviation and commercial aircraft.

[15] ATL served about 90 million passengers in 2007 (Table 12.1), with a fleet mix of mostly narrow-body jets.

for aircraft moving above highways, airport access roads, or rail lines. Few generalizations can be made about taxiway systems as their configuration largely depends on the particular characteristics of each airport layout. It is worth noting, however, that taxiway systems are sometimes overdesigned, especially at newer airports that often feature full-length, dual taxiways running in parallel to every runway and/or a large number of high-speed runway exits. Such features are often excessive for the needs of the available air traffic. At the opposite extreme, many airports, especially older ones with limited land areas, have taxiway systems with "bottleneck" spots, which cause significant delays to taxiing aircraft.

Finally, *apron areas* and *aircraft stands* constitute the third major element of airport airsides. The shape and dimensions of apron areas that are adjacent to passenger or cargo terminals are largely determined by the configuration of the terminals themselves. Aircraft stands can be *"contact"* (i.e., adjacent to terminals and served by "jetways" or "aviobridges") or *"remote"*, necessitating a fleet of ground vehicles for the transport of passengers to/from the terminal. Aircraft stands are characterized by their dimensions, which determine the types of aircraft that can be accommodated at each stand. Stands can be for *shared use* by all airport users or for *exclusive use*, i.e., reserved for a specific airline or group of airlines or a specific ground services provider ("handling company").

12.3.3 Typology of Passenger Buildings

Passenger buildings or "terminals" and the services they provide are the primary shapers of public perceptions about any individual airport. Typically, passenger buildings are also the most expensive facilities to build, operate and maintain at an airport. The cost of a new, fully equipped, very large terminal building can now exceed $1 billion.[16] Finally, passenger buildings are very important generators of airport revenue from space rentals and, especially, from *commercial activities*, such as retail stores, duty-free shops, restaurants, etc. Revenues from such activities have become a critical component of airport economics (Section 12.5).

Passenger buildings are classified according to an informal typology that focuses on two important characteristics. The first is the geometric layout of the building and identifies four generic *configurations*, shown schematically in Figure 12.4: *linear, finger* (or *pier*), *satellite* and *transporter*. In practice, only a minority of passenger buildings, mostly new ones, are "pure" applications of one of these four configurations. Most buildings actually fall in a fifth category, i.e., they are *hybrids* of two or more of the four generic configurations (bottom of Figure 12.4).

The second part of the typology distinguishes between *centralized* and *decentralized* (or *dispersed*) processing. This refers primarily to the way in which *departing* passengers are processed. Centralized processing provides a single point of access for departing passengers, such as a central lobby area where the check-in counters for all of the building's gates are located. From this central lobby passengers then proceed to the other parts of the building. Note that airports that have multiple terminals (see, e.g., New York/LaGuardia in Figure 12.1) are, by definition, "decentralized." However, individual terminals at these airports may be "centralized," in the sense that passengers departing from each one of these buildings may have a single access point to that building.

---

[16] The Terminal 5 complex at London/Heathrow cost the equivalent of around $9 billion, and the new baggage handling system at Amsterdam/Schiphol Airport about $500 million, by itself.

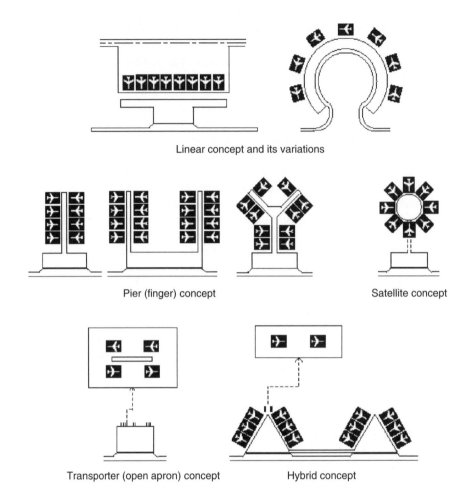

Linear concept and its variations

Pier (finger) concept   Satellite concept

Transporter (open apron) concept   Hybrid concept

Figure 12.4 Standard and hybrid configurations of passenger building and aprons
Source: ICAO (1999)

The combination of geometric configuration and centralized or dispersed processing is sometimes referred to as a *concept* (e.g., "centralized finger concept"). Figures 12.1–12.3 illustrate three such concepts. The central passenger terminal of LaGuardia (Figure 12.1) has four straight fingers. Atlanta International, on the other hand, offers an example of a satellite configuration with centralized processing. The Main Terminal (Figure 12.3) provides a single access point for originating passengers, who are then distributed to their departure gates in one of the satellite Terminals A through E or at the Main Terminal. The satellite terminals are connected with the Main Terminal by an underground tunnel that includes a *people mover* system. This configuration is very effective for Atlanta's traffic, which consists roughly of 75% connecting passengers and 25% originating/terminating: most connecting passengers never see the Main Terminal, but simply transfer between gates at the satellite buildings – quite often without even having to move from one satellite building to another. The configuration is also effective for the airline users, as the space between the satellite buildings allows for easy aircraft circulation into and out of the contact gates, while all the buildings are close to both pairs of runways.

Munich International's concept is quite different. The original Terminal 1, shown in the middle of Figure 12.2 as a thick solid line adjacent to the apron and perpendicular to the runways, is a long linear building with aircraft stands located on the apron side and a roadway on the opposite side. It is also decentralized, in the sense that there are four distinct gate modules (also seen in Figure 12.2) along the roadside façade of the building. Departing passengers must use the gate module that serves the airline they are using or the particular destination they are flying to. This concept was thought to be particularly convenient when the airport was designed in the mid-1970s, because it would minimize walking distances between the roadside and aircraft gates for originating and terminating passengers. It has, however, proved quite ineffective and costly for an airport that gradually evolved into an important connecting point for Lufthansa: walking distances between gates for transfer passengers are long, while decentralized processing at each of the modules means significant duplication of services at the different processing points. In addition, it is difficult to balance passenger loads at the four modules: there may be times when one or more of the four are overloaded, while others are empty. In response to these problems, the original building (now Terminal 1) has been supplemented with a new larger building (Terminal 2), developed jointly by Lufthansa and the local airport authority. Terminal 2 uses a *centralized*, linear concept with a more compact gate arrangement and Lufthansa has moved all of its operations to that building.

12.3.4   Assessing Passenger Building Concepts

The evaluation and comparison of alternative passenger building concepts is a complex task. There are several stakeholders to be considered and many attributes to be concerned with. The most important among the stakeholders are the airport operator, the passengers, the airlines and the concessionaires in the building. Their objectives and priorities do not necessarily coincide. For example, a decentralized linear building, such as Munich's Terminal 1 described above, typically offers short walking distances and quick access to gates for departing passengers, as well as easy aircraft maneuvering in and out of the contact stands for the airlines. However, this concept might be far from ideal for concessionaires (it reduces passenger flows through the building), for the airport operator (it typically requires longer buildings and extensive duplication of facilities and services at each set of gates) and for transfer passengers and hubbing airlines (it requires long walking distances between connecting flights).

The list of attributes to be considered when evaluating the level of service provided by any particular passenger building can be long, with different attributes being of interest to different stakeholders. Some of these are quantifiable and can be measured easily, while others are more qualitative. An incomplete list would include the following:

- Space availability, as measured by area per building occupant.
- Waiting times at processing facilities in the building.
- Walking distances.

- Minimum connection times for transfer passengers.[17]
- Ease with which aircraft can move in and out of contact gates.
- Suitability for development of commercial activities and thus for raising non-aeronautical revenues.
- Controllability of the building in terms of security (e.g., can passenger flows be channeled through a small number of checkpoints?).
- Sense of orientation and ease of finding one's way – it is well known that passengers have a lot of difficulty finding their way through some airport buildings.
- Flexibility of the building with respect to modifying the use of its various spaces, if future conditions so warrant.
- Modularity and expandability of the building.
- Ambience and architectural quality.

All of these attributes must, of course, be weighed against the fixed and variable costs associated, respectively, with constructing the building and with operating and maintaining it. Note, as well, that many of the attributes may require several different "metrics" to describe them. For example, the average walking distance, the variance (a measure of variability) of the walking distance and the maximum walking distance may all be useful in assessing a passenger building's level of service.

It should also be obvious from the above that the "best" concept will change from airport to airport, depending on local conditions. Three particularly important factors are:

1. The volume of passenger traffic: For example, finger terminals may work well for moderate volumes of traffic but, as the numbers of flights and passengers increase, the required length of the fingers and the resulting passenger walking distances may become excessive (as, for example, at Chicago O'Hare or at Amsterdam Schiphol).
2. Share of connecting traffic: Some concepts that work well for originating and terminating passengers (e.g., decentralized linear buildings) may be poor choices for transfer passengers (as, for example, at Munich Terminal 1 or at Dallas/Ft. Worth).
3. Seasonality of the traffic: When traffic is highly seasonal, inexpensive solutions may be sought for the peak periods. For example, many airports use remote stands (the transporter concept) to serve excess traffic during their peak season, rather than build additional contact stands that will be underutilized for much of the year. When traffic volumes are roughly constant throughout the year, then the "remote stand strategy" may become expensive and may be perceived by passengers as lowering the overall level of service.

12.3.5 Ground Access

The ground access system is another critical component of airports. At the busiest airports, these systems can be complex and expensive, involving combinations of highway and rail access, usually supported by multiple large parking garages and (occasionally elaborate)

---

[17] This is the minimum required time between the arrival of a flight and the departure of a different flight that a passenger can connect with. If, for example, the minimum connection time at an airport is declared to be 45 minutes, airlines cannot advertise two flights separated by fewer than 45 minutes as "connecting."

train stations located under passenger terminals. A few basic considerations and facts concerning such systems will be summarized here.

The local origins and destinations of airport users and airport employees are, as a rule, widely dispersed in sprawling metropolitan areas. As a result, and with very few exceptions, major airports depend primarily on highways and roadways and on a combination of private automobiles, rental cars, taxis, buses, trucks and specialized ground vehicles to transport the bulk of passengers, airport-site employees, cargo and various supplies to/from the airport. The private automobile, in particular, has held a dominant position among the various modes of ground access to airports in most places, most notably in the USA.

Investment in and use of rail access varies widely around the world. Most airports in the USA, either by choice or due to economic and political realities, have placed limited reliance on rail access. With the exception of Washington Reagan, where roughly 15% of passengers use the Metro rail system, the share of rail access is well under 10% at all US airports (de Neufville and Odoni, 2003). By contrast, it is in the 20 to 40% range at many of the busiest airports in Europe and East Asia. In general, the degree of success of rail as a means of airport access depends on several factors such as: local and national culture vis-à-vis use of the railroad; the existence of an extensive national or regional rail network and the connectivity of the airport's rail link to that network; and the quality and cost of alternative means of airport access.

When it comes to highway and road access, it is essential to realize that employee vehicle traffic, as well as cargo and supply vehicle traffic, are very significant: each typically accounts for at least 20% of total daily road traffic to/from the airport. Consider, for example, employee traffic. A very rough "rule of thumb" is that the relationship between the annual number of airport passengers and the number of airport-related employees is 1000 employees for every 1 million passengers.[18] Thus, about 20 000 airport-related employees can be expected to work at or near an airport with 20 million annual passengers. If 70% of these employees come to the airport on any given day and if 50% of those use a private automobile, this means about 7000 employee car trips *each way* to and from the airport. Moreover, the trips will be concentrated around the peak commuting hours of the day, thus placing a further strain on the available highway and road infrastructure. Extensive employee parking facilities will also be required. Fortunately, these facilities do not have to be located close to the terminal buildings and can be served through a shuttle bus system.

Automobile parking facilities for airport passengers and visitors are important because they require sizable investments and consume valuable, centrally located space. Major airports are also increasingly reliant on revenues from these facilities (Section 12.5.2). The total number of parking spaces provided varies widely from airport to airport, with an average figure of around 500 and a range of 200 to 1200 per million annual passengers (de Neufville and Odoni, 2003). Parking facilities are subdivided into: short-term parking areas, primarily for those picking up arriving passengers, with typical occupancy times of 30 minutes to 3 hours per vehicle; main parking areas, mostly for business travelers, often housed in multi-story, centrally located buildings with occupancies of the order of 8 hours to a few days per vehicle; and long-term parking with vehicle stays of several

---

[18] This relationship varies considerably across regions of the world.

days to a week or longer. Like employee parking, long-term parking areas, as well as parking areas for car rentals, are often remotely located and served through shuttle buses. Car rentals represent another fast-growing activity and an increasingly important source of revenue for airport operators.

With growing numbers of passengers and air freight volumes, the ground access systems of busy airports are becoming very complex and expensive and often the cause of confusion and anxiety for airport users. They pose a true challenge to airport designers and operators.

12.4 Capacity, Delays and Demand Management

The future growth of the global air transportation system depends to a significant extent on the availability of adequate airport capacity. In this section, the capacity of the airside elements (runways, taxiways and aprons) is defined and discussed, followed by a qualitative review of issues related to air traffic delays and congestion at airports. The section concludes with a review of the more opaque notion of the capacity of passenger terminals. As will be seen, this capacity can only be computed with reference to a desirable level of service.

12.4.1 Airside Capacity

The constraints imposed by the capacity of runway complexes at major airports are certainly among the most important challenges facing air transport operations, both in the long run and on a daily basis. From a long-run perspective, growth in demand has generally outpaced increases in airport capacity over the past 40 years, often by a considerable margin. In the USA, this led to delays of critical proportions in the late 1990s and in 2000. After September 2001, a significant drop in US domestic passenger demand and air traffic operations took place, but the respite this drop provided has proved to be only temporary. At the most important European airports and at some Asian airports the shortfall of capacity is even more severe, but delays are kept under partial control by restricting the number of scheduled operations through the system of "slot coordination" (Chapter 2). Barring a drastic change in the landing and takeoff performance of commercial aircraft, the runway complexes of major airports will probably continue to be among the scarcest resources of the international air transport system. New runways are very expensive to build, require great expanses of land and, most important, have environmental and other impacts that necessitate long and complicated approval processes with uncertain outcomes.

On a daily basis, the variability of the capacity of runway complexes is the primary cause of air traffic delays. With the same amount of demand, delays at an airport can be modest one day and extremely long the next. Adverse weather conditions, strong winds and other largely random events result in the frequent instances when air travel times become unpredictable and excessive due to delays, disruption of flight schedules, flight cancellations, diversions of flights to different airports, and missed flight connections.

The capacity of a runway (or of a system of simultaneously active runways at an airport) is defined here as *the expected ("average") number of movements (landings and takeoffs) that can be performed per unit of time, typically 1 hour, in the presence of continuous demand and without violating air traffic control (ATC) separation requirements*. This is often referred to as "maximum throughput" capacity because of the assumption that the runway is continuously busy and thus operates at its maximum potential. Note that this definition recognizes that the actual number, N, of movements that can be performed during an hour on a runway (or a given set of simultaneously active runways) is a random variable, i.e., it can take on a range of values with certain probabilities that depend on the specific conditions at hand. The capacity, C, is then defined as being equal to $E[N]$, the expected (or "average") value of N.

Chapter 13 provides more detailed information on the factors that determine the capacity of individual runways or of runway systems. Mathematical models[19] and simulation models have been instrumental in developing an understanding of the "physics" of airport capacity and in computing the capacity of a runway system under any given set of conditions. Such models are also extremely valuable in assessing the impacts of proposed procedural or technological changes on airport capacity, as well as in airport planning. General-purpose simulation models of airside operations first became viable in the early 1980s and have been vested with increasingly sophisticated features since then.

In addition to potentially being highly variable from day to day at any single airport location, capacities span a wide range of values across airports worldwide. Even if one restricts comparisons to countries with reasonably advanced ATM systems, the capacities of runways used for mixed (landings and takeoffs) operations range from as low as 20+ per hour *per runway* to as many as 60, depending on such critical factors as the air traffic control separation standards in use, the mix of aircraft types using the airport and, more generally, the overall performance of the ATM system (see also Chapter 13). At airports with a large percentage of wide-body jets in the traffic mix, the upper limit is around 48–50 operations per hour. Airports in the USA are generally near the high end of these ranges, due primarily to using the tightest separation standards and having one of the most proficient controller workforces in the world. Another indication of the capacity advantage that US airports enjoy is that about 25 multi-runway airports in the USA routinely achieve capacities of more than 100 movements per hour under favourable weather conditions[20] (good visibility, light winds) whereas only four airports outside the USA (Paris CDG, Amsterdam, Madrid and Toronto) exceed, as of 2008, the 100 per hour threshold.

Turning briefly to the other elements of airside, the capacity of *taxiway systems* depends greatly on local conditions and on the geometric layout of the airfield. At major airports, the capacity of the taxiway system almost always exceeds the capacity of the runway system by a considerable margin. Delays sustained at specific bottleneck spots on the taxiways are typically much smaller than the delays experienced due to the capacity limitations of the runway system. However, some exceptions may exist at older airports with limited land area or inadequate taxiway systems.[21] Taxiway capacity problems are thus airport specific and must be resolved in the context of local conditions.

In the case of *aprons*, it is important to distinguish between the *static* and the *dynamic* capacity. The former refers simply to the number of available stands, i.e., the maximum number of aircraft that can be stationed on the apron simultaneously. As stands can be

---

[19] These can range from simple models for cases involving, for example, a single runway or two independent parallel runways, to more advanced ones for more complex layouts.

[20] Dallas/Ft. Worth (DFW) can perform up to 300 movements in 1 hour and seven other airports have VFR capacities of 150 per hour or higher.

[21] For example, many airports in India today have woefully inadequate taxiway systems.

of different sizes (e.g., some can accommodate the largest aircraft and others cannot), static capacity is often specified with reference to the number of stands available for each size of aircraft (see also Tables 12.2 and 12.3). The dynamic capacity, on the other hand, indicates the number of aircraft that can be served at the apron per unit of time, typically per hour, and is the more important measure for operating and planning purposes. The dynamic capacity takes into consideration the stand occupancy times: for a given static capacity, short stand occupancies result in a high dynamic capacity and vice versa. The determination of the dynamic capacity is often difficult, not only because of the differences in the sizes of stands, but also because of the existence of a large number of constraints and conditions on which aircraft are eligible to use the various stands.[22] The most critical question in this respect is whether the dynamic capacity of the aprons is consistently greater than the capacity of the runway system, so that the apron system does not constrain runway operations. This is indeed the case today at practically all the major airports. Should the need arise, however, increasing apron capacity to keep up with growing numbers of runway movements can be very difficult at space-constrained airports.

12.4.2 Airside Delays and Mitigation

A natural consequence of limited airside capacity is air traffic congestion and delays. The most widely reported statistic on delays is the number of flights that are "on time."[23] A widely used practice internationally is to classify a flight as being "on time" if it arrives at its destination not later than 15 minutes after its scheduled time of arrival. In 2007, a very large percentage – roughly 26%, up from 22% in 2005 – of all flights by major airlines in the USA were "late" by this definition.[24] The percentage of late flights was even higher at some of the most critical hub airports, such as Atlanta, Chicago O'Hare and Newark. It is important to realize that these on-time statistics seriously underreport the true extent of delays, because airlines already introduce an allowance for delay ("schedule padding") when they prepare their schedules. For example, if the delay-free gate-to-gate time from Airport A to Airport B is 90 minutes, an airline may schedule its flights for 100 minutes, thus allowing a buffer for 10 minutes of delay for each A-to-B flight on the basis of historical flight records – see ElAlj (2003) for an extensive study. Thus, if that flight arrives at Airport B, on a given day, 17 minutes after its scheduled arrival time, it will have actually incurred a delay of 27 minutes relative to the true delay-free gate-to-gate time. A 27-minute delay on what could have been a 90-minute flight implies, of course, an enormous cost for the airline.

With continuing growth in the number of flights, airside congestion and delays will be persistent problems at major airports worldwide and may reach crisis proportions in some regions. A few important qualitative points about the characteristics of airside delay are the following. To begin with, delays may occur even during periods when total airside demand is lower than airside capacity. For example, if the average demand (as

---

[22] For example, certain groups of stands may be reserved for only certain airlines or only certain types of flights (e.g., long-range ones or those arriving from or departing for a particular set of countries).

[23] In the USA, airlines are required by law to make public every month their on-time performance record.

[24] This marked another historical high, following a brief period of improvement in 2001–2003, primarily due to low traffic during that period.

measured by the number of requests for landings and takeoffs) at a given airport over any period of 4 hours or so is more than 80–85% of the airport's good-weather capacity, that airport is likely to be highly prone to delays due to variability in weather (and, thus, in runway capacity) and to the intentional or unintentional "clustering" of demands for runway operations over time.[25] Moreover, at that level of utilization, delays are extremely sensitive to changes in demand and/or capacity: a 1% increase in the average demand may easily lead to a 10% increase in average delay, when the airport operates at or very near its maximum capacity. The dynamic behavior of air traffic queues and delays is complex, as well: for instance, delays that occur in the morning often propagate through much of the day and may severely affect operations many hours later. Advanced computer-based tools are typically needed to obtain good estimates of delay-related measures at a busy airport over time.

In the long run, both the expected length and the variability[26] of airport delays increase nonlinearly with increases in the runway system's utilization ratio, i.e., the ratio of the demand rate to the (maximum throughput) capacity. Delays will be very long and highly variable from day to day at runway systems that operate with utilization ratios in excess of the 0.85–0.9 range during the most active 15–18 traffic hours of the day.

In addition to its direct effects on passenger level of service and on operating costs, airport congestion poses some difficult strategic questions for airlines. For example, one way to increase schedule reliability is to plan for longer ground turnaround times for aircraft at the most congested airports. In this way, the schedule can "absorb" some of the delays to aircraft that arrive late at these airports. However, this approach obviously has negative consequences for aircraft and crew utilization.[27] Another strategic alternative is to increase reliance, in the long run, on larger aircraft and correspondingly reduce flight frequencies in certain markets. But, in a highly competitive environment, this may risk losing market share (Chapter 3). Finally, some airlines, especially many of the low-cost carriers, have adopted the strategy of bypassing, when possible, the most congested airports, either by providing non-stop flights without connecting at "hubs" or by utilizing secondary, less congested airports in certain metropolitan regions.

In addition to the adjustments that airlines can make themselves, aviation authorities, airport operators and ATM service providers have at their disposal two general approaches to relieving airport congestion. One, obviously, is to increase capacity. The most effective way to this end is through the construction of additional runways, either at existing airports or at entirely new ones. However, as already stated, this is often extremely difficult to do. The capacity of existing runway systems can also be increased through enhancements which may involve physical improvements (e.g., construction of high-speed runway exits) or may be ATM related (e.g., reduced separation requirements between aircraft operating on the same runway or on different runways, improved sequencing of landings and takeoffs, etc.). The latter are discussed further in Chapter 13.

The second approach is through measures that affect airport demand. Measures of this type can be of a strategic or a tactical nature. On the strategic side, the general area of *demand management* has been discussed in Chapter 2. Demand management refers to any set of administrative and/or economic measures and regulations aimed at *restraining*

---

[25] For example, for marketing purposes, many departures are often scheduled on the hour or the half hour.

[26] As measured by the variance of the delay.

[27] Chapter 9 discusses approaches toward the construction of robust and "passenger-friendly" schedules.

or otherwise affecting the demand for access to an airport during certain parts of the day and/or of the year, when congestion would otherwise be experienced. It is a "proactive" approach in the sense that these measures and regulations are set in advance and for extended periods of time (e.g., for six months at a time). There is growing pressure internationally to replace or supplement existing, purely administrative demand management procedures with "market-based" measures, such as congestion pricing, slot auctions and trading of slots (Chapter 2). On the tactical side of approaches that affect airport demand, *air traffic flow management* (ATFM) refers to a set of measures, which may be taken on a *daily, as-needed* basis by ATM service providers (e.g., the FAA in the USA and EUROCONTROL in Europe) in response to the specific demand/capacity relationships that prevail during each day. Current ATFM practices are described briefly in Chapter 13.

12.4.3 Landside Capacity and Level of Service

The task of estimating the capacity of a passenger terminal is quite complex. It is also often the cause of considerable confusion and controversy, mostly due to inadequate understanding of the assumptions underlying such estimates. The capacity is usually stated in terms of millions of annual passengers (MAP). However, in order to compute this number, one must first determine the number of passengers that can be handled during an hour of peak demand (the "design hour") and then convert the hourly figure into an annual estimate. The conceptual issues involved and the methodology used for this purpose will be described briefly in this section. These topics are discussed in detail in de Neufville and Odoni (2003) and, in a less critical spirit, in IATA (2004).

The notion of the "capacity" of any passenger building is different from the notion of the capacity of, say, a container. The capacity of a container, as measured by the volume it can hold, is a fixed quantity. The capacity of a passenger building, on the other hand, depends on decisions about a desirable "level of service" (LOS), i.e., about how much crowding and delay can be tolerated. An area of $100\,\text{m}^2$, for example, can be said to have a capacity of 50 passengers or of 100 passengers, depending on whether 1 passenger per $2\,\text{m}^2$ or 1 passenger per $1\,\text{m}^2$ is deemed to be the acceptable limit for the "density" of space occupancy in the passenger building. Some LOS standards of this type have been established and are being met with increasing international acceptance.

The most widely used among them are the space-related standards proposed by IATA. By analogy to highway LOS standards, IATA defines six different LOSs, A through F, which are described in qualitative terms in Table 12.4, with reference to the flow of passengers and other users within the building and to the delays encountered in the process. Table 12.5 summarizes the space per passenger (or other occupant of a given space) that IATA associates with each LOS. Note that (i) the space requirements depend on the activity that takes place in the relevant area of the building[28] and (ii) IATA has *not* specified an analogous set of LOS standards for the amount of time that passengers spend waiting for service (such as a statement like "an average waiting time of between 2 and 5 minutes for check-in should be considered LOS B"). However, some major airport operators (e.g., the British Airports Authority and Aéroports de Paris) as well as some

---

[28] The space requirements shown in Table 12.5 have been revised recently by IATA with many details added, such as making space dependent on whether baggage carts are also circulating in an area (IATA, 2004). The six LOSs and the fundamental concepts are still the same, however.

Table 12.4 Definition of level of service standards for passenger buildings

Level of service	Description of standard		
	Quality and comfort	Flow condition	Delays
A	Excellent	Free flow	None
B	High	Stable, steady	Very few
C	Good	Stable, steady	Acceptable
D	Adequate	Unstable, stop and go	Barely acceptable
E	Inadequate	Unstable, stop and go	Unacceptable
F	Unacceptable	Cross-flows	Service breakdown

Sources: de Neufville and Odoni (2003); adapted from IATA (2004)

Table 12.5 Space to be provided for passengers (in m^2 per passenger)

Activity	Situation	Level of service standard					
		A	B	C	D	E	F
Waiting and circulating	Moving about freely	2.7	2.3	1.9	1.5	1.0	Less
Bag claim area (outside claim devices)	Moving, with bags	2.0	1.8	1.6	1.4	1.2	Less
Check-in queues	Queued, with bags	1.8	1.6	1.4	1.2	1.0	Less
Hold room; govt. inspection area	Queued, without bags	1.4	1.2	1.0	0.8	0.6	Less

Sources: de Neufville and Odoni (2003); adapted from IATA (1995)

airlines in the USA and elsewhere have developed their own LOS standards for passenger waiting times in airport terminals. These standards, however, are far from uniform.

A second important concept is that of "dwell time," the length of a person's stay in a specified part of the building. Dwell time depends on the kind of activity taking place in the area. For example, dwell times in the departure lounges or in shopping areas are usually much longer than in passport control areas. If the flow of departing passengers in a terminal is 1000 per hour during peak hours and if the passengers stay in the check-in area for an average of 1 hour each, space is needed for 1000 simultaneous occupants of the check-in area. If they stay for only 15 minutes, space is needed for only 250 occupants. Thus, the capacity of the space in each part of a terminal (and of the terminal as a whole) depends as much on dwell time as it does on the selected LOS standards for space available per passenger (as shown, for example, in Table 12.5). The number of people[29] per hour that can be accommodated in an area of A m^2 in the terminal, when the LOS space standard selected requires s m^2 per person and the average dwell time per person is D hours, can then be approximated through the expression:

$$\text{Capacity per hour} = \frac{A}{s \cdot D}$$

---

[29] Note that, in some parts of the terminal, the occupants of the space may include non-passengers who are waiting to greet arriving passengers, or are accompanying departing passengers, or are simply visiting the airport on their own.

For example, if an area of $400\,m^2$ is provided in front of passport control, LOS D is desired during peak hours (i.e., $0.8\,m^2$ per person from Table 12.5) and a dwell time of 10 minutes (or 1/6 hours) is expected, the approximate hourly capacity of that area is $3000(= 400 \cdot 6/0.8)$ passengers.

It can now be seen that, to compute the hourly capacity of any element of the terminal building, one needs to specify the desired LOS for the terminal during the peak hours of the year (e.g., LOS C), the associated LOS standard (e.g., Table 12.5 or some other set), as well as a reasonable set of dwell time assumptions. To then compute the capacity of the *entire* building, one must: (i) estimate the hourly capacity of every major element of the terminal on the departures side (main departures hall for well-wishers and passengers, check-in, emigration passport controls, security controls and waiting areas, such as gate lounges) and on the arrivals side (immigration passport controls, baggage claim, customs, and main arrivals hall for greeters); (ii) estimate the capacity of the corridors, stairs, escalators and lifts that connect these areas, as well as of the inbound and outbound baggage handling system; and (iii) thus identify the most constraining elements ("bottlenecks") on the arrivals and the departures side. The capacity of these most constraining elements determines the overall hourly capacity of each side.

The conversion of the hourly capacity estimate, once it has been determined, to an estimate of the annual capacity of a passenger terminal is usually done through a simple multiplication by a constant – a "conversion factor" – that reflects certain important characteristics (current or expected in the future) of airport demand, such as the typical daily profile of demand at the terminal, the variability of demand by day of the week, and the seasonality of demand. For instance, assume that, at a particular terminal: (i) 10% of a typical peak summer day's total demand occurs during the peak hour of the day; and (ii) the total demand on a typical peak summer day is about 20% greater than the total demand during the average day of the year. Then, the conversion factor will be $10 \cdot 365/1.2$ or about 3040. In other words, the hourly capacity of the terminal, as computed previously, is multiplied by 3040 to arrive at an estimate of the annual capacity: if the hourly capacity has been computed as 3500 per hour, the annual capacity of the building is approximately $3500 \cdot 3040 \approx 12$ million. Note that for large terminals, like this hypothetical one, the annual capacity is usually stated to the nearest integer, in millions, to reflect the fact that the estimate is a rough one. Typical values for the factor that converts hourly capacity to annual capacity range from 2500 for highly seasonal smaller airports with a sharp hourly peak in their daily schedule of demand to 3500 for very busy airports with limited traffic seasonality and several peak and near-peak hours in their daily schedule.

As already mentioned, estimates of passenger terminal capacity are often the subject of confusion and controversy. One of the reasons is that they are not particularly reliable in the first place. The estimates are typically prepared when a building is being designed and are thus based on projections of future conditions. Many of the assumptions that go into these projections often prove wrong. For example, dwell times can change dramatically as a result of new security requirements, as has happened in recent years. Processing times at the various elements of the terminal may increase or decrease as a result of new technologies or procedures (e.g., electronic ticketing, self check-in, hand baggage screening). The patterns of daily, weekly or seasonal use of the airport may also change due to service by new airlines, new routes or imposition of night curfews on operations.

A second reason is that major terminals often end up serving many more annual passengers than they were designed for. Typically, major terminals are designed to provide LOS C during peak hours in a "target year," usually 5–10 years from their opening day. Note this means that for the great majority of the time, during that target year, the terminal will provide LOS A or B to its passengers. In fact, this will still be the case for several more years beyond the target year, i.e., after the LOS during peak hours will have fallen below LOS C and some elements of the terminal will be operating at LOS D ("Adequate") during these hours. Passenger terminals at some of the world's busiest and best airports are, in fact, currently operating at Level D, or even worse, in peak conditions – an occurrence common at older terminals, where traffic has increased gradually over many years. Such terminals may thus be operating at traffic levels significantly exceeding (sometimes by 30% or more) their "design capacity." However, despite being crowded at peak times, they still provide an adequate LOS during the great majority of the time and in a cost-effective way.

Finally, a third reason for the confusion is that improper cross-airport comparisons are often made between the capacities of terminals. For instance, if two terminals of similar size and configuration are constructed at two different airports, A and B, it is not unusual to find out that the annual capacity estimate for the terminal at A is very different from the estimate for the terminal at B. When this is pointed out, one or another of the two airports may be subjected to considerable criticism, e.g., for supposedly using the available space inefficiently. The critics often forget that, as explained in this section, the capacity of any passenger building depends strongly on the local circumstances, as well as on the selected LOS targets.

12.5 Institutional, Organizational and Economic Characteristics

This final section of the chapter begins by summarizing the typical ownership and management arrangements at the world's busiest airports. This leads to a discussion of the critical topic of user charges at airports and of the sources of airport revenue, including the ongoing controversy about the treatment of revenues from aeronautical and from non-aeronautical (i.e., commercial) activities. The chapter concludes with a review of the different approaches to financing capital projects at airports.

12.5.1 Airport Ownership and Management

The institutional, organizational and economic characteristics of airports not only vary widely, but are also undergoing rapid changes, stimulated in large part by the global trends toward airport privatization and toward airline deregulation (Chapter 2).

Many different "models" of airport ownership and management exist. The most traditional one places airport management in the hands of an agency of the national government, e.g., a national civil aviation authority (CAA). It is widely accepted, however, that these agencies often fail to meet the need for efficiency, innovation and responsiveness to an increasingly competitive and fast-changing airline industry. Thus, there is a worldwide movement toward decentralizing the governance of large airports, either through transitioning to local government control or, increasingly, by adopting some version of the *airport authority* concept. An airport authority is a corporate entity, owned by government or by private investors, or by a combination of the two, which acts as an autonomous and flexible airport operator. This entity is typically given the right to act as de facto owner, and to manage and operate one or more airports for a specified period of time, the *concession period*, e.g., for 25, 30 or 50 years. Provisions may also exist for extending the concession period, subject to performance.

In the USA, where the federal government does not operate any airports, practically all the major ones are owned by a state or a local government or by a combination of the two. Many of these airports (e.g., Honolulu International, Chicago O'Hare, Denver, Los Angeles) are operated by divisions of state or city governments, while others (e.g., Boston, San Francisco, as well as the three large commercial airports of New York City as a group) are operated by an airport authority. Elsewhere in the world, the trend during the past 20 years has definitely been in the direction of establishing airport authorities to run major airports.

The organizational structure of airport authorities typically includes units with responsibilities in the areas of: legal affairs; financing and financial management; planning; public relations; administration and human resources; environmental affairs; engineering and technical; commercial activities; and airport operations. As an airport grows in size, its organizational structure also tends to become increasingly complex. This is particularly true of airport authorities with responsibility for more than one airport (e.g., the British Airports Authority or the Port Authority of New York and New Jersey) or those that engage in extensive activities outside the core airport business, as is currently the case with several European and Asian entities.

12.5.2 Airport User Charges

Operators of major airports derive the overwhelming portion of their income from a wide variety of aeronautical and non-aeronautical user charges and other fees.[30] International practices, however, vary considerably when it comes to what user fees are imposed and how they are computed.

The process of developing a system of airport user charges is quite complex. It requires specification of policy guidelines (e.g., the largest airports typically aim at recovering all costs, including depreciation, plus earning a reasonable rate of return on capital investments), definition of revenue centers and cost centers at the airport, development of a detailed cost base, allocation of costs to revenue centers, development of a pricing methodology, and consultation with airport users. Ideally, a system of charges at an airport should have all of the following attributes: transparency; generation of adequate revenues to achieve its economic objectives (see below); reasonableness of charges on an absolute basis and in comparison to other airports with similar characteristics; promotion of the efficient use of airport facilities, especially when it comes to congested airports; and flexibility, so that charges can be revised easily in response to changing conditions. In practice, the charging systems of many airports often fall far short of achieving some of these goals.

---

[30] This includes various airport-related taxes and fees (e.g., "passenger facility charges" in the USA). Some airports also receive outright government grants; however, in the case of major airports, the proportion of income received from these grants has been steadily diminishing in recent years, as the emphasis worldwide has shifted to economic self-reliance.

User charges are classified as *aeronautical* and *non-aeronautical*. The former are for services and facilities related directly to the processing of aircraft and their passengers and cargo, while the latter are for the use of *ancillary* (i.e., supplementary) services, facilities and amenities. The principal types of *aeronautical charges* are:

1. The *landing fee* that aircraft operators pay for use of the airfield (runways and taxiways).
2. The *terminal area navigation fee*, a charge for air traffic management services on and near the airport; this is usually a part of the landing fee, but it is charged separately in some countries.
3. The *aircraft parking and hangar fee* for the use of contact and remote apron stands and, if applicable, of hangars; many airports also charge a separate fee for overnight aircraft parking.
4. The *airport noise charge*, which is imposed separately at some airports, currently mostly European ones, to cover the cost of noise monitoring systems and of noise mitigation measures.
5. Various *passenger service charges* intended to cover costs directly related to the use of passenger buildings and of passenger processing services therein.
6. The *cargo service charge* that some airport operators use to cover parts or all of the costs of cargo processing facilities and services.
7. The *security charge* that pays for (part or all of) aviation security equipment and services at the airport.
8. The *ground handling charge* for the servicing of aircraft and the processing of passengers and their bags. Ground handling charges are often subdivided into charges for: *ramp handling*, i.e., services provided on the apron (the "ramp") such as loading and unloading of aircraft, baggage handling and sorting, cleaning of aircraft, passenger transport to/from remote stands, aircraft deicing, etc.; and *traffic handling*, i.e., the processing of passengers (check-in, ticketing, boarding, etc.) and of bags, as well as the provision of information services, preparation of various handling documents, etc. Ground handling services can be provided in four different ways: by the airport operator; by the airline itself ("self-handling"); by one airline to another; and by specialized companies ("ground handlers") that obtain a license to operate on an airport's premises. When the airport operator itself provides the handling services, it collects handling charges from the airlines it serves. In the case of third-party handling (one airline to another or handling by a specialized company) the airport operator collects a license fee or a percentage of handling revenues from the service provider. Self-handling airlines usually pay no airport ground handling charges, but obviously incur all relevant expenses themselves.
9. *Concession fees for aviation fuel and oil* collected either from on-site licensed companies that supply fuel and oil to the airlines and other aircraft operators or directly from the airlines and aircraft operators. In the latter case, the airport operator itself may buy the fuel from external suppliers and resell it to the ultimate users, adding its own fee to the cost.

In addition to the above, fees are typically collected to cover the cost of a number of government services (e.g., for passport control, customs and health inspection) provided at airports. These fees, however, do not accrue to the airport operator.

Non-aeronautical charges span a broad and seemingly ever-expanding range of possibilities that may include: commercial concession fees paid to the airport operator by duty-free and retail shops, bars and restaurants, bank and currency exchange branches, and other such businesses contracting to operate on the airport's premises; tolls for automobile parking and rentals; rentals for airport land, space in buildings, advertising space, and assorted equipment; and fees for the provision by the airport operator of engineering services and reimbursable utilities to airport users. Another category of non-aeronautical revenues is derived from a number of *off-airport activities* that a growing number of (mostly European and Asian) airport operators are increasingly undertaking. Such activities include: provision of consulting services to other airports; provision of educational and training services; management contracts at other airports; real-estate ventures outside the airport's premises; and equity investments in other airports in the context of various airport privatization ventures.

One of the most striking and consistent trends in the airport sector over the past 25 years has been the growing importance of non-aeronautical revenues. Major airports worldwide currently derive roughly as much revenue, on average, from non-aeronautical charges as from aeronautical ones. This is a dramatic change: until the late 1980s, aeronautical revenues were the dominant source of airport revenue. Of the many reasons for this trend, the most fundamental is that airports cater to growing numbers of relatively affluent people who spend, often not by choice, increasing amounts of time on airport premises.

Table 12.6 shows the amounts of revenue obtained from all sources by the 32 busiest, in terms of annual passengers, US airports in 2006. The table lists both operating and non-operating revenues. The latter consist of direct grants from the government, income obtained through the "passenger facility charge" (PFC), which is essentially a tax imposed on air passenger tickets, interest on deposits and other miscellaneous items. Among the sources of operating revenues, the largest three were: landing fees; aeronautical charges for use of passenger and cargo buildings (e.g., space rentals by airlines, gate leases, etc.); and fees for automobile parking. Note that the combined revenue from automobile parking and car rental facilities and services is as high as the revenue from any single aeronautical source! Roughly 56% of operating revenue comes from aeronautical charges and fees and 44% from non-aeronautical. Operating and non-operating revenue comprise, respectively, 72% and 28% of total revenue. By far the two main sources of non-operating revenue are grant receipts from the federal government's Airport Improvement Program (AIP) and the passenger facility charges. The PFC, which was initiated through federal legislation in 1992, has gradually become one of the most important sources of revenue for US airports and, in 2007, adds about 3% to the average cost of a domestic air ticket sold in the USA (Karlsson *et al.*, 2007).

The picture is similar, with a few differences, if one looks at all the airports with some scheduled airline service in the USA, approximately 500 in total. The breakdown of operating revenue was 53% from aeronautical charges vs. 47% from non-aeronautical in 2006, thus suggesting that smaller airports rely even more than the busiest ones on non-aeronautical revenues. The most important source of operating revenue for these smaller airports is automobile parking and car rental concessions, which, in combination, account for about 28% of total operating revenue. An important difference with the busiest airports is that non-operating revenues comprise 32% of total revenue (vs. 28% for the busiest airports). This is because smaller airports rely heavily on federal AIP grants: the

Table 12.6 Sources of revenue for the 32 US airports with largest number of annual passengers in 2006

Revenue source	Revenue ($000)	Percentage of total operating revenue	Percentage of total revenue
Terminal rental charges	2 298 351	25	18
Landing fees	1 954 598	22	16
Cargo and hangar rentals	260 011	3	2
Fuel sales	105 240	1	1
Other	398 246	4	3
Total aeronautical revenue	5 016 447	56	40
Land and non-terminal facilities	294 835	3	2
Terminal concessions	922 634	10	7
Rental cars	712 149	8	6
Parking	1 552 794	17	12
Other	552 804	7	5
Total non-aeronautical revenue	4 035 215	44	32
Total operating revenue	9 051 662	100	72
Interest income	661 542		5
Grant receipts	924 221		7
Passenger facility charges	1 747 435		14
Other	165 661		1
Total non-operating revenue	3 498 859		28
Total revenue	**12 550 521**		**100**

Source: Data obtained from FAA Form 5100–127

size of AIP grants to all airports is about the same (in fact, a little greater) as the funds collected through PFCs. By contrast (Table 12.6) total revenue from the PFC for the 32 busiest airports is twice as large as the revenue from federal grants.

When it comes to major airports worldwide, the situation is entirely analogous – although comprehensive and internally consistent statistics are much harder to obtain. Based on a number of international surveys, operating revenues from aeronautical sources are about equal on average to those from non-aeronautical sources at major airports in Europe and Asia–Pacific, as in the USA. If anything, revenues from non-aeronautical sources seem to be slightly greater. A wide range of values can be found, however. Such airport operators as the British Airports Authority, the Paris Airports Authority (ADP), Singapore and Hong Kong now obtain more than 60% of their operating revenues from non-aeronautical sources. By contrast, the aeronautical revenues of the Milan Airports Authority (SEA) represented until recently more than 70% of total operating revenues. The reason is that SEA maintained a monopoly on ground handling services at the Malpensa and Linate Airports and derived a large portion of its revenues from these services.[31] It is also noteworthy that, in contrast to US airports, commercial revenues from retail shops and restaurants usually far exceed those from parking and car rentals at major European and Asian airports. Overall, as noted

---

[31] Until the late 1990s, several airport operators in Italy and in Germany provided ground handling services on a monopolistic basis at their own airports. They also employed a large workforce for this purpose.

Table 12.7 Revenue and profitability of airports and
airlines (calendar year 2004 or fiscal year 2004–2005)

	Top 100 airports (in revenue)	Top 150 airlines (in revenue)
Total revenue	$42 billion	$392 billion
Net operating result	$8.2 billion	$9.6 billion
Operating margin	20.9%	2.6%
Net result	$4.1 billion	−$3.1 billion
Net margin	11.3%	−1.3%

Source: Airline Business Magazine, December 2005

earlier, the balance globally has been shifting toward increased reliance by airports on
non-aeronautical sources.

With respect to overall financial performance, the busiest airports in the world are
typically highly profitable. As indicated by Table 12.7, the total revenue at the 100 airports
in the world with the highest annual turnover in 2004 was only about 10% of the total for
the 150 top-revenue airlines ($42 billion vs. $392 billion). Yet, the net operating result
(= operating revenues − operating costs) was roughly equal to that of the airlines ($8.2
billion vs. $9.6 billion) and amounted to 20.9% of operating revenue for the airports
vs. only 2.6% for the airlines. Similarly, the overall net margin of the airports for the
year was 11.3%, while the airlines incurred a loss of 1.3%. This striking difference in
financial performance is one of the root causes of persistent tension between airports and
the airlines, with airlines contending that airport operators charge excessive aeronautical
fees, taking advantage of their quasi-monopolistic status.

12.5.3 Economic Regulation, Single Till and Dual Till

Airport aeronautical charges are subject to some form of regulation in most countries.[32]
This practice stems from a desire to "contain" the cost of aeronautical facilities and services to their users and to protect the airlines and their passengers from potential abuses
resulting from airport privatization or from the quasi-monopolistic position that airports
occupy when it comes to serving origin–destination traffic (see also Chapter 2). The ICAO
Council has stated that airport operators may recover the *full cost, "but no more", of aeronautical facilities and services* (ICAO, 1992). Full cost includes the cost of operations,
maintenance, management and administration, as well as interest on capital investment,
depreciation of assets, and, when conditions permit, a fair return on investment. By contrast, the non-aeronautical revenues of airports are largely – but, in some countries, not
fully – unregulated.

The two most common approaches used by economic regulators of airport aeronautical
charges internationally are: (i) specifying upper limits on the rates of return on investment
that airports can earn; and (ii) restricting the annual rate of increase of unit charges. For
example, until recently, a 7.5% target rate of return on net capital assets was specified

---

[32] In the most extreme (and hardly unusual) form of regulation, the government – not the airport operator – sets
the charges by itself and revises them periodically.

by UK regulators of the British Airports Authority, while limits were also placed on the annual rate of increase of the BAA's aeronautical charges.[33] In other cases (e.g., Vienna) the limits on the annual rates of increase in aeronautical charges have been tied to traffic growth rates – the higher the growth rate, the lower the limit. A variety of other regulatory schemes, with a similar general spirit, are in effect at many major airports around the world, especially the ones with some degree of privatization. Nonetheless, the setting of airport user charges continues to be a controversial topic and a constant irritant in the relationship between airport operators and airport users, leading to numerous major disputes over the years and occasionally to litigation. This is not surprising in view of the vagueness of existing regulatory guidelines on this subject at the international level.

A critical question for regulators is whether aeronautical charges should be affected in any way by airport earnings on the non-aeronautical side. Airlines have argued in favor of the so-called *"single till"* approach. Under single till, the *total* operating revenues of the airport, i.e., the sum of aeronautical and non-aeronautical revenues, must be considered when computing the aeronautical fees that the airport operator may charge under the constraints imposed by the economic regulator, such as any limits on the rate of return on capital invested in aeronautical facilities. What this means, in effect, is that the net revenues from non-aeronautical services (which are usually highly profitable) end up offsetting a large portion of aeronautical costs. This, in turn, means that the airport needs to collect less revenue from aeronautical charges in order to reach its regulated rate of return on aeronautical investment (or other economic target). Thus, in the great majority of cases, the single till approach will lead to reduced charges for aeronautical services. Non-aeronautical services end up subsidizing aeronautical ones. For example, the proceeds from duty-free sales will help reduce the landing fee to be charged under this scheme.

In contrast, airport operators generally support the *"dual till"* approach. Under dual till, the aeronautical side of the airport's business is treated separately from the non-aeronautical one. The airport operator will set charges on the aeronautical side to achieve the regulator-specified economic targets *solely though revenues from aeronautical facilities and services*. At the same time, the airport operator is usually largely free to seek the maximum possible profit on the non-aeronautical side. Thus, in most cases, the dual till approach will result in higher aeronautical user charges and higher overall profits for the airport operator than under single till. Full cost recovery plus a fair return on investment is achieved on the aeronautical side, while all the profits from the unregulated non-aeronautical facilities and services accrue to the airport operator as well.

The controversy regarding single till vs. dual till is largely unresolved at this time. Single till is applied at some airports and dual till at others. In the USA, the approximate counterparts to single till and dual till are the *"residual"* and *"compensatory"* systems, respectively. Under the former, airlines pay only for the difference between (i) the airport's total revenue target in each year and (ii) the revenue from all non-aeronautical sources and from general aviation.[34] There is, however, an important difference between the singe till and the residual systems. To benefit from the residual system, US airlines must take on

---

[33] These limits differed across BAA airports, being tightest at the two busiest airports, Heathrow and Gatwick. They were stated with reference to RPI, the Retail Price Index (e.g., "the maximum annual increase in charges is limited to $x\%$ below the RPI").

[34] Simply put, if the airport needs a total revenue of $100 to meet its financial targets in a year and non-aeronautical activities produce a revenue of $70 in that year, the airport will raise the remaining $30 from aeronautical charges.

a significant financial risk, by signing long-term use agreements with the airport operator under which they underwrite the service of debt issued by the airport (Section 12.5.4). Thus, the airlines essentially agree to cover any shortfall that may occur in the future in servicing airport debt. Under the compensatory system, by contrast, the airport operator assumes the full financial risk associated with servicing its debt, while the airlines and general aviation pay for the full cost of aeronautical facilities and services. Airport hubs, dominated by one or two airlines (e.g., Minneapolis/St. Paul, Atlanta, Cincinnati), usually operate under the residual system, while mostly origin–destination airports that do not have to rely heavily on any single airline (e.g., New York's airports, Boston) operate under the compensatory system.

12.5.4   Financing Capital Projects

The financing of large-scale infrastructure development or improvement projects is always a central concern of airport owners and operators. Airport capital investments can be financed in many different ways ranging from grants from national governments to revenue bonds issued and serviced by airport operators. The alternatives can be classified into the following broad categories:

1. outright government grants;
2. special-purpose user taxes, such as the passenger facility charge (PFC) that is added to airfares in the USA or the "head taxes" that air passengers pay in many countries upon arrival or prior to departure;
3. low-cost loans provided by national or international development banks such as the World Bank or the European Investment Bank;
4. retained portions of the profits generated by the airport operator;
5. loans from commercial banks or other sources at financial market rates;
6. general obligation bonds, which are secured through the full taxing power of the issuing government entity (national, regional or local): should revenues from the airport prove insufficient to service obligations to bondholders, taxpayers at large must cover the shortfall;
7. revenue bonds issued directly by an airport authority, which is solely responsible for servicing obligations to bondholders through airport revenues;
8. project-specific revenue bonds, whose servicing relies solely on revenues from a particular airport project (e.g., a new terminal building): such bonds are usually issued by an airport authority, often in combination with other investors; and
9. private financing obtained by granting specified rights to airport revenues or leasing rights, as in the case of build, operate and transfer (BOT) contracts for the development of a single facility (e.g., an automobile parking garage, or a passenger terminal) or of a complex of facilities (e.g., an entire airport in some instances).

The financing alternatives available in each case depend on the characteristics of the airport and on national statutes and economic policies and practices. For example, alternative (1) is still the most common type of airport financing in many, usually developing countries, which also rely heavily on (2) and (3). In the USA, alternatives (2), (7) and (8) and, to a lesser extent, (9) are the predominant sources of financing capital projects at major airports. Alternative (1), which played an important role up to the 1980s through

Table 12.8 Financing of Athens International Airport which opened in 2001. Amounts indicated are millions of US dollars (1996 prices)

1. European Investment Bank	1128	47%
2. Consortium of commercial banks	360	15%
3. Airport development fund (Greece)	288	12%
4. European Union grants	264	11%
5. Greek State grants	168	7%
6. Share capital (55% Greek State, 45% German consortium)	144	6%
7. Secondary debt (commercial rates)	48	2%
Total	**2400**	**100%**

the FAA's AIP, is now a secondary source of funding for the busiest airports as federal grants for airport development are now mostly directed to smaller airports. In Western Europe and in the Southeast Asia, Pacific Rim and Oceania regions, the importance of alternatives (2), (4) and (8) is clearly increasing and that of (1) decreasing. Airport revenue bonds, (alternatives (7) and (8)) were an unusual practice outside the USA until the 1980s, but now constitute an increasingly common way of financing airport capital projects in these parts of the world.

Table 12.8 illustrates some of these points by listing the sources of financing for the new Athens International Airport, which opened near the town of Spata in March 2001 and cost the equivalent of $2.4 billion (1996 prices). It can be seen that: approximately 30% of the project (Items 3, 4 and 5 in the table) was financed through outright grants from the Greek State, from the European Union and from an airport users tax (Item 3); 47% came from a low-interest loan from the European Investment Bank (Item 1); 6% from the airport's shareholders (Item 6); and 17% from funds borrowed at commercial rates (Items 2 and 7).

The ability of airport operators to obtain favorable financing terms depends in large part on assessments performed by credit rating agencies. Prominent among them, when it comes to airports, are Moody's Investors Services, Standard and Poor's, and Fitch ICBA. These three organizations also issue, on a regular basis, informative reports on the creditworthiness of the operators of many major airports around the world, using mostly similar sets of credit rating criteria.

References

ACI – Airports Council International (2008) *Worldwide Airport Traffic Statistics: December 2007*, posted 12 March 2008, www.aci.aero.

de Neufville, R. and Odoni, A. (2003) *Airport Systems: Planning, Design, and Management*, McGraw-Hill, New York.

ElAlj, Y. (2003) "Measuring the True Delays in the Air Traffic Control System," SM thesis, Massachusetts Institute of Technology.

FAA – United States Federal Aviation Administration (2007) *Airport Design*, Advisory Circular 150/5300-13, incorporates Changes 1 through 13, US Government Printing Office, Washington, DC.

IATA – International Air Transport Association (1995) *Airport Development Reference Manual*, 8th edition, Montreal, Canada.

IATA – International Air Transport Association (2004) *Airport Development Reference Manual*, 9th edition, Montreal, Canada.

ICAO – International Civil Aviation Organization (1992) *Statements by Council on Charges for Airports and En Route Navigation*, ICAO Doc. 9082/4, updated in 2001, Montreal, Canada.

ICAO – International Civil Aviation Organization (1999) *Airport Planning Manual, Part 1, Master Planning*, ICAO Doc. 9184, Montreal, Canada.

ICAO – International Civil Aviation Organization (2007) *Aerodromes, Annex 14 to the Convention on International Civil Aviation, Volume I: Aerodrome Design and Operations*, 3rd edition, Montreal, Canada.

Karlsson, J., Odoni, A., Geslin, C., and Yamanaka S. (2007) "Airline Ticket Taxes and Fees in the United States and European Union," in D. Lee (ed.), *Advances in Airline Economics*, Elsevier, Amsterdam, Vol. 2, pp. 255–274.

13

Air Traffic Control

R. John Hansman and Amedeo Odoni

13.1 Introduction

Air traffic control (ATC) is a critical factor in the operations of every airline.[1] In most of the world, all the phases of a scheduled airline flight from taxi-out and takeoff to landing and taxi-in require the approval of ATC. In order to execute their schedules efficiently airlines must understand the functioning of the ATC system and the constraints it imposes. In addition, the policies, procedures and costs of the ATC service providers in an airline's regions of operation can be an important factor in shaping an airline's strategies and business plan.

The purpose of ATC is to ensure the safe and efficient flow of air traffic. This is accomplished through four basic ATC services. Safety is supported through *separation assurance*, whereby air traffic controllers are responsible for keeping aircraft separated from each other, as well as from other hazards such as terrain or the wake vortices of other aircraft. ATC provides *flight information* services to aircraft, such as weather reports and updates on airport conditions. ATC also has an important role in notifying and alerting appropriate agencies about aircraft in need of *search and rescue*. Finally, efficiency is supported through *congestion management*, with ATC organizing traffic flows into congested airports and airspace.

This chapter provides an introduction to the principal features of ATC systems and to some related current developments. It begins (Section 13.2) by reviewing the generic elements of an ATC system. Although the equipment and technologies may vary across

---

[1] The terms "air traffic management" (ATM) and "air navigation services" (ANS) are also used instead of "air traffic control." Some attempts have been made to draw fine distinctions among these three terms. For example, proponents of the use of "air traffic management" (a more recent term) have argued that the "ATM system" consists of two subsystems: "air traffic control," which has the tactical mission of providing separation assurance between aircraft, and "traffic flow management," which is concerned with the more strategic task of congestion management. However, most people, including specialists, use "ATC," "ATM" and "ANS" interchangeably, with ATC being by far the most common. "ATC" will also be used throughout this chapter.

The Global Airline Industry P. Belobaba, C. Barnhart and A. Odoni
© 2009 John Wiley & Sons, Ltd

the globe, all ATC systems must provide for such fundamental tasks as traffic surveillance, communications, navigation, separation assurance, and information gathering. The chapter then describes how most ATC systems organize and structure the airspace for which they are responsible (Section 13.3), the way ATC typically operates (Section 13.4) and the use of Standard Operating Procedures to increase safety, reduce workload and enhance the predictability of traffic patterns (Section 13.5). Coverage then turns to capacity limitations (Section 13.6), with emphasis on the main bottlenecks of the ATC system, namely the runway complexes of major airports and associated terminal area airspace. The principal factors that determine the capacities of single and multiple runway systems are identified and the reasons for the high variability of these capacities are explained. The chapter concludes with a review of traffic flow management, which provides the means by which ATC organizes traffic at the macroscopic level for the purpose of managing congestion and minimizing the impacts of capacity shortfalls. Other relevant material is contained in Chapter 8, which describes interactions with ATC during every phase of a flight, as well as in Chapter 12, which provides further information on runway capacities at major airports. For far more extensive coverage of ATC, the reader is referred to Nolan (1999).

13.2   The Generic Elements of an ATC System

The generic elements of any ATC system are shown in Figure 13.1. The air traffic controller observes the ATC traffic situation through a *surveillance* system. The controller issues commands ("clearances") to aircraft through a *communications* system and the aircraft fly the cleared route using a *navigation* system. Other important technical elements include *flight and weather information systems*, which provide pilots and controllers with up-to-date information. Each of these will be discussed briefly below.

13.2.1   Communications Systems

Most ATC communications take place currently on voice radio channels in the very high-frequency (VHF) band reserved for aviation use. Due to the nature of voice radio

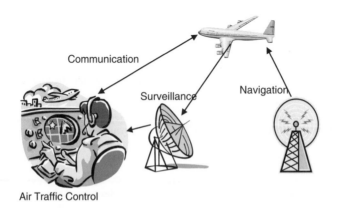

Figure 13.1   The generic elements of ATC

communications, only one transmission can be conducted at a time. Multiple radio transmissions result in "blocking" of the channels. Consequently, confirmation by "readback" is required for ATC clearances and the communications channels limit the number of aircraft that can be managed on a single frequency. The VHF channels are spaced by 25 kHz in most of the world, but frequency congestion is forcing some regions (e.g., parts of Europe) to move to 8.33 kHz channel spacing.

VHF communications are very reliable, but only transmit along a "line of sight" so range is limited (typically to less than 100 nautical miles) and a network of ground stations is required to provide communications coverage over large regions. In oceanic regions it is necessary to use high-frequency (HF) "shortwave" radios which can communicate "over the horizon" by reflecting off the ionosphere. The quality of HF radio is poor and most flight crews do not continuously monitor HF signals except when alerted to do so by ATC through an HF SELCAL message which activates a light and audio alarm in the cockpit. Some aircraft are equipped for satellite-based voice communications (SATCOM) over oceanic regions. However, SATCOM has not yet been implemented in most ATC facilities due to cost considerations.

The ability to exchange data between the aircraft and the ground for ATC purposes is still surprisingly limited due to difficulties in developing common communications standards. The most widely used system is ACARS (Aircraft Communications Addressing and Reporting System) which is a low-bandwidth VHF system originally developed for airline communications and limited to alphanumeric text. Several higher-performance voice data link (e.g., VDL Modes 2–4) systems have been developed for Controller–Pilot Data Link Communication (CPDLC), but implementation has been blocked by costs and the lack of agreement on technical standards. Satellite-based CPDLC systems are used in some aircraft, but equipment and message costs for these systems remain fairly high and use is normally limited to aircraft operating over oceanic or remote regions. General broadcast satellite services such as XM or Sirius are also used to send non-aircraft-specific data, such as weather information, to aircraft in flight.

13.2.2 Navigation Systems

The navigation system is critical in ATC as it defines the underlying structure of airways that controllers use to organize traffic and issue commands. The navigation systems used while "en route" are often different from those used while on "approach" to an airport. The en route systems must be capable of longer-range coverage, while the approach systems must have a higher precision to avoid terrain and obstructions at low altitude.

En Route Navigation Systems

The fundamental en route navigation aid used worldwide for the past 50 years has been the VHF omnidirectional range (VOR) system. The VOR system consists of ground-based transmitters which allow aircraft to navigate on specific "radials" to and from the VOR stations. Some stations and aircraft are also equipped with distance measuring equipment (DME) which allows the distance to the VOR station to also be measured. Like VHF radio communications, the VOR system is limited to "line of sight," so a network of VOR stations is required in extended regions. These stations are often located at or near airports or on high ground and are spaced so as to give good navigational coverage along key

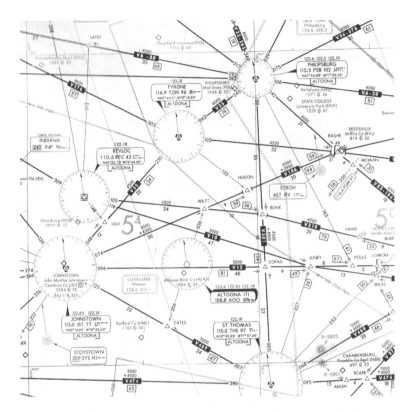

Figure 13.2 Example of an IFR en route chart with low-altitude "Victor" airways connecting VOR stations (not to be used for navigation)

routes. The VOR network provides the framework for the low- and high-altitude airway structure which can be seen in navigational charts, such as that in Figure 13.2. VOR stations are depicted as compass roses. The high-altitude "Jet" airway structure connects a network of high power VOR stations while the low-altitude "Victor" airway structure connects a denser network of lower-power stations.

A number of additional navigation systems have emerged to complement VOR navigation in recent years. These include satellite-based navigation systems, such as the Global Positioning System (GPS) and Inertial Navigation System (INS). These systems are not limited to supporting radial navigation between pairs of ground stations, but make it possible to navigate directly between any pair of points on the earth. In addition, they are not limited by the range of the ground stations and are therefore very useful in oceanic regions which are out of range of the VOR network.

The development and fast-growing use of GPS and other satellite-based navigation systems has spurred a movement toward the Required Navigation Performance (RNP) approach, which allows the use of any navigation system that meets a required RNP standard, without specifying the navigation system that provides that standard. RNP levels will depend on the flight application: RNP 1 requires horizontal navigation accuracy of less than 1 nautical mile, whereas RNP 0.3 would require better than 3/10 nautical mile accuracy.

Vertical navigation is accomplished through barometric altimetry. Aircraft altitude is measured through barometric pressure, with lower pressures indicating higher altitude. In order for an aircraft's altimeter to give an accurate reading it must be referenced to the local surface pressure which changes with weather. The "altimeter setting" is measured at weather reporting stations or airports and is transmitted to the aircraft by ATC. Above a "transition altitude" well above terrain (18 000 ft in the USA) all altimeters are set to the standard atmospheric surface pressure (1010 mb, 29.92 in Hg) to minimize the number of altimeter setting changes and the chance of aircraft being on different altimeter settings in the same area. Altitudes referenced in this way are termed "flight levels." FL360, for instance, corresponds to an indicated altitude of 36 000 ft at the standard altimeter setting.

At low altitudes the minimum vertical separation between aircraft is 1000 ft. Above FL290 the flight levels have historically been separated by 2000 ft because of the smaller pressure differences between flight levels at high altitude. With improved altimetry systems, a Reduced Vertical Separation Minimum (RVSM) of 1000 ft at all levels has been implemented in some regions, including the USA and most of Europe.

Approach Navigation Systems

Runways are classified into non-instrument and instrument. A *non-instrument* (or *visual*) runway is intended for the operation of aircraft using only visual approach procedures. An *instrument* runway permits instrument approach procedures in low-visibility conditions. Instrument runways are further subdivided into *non-precision approach* and *precision approach*.

For precision approaches, the Instrument Landing System (ILS), shown schematically in Figure 13.3, is the basic system used worldwide. The ILS consists of a "Localizer," a lateral navigation beam aligned with the runway centerline, and a "Glideslope," a vertical navigation beam aligned with a standard descent angle (typically 3°) to the runway touchdown zone. In addition a set of radio marker beacons at fixed distances from the runway are required, as well as approach lighting to guide the pilot to the runway threshold. Figure 13.4 shows an example of a typical ILS approach procedure. There are several ILS categories depending on precision. The basic Category I ILS requires a minimum visibility of 1/4 mile and a ceiling of 200 ft (unless terrain obstructions dictate higher minima). Category II and Category III allow lower minima (Table 13.1), but require additional ground and aircraft equipment (e.g. a radar altimeter) and crew training.

For non-precision approaches, less accurate navigation aids, such as a VOR or a non-directional beacon (NDB), are used to provide directional guidance for maneuvering and alignment with a straight-in approach. These non-precision approaches have higher ceiling and visibility minima and do not provide specific vertical guidance, but rely on a series of step-down maneuvers during the approach. These "dive and drive" non-precision approaches have a much higher accident rate than ILS approaches (Enders *et al.*, 1996).

GPS can also be used for approach navigation. The basic civilian GPS system has sufficient accuracy for non-precision approach procedures and, in many cases, GPS can be used in lieu of VOR or NDB for existing non-precision approach procedures. GPS has also enabled low RNP approaches (e.g., RNP 0.3 or below) which can give vertical guidance for non-precision approaches and can be used in areas with difficult terrain. An example of an RNP approach into Juneau, Alaska is shown in Figure 13.5. For this

FAA Instrument Landing Systems

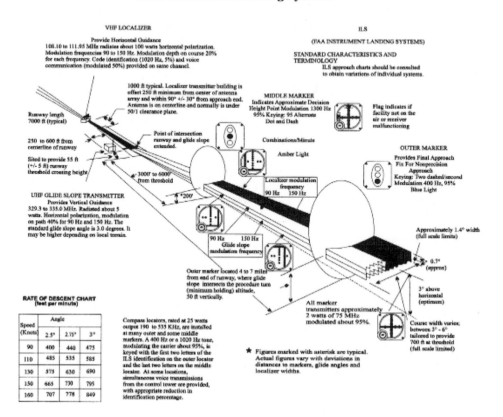

Figure 13.3 Instrument Landing System (ILS)
Source: FAA Airman's Information Manual

Table 13.1 Precision instrument approach categories

	Decision height	Visibility or runway visual range (RVR)
Category I	60 m (200 ft)	Visibility: 800 m (0.5 mile) *or* RVR: 550 m (1800 ft)
Category II	30 m (100 ft)	RVR: 350 m (1200 ft)
Category III-A	0 m	RVR: 200 m (700 ft)
Category III-B	0 m	RVR: 50 m (150 ft)
Category III-C	0 m	RVR: 0 m

Source: ICAO (1999)

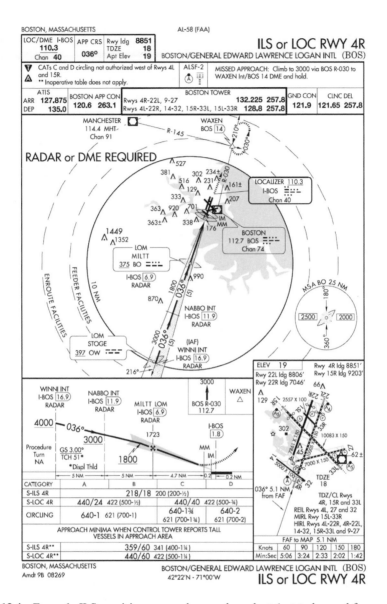

Figure 13.4 Example ILS precision approach procedure chart (not to be used for navigation)

approach the minimum decision altitude is 1238 ft at an RNP level of 0.3 and lowers to a decision altitude of 336 ft at an RNP level of 0.15.

To use GPS as the primary navigation source for precision approaches, it is necessary to augment the basic civilian GPS in order to improve accuracy and provide the ability to detect problems with the GPS position measurement ("integrity"). The accuracy of GPS measurements is degraded due to the slowing of the GPS timing signals as they travel from the satellites through the ionosphere. Two competing approaches to GPS augmentation exist. In the Ground-Based Augmentation System (GBAS) a reference

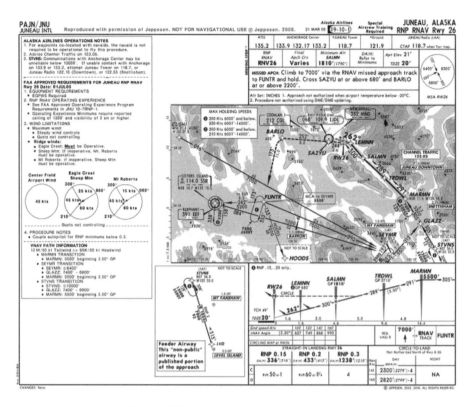

Figure 13.5 RNP approach procedure chart
Source: Courtesy of Alaska Airlines and Jeppesen (not to be used for navigation)

station with a GPS receiver at a known location is used to measure any GPS errors and to transmit corrections to the local area. This system can have accuracy sufficient for precision approaches at the Category II or III levels but requires a reference station near each airport. The Satellite-Based Augmentation System (SBAS) uses a series of widely spaced reference stations to develop a simple model of the ionospheric corrections which can be broadcast to aircraft from a satellite. SBAS can give coverage over a wide area but its vertical accuracy is limited to roughly Category I precision approach levels. The US SBAS system is the Wide Area Augmentation System (WAAS) and is currently operational. Other SBAS systems are currently in development in other regions of the world (EGNOS in Europe, MSAS in Japan, GAGAN in India).

13.2.3 Surveillance Systems

The surveillance system is the means by which the controllers monitor the air traffic situation. This system is critical for all ATC operations other than at control towers in good visibility, when the controllers can directly observe the air traffic.

The most basic form of aircraft surveillance is position reporting where the crew communicates its position, altitude and intent information through radio communications to ATC. Controllers traditionally track the reports by moving physical "shrimp boat" markers of aircraft position on a map or by organizing flight progress strips in a pattern which relates to the air traffic situation. In modern systems, the reports can be presented on electronic maps or flight strip displays. Position reports are the primary means of surveillance when no other surveillance systems are available. This includes most of oceanic airspace, remote regions such as Alaska, Northern Canada or Western China, and developing regions with low flight density, such as parts of Africa and South America. Position reporting also serves as the "backup" in the event of failure of other surveillance systems.

Radar comprises the fundamental aircraft surveillance system currently used for most domestic ATC systems. There are two types of ATC surveillance radars. *Primary* radars measure range by the round-trip time of an interrogation pulse reflected off the metal surfaces of an aircraft (the "skinpaint"). *Secondary* radars (ATC Radar Beacon System, ATCRBS) require that aircraft be equipped with a transponder which receives and retransmits an interrogation pulse along with an identification code for the responding aircraft and other aircraft data. Each aircraft is assigned a four-digit transponder code which is used to identify the specific aircraft on the radar display (Figure 13.6).

The direction (or "azimuth") of the aircraft is determined by the direction the antenna is pointing to when the aircraft is detected. In order to scan all directions, ATC surveillance radars typically rotate at a constant speed which determines the update rate of the radar display. Short-range airport surveillance radars (ASRs) used in terminal areas typically require slightly more than 4 seconds per update. Long-range air route surveillance radars (ARSRs) used for en route control typically update once every 12 seconds. Because of the relatively low update rates, most ATC radar displays present tracks of recent radar "hits" so that the controller can estimate the direction of the aircraft as shown in Figure 13.6. Radar processors can also integrate these "hits" with tracking software to estimate aircraft speed and direction. However, the low update rates constitute a significant limitation, as it takes several "hits" and thus as much as a few minutes to determine if an aircraft has changed course or direction.

Altitude is not directly measured by ATC radars, but is reported through the ATCRBS transponder reply. The barometric altitude of the aircraft (in 100 ft increments) is measured by an on-board "altitude encoder" which is integrated with the transponder. Altitude data are received with the aircraft identification code and integrated as part of the "data block" on the radar display (Figure 13.6).

Radars are also used for surveillance of traffic on the airport surface. Surface surveillance radars (SSRs) often have "blind spots" due to blocking of the radar signal by buildings or obstructions. Integrated sensor systems, such as ASDE-X, combine radar with multi-lateration whereby aircraft position is measured through an array of receivers which detect aircraft transponder signals. Multi-lateration can also be used for surveillance of airborne aircraft.

Automatic Dependent Surveillance (ADS) is an emerging surveillance approach whereby aircraft automatically transmit position reports and intent data. This is essentially an automated version of the manual position reporting approach described earlier. A number of different ADS systems exist. ADS-A (Addressed) transmits position reports to the ground when requested by ATC. ADS-C (Contract) transmits position reports to ATC at defined periodic intervals or upon the occurrence of specific conditions (e.g., crossing a reporting point). Both ADS-A and ADS-C are being used in parts of oceanic or remote

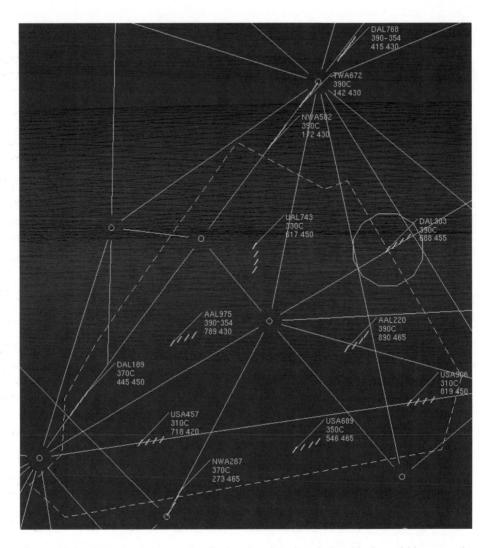

Figure 13.6 Example en route radar display showing aircraft data blocks and history tracks

airspace (South Pacific, North Atlantic) through satellite or other communication links
when available.

ADS-B (Broadcast) transmits aircraft position and state information at high update rates
(1 second) in signals which can be received by other equipped aircraft in the area, as well
as ATC. The high quality of the surveillance information, the high update rate and the
low cost of the ADS-B ground stations (compared to radar) make ADS-B an attractive
alternative to conventional ATC surveillance. ADS-B surveillance is being implemented
in Australia, the USA and Europe and is planned for many other regions. Because ADS-B
also enables equipped aircraft to directly observe each other through Cockpit Display of
Traffic Information (CDTI), it is expected that ADS-B will enable new, more efficient
procedures based on airborne self-separation.

13.2.4   Flight and Weather Information Systems

The ATC system receives critical support from several types of information systems. Aircraft flight plans are managed in a central flight data processing system (the "host") which accepts and distributes aircraft flight plans to the various ATC facilities along the expected route of flight. An increasing number of computer-based decision support tools are being developed to help controllers optimize the flow of traffic in constrained airspace. Finally, extensive weather information systems generate and disseminate general and airport-specific forecasts and observations, as well as winds aloft forecasts, hazardous weather alerts and pilot reports (PIREPS).

13.3   Airspace and ATC Structure

Civilian airspace is designated as "controlled" or "uncontrolled." In controlled airspace, traffic is supervised and managed by ATC, so that aircraft may fly in low-visibility Instrument Meteorological Conditions (IMC). Aircraft are not directly managed by ATC in uncontrolled airspace, which is normally restricted to low altitudes or remote regions where the traffic density is low. In most countries, some airspace is reserved for military operations, which civilian aircraft may be prohibited from entering or may require coordination with military authorities before using.

Figure 13.7 shows the general ATC control structure in the USA. Similar structures are used, with variations, in most of the world. Some elements may be combined in regions or countries with low traffic density. The different control regions are discussed briefly below.

At airports, Surface ("Ground") Control manages aircraft and other vehicles on the airport surface as they taxi to/from the runway from/to the ramp (or "apron") areas. Local ("Tower") Control is responsible for aircraft taking off and landing. The tower controls the active runways and the local maneuvering airspace typically as far as a radius of about

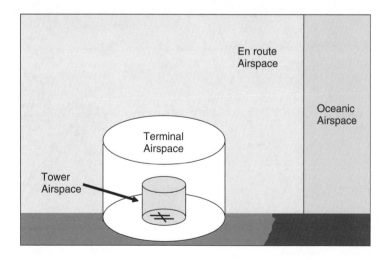

Figure 13.7   Typical structure of airspace

5 miles from the airport and up to 2500 or 3000 ft above the airport surface. For some large and busy airports, such as Atlanta, Local Control may be split into sectors with different controllers managing each runway or sets of runways (Chapter 12). It should be noted that many small airports do not have an active control tower and rely on procedural "rules of the road" for managing traffic. However, control towers (as well as a crash rescue capability) are normally required for airports with scheduled air service in most of the world.

Terminal Area (or "Terminal Airspace") "Approach" or "Departure" Control manages aircraft in the descent and initial approach phases, as well as during the phases of departure and climb-out from major airports. In the USA, the terminal airspace associated with Terminal Radar Approach Control (TRACON) facilities typically extends 50 miles from and up to 18 000 ft above the central airport. These dimensions are also typical of terminal airspace at major airports around the world. TRACON facilities typically use short-range, high-update-rate (4 seconds) ASRs. The airspace in the TRACON is subdivided into a number of sectors which are configured to support the arrival and departure flows depending on the airport configuration (i.e., the set of active runways in use – see Section 13.6 and Chapter 12). In some high-density metropolitan areas (e.g., New York, Washington, San Francisco, Los Angeles) there may be multiple high-activity airports and a combined TRACON is used to coordinate the arrival and departure traffic. An example of the New York TRACON traffic flow showing the trajectories of aircraft flying into the four major airports in the metropolitan area is shown in Figure 13.8.

En route ("Center") Control is responsible for traffic above and between terminal airspace. The areas of responsibility of the 20 Air Route Traffic Control Centers

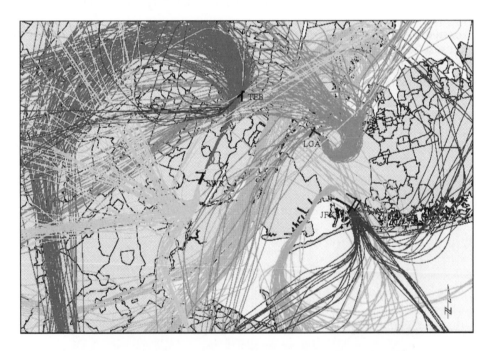

Figure 13.8   New York arrival and departure trajectories (JFK, LGA, EWR, TEB)
Source: Courtesy of the Port Authority of NY & NJ

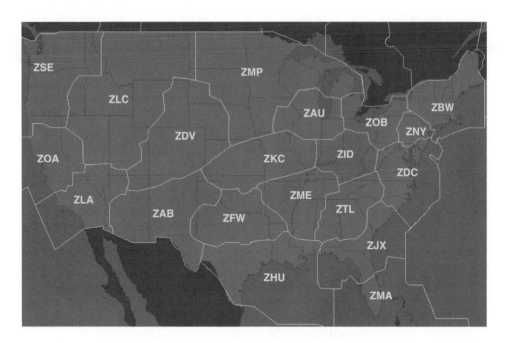

Figure 13.9   Continental US Air Route Traffic Control Centers

(ARTCCs), which manage en route traffic over the continental USA, are shown in Figure 13.9. ARTCC facilities typically use the longer-range, slower-update-rate (12 seconds) ARSRs. The airspace in the Center is split into three-dimensional sectors based on the traffic flows. The geographical configuration of low-level en route sectors is shown in Figure 13.10.

In regions with high traffic densities, such as Europe and the USA, an additional layer of traffic control has been added to coordinate, on a more aggregate scale, traffic flowing across the various ATC facilities. This layer is referred to as the air traffic flow management (ATFM) system and will be described further in Section 13.7.

The control of aircraft in international airspace between countries and over "Oceanic" regions is coordinated by the International Civil Aviation Organization (ICAO) which assigns control responsibility through a structure of Flight Information Regions (FIRs). An example of the FIRs which manage air traffic over the North Atlantic is shown in Figure 13.11.

An additional consideration concerns the procedures for aircraft crossing international borders. It may be necessary for aircraft to be identified in an Air Defense Identification Zone (ADIZ) before entering sovereign domestic airspace. These procedures can vary by country and may sometimes complicate ATC operations.

13.4   ATC Operations

ATC, as currently practiced, is a human-centered contract process in which controllers and flight crews (or dispatchers) negotiate for access to airport or airspace resources. The contract is the ATC "clearance," which is *executed* by the flight crew and *monitored* by

Figure 13.10   Low-altitude sectors in the USA

the controllers. In the event of a need for a change, the clearance is renegotiated and an amended clearance is issued.

The basic ATC control loop is shown in Figure 13.12. The air traffic controller observes the traffic situation through his or her surveillance system. The controller issues commands to the aircraft, in the form of either clearance amendments or "vectors." Vectoring is a process whereby the controller gives steering commands to the pilot in the form of heading, speed and altitude. Controllers normally employ vectoring when they exercise tight control over the maneuvering of aircraft. This often occurs in the terminal area, where aircraft are being maneuvered onto final approach, but may also occur in En route Control areas, when aircraft are being maneuvered around traffic or weather.

The precision of ATC is constrained by the performance of the surveillance system and the limitations of the ATC control loop shown in Figure 13.12. These constraints stem from the relatively slow update rates and uncertainty concerning aircraft positions in radar-based systems, coupled with communication delays on the voice channels and the variability of pilot response. It can take several radar updates to determine if a pilot has responded correctly to a command. As a consequence, controllers cannot reduce the separation between neighboring aircraft below certain minima. These are the ATC-specified "minimum separation standards," which limit the capacity of the ATC system. The current minimum radar separation standard in most terminal airspace is 3 nautical miles for aircraft flying at the same altitude. This is increased to 5 nautical miles in en route airspace, due to the lower precision and the slower update rate of en route radars (ARSR). In most of the world, as noted earlier, vertical separation standards are 1000 ft increasing to 2000 ft above FL290, except in the USA, Europe and the North Atlantic where

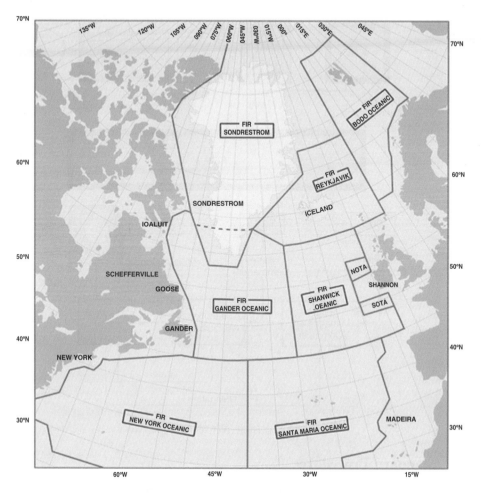

Figure 13.11 North Atlantic FIR boundaries
Source: Gaudet (2008)

the Reduced Vertical Separation Minimum (RVSM) is 1000 ft at all altitudes because all aircraft above FL290 must be equipped with high-accuracy altimetry systems.

In those parts of oceanic airspace where surveillance is based solely on position reporting and the communication links are unreliable, it is necessary to have large separation minima due to potential communication delays. Current longitudinal separation is typically 60 miles, but can be lower if good communications and navigation capabilities are available.

Finally, hazardous wake vortices behind aircraft constitute an additional factor driving separation standards in the terminal area. The separation requirements, on arrival and departure, depend on the relative weight of the leading and trailing aircraft, as the strength of the wake vortex effect generated by an aircraft generally increases with the aircraft's weight. These requirements will be discussed in more detail in Section 13.6.

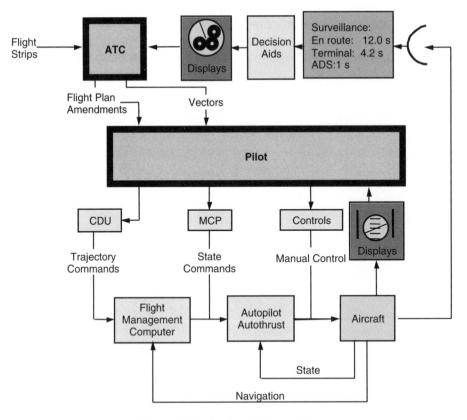

Figure 13.12 Basic ATC control loop

13.5   Standard Procedures

Standard Operating Procedures (SOPs) are defined for most ATC systems. The SOPs specify the normal operating procedures, as well as standard routings and communications procedures. An example is the standard altitudes for flights. The SOP is for aircraft flying westbound to be assigned cruise altitudes in even thousands of feet (e.g., 12 000 ft), while aircraft with eastbound headings are assigned odd altitudes. This avoids head-on encounters and gives the controllers more time to deal with traffic conflicts at the same altitude. Another standard procedure is the holding pattern, with a racetrack-like trajectory around a reference point, which can be used to delay aircraft in the air when it is not possible to hand off the aircraft into the next sector due to traffic, weather or some other reason. Some of the standard routings associated with SOPs are published, such as Standard Instrument Departures (SIDs) – see Figure 13.13 – and Standard Terminal Arrival Routes (STARs).

One important SOP involves priority for service and equity of service. The US system – and practically all other advanced ATC systems – generally runs on a first-come, first-served (FCFS) basis, where each ATC facility deals with aircraft or requests for service in the order that they arrive. Some exceptions may take place in the course of daily operations for the purpose of increasing processing efficiency. Exceptions

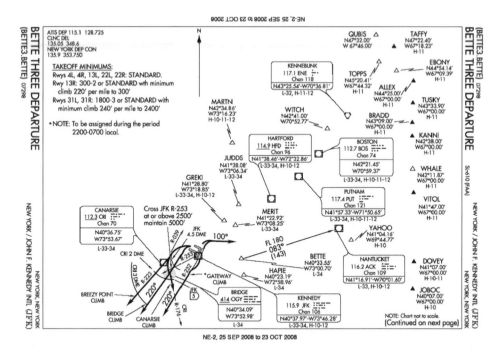

Figure 13.13 Example Standard Instrument Departure for JFK (not to be used for navigation)

also include high-priority cases, such as aircraft with a declared emergency, medical evacuation ("Lifeguard") flights and high-priority military aircraft, such as Air Force 1.

As the ATC system has grown, it has adopted specialized procedures to deal with local conditions. An example can be seen in the flight paths into and out of the key airports in the busy New York terminal airspace shown in Figure 13.9. In order to minimize the need for coordination between airports, arrival and departure trajectories into and out of each airport are procedurally separated. This allows for a high rate of operations at the New York airports, but makes it difficult to authorize a non-standard flight path. Special procedures may also be in place for transitions between two ATC facilities. These are described in Letters of Agreement (LOA), which often define the locations where handoffs between neighboring facilities take place, as well as handoff procedures and maximum handoff rates.

One important class of procedures applies to emergencies or equipment failures. For example, if an aircraft loses its communications capability, the standard procedure is for the aircraft to fly the last assigned clearance and routing. If the aircraft is in a holding pattern, it will have been given an "Expect Further Clearance Time" (EFCT) and will be assumed to leave the holding pattern at that time. ATC is expected to clear other aircraft out of the way.

13.6 Capacity Constraints

ATC and airport capacity constraints are becoming a limiting factor in many regions of the world. The principal "bottlenecks" are the runway systems of the world's busiest airports

Table 13.2 For purposes of specifying minimum ATC separations on landing and takeoff, aircraft are classified into a small number of classes, according to their maximum takeoff weight. ICAO and the FAA use somewhat different classifications and class names

ICAO classification:	
Heavy (H):	MTOW greater than 136 tons (~300 000 lb)
Medium (M):	MTOW between 7 tons (~15 400 lb) and 136 tons (~300 000 lb)
Light (L):	MTOW under 7 tons (~15 400 lb)
FAA classification:	
Heavy (H):	MTOW greater than 255 000 lb (~116 tons)
Boeing 757:	The B757, whose MTOW places it at the borderline between the FAA's L and H classes, constitutes an aircraft class by itself, due to its severe wake vortices
Large (L):	MTOW between 41 000 lb (~19 tons) and 255 000 lb (~116 tons)
Small (S):	MTOW less than 41 000 lb (~19 tons)

and the terminal airspace around them. The taxiways, ramps and gates are normally designed to match or exceed the runway system's capacity. En route capacity limitations are also often severe, especially in certain parts of Europe and, with increasing frequency, the USA (see Section 13.7). This section reviews airport and terminal airspace capacity limitations and their implications for air traffic delays.

The capacity of a *single* runway is largely determined by the minimum separation requirements between consecutive aircraft operations. For purposes of separation on landing and on takeoff, aircraft are subdivided by ATC into a small number of classes, usually three or four, according to their maximum takeoff weight (MTOW). The classes defined by ICAO (and used in many countries around the world) and the FAA (used in the US) are shown in Table 13.2. The minimum separation requirements are then specified *for every possible pair of classes* (e.g., "Heavy" aircraft followed by "Light") *and for every possible sequence of movements* ("arrival followed by arrival," "departure followed by departure," "departure followed by arrival" and "arrival followed by departure"). Table 13.3 shows the separation requirements that apply to the arrival-followed-by-arrival case as specified by ICAO and the FAA for operations under Instrument Flight Rules (IFR). Note that the two sets of requirements are quite similar, but also include some differences. For example, in the USA, the Boeing 757 has a special designation due to its intermediate MTOW (255 000 lb) and the fact that it has been the source of a number of reported wake vortex encounters. Note also the larger separations (4, 5 and 6 nautical miles) required when lighter aircraft trail heavier ones: these are the "wake-vortex separations" mentioned earlier in this chapter. Because of its large size, the Airbus 380 has been designated internationally as a special category of aircraft for now. Conservative separation standards have been set for this aircraft, until more data on its wake-vortex characteristics becomes available. At the time of this writing, the separations on final approach when the leading aircraft is an Airbus 380 have been set by ICAO to be equal to the requirement behind "Heavy" aircraft *plus 2 nautical* miles, i.e., 6, 7 and 8 nautical miles instead of the 4, 5

Table 13.3 Minimum separation requirements for same-runway arrivals under Instrument Flight Rules (IFR)

ICAO separations

A. Throughout final approach, consecutive aircraft must be separated by at least the distance (in *nautical miles*, nm) indicated by the matrix below:

		Trailing aircraft		
		Heavy	Medium	Light
Leading aircraft	Heavy	4	5	6
	Medium	3	3	4
	Small	3	3	3

B. The trailing aircraft cannot touch down on the runway before the leading aircraft has exited the runway

FAA separations

A. Throughout final approach, successive aircraft must be separated by at least the distance (in *nautical miles*) indicated by the matrix below:

		Trailing aircraft		
		H	L + B757	S
Leading aircraft	H	4	5	5/6*
	B757	4	4	5
	L	3	3	4*
	S	3	3	3

*Separations indicated with an asterisk are distances required at the time when the leading aircraft is at the threshold of the runway

B. The trailing aircraft cannot touch down on the runway before the leading aircraft has exited the runway

and 6 nautical miles, respectively, shown in Table 13.3. It is possible that this will change in the future.

In addition to minimum separation requirements, numerous other factors might affect significantly the capacity of a *single* runway. Some important ones include:

- The performance characteristics of the aircraft using the runway, such as speed on final approach, which influence runway occupancy time and the time separation between consecutive landings. In most cases the runway must be clear before the next landing or takeoff.

- The geometric characteristics of the runway, such as the availability of well-spaced high-speed exits that may reduce significantly runway occupancy times on landing.
- The mix of aircraft types using the runway: a homogeneous mix of aircraft (e.g., an almost all-"Large" mix) will result in higher capacities than when the runway is used by a diverse mix.
- The types of movements (arrivals only, departures only, mixed operations) that are assigned to the runway during a time interval.

As already noted in Chapter 12, the capacity of a single runway will vary from airport to airport and from country to country, but an upper limit is about 60 aircraft per hour, with more typical values being in the range of 40–50 movements per hour for runways which are used for mixed operations. Approaches to computing the capacity of a runway, taking all the above-listed factors into consideration, are reviewed in de Neufville and Odoni (2003).

Two or more runways typically exist at the great majority of the busiest airports. Two additional factors enter the computation of the capacity of such multi-runway airports:

- The geometric layout of the various configurations, e.g., whether two simultaneously active runways intersect or are parallel to one another.
- The degree of *dependence* between operations on different runways, i.e., the extent to which operations on a runway may constrain operations on one or more other active runways. For example, in the case of a pair of intersecting runways, the location of the intersection relative to the points where takeoffs are initiated or where landing aircraft touch down greatly affects the degree to which operations on the two runways interact. Similarly, in the case of two parallel runways, the degree of dependence is determined by the distance between the centerlines of the runways, as discussed in Chapter 12. In the best case, movements on two different simultaneously active runways are mutually independent.

Multi-runway airports may have many alternative *runway configurations*, i.e., sets of simultaneously active runways with which the airport may operate at different times. An often-used configuration for Boston Logan International Airport (BOS) is shown in Figure 13.14. In the configuration shown, three of the five existing runways at BOS are simultaneously active, with one runway used solely for arrivals, another solely for departures, and the third for mixed operations. In general, the use of configurations is dictated by weather and/or noise considerations. It is desirable to have aircraft land and take off into the surface wind. In light winds or "calm" conditions, the choice of configuration is based on such criteria as maximizing capacity, minimizing noise impacts on neighboring communities, or retaining the opportunity to switch easily to some other favorable configuration under a forecast weather change. For example, BOS may operate with a single runway during the period of 23:00–06:00, when traffic is very light and noise minimization is extremely important, and with one, two or three runways (as in Figure 13.14), depending on weather conditions and traffic, during the rest of the day.

The overall airport performance can be represented as a Pareto frontier of achievable arrival and departure rates, such as those shown in the example of Figure 13.15. These frontiers are often referred to as *runway configuration capacity envelopes* (RCCEs) and provide a convenient way for displaying the capacity of any given runway configuration. Figure 13.15 shows two typical RCCEs for a hypothetical runway configuration M at

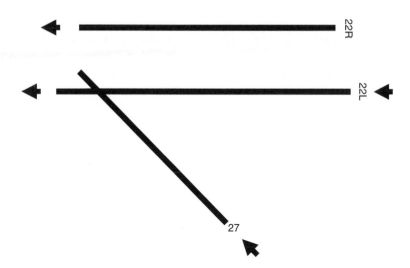

Figure 13.14   An often-used runway configuration at Boston's Logan International Airport (BOS). In this configuration, runway 22L is used for both landings and takeoffs, runway 22R for takeoffs only and runway 27 for landings only. Note that 22R and 22L are close parallel runways and that 27 and 22L intersect. As a result, operations on the three runways must be carefully coordinated by ATC.

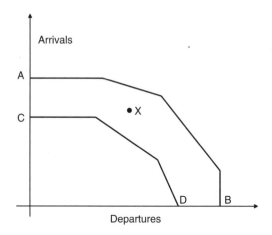

Figure 13.15   Two runway configuration capacity envelopes (RCCEs) describing the available departure and arrival capacities of a hypothetical runway configuration M in VMC (envelope AB) and IMC (envelope CD)

some given airport. The horizontal and vertical axes indicate the number of departures and arrivals, respectively, per unit of time (typically per hour or per 15 minutes). The boundary connecting points A and B in Figure 13.15 represents the RCCE of configuration M under Visual Meteorological Conditions (VMC), while the one connecting points C and D is the RCCE under Instrument Meteorological Conditions (IMC). Any point inside an RCCE corresponds to a *feasible* combination of departures and arrivals, and any point outside is *infeasible*. For example, point X in Figure 13.15 is feasible in VMC, but infeasible in IMC. In the latter case, configuration M does not have sufficient capacity to accommodate simultaneously the number of arrivals and of departures associated with X during a single unit of time, resulting in the queuing of aircraft.

Note that, as suggested by Figure 13.15, the airport capacity is generally lower for IMC operations than for VMC: for any particular runway configuration the RCCE for IMC operations is contained "inside" the RCCE for VMC operations. In addition, the capacity will vary with the airport configuration in use. As already noted, the number of possible runway configurations can be large for multi-runway airports. BOS, with five runways, typically employs more than 20 different configurations during a year, each consisting of one, two or three simultaneously active runways. Because most of these configurations are associated with two RCCEs, one for VMC and the other for IMC, the total number of RCCEs that need to be considered for BOS is of the order of 40!

The resulting variability of airport capacity creates a dilemma for airports and airline schedulers. Should arrivals and departures be scheduled assuming good weather conditions or should a "worst-case scenario" of poor visibility and adverse wind directions be assumed? In airports where the number of departure and arrival "slots" is restricted a priori,[2] the conservative IMC rates are normally used as guidelines for scheduling operations in order to ensure that sufficient airport capacity is available under all weather conditions. At airports without slot restrictions, airlines are free to schedule at rates above the IMC capacities or, even, the VMC capacities. As a result, they may sustain long delays at times when the actual capacity falls below the level of demand.

The impact of capacity variability can be seen in the simple example shown in Figure 13.16. In this example the nominal VMC capacity of a hypothetical "hub" airport is 50 flights per hour and the airlines have chosen to "over-schedule" during the four "banks" (or "waves") of the day, in order to maximize the number of connections at the airport. Even though demand exceeds capacity at times, the low-activity periods between banks allow the delayed flights to operate on time or with modest delays when the capacity is at nominal levels. If, however, the capacity of the airport were to drop to 30 flights per hour due to adverse weather conditions, then all flights after the initial bank would be delayed and the delays would build up throughout the day, as shown in the lower half of Figure 13.16.

13.7 Congestion and Air Traffic Management

As air traffic grows, demand can exceed capacity at key points of the air transportation network and at critical times. These local overloads create delays which propagate to other parts of the air network, amplifying congestion as increasing numbers of local capacity constraints come into play. Moreover, the average delay generally increases faster than linearly with traffic. This can be seen in a set of US delay data[3] shown in

---

[2] As noted in Chapter 2, most of the busiest commercial airports outside the USA fall in this category. In addition, a few airports within the USA (New York LaGuardia, New York JFK, Washington Ronald Reagan National and Chicago O'Hare) still operated at the time of this writing with slot limits – which, however, may be removed in the future.

[3] This particular set of data records the minutes of delay per month sustained by scheduled flights in sectors where the flight experienced a delay of more than 15 minutes (within the single sector). Thus, the amounts of delay shown in Figure 13.16 represent only a small part of all the delays sustained by scheduled flights in the USA.

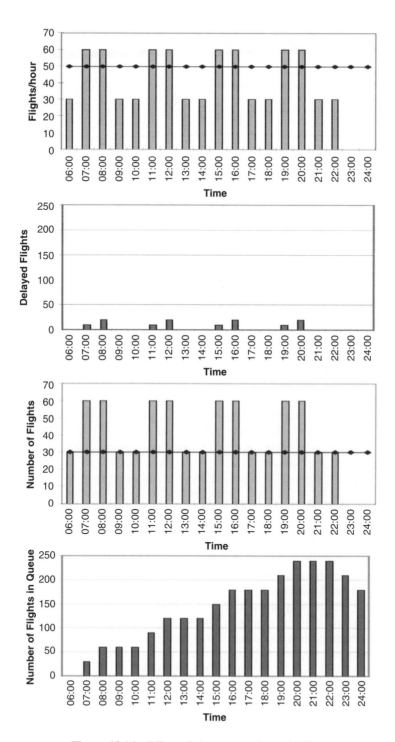

Figure 13.16 Effect of airport capacity variability

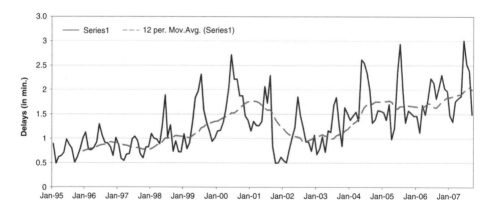

Figure 13.17   US flight delays
Source: FAA Operations Network – OPSNET

Figure 13.17. Beginning in 1998, a pattern of summer delay spikes has developed, due to convective weather and increased seasonal traffic demand. The delays moderated briefly due to reduced demand after the September 11, 2001 attacks, but have returned with increasing traffic levels. Air traffic delays experienced elsewhere have displayed similar patterns over time to those in the USA.

One of the main functions of ATC is to manage the flow of air traffic when congestion does occur. Because aircraft cannot stop in midair, they can only be delayed briefly in holding patterns or they can be rerouted when airborne or held on the ground prior to departure. ATC must manage the process by assuring that no downstream sector or airport exceeds the Operationally Acceptable Level of Traffic (OALT). The OALT may be determined by airspace capacity, defined as the maximum number of aircraft allowed in a specific sector, or by airport capacity.

Airspace capacity is typically determined by the capacity of the ATC system's sectors. The maximum capacity of a sector is limited by controller workload and by the complexity of air traffic patterns. Typical maximum traffic counts in individual en route sectors are 10–20 simultaneously present aircraft in a radar-supported environment. The traffic flow rates may also be limited by restrictions on downstream sectors. This can be significant at international border crossings or when several sectors are feeding traffic into a merging sector near a high-volume destination.

At the airport level, the OALT is often specified through the Airport Acceptance Rates (AARs). The AARs for any single airport provide a forecast of the arrival capacity per hour (or other time unit) of that airport over the next several hours. For example, a 4-hour AAR schedule for BOS may be the sequence [40, 33, 33, 40] indicating that BOS is expected to be in a position to handle 40 arrivals during the first hour of the period, 33 during the second and third, and 40 during the fourth. As indicated in the previous section, the arrival capacity of a system of runways depends on numerous factors and is subject to great uncertainty under variable or adverse weather conditions. In such circumstances, it is often extremely difficult to predict the capacity of an airport even 2 or 3 hours in advance.

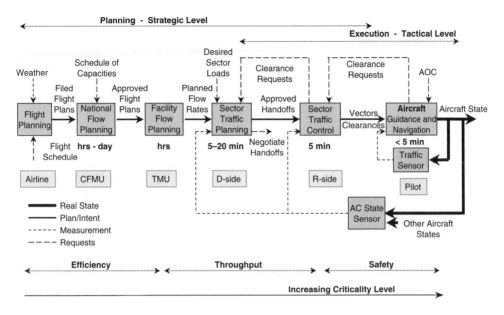

Figure 13.18 ATC strategic coordination architecture (adapted from Haroldsdottir *et al.*, 1998)

The management of capacity and delays has become one of the most important tasks of ATC in high-demand regions and is referred to as traffic flow management (TFM). Figure 13.18 (Haraldsdottir *et al.*, 1998) illustrates the strategic coordination architecture utilized in the US for TFM purposes. As indicated, TFM is carried out in a hierarchical fashion: depending on the situation at hand, overloads may be dealt with at the local, regional or national level. The tactical ATC control loop discussed earlier (Figure 13.12) is on the right side of the figure. The R-side (Radar) controller manages aircraft on his or her radar at a tactical time scale of minutes. The R-side controller is supported by a more "strategic" D-side (Data) controller who coordinates handoffs with other sectors and is responsible for sector traffic planning at the 5- to 20-minute time scale. Coordination at the en route Center level, i.e., on a regional scale, is accomplished by the Traffic Management Unit (TMU), which coordinates traffic management across sets of sectors within the same Center, as well as across neighboring Centers. The TMU is responsible for facility flow planning at timescales of 30 minutes to 2 hours. If a serious overload problem is forecast or if the traffic is projected to exceed the OALT in a sector, the TMU can impose flow restrictions on traffic entering a Center or a part of a Center. Flow restrictions typically take the form of "miles-in-trail" (MIT) or "time-in-trail" (TIT) requirements at defined waypoints. MIT and TIT mean that extended separations are required between consecutive aircraft (e.g., 30 nautical miles or 5 min) flying past these waypoints.

When traffic overloads cannot be managed at the local or Center level, flow management interventions of a national or quasi-national scope may be implemented. In the USA and in Europe flow planning of such extensive scope is accomplished by a Central Flow Management Unit (CFMU) that evaluates the actual capacity of the system on a time horizon of several hours to a day and can impose strategic restrictions to minimize overloads in the system. The CFMU in the USA is in the Air Traffic Control System Command Center (ATCSCC) near Washington, DC. The European CFMU is operated by EUROCONTROL in Brussels and coordinates air traffic across the 38 Member States of the European Civil Aviation Conference (ECAC). The role of the CFMU in the USA is viewed as mostly reactive, intervening only as called for by weather conditions or other circumstances. In Europe, by contrast, all aircraft for which a flight plan is received must be cleared by the CFMU, i.e., receive a "departure slot" before they can leave their parking stand.

The CFMU projects traffic loads based on airline schedules and filed flight plans and compares the projected traffic demand to the expected capacity of key elements of the system, such as major airports or congested en route airspace. Because the capacity often depends on weather, the weather forecasting capability of the ATC system plays a very important role in determining the effectiveness of CFMU interventions. Particularly critical are forecasts of convective weather, which can block routes or critical fixes en route, and of low-ceiling, low-visibility thunderstorms, snow or ice near major airports which can reduce the AAR. Other factors affecting capacity, such as navigational equipment failures or runway closures due to snow removal or maintenance, are also considered. Once areas of potential overload are identified, strategies to manage the traffic are generated. The strategies employ three principal types of interventions: *ground holding*, i.e., intentionally delaying an aircraft's takeoff for a specified amount of time; *rerouting*, i.e., changing or restructuring some flight routes to modify the distribution of traffic flows; and *metering*, i.e., controlling the rate at which traffic crosses some specified spatial boundaries by adjusting the spacing between aircraft. Ground holds are typically employed when a serious overload is expected at a specific airport and are imposed on aircraft with that destination. Rerouting is often used when convective weather is blocking a route or if there is a traffic overload in a specific sector. Metering is used to manage aircraft flow rates into specific sectors.

In recent years, the ATCSCC in the USA has increasingly coordinated its interventions with airlines, airports and ATC facilities, in a process which is known as collaborative decision making (CDM). The objective of CDM is to share information among stake-holders and thus provide a "common knowledge base" on which to base TFM decisions. Prior to CDM, the CFMU would often base its interventions on obsolete information. For example, when a major snowstorm reduced capacity at hub airports, such as Chicago O'Hare, the airlines would proactively cancel many flights. However, the CFMU did not have this information and thus could not assess accurately the extent of the projected overload. In the absence of information about flight cancellations, the overload would typically be overestimated and the CFMU would often impose unnecessary restrictions on traffic. To optimize the performance of the TFM system, CDM employs a variety of information-sharing techniques including a web-based Flight Schedule Monitor, periodic teleconferences with the airlines, and rescheduling methods aimed at ensuring fairness in the allocation of the limited available capacity among the airlines and other airspace users. A detailed overview of the air traffic flow management systems in the USA and in Europe, as well as a description of how CDM currently operates in the USA, can be found in Ball *et al.* (2007).

13.8   Future ATC Systems

In order to provide the potential for growth of the air transportation system, various efforts are under way to improve the capacity, efficiency and environmental performance of the ATC system. In mature ATC regions, such as the USA and Europe, there is little opportunity to expand the system by adding new airports or runways, but there are extensive planning efforts (NextGen in the USA, SESAR in Europe) in progress to define the long-term future of the ATC system. These plans are still in their developmental phase, but generally share many elements including increased use of satellite navigation systems, surveillance systems based on ADS-B, use of time as a control parameter in trajectory-based clearances, broad information sharing through system-wide information management (SWIM), moving controllers to a more supervisory role and shifting some ATC functions to the cockpit. In developing ATC regions, such as China, India and the Middle East, much of the focus has been on building new airport and runway capacity although these regions can also quickly incorporate and benefit from new ATC operational concepts and technologies.

References

Ball, M.O., Barnhart, C., Nemhauser, G., and Odoni, A. (2007) "Air Transportation: Irregular Operations and Control," in C. Barnhart and G. Laporte (eds.), *Handbook in Operations Research and Management Science, Volume 14*, Elsevier, Amsterdam, pp. 1–73.

de Neufville, R. and Odoni, A. (2003) *Airport Systems: Planning, Design and Management*, McGraw-Hill, New York.

Enders, M.J., Dodd, R., Tarral, R., and Khatwa, R. (1996) "Airport Safety: A Study of Accidents and Available Approach-and-landing Aids," *Flight Safety Digest*, Vol. 15, No. 3, March.

Gaudet, M. (2008) "Harmonization of Aviation User Charges in the North Atlantic Airspace," SM thesis, Civil and Environmental Engineering Department, Massachusetts Institute of Technology.

Haraldsdottir, A., Schwab, R., and Alcabin, M. (1998) "Air Traffic Management Capacity-Driven Operational Concept Through 2015", 2nd USA/Europe Air Traffic Management R&D Seminar.

ICAO – International Civil Aviation Organization (1999) *Aerodromes: Annex 14 to the Convention on International Civil Aviation, Volume I: Aerodrome Design and Operations*, 3rd edition, Montreal.

Nolan, M. (1999) *Fundamentals of Air Traffic Control*, 3rd edition, Brooks/Cole Wadsworth, Pacific Grove, CA.

14

Air Transport and the Environment

Karen Marais and Ian A. Waitz

14.1 Introduction

Aviation is a critical part of national economies, providing for the movement of people and goods throughout the world and enabling worldwide economic growth. However, along with the growth of aviation have come concerns regarding noise, air quality, water quality and impacts on climate. While aircraft have become more fuel efficient and less noisy over the last 35 years, most projections for the rate of growth of air transport exceed projections for the rate of technological advancement for noise and emissions such that the environmental consequences of aviation may increase.

At the same time, public awareness of environmental issues, and political pressure to manage environmental impacts, have increased dramatically. Three-quarters of delays experienced by runway expansion projects at the 50 busiest US airports have been attributed primarily to environmental concerns. Of these 50 airports, 12 have had at least one expansion project cancelled or indefinitely delayed due to environmental issues (GAO, 2000). Environmental constraints are therefore likely to pose a fundamental limitation on aviation growth in the twenty-first century (NSTC, 1995; Waitz *et al.*, 2004).

Aviation affects the environment at the local, regional and global levels. Water quality around airports is adversely affected by runoff from aircraft and airfield deicing operations, as well as other sources such as fuel leaks, spills, and solid and liquid waste treatment and disposal. Because water sources are often connected to each other, the adverse impacts of local deterioration in water quality may be felt in regions far removed from airports. Section 14.3 focuses on one of the largest sources of airport runoff pollution in cold climates, aircraft and airfield deicing.

Noise from aircraft causes sleep disturbance, interrupts speech and adversely affects property values around airports. As a result, local communities often vigorously oppose airport expansion plans. Section 14.4 discusses the sources, effects and control of aviation noise.

The Global Airline Industry   P. Belobaba, C. Barnhart and A. Odoni
© 2009 John Wiley & Sons, Ltd

At the local and regional level, emissions from aircraft, airport traffic and stationary airport sources adversely affect air quality and therefore health. In the USA many counties containing airports do not meet federal Clean Air Act standards (Pub. L. 101–549, Nov. 15, 1990, 104 Stat. 2399). Section 14.5 discusses the sources, impacts and control of aviation emissions at the local and regional level.

At the global level, aircraft emissions contribute to climate change by increasing the levels of greenhouse gases such as CO_2 and H_2O in the troposphere and stratosphere. Aircraft NO_x emissions have indirect effects through production of tropospheric ozone (a warming effect) and through removal of methane (a cooling effect). In addition, contrails from aircraft engines directly and indirectly (through the formation of cirrus clouds) increase cloud cover and therefore alter the atmosphere's radiative forcing (tending to produce a net warming effect). Section 14.6 discusses the impact of aviation emissions on climate change.

There are several challenges to limiting the environmental impact of aviation. Aviation growth is correlated with economic growth. Placing inappropriate constraints on aviation may have negative consequences for local, national and world economies. But allowing environmental impacts to go unaccounted in consumer and producer behavior (i.e., allowing environmental externalities) also produces negative economic impacts. Thus, a balance must be struck. Balancing society's objectives for mobility and environmental quality is particularly challenging because aviation's environmental impacts are highly interrelated. For example, quieter engines may be heavier and therefore less fuel efficient, leading to an increase in emissions. And there are inevitable trade-offs among safety requirements and environmental and business goals. For example, aircraft are required to carry fuel reserves to ensure they can safely reach their destination – and beyond if necessary. These reserves are rarely used, but they add to the aircraft's weight thereby increasing fuel use (and thus increasing operating costs and environmental impacts). To effectively balance the need for mobility with the demand for environmental protection, actions in the commercial air transport arena must address a wide range of scientific, design and policy problems that require joint attention to noise, air quality and climate issues. Lastly, the timescales involved in changing aviation technology and hence environmental impact are very long. The introduction of aircraft and engine design changes to affect noise and emissions performance therefore has limited impact in the short term. New aircraft development typically takes 5 years or more to be proven commercially acceptable and certified, production runs on successful commercial aircraft may last 15 to 20 years, and aircraft service lifetimes average 25 to 35 years (NRC, 2002). Thus it can take up to 40 years to turn a fleet over to a new technology. In addition, the high capital costs of aircraft coupled with the high residual value of aircraft provide airlines with a strong disincentive to prematurely phase out or retire aircraft. For example, the cost of premature retirements to enable noise reduction has been estimated at between $5 billion and $10 billion in the USA alone Morrison *et al.*, 1999; GAO, 2000).

This chapter provides an introductory overview of civil aviation's impacts on water quality, community noise, local air quality, and climate change, and how these impacts can be mitigated. We begin with a review of the local, national and international regulatory bodies that address the environmental impact of aviation.

14.2   Limiting Aviation's Environmental Impact: The Role of Regulatory Bodies

There are three main ways to limit the impact of aviation on the environment: operational changes, technological changes and policy changes (that may or may not directly impact operations and technology). Operational changes include limiting flight hours (e.g., noise curfews) and requiring aircraft to fly in narrowly defined flight tracks. Technological changes include quieter or "cleaner" engines that produce less emissions. Policy changes include land-use controls around airports and financial measures such as landing fees, emissions taxes and emissions trading. While environmental goals represent only a fraction of the many interdependent safety and performance objectives in aircraft and engine design, the influence of environmental controls has been growing. Further, there have long been strong incentives for improved fuel efficiency. More fuel-efficient aircraft emit less CO_2 and therefore have less impact on global climate. Finally, passengers in general do not consider environmental performance important in selecting an airline – concerns such as on-time performance, availability of direct flights, and ticket price tend to trump environmental performance. Even when passengers do wish to select environmentally friendly airlines or aircraft, the necessary information is not easily available to those unfamiliar with the industry.

While environmental performance ranks behind safety and economics in priorities in the marketplace, there is a growing awareness of the important link between environmental constraints and aviation economics (e.g.,Waitz *et al.*, 2004; JPDO, 2006). This awareness is fueling heightened attention across the industry to addressing aviation's environmental impacts. It falls to local, national and international regulatory bodies to seek policies that balance society's desires for mobility and environmental quality.

The International Civil Aviation Organization (ICAO) is a United Nations body that attempts to harmonize international regulatory standards for aircraft noise and emissions by recommending appropriate standards that regulatory bodies in member nations around the world can adopt (NRC, 2002). With regard to the environmental impact of aviation, ICAO strives to limit or reduce the number of people affected by significant aircraft noise, the impact of aviation emissions on surface air quality, and the impact of aviation greenhouse gas emissions on the global climate (ICAO, 2005a). Within ICAO, the Committee on Aviation Environment Protection (CAEP) is responsible for setting standards relating to noise and emissions. However, this process has not led to a common set of rules around the world because of the actions of local and regional groups. Airports and airlines are subject to different standards imposed by local and regional bodies. For example, an aircraft may be subject to different operating restrictions based on its noise level at its origin and destination airports. These differing standards may lead to conflict because of the interdependencies between noise and emissions. Aircraft designed to meet stringent noise standards at one location may compromise emissions more generally.

Most nations have one or more governmental agencies that are responsible for managing the impact of aviation on the environment. In the USA, the Environmental Protection Agency (EPA) is responsible for the establishment and enforcement of US environmental protection standards. The Clean Air Act of 1973 (CAA), which was last amended in 1990, provides for minimum air quality standards for certain pollutants (the National

Ambient Air Quality Standards, or NAAQS) and requires states to implement a plan (state implementation plan, SIP) to achieve or exceed these minimum standards (Pub. L. 101–549, Nov. 15, 1990, 104 Stat. 2399). For aircraft, the enforcement responsibility for clean air standards is borne by the Federal Aviation Administration (FAA). The CAA preempts states from setting different aviation emissions standards (EPA, 2005). Since publication of the initial emissions standards in 1973, the FAA has worked with ICAO on the development of international aircraft emission standards (EPA, 2005). Finally, NASA is responsible for increasing the range of environmental mitigation options that are technically feasible through basic research (NRC, 2002).

The FAA also bears the responsibility for setting and enforcing aviation noise standards on behalf of the EPA. The Aircraft Noise Abatement Act of 1968 (49 U.S.C. 44709, 44715) tasked the FAA with developing and enforcing safe standards for noise generated by aircraft. Several subsequent acts have been used to promulgate additional aviation noise restrictions and distribute funding for mitigation (typically sound insulation for homes and land purchase) around airports. In particular, the Aviation Safety and Noise Abatement Act of 1979 provides assistance to airport operators to prepare and carry out noise compatibility programs (Pub. L. 96–193).

In Europe, the European Union monitors and limits emissions by means of Directives issued by the Environmental Council (European Commission, 2006a). The EU Directives aim to harmonize monitoring strategies, measuring methods, calibration and quality assessment methods across the EU. Previously, emissions standards were set at national levels. Aircraft noise levels in the vicinity of airports are regulated by the Joint Aviation Authority (JAA). In addition to these governmental bodies, the communities around airports are having an increasing influence on airport and aircraft operations. This influence is typically produced through public reaction to noise assessments and environmental impact statements that are required prior to airport development projects or airspace redesign.

14.3   Airport Water Quality Control

Airports affect the quality of water in surrounding waterways, rivers and wetlands primarily through surface discharge of contaminated water. The main sources of contaminants are (BAA, 2003):

- **Aircraft deicing** is the dominant source of biological oxygen demand (BOD) from airport activities, and is discussed in detail in this chapter.
- **Aviation fuel spills** can occur as a result of human error, faulty valves or fuel venting. Refueling spills tend to peak during warm periods when the sun's heat on aircraft wings can lead to overflows from fuel vents on full fuel tanks.
- Inadequate storage, handling or disposal of **oils and other chemicals** can also contribute to pollution. In the UK, The Control of Pollution (Oil Storage) (England) Regulations 2001 require containers to be strong enough to hold the oil without leaking or bursting and to be positioned in a place where they will not be damaged by vehicle collisions.
- Airports must also control plants and pests to ensure that runways and aircraft pavement areas are kept free of weeds for aircraft safety and to preserve the infrastructure. If airports are located in conservation areas they may also be required to manage invasive species populations. **Pesticides and herbicides** may run off into the groundwater and storm water drains and adversely affect plant and animal life.

- Water that has been used in **firefighting operations** may contain high concentrations of hydrocarbons and heavy metals released from burnt materials.
- Surface runoff from **ground transport** areas, including car rental bases, bus stops and bus stations, car parks and roads, can contain a wide range of potentially polluting chemicals such as oils, fuel and heavy metals.
- Leaks and spills from the **fuel farm and fuel supply infrastructure**.
- **Construction and contractors' compounds**.
- Aircraft are washed in three ways: dry wash, wet cosmetic wash and wet maintenance wash. Effluent from **wet washes** can contain heavy metals and hydrocarbons. Runoff from **washing of ground support vehicles** can contain oils, grease, detergents and solvents as well as heavy metals such as lead and cadmium. This runoff can be significant; for example, there are approximately 17 000 vehicles belonging to the airport, airlines and contractors on the airfield at Heathrow. Residue from **washing of aircraft stands** using a mix of soap and water can produce a small BOD load.

In regions where temperatures drop below the freezing point, aircraft surfaces must be deiced before takeoff to ensure that wing control surfaces can function and that the aerodynamic properties of the wing are not changed by ice. Ice layers of only a few millimeters can adversely affect aircraft aerodynamic performance (Valarezo *et al.*, 1993; Lynch and Khodadoust, 2001). Deicing and anti-icing of aircraft are therefore essential for safety. According to the US EPA, it takes on average 500–1000 gallons (2250–4500 liters) of deicing fluid to deice a large commercial aircraft (EPA, 1995). Over a year, a medium-sized airport may use more than 264 000 gallons of deicing fluid (Betts, 1999). In addition, iced-over or snow-covered runways, taxiways and aprons must also be deiced, using a combination of mechanical methods, deicing fluids and solid chemicals, to ensure safe landings, takeoffs and taxiing.[1]

Deicing and anti-icing of aircraft are most commonly done by applying aircraft deicing fluids (Type I ADFs) and aircraft anti-icing fluids (Type IV AAFs) (Switzenbaum *et al.*, 2001). ADFs and AAFs consist mainly of ethylene or propylene glycol mixed with several additives, such as corrosion inhibitors, flame retardants, wetting agents, pH buffers, dispersing agents and thickeners. These additives protect the aircraft and allow the glycol to cling to aircraft surfaces for longer periods of time, reducing the need for repeated deicing while aircraft await takeoff. The exact compositions of different ADFs and AAFs are trade secrets.

Clearing of airfield surfaces is achieved using chemical deicing agents (e.g., ethylene or propylene glycol, urea, potassium acetate, sodium acetate, calcium magnesium acetate, or an ethylene glycol-based fluid known as UCAR) to loosen snow and ice, which can then be removed by mechanical means (snowplows and brushes). Sand and salt (sodium or potassium chloride) are used less frequently because they may cause damage to aircraft.

14.3.1 Effects of Deicing Fluids

Deicing chemicals become an environmental concern when they run off aircraft and airfield surfaces and disperse into surface water. Approximately 75–80% of Type I ADF

---

[1] This section draws primarily on documents from the United States Environmental Protection Agency (EPA, 2002) and Transport Canada (2006), and on Switzenbaum *et al.* (2001).

is deposited on the deicing area pavement, and a further 15–20% of Type I and IV fluids is sloughed from aircraft surfaces during taxi and takeoff (Switzenbaum *et al.*, 2001). A portion of this fluid will eventually run off into the airport surroundings, and find its way into the groundwater and storm water drains if it is not captured. Groundwater pollution is particularly serious because it persists even when the source of pollution is removed (BAA, 2003). It can take many years before uncontaminated water flushes the pollution out. Although some glycol has been found in the air and in groundwater, deicing runoff is primarily dispersed into surface water by storm water discharges (Transport Canada, 2006). This runoff is exacerbated by rainfall, when accumulated product is simultaneously washed into airport surroundings. The impact of these chemicals depends on the weather conditions (e.g., levels of rainfall, temperature) and flow rates in rivers if present.

Surface and groundwater contamination directly affect plant and animal life and may eventually reach the drinking water supply. The EPA estimates that each year in the USA 21 million gallons of ADF is discharged into surface waters from airport and aircraft deicing operations (EPA, 2002). An additional 2 million gallons is discharged into publicly owned treatment works.

While ethylene and propylene glycol have low aquatic toxicity and do not persist in the environment for more than 3–20 days, they serve as a biological substrate ("food") for microorganisms (Miller, 1979; Transport Canada, 2006). These organisms consume oxygen in proportion to the amount of substrate present, and thereby deplete oxygen levels in water. The amount of oxygen these organisms take from the water is indicated by the BOD. High BOD levels kill aquatic plant and animal life. Thus, even though glycol is biodegradable, it can nevertheless result in environmental deterioration when discharged in sufficiently large volumes. In Canada the Canadian Environmental Protection Act (CEPA) limits glycol discharge resulting from aircraft deicing at airports to no more than 100 mg/l at the discharge point into receiving or surface water (Transport Canada, 2006).

Ethylene and propylene glycol are toxic to mammals and may cause or worsen neurological, cardiovascular and gastrointestinal health problems (EPA, 2002). Glycols may also lead to birth defects and can cause death directly when ingested in large doses. In the USA ethylene glycol is classified as a hazardous air pollutant, with the attendant requirements for monitoring and control (see Section 14.5).

The additives in ADFs can have high aquatic toxicity, and may increase the toxicity of glycols (Pillard, 1995). The primary additives from an environmental viewpoint are the triazoles, which are primarily used as yellow-metal corrosion inhibitors (Switzenbaum *et al.*, 2001). These corrosion inhibitors react readily with each other and with glycols to produce toxic by-products. Wetting agents, flame retardants, pH buffers and dispersing agents also have high aquatic and mammalian toxicities.

Liquids and chemicals used to clear airport surfaces also affect soil and water quality adversely. The impact of these chemicals is highly variable and depends on diverse factors, including drainage systems, topography, temperature, precipitation and the type, dilution and storage conditions of the chemicals used (Ramakrishna and Viraraghavan, 2005). There is still a significant lack of understanding of the health and environmental effects of surface deicing chemicals, other than sodium and chloride, and therefore little legislation limiting use of these chemicals. Sodium increases blood pressure and can contribute to cardiovascular, kidney and liver disease. In the USA, a Drinking Water Equivalent guidance level of 20 mg/l is considered as protective against the non-carcinogenic adverse health effects of sodium (EPA, 2002). Chloride corrodes pipes and adds a salty taste to water. Drinking water in the USA is required to have no more than 250 mg/l chloride (EPA, 2002).

Managing Airport Water Quality

The impact of deicing activities on the environment can be reduced by using lower-toxicity chemicals, using lower amounts of chemicals, or by controlling runoff (EPA, 2002). Alternative ADFs with lower toxicity, lower biological oxygen depletion and higher biodegradability include potassium acetate, sodium acetate, sodium formate and potassium formate. For example, Heathrow uses acetate-based fluid (potassium acetate) for airfield deicing (BAA, 2003). However, glycol-based fluids must still be used in extreme cold and wet weather periods when a fluid with a higher viscosity is needed.

The necessary glycol concentration depends on temperature – higher temperatures require lower levels of glycol. Storing multiple concentrations of glycol, or using systems that allow deicers to set the glycol concentration in ADF, can reduce the total level of glycol needed, and therefore the total runoff. At temperatures above $-2°C$, hot water alone may be sufficient for deicing (Switzenbaum et al., 2001).

ADF use on aircraft, as well as runoff, can be reduced by applying ADFs using computerized spraying systems that work in a similar manner to automatic car washes (EPA, 2002). These systems can reduce the total volume of fluid used and the time needed for deicing, but their cost may be prohibitive for small airports. In some cases, boom- or truck-mounted deicing may be needed for additional deicing in areas not easily reached by the automated system. Using boom-mounted deicers can also reduce the volume of ADF used by allowing operators to spray more closely to the aircraft surface, reducing overspray. Mechanical deicing technologies such as deicing boots, electrical resistive (using electrical resistors installed on the wing to generate heat) or infrared heating can also be used to reduce or eliminate the need for ADFs.

It is difficult for deicing operators and aircraft crews to detect ice on aircraft. Pilots use guidelines based on weather, type of deicing fluid used and the time elapsed since the last application to determine if repeated deicing is needed after initial at-gate deicing (Switzenbaum et al., 2001). Current practice is therefore to deice aircraft whenever there is any suspicion that ice might have formed. Ice detection systems that rely on magnetostrictive,[2] electromagnetic and ultrasonic devices can be used to detect ice and thus reduce or eliminate the need for deicing on an aircraft-specific basis (EPA, 2002). These systems may be too expensive for small aircraft, but may result in time and cost savings for operators of large aircraft.

The need for deicing of airfield surfaces can be reduced by pre-treating pavement prior to storms or icing conditions using agents such as potassium acetate (EPA, 2002). The FAA estimates that such treatments can reduce the need for pavement deicing and anti-icing agents by 30 to 75%. Deicing agent use can also be reduced by using mechanical methods to remove snow.

---

[2] Ferromagnetic metals such as iron, cobalt and nickel are magnetostrictive. These materials change shape when placed in a magnetic field. The converse is also true: extending or compressing these materials changes their magnetic properties.

Runoff of applied ADFs can be limited by use of deicing pads that capture runoff. Deicing pads are often located away from gates and used by multiple airlines, though the trend is to use a combination of at-gate and pad deicing. At-gate deicing is preferred by airlines, primarily because it interferes less with scheduling and allows them to use the same personnel to do the deicing and other gate tasks such as loading and unloading of aircraft (Switzenbaum *et al.*, 2001). Deicing pads may also become congested during inclement weather or peak traffic periods. However, deicing pads free up gates, thus allowing more intensive gate use. When deicing pads are located close to runways, they can also reduce the need for AAFs and repeated deicings. Glycol recovery vehicles and vacuum sweeper trucks can be used to collect ADFs, as well as snow, slush and salt from airfield surfaces. These vehicles may increase any congestion already present due to other airport ground vehicles and also add to local air pollution (see Section 14.5).

If deicing runoff is collected, it can be reused or disposed of in ways that have varying degrees of environmental impact. There are three primary options for treatment and disposal of deicing fluids: off-site treatment, on-site treatment and recovery (Switzenbaum *et al.*, 2001). Off-site treatment and disposal involves collecting runoff and then transferring it to off-site treatment facilities. On-site treatment uses aerobic or anaerobic facilities at the airport to treat and dispose of runoff. Recovery options include filtration, reverse osmosis and distillation to remove glycol from runoff. Recovery of glycols for reuse is, however, difficult because of quality control concerns. It is also expensive and may therefore only be applicable at large airports (EPA, 1994).

Aerobic systems are widespread in diverse municipal and industrial applications. These systems convert toxic chemicals into harmless end products quite effectively, but can generate large amounts of sludge for disposal. The simplest approach is detention basins or constructed wetlands that are used to "settle" runoff and reduce its BOD before it is allowed into surface and groundwater. For example, balancing reservoirs at Heathrow are used to allow the microbial breakdown of glycol to take place and levels of dissolved oxygen to recover before water is discharged into waterways (BAA, 2003).

Anaerobic processes use microorganisms to ferment toxic wastes into methane and carbon dioxide in the absence of oxygen. These processes have several advantages over aerobic processes, including lower sludge production, no need for oxygen, lower nutrient requirements, and the production of methane, which can be used for heating (McCarty, 1964). Disadvantages include slower processing of waste, the production of hydrogen sulfide, which has an unpleasant odor, and the disposal of methane, if it cannot be used.

14.4   Noise

Aviation noise results from a combination of the acoustic energy generated by unsteady fluid mechanical processes within an engine, unsteady interaction of the exhaust jet with the surrounding air and unsteady flow produced by the airframe. There are several different types of noise that have different types of impacts. Takeoff and approach noise are the primary source of community noise complaints. Taxi and engine run-up noise also result in community noise complaints. In quiet areas such as national parks, flyover noise from aircraft at cruise altitudes can be annoying and therefore lead to noise complaints (e.g., FICAN, 1992). Finally, sonic booms and hyperbooms (low-intensity thermospherically refracted remains of sonic booms) are not tolerated by most communities. For example,

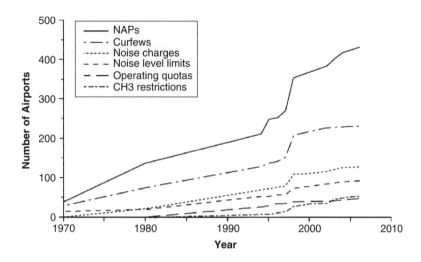

Figure 14.1   Growth in airport noise restrictions (NAPs = Noise Abatement Policies)
Source: Boeing (2006a)

community resistance to sonic booms led ICAO to place restrictions on the Concorde that prevented overland supersonic flight. Therefore the Concorde flew only between New York City and Heathrow or Paris. Current efforts to develop supersonic business jets are focusing on reducing sonic booms to a degree that communities would accept overland flight (see Section 14.4.5).

Over the past several decades there has been a 95% reduction in the number of people exposed to high levels of aviation noise in the USA, despite the six-fold growth in aviation (Waitz *et al.*, 2004). This dramatic reduction was made possible by improvements in technology. But resistance to noise is increasing worldwide, as evidenced by the increasing number of non-governmental organizations (NGOs) devoted to reducing aviation noise (NRC, 2002). Figure 14.1 shows the growth in noise-related operating restrictions in the USA over the past four decades (Boeing, 2006a). Communities surrounding airports react strongly to aviation noise and may delay or cancel airport expansion plans or impose operating restrictions (e.g., flight path restrictions and curfew hours during which operations are not permitted).

The failed runway addition at Logan International Airport in Boston illustrates how community objection to noise can stymie airport expansion efforts. In the early 1970s airport authorities attempted to add a new runway (33R/15L) parallel to an existing runway (33L/15R) (NRC, 2002). Construction was already well under way when community members blocked the bulldozers preparing the runway and prevented completion of the runway. The court subsequently forbade the construction of this or other additional runways. This injunction was finally lifted in November 2003. Runway 33R/15 L exists, but is only 2557 feet (779 m) long and therefore too short to land large jet aircraft. Like other airports, Logan has experienced a dramatic growth in operations, from 300 000 takeoffs or landings in 1975 to 500 000 in 2000. To support this increase in traffic, the airport proposed an alternative 5000-foot runway to service commuter and light aircraft. This runway will allow aircraft to approach primarily over Boston Harbor, rather than over the densely populated communities nearby. This runway was also opposed by the local community but the state's Supreme Judicial Court overturned the last legal objections in 2004. The runway, referred to as runway 14/32, was opened for operations in November 2006.

This section first discusses the effects of noise on local communities and the environment. Next, techniques for measuring noise are presented. These measurements are used to show noise trends, as discussed in Section 14.4.3, and form the basis of noise regulation policies, as discussed in Section 14.4.4. This section concludes with a discussion of emerging issues in noise, such as the potential impact of low-frequency noise.

14.4.1 Effects

Community actions such as those experienced at Logan International Airport contribute to airport congestion, increase travel delays, and may therefore increase airline capital and operating costs and hence ticket prices (NRC, 2002). In the USA, noise is the biggest concern for airport officials at the 50 busiest US airports, as shown in Figure 14.2.

At sufficiently high levels, noise leads to auditory damage. Hearing loss is a concern for airport support personnel who work in close proximity to aircraft idling on the ground or taking off and landing. Residents of communities surrounding airports do not suffer hearing loss as a result of aircraft noise, but noise nevertheless results in a variety of adverse physiological and psychological responses. Specifically, noise disturbs sleep and interferes with speech, both of which may lead to reduced productivity in learning and work (e.g., Shield and Dockrell, 2008; FICAN, 1992; Matsui *et al.*, 2004; Raschke, 2004; Matheson *et al.*, 2003; Haines *et al.*, 2001). Studies have linked noise to non-auditory health effects, such as hypertension in children and adults (Jarup *et al.*, 2008) and heart disease (Ising and Kruppa, 2004).

Noise can also be expected to adversely impact wildlife. Studies have attempted to determine the impact of aircraft noise on over 100 different species of domestic and wild

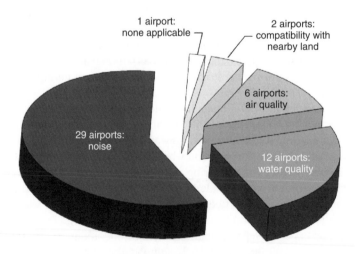

Figure 14.2   Environmental issues that most concern officials at the 50 busiest US airports
Data source: GAO (2000)

mammals, birds and marine mammals (Waitz *et al.*, 2004). In the absence of definitive data on the effect of noise on animals, the US National Research Council has proposed that the same protective noise criteria be used for animals as for humans (NRC, 1995).

There is strong evidence that airport noise reduces property values (e.g., Nelson, 2004; Schipper *et al.*, 2001). The effect of noise on property values is usually estimated using the Noise Depreciation Index (NDI), which is defined as the percentage reduction in housing capital for an increase in 1 dB of perceived noise. Typical NDIs range from 0.5% to 1.5% change in housing value per dB of perceived noise for noise levels above about 50 dBA. A house in a region of high aviation noise may experience 10% to 15% devaluation compared to a house in a region with low or no aviation noise.

14.4.2 Noise Measurement

There are a variety of metrics available for assessing the noise and emissions performance of aircraft and the transportation systems in which they operate. Some metrics are appropriate for understanding trends in technology, while others are more useful for evaluating environmental impact; others are useful for communicating with the public (cf. Eagan, 2007). The most widespread measure of adverse reactions to living in noisy environments is annoyance, which is a generalized and subjective measure that, by definition, overlaps with sleep disturbance and speech interference. There are several well-established procedures and metrics for relating instrument sound measurements to human annoyance. These procedures and metrics take account of the non-uniform response of the human ear in both frequency and amplitude, sensitivity to tonal versus broadband noise, and how annoyance with a particular source relates to the background noise level. For a single aircraft operation these effects are usually represented by the effective perceived noise level (EPNL), measured in decibels. EPNL forms the basis of noise certification standards for commercial aircraft set under the Noise Control Act (NCA) and subsequent amendments.

Figure 14.3 shows noise levels for commercial aircraft presented in terms of EPNL for a measurement taken during takeoff 450 m to the side of the runway (except for some four-engine aircraft, for which measurements were taken using the earlier 650 m standard). There has been a general decline in noise levels from 1960 to 1995.

When assessing the impact of an airport and all the associated aircraft operations on the local community, it is useful to consider a summative measure of the noise produced by flight operations over some period of time, such as the day–night noise level (DNL). The DNL has been shown to correlate well with community annoyance from aircraft noise. DNL is calculated as the A-weighted sound energy (i.e., the averaging accounts for unequal loudness perception across the audible frequency spectrum) averaged over a 24-hour period. A 10 dB penalty is added for nighttime events, under the assumption that nighttime events are twice as annoying as daytime events because of the increased potential for sleep disturbance and because background noise tends to be lower at night (cf. Jarup *et al.*, 2008). Table 14.1 summarizes community response to noise as characterized by DNL (FICAN, 1992). For comparison, the range of exposure to noise in urban areas is typically 58 to 72 dB, and in suburban and wilderness areas the ranges are 48 to 57 dB and 20 to 30 dB, respectively.

While the DNLs correlate well with noise annoyance in most areas, they do not take into account annoyance that occurs when aviation noise is experienced in relatively quiet areas, such as small communities, national parks or farmland. In such areas, even the

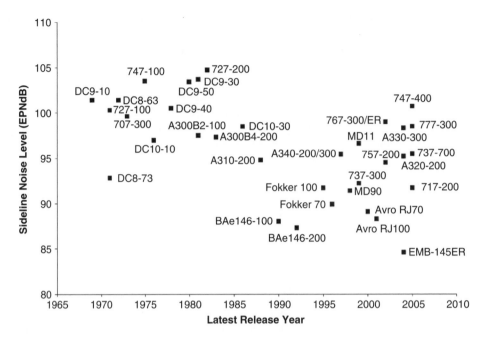

Figure 14.3 Effective perceived noise level for commercial aircraft as per certification data
Data source: FAA 2001

Table 14.1 Residential response to noise levels described by DNL

DNL (dB)	Hearing loss	Annoyance		
		Percentage of population highly annoyed	Average community reaction	General community attitude toward area
75 and above	May begin to occur	37	Very severe	Noise is likely to be the most important of all adverse aspects of the community environment
70	Will likely not occur	22	Severe	Noise is one of the most important adverse aspects of the community environment
65	Will not occur	12	Significant	Noise is one of the most important adverse aspects of the community environment
60	Will not occur	7	Moderate to slight	Noise may be considered as an adverse aspect of the community environment
55 and below	Will not occur	3	Moderate to slight	Noise may be considered as an adverse aspect of the community environment

Data source: FICAN (1992)

relatively low noise levels of aircraft at 35 000 ft may cause annoyance (e.g., Lim *et al.*, 2008). For example, Denver International Airport was constructed at least 10 miles (16 km) from residential areas (NRC, 2002). When the airport began operating, complaints were received from Boulder (40 miles away), Ward and Lakespur (60 miles away) and Peyton (85 miles away), among other areas.

Noise annoyance is also related to an airport's ability to restrict deviations from the "normal noise experience" (NRC, 2002). For example, Atlanta's Hartfield International Airport, one of the world's busiest airports, receives relatively few noise complaints, in part because aircraft keep to narrowly defined flight tracks, thereby minimizing the variation in the noise level that local communities have come to expect. Similarly, changes in flight tracks may result in relatively more complaints from newly impacted communities.

Another approach to measuring noise annoyance is according to the number of noise complaints received by an airport. Higher levels of annoyance should result in higher levels of complaints, but this is not always the case. Complaints are mediated by factors such as personality, ease of making complaints and expectation of response (Maziul *et al.*, 2005). Relying on complaints as a measure of annoyance may therefore lead to poor estimates of annoyance levels.

14.4.3 Noise Trends

Figure 14.4 shows the historical evolution of commercial air transport-related noise exposure in the USA as well as future projections developed by the FAA. The latest estimates indicate that approximately 500 000 people reside within the 65 dB DNL contours of commercial airports in the USA (NRC, 2002). Over the past 30 years or so, the

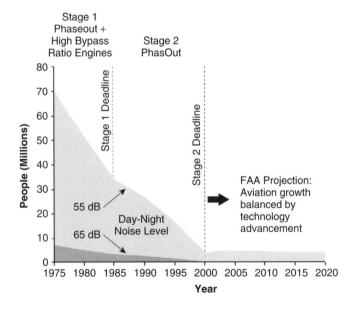

Figure 14.4 Estimated number of people exposed to commercial aircraft noise in the USA
Source: FAA (1997)

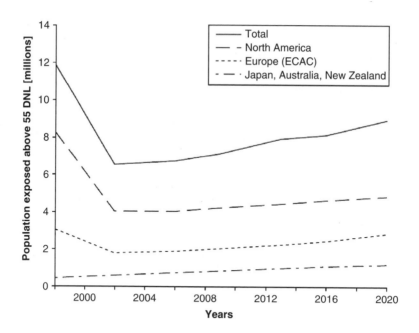

Figure 14.5   Predictions of noise exposure trends, 2004
Source: IATA (2004)

number of people living within the 55 dB DNL contour has been reduced by 95% (NRC, 2002). These large reductions in population exposure are due primarily to operational and engine technology improvements. Lower-noise aircraft operations were enabled by advances in aircraft communication and navigation capabilities and in surveillance and air traffic management technology. High-noise aircraft were phased out in two stages through regulatory action enabled by the availability of improved engine technology. A large reduction in noise in the late 1960s and throughout the 1970s was brought about by the introduction of the turbofan engine. This engine was introduced primarily for its increased fuel efficiency – decreased noise was an ancillary benefit (NRC, 2002). More evolutionary changes in the 1980s and 1990s, including higher bypass ratio engines and improved acoustic liner technology, contributed to a further, though less dramatic, decrease in noise.

Recent predictions from CAEP place the total North American, European, Australian, New Zealand and Japanese populations exposed to aviation noise at about 9 million people, as shown in Figure 14.5 (ICAO, 2004). The dramatic decrease shown in noise exposure between 1998 and 2002 is attributed to the accelerated retirement of older, noisier, aircraft.

The principal limitations to further reductions in noise exposure are the increase in traffic due to increased demand, and long transition periods to new technology, as discussed at the beginning of this chapter. Over the next 20 years, FAA and CAEP estimates suggest that the number of people affected by noise (i.e., people falling within the 55 dB DNL contour) will increase because improvements in technology are expected to be more than offset by an increase in operations.

Despite these difficulties in limiting noise exposure, there is potential for technological and operational improvements to reduce aircraft noise (cf., Casalino *et al.*, 2008). For example, the European Union has established research programs whose aim is to reduce perceived noise from new aircraft to one-half of the average 2001 levels by 2020 (ACARE, 2001). Research is also ongoing at NASA, but most recently at a lower funding level than has historically been the case. Research within the FAA currently focuses on the development of better metrics and tools to assess aviation noise impacts, and on the development and implementation of operational procedures to mitigate aviation noise.

14.4.4 Noise Controls

It is widely recognized that a multi-pronged approach to noise reduction is necessary, with the greatest near-term opportunities existing with operational procedures, and reductions in source noise (airframes and engines) being required in the long term for further progress. In 2001, CAEP developed a comprehensive series of recommendations to reduce the environmental impact of aircraft noise in a cost-effective manner. The committee also endorsed a balanced approach to noise mitigation, focusing on: (i) noise reduction at the source; (ii) land-use planning and management; (iii) noise abatement operational procedures; and (iv) operating restrictions on aircraft use (ICAO, 2005a). Part of this approach is a new noise standard that is 10 dB lower than current standards on a cumulative basis (i.e., the sum of the noise levels for the three certification points is 10 dB lower than current standards).

National regulatory bodies such as the FAA generally follow noise restrictions established by ICAO and CAEP in developing noise standards. Aircraft are divided into stages (in the USA), or chapters (in Europe), with noise decreasing at higher stages/chapters. This classification scheme allows successively stricter noise controls to be applied by defining higher stages as noise reduction technology improves. Stage 1 aircraft have been phased out of commercial operation since 1984 (Boeing, 2006b). Unmodified Stage 2 aircraft over 75 000 lb have been staged out since 1999. Stage 3 aircraft are required to meet separate standards for runway takeoffs, landings and sidelines, which range from 89 to 106 dBA according to the aircraft's weight and number of engines. Stage 4 noise limits became active in January 2006 and require that in addition to meeting Stage 3 limits, noise in each flight phase be cumulatively 10 dB lower than the existing Stage 3 limits. The Stage 4 standard applies to newly certified aircraft and to Stage 3 aircraft that are recertified to Stage 4.

While federal and industry investments can be applied to reduce aircraft noise, local authorities typically control land-use decisions near airports. There are many examples where national and international land-use guidance designed to mitigate impacts has not been followed by local authorities, thus exacerbating noise issues (GAO, 2000). For example, in the US several airports were relocated to initially sparsely populated areas (e.g., Dallas–Fort Worth International Airport, Naval Air Landing Field Fentress and Denver International Airport) in order to limit noise complaints. But local land-use decisions resulted in increased construction of housing and other facilities near the airports and a concomitant increase in noise complaints.

While some communities have taken active roles in addressing land-use issues near airports (e.g., by establishing building codes and guidelines for sound insulation of new homes, and by providing interactive tools and property locators to enable communities to better understand noise levels in particular locations), inconsistencies or conflicts often exist between national and international aviation policy and local land-use decision making. In the USA, most federal funding available for addressing noise is being used for noise mitigation at selected airports, primarily by means of sound proofing and insulation of buildings in high-noise areas (currently $0.3 billion per year derived from ticket taxes) or by purchasing land to extend airport boundaries around high-noise areas. Much smaller amounts of federal funding are available for research and development to promote the technological advances that are ultimately required for long-term progress (NRC, 2002).

Local actions can promote noise control when they focus on reducing noise at the source. Many European airports have their own noise rules and provide financial incentives to airlines to use quieter aircraft by penalizing noisy aircraft. Noise rules are commonly based on margins relative to Stage/Chapter 3 standards (Aviation International News, 2004). For example, in France, Lyon's Saint Exupéry Airport places a curfew on aircraft within 5 dB of Chapter 3 limits from 22:00 to 06:00 (excluding operations mandated for safety reasons).

14.4.5 Emerging Issues

This section discusses two emerging issues in aviation community noise impacts: low-frequency noise and low-boom supersonic aircraft.

Low-Frequency Noise

Recently interest has resurged in the potential impacts of low-frequency noise from aircraft on local communities. Low-frequency noise is acoustic energy at frequencies below 50 Hz, thus falling below the noise spectrum typically considered in human annoyance metrics such as EPNdB and DNL. Low-frequency noise issues have been raised at several airports in the USA, most notably Minneapolis–St. Paul International Airport, as well as at San Francisco International Airport, Baltimore–Washington International Airport, Boston Logan International Airport and Los Angeles International Airport.

Several studies have identified links between low-frequency noise, sleep disturbance and annoyance (e.g., Berglund *et al.*, 1996; Persson, 2004). However, these studies focused on non-aviation sources of low-frequency noise; to date little has been published directly linking low-frequency noise from commercial aviation to sleep disturbance and annoyance (e.g., Schust, 2004; Leventhall, 2004; Findeis and Peters, 2004). Low-frequency noise has not been shown to have a direct adverse effect on health, although sleep disturbance may have adverse health effects, as discussed previously. Low-frequency noise also contributes to noise annoyance by causing structures such as buildings to rattle audibly, though it does not cause structural damage (FICAN, 2002; Hubbard, 1982).

There is no widely agreed upon metric for measuring low-frequency noise and predicting noise-induced rattle (FICAN, 2002). Further research is needed to assess the impacts of aviation-induced low-frequency noise and develop mitigation strategies if necessary.

Low-Boom Supersonic Aircraft

Community resistance to the sonic booms generated by supersonic aircraft has limited the use of supersonic commercial aircraft. For example, Concorde flights were discontinued in

2003 due in large part to operating restrictions that made it difficult to operate the Concorde profitably. Recently there has been a resurgence in interest in developing a new generation of "low-boom" supersonic commercial aircraft. These "low-boom" aircraft would have significantly reduced sonic booms that would not preclude overland operation. Research efforts to reduce sonic booms are focused on airframe modifications that will result in a boom pressure signature that is less annoying than the typical "N-wave" created by most supersonic aircraft. However, sonic booms are not the only environmental concern for supersonic flight. Supersonic aircraft also tend to be less fuel efficient than subsonic aircraft and therefore generate more emissions.

14.5 Surface Air Quality

Although noise is currently the primary environmental constraint on airport operations and expansion, many airports either put air quality concerns on an equal footing with noise or anticipate they will be on an equal footing soon (GAO, 2000). While the connection between noise and human health is somewhat unclear, emissions are known to have a direct impact on human sickness to lead to an increased risk of premature death. In addition, the link between emissions and climate change (which in turn affects human and ecosystem health) is becoming ever clearer with continuing research (see Section 14.6). Emissions impacts occur over several different timescales. Air quality is immediately affected and varies on a daily basis with emissions volumes, while health impacts may take longer to emerge and tend to persist for longer periods. The impact of emissions on climate change persists for decades or centuries (see Section 14.6).

In the USA, air quality has steadily improved since the 1970s as a result of the CAA and related amendments, which have led to reductions in pollution from most sources (EPA, 1999a; EPA, 2002). However, many of the technologies employed for land-based sources are not applicable to aircraft because of the more severe weight, volume and safety constraints. Thus, although aviation is a small overall contributor to air quality impacts, some aircraft emissions are growing against a background of generally decreasing emissions from other sources (Waitz *et al.*, 2004). Airports are one of the largest single sources of pollution in some US counties, as shown for example by Figure 14.6, which shows the contribution of aviation to NO_x emissions in 10 major cities.

Like most fossil fuel combustion sources, aircraft engines emit several different chemical species that have an impact on health and ecosystems, including carbon dioxide (CO_2), water vapor (H_2O), nitrogen oxides (NO_x), unburned hydrocarbons, such as formaldehyde and benzene (UHC),[3] carbon monoxide (CO), sulfur oxides (SO_x), other trace elements including the extended family of nitrogen compounds (NO_y) and hazardous air pollutants (HAPS), and both volatile and non-volatile particulate matter (PM), primarily PM2.5 (particulate matter smaller than 2.5 micrometers). *Primary* PM includes dust, dirt, soot, smoke and liquid droplets directly emitted into the air. Particles formed in the atmosphere by the condensation or transformation of emitted gases such as NO_x, SO_2 and UHCs are also considered particulate matter, and are referred to as *secondary* PM. Recent estimates suggest that these sources for secondary PM can be quite significant, with impacts equal

---

[3] Volatile organic compounds (VOCs) are a class of hydrocarbons emitted from certain solids or liquids. In the aviation context, VOCs are released both by unburned fuel and also as a by-product of fuel combustion.

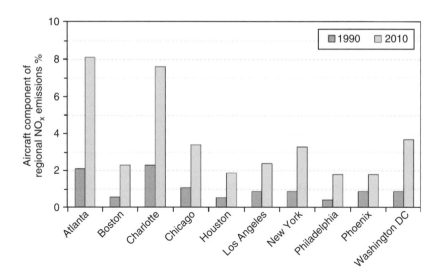

Figure 14.6 Aircraft component of total regional emissions, 1990 and projections for 2010
Data source: EPA (1999b)

to or greater than those of the direct PM emissions (see. e.g., Rojo, 2007). In addition, fuel venting directly introduces hydrocarbons into the atmosphere. Aircraft engine emissions consist of about 70% CO_2 and just under 30% H_2O, with the remainder of pollutants making up less than 1% of the total (FAA, 2005). The bulk (about 90%) of emissions is emitted at high altitudes, where the emissions contribute to atmospheric changes that can affect climate (see Section 4.6). About 10% of emissions occur at surface levels during ground operations, landing and takeoff. Emissions at this level are known to directly impact surface air quality. There is also emerging research that suggests cruise emissions may impact surface air quality.

The relative proportion of each emission is influenced by different factors. CO_2 and H_2O are products of hydrocarbon combustion and the amount of these gases emitted is therefore directly related to the *amount* of fuel consumed, which is in turn a function of aircraft and engine fuel efficiency, as well as the length of time that an aircraft's engines or auxiliary power unit (APU) are running. The total amount of sulfur species emitted is related to the concentration of sulfur in the fuel (controlled by a fuel *specification*) and to the amount of fuel burned. Emissions of NO_x, NO_y, non-volatile and volatile PM, CO and UHC are related to the *manner* in which fuel is combusted and to post-combustion chemical reactions occurring in the engine. These emissions are therefore a function of engine design in addition to overall fuel burn.

NO_x emissions can be difficult to reduce because of the high temperatures and pressures used to increase efficiency and thrust per unit mass flow. Thus, some trade-offs exist between NO_x emissions and CO_2 and H_2O emissions. In terms of the combined effects of aviation emissions on climate and air quality, it is not obvious whether increased NO_x should be traded in return for decreased CO_2 and H_2O, or vice versa. Finally, PM and UHC emissions are also affected by fuel composition. Table 14.2 summarizes the principal factors determining emissions rates.

Table 14.2 Chemical species resulting from engine combustion and their health effects

Chemical species	Principal determining factors	Health impact
CO_2	Fuel consumption	No direct health impact
H_2O		
UHC	Engine design, fuel composition	Contributes to smog and ozone formation at surface levels
VOC	Engine design, fuel composition	Temporary health effects such as nausea, fatigue and throat irritation. Certain VOCs are categorized as hazardous air pollutants and are carcinogenic
SO_x	Fuel consumption, fuel composition	Sulfur dioxide: High concentrations of SO_2 adversely affect breathing and may aggravate existing respiratory and cardiovascular disease. Asthmatics, sufferers of bronchitis or emphysema, the elderly and children are the most sensitive to SO_2. Moreover, SO_x undergoes chemical and microphysical processes in the atmosphere eventually forming ammonium sulfate aerosols which contribute to ambient particulate matter concentrations (with health affects as described for PM)
NO_y	Engine design	Contributes to smog and ozone formation. NO_2 can irritate the lungs, cause bronchitis and pneumonia, and lower resistance to respiratory infections. NO_x also undergoes chemical and microphysical processes in the atmosphere, eventually forming ammonium nitrate aerosols which contribute to ambient particulate matter concentrations (with health effects as described for PM)
NO_y		Contributes to smog and ozone formation
CO		Contributes to smog and ozone formation. CO is taken up by the body like oxygen and therefore limits the delivery of oxygen to the body's organs and tissues. Exposure to high levels of CO can impair visual perception, manual dexterity, learning ability and the performance of complex tasks. CO poses the most serious threat to those who suffer from cardiovascular disease, particularly those with angina or peripheral vascular disease

(continued overleaf)

Table 14.2 (*continued*)

Chemical species	Principal determining factors	Health impact
Primary PM Secondary PM	Fuel combustion, engine design Secondary pollutant (formed from SO_x NO_x, VOC, ammonia and other compounds)	Contributes to smog formation. PM has adverse effects on the respiratory system, is a carcinogen, and may lead to premature death. It damages lung tissue, aggravates existing respiratory and cardiovascular disease, and produces alterations in the body's defense systems against foreign materials. Individuals with chronic obstructive pulmonary or cardiovascular disease or influenza, asthmatics, the elderly and children appear to be most susceptible to the damaging effects of PM
Ozone	Secondary pollutant (resulting from NO_x, VOC, and in some cases CO, interaction)	Ozone causes health problems by damaging lung tissue, which reduces lung function and sensitizes the lungs to other irritants. Ozone affects not only people with already impaired respiratory systems, such as asthmatics, but also healthy adults and children (EPA Green Book, 2006)

Source: [EPA, 2006]

The *direct* emissions mentioned above also contribute to the formation of additional chemical species, or *secondary* emissions, such as aerosols, some types of particulates, and ozone (O_3) at surface levels and in the troposphere. While ozone in the stratosphere is desirable, protecting us from ultraviolet radiation, it is also a pollutant that contributes to smog at the lowest atmospheric levels, and to climate warming in the troposphere. Airports typically contribute 2–5% of the NO_x to US county-level inventories (FAA, 2005).

Aircraft are not the only sources of aviation-related emissions. Airport support operations, such as airport shuttle busses, stationary power sources, construction equipment and passenger traffic in and out of the airport, also contribute to an airport's total emissions budget.

14.5.1 Effects

The chemical species mentioned above affect the environment at local and global levels directly and also through reactions with other chemicals in the atmosphere. Table 14.2 summarizes the local health and environmental impacts of aviation-related emissions. Refer also to NRC (2002) for an extensive discussion of the impacts of each chemical species. Section 14.6 discusses the impact of emissions on global climate.

There is still uncertainty about the relative importance of aviation emissions in terms of health impacts. However, two recent studies indicate that the largest impacts can be ascribed to secondary PM formed from SO_x emissions, followed by secondary PM formed from NO_x, and then primary PM emissions (Rojo, 2007; Sequeira, 2008).

14.5.2 Emissions Measurement

As with noise, there are several ways of characterizing aircraft emissions. Trends in technology with regard to emissions performance of aircraft and the transportation systems in which they operate are typically measured in terms of the amount of pollutant released per some measure of aircraft operation. For example, NO_x emissions are a strong function of engine pressure ratio and the overall rated thrust output of the engine. Technology performance is typically evaluated in terms of the mass of NO_x emitted per unit thrust, and placed on a sliding scale in terms of engine pressure ratio. This metric is the current basis for NO_x certification standards for engines.

The impact of aviation on air quality is more appropriately measured using some cumulative measure of emissions over time. For example, the impact of NO_x emissions on local air quality is directly related to the total mass of NO_x emitted by all aircraft flying in a particular region (e.g., kg NO_x per day).

Population exposure to pollutants is measured by monitoring the ambient air quality, which is characterized by the concentrations of pollutants in the air. For example, ozone presence is measured in parts per million (ppm). Measurements are typically taken several times in a given time period to ensure that changes in levels during the time period are observed.

The total impact of pollutants on populations is more difficult to measure because the effects of pollutants are dependent on population demographics, may take years to appear, and are difficult to disconnect from other sources of negative health impacts (EPA, 1999a). Nonetheless, a wide range of epidemiological evidence has been collected which has formed the basis for concentration–response relationships used in the USA and Europe (BenMAP, 2005; European Commission, 2005); these relate ambient concentrations of pollutants to increased risks for various health consequences including premature mortality.

14.5.3 Emissions Trends

Aviation currently uses approximately 2–3% of fossil fuels in the USA, or about 14% of fossil fuel used in the transportation sector (USDOT, 2008). Aviation's share of fossil fuel use may increase in the US and Europe as other industries turn increasingly towards alternative energy, such as hybrid car engines and wind power – although alternative fuels are also being investigated for application to commercial aviation (e.g., IPCC, 1999, Chapter 7).

Technological advancement has significantly reduced aircraft engine fuel consumption and emissions – fuel consumption per passenger kilometer has decreased by 70% over the past four decades, as illustrated by Figure 14.7 (ICAO, 2005a; Lee et al., 2004) (in the figure we show fuel energy used to move a passenger a unit distance in terms of megajoules per revenue passenger kilometer). Aircraft operations – airports served, stage lengths flown and flight altitudes – also have a significant impact on fuel efficiency, as shown in Figure 14.8 (Babikian et al., 2002).[4] Regional aircraft generally fly shorter stage

---

[4] The y axes on the figure use "energy intensity," which is closely related to fuel efficiency. Energy intensity is a measure of how much energy is required to move one passenger 1 kilometer. It is typically measured in megajoules per ASK or RPK.

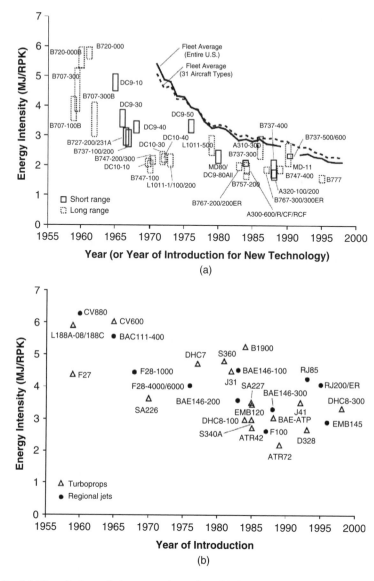

Figure 14.7 (a) Historical trends in energy intensity of US large commercial fleets; (b) historical trends in energy intensity of US regional fleets

Source: Lee *et al.* (2004), reproduced with permission

lengths than large aircraft, and therefore spend a greater fraction of their engine-on time in non-optimum, non-cruise stages of flight like taxiing.

Emissions can be reduced by using improved engines that release smaller amounts of pollutants per pound of thrust, improved airframes that require less thrust to transport a given load a given distance at a given speed, and improved operational procedures that enable more efficient use of aircraft (NRC, 2002). Over the past 50 years or so,

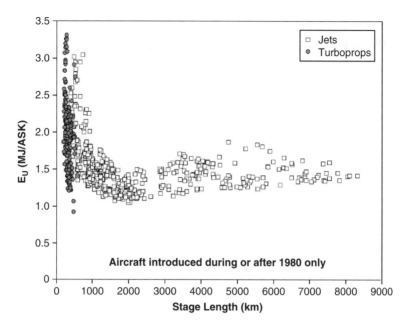

Figure 14.8   Variations in energy intensity with stage length
Source: Reprinted from: *Journal of Air Transport Management*, 8:6, R. Babikian *et al*, The Historical Fuel Efficiency Characteristics of Regional Aircraft from Technological, Operational and Cost Perspectives, 389–400, 2002, with permission from Elsevier

there have been major advances in engine turbine design as a result of concerted efforts between engine manufacturers and government agencies in the USA and Europe. These advances have increased fuel efficiency as well as engine performance, durability and reliability. Approximately 60% of the improvement in fuel efficiency can be attributed to advances in engine technology, with the remainder attributable to improvements in aerodynamics and changes in operations (most particularly increased load factors) (Lee *et al.*, 2001). Over the same period there were also significant advances in airframe design and fabrication enabled by the development of strong lightweight materials and structures and computer-aided design tools that optimize aerodynamic efficiency. Because several species of emissions (primarily CO_2 and H_2O) are directly related to fuel consumption, increases in fuel efficiency have the fortunate side effect of reducing those species as well.

In contrast, NO_x emissions, which depend mainly on the temperature and pressure in an engine, are the most difficult to limit because modern aircraft engines operate at high temperatures and pressures to increase fuel efficiency and thrust per unit engine weight. Figure 14.9 shows that while emissions of all other species have declined over time, NO_x emissions continue to pose a challenge. Combustors that reduce NO_x by as much as 60% below international standards are now available (NRC, 2002). But some estimates suggest that engines with these combustors will be heavier and more expensive to operate than those with simpler combustors that reduce NO_x to about 35% below current standards (CAEP/6-IP13). In the absence of strong incentives or regulations, airlines are

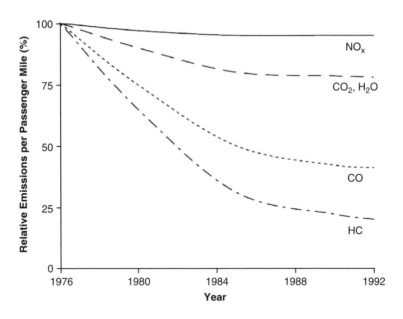

Figure 14.9   Relative aircraft emissions from 1976 to 1992
Data source: FAA (2005)

unlikely to select the more expensive combustors. However, there are alternatives for reducing NO_x that do not require trade-offs with fuel efficiency: various improvements in combustor technology and airframe aerodynamics and weight have led to reductions in NO_x emissions without negative effects on fuel efficiency (Waitz *et al.*, 2004; ICAO, 2005a).

The contrast between the dramatic CO_2 and H_2O reductions and the less significant NO_x reductions illustrates that environmental improvements are most likely to occur when they are directly coupled to economic performance, such as fuel efficiency. There are, however, many opportunities for technological and operational improvements to reduce emissions of NO_x, UHC, CO and PM. For example, studies indicate that changes in air navigation and air traffic management could yield as much as a 12% reduction in fuel use worldwide (IPCC, 1999). The European Union (EU) hopes to achieve an 80% reduction in NO_x emissions by 2020, through a combination of increased fuel efficiency and engine design advances (ACARE, 2001). Research and development programs are also being pursued within NASA, although currently at a lower funding level than has been the case historically. The FAA is developing several promising operational opportunities for reducing fuel burn and emissions such as modified takeoff and landing procedures, and modernization of the air traffic management system to reduce en route and ground delays. These options for reducing emissions present major engineering, safety and cost challenges that must be overcome before they can be implemented in airline fleets.

Long development times and service lifetimes mean that the benefits of improved technology are achieved very slowly at airline, national and international fleet levels. In addition, fuel efficiency may vary by as much as 30% about average levels as a result of variations in operating conditions such as load factor, flight speed, altitude and routing

(Lee *et al.*, 2004). Based on past trends, improvements in newly designed aircraft are expected to reduce fuel consumption per revenue passenger kilometer by about 1% a year over the next 15 to 20 years (NRC, 2002). Such increases in efficiency and the associated reductions in emissions would not be sufficient to offset the expected increase in traffic of 3–5% a year over the same time period (IPCC, 1999; Lee *et al.*, 2001; NRC, 2002).

14.5.4 Emissions Controls

CAEP has established emissions measurement procedures and compliance standards for soot (PM), UHCs, CO and NO_x to limit emissions around airports since 1981 (EPA, 2005). ICAO has approved four terms of reference for setting standards: standards must be technologically feasible, economically reasonable, environmentally beneficial, and balanced to account for interrelationships between emissions and noise (EPA, 2005, quoting ATA). This approach aims to develop realistic regulations and standards that take account of current, medium- and long-term engine technology availability, while weighing the relative benefits of reductions in CO_2 vs. NO_x and providing a compromise between community benefit and implementation costs.

Because most operations in the vicinity of airports involve taking off and landing (as opposed to flying at cruise altitude), ICAO defines a landing and takeoff cycle to characterize the operational conditions of an aircraft engine within the environs of an airport. The emissions standards are applied to all newly manufactured turbojet and turbofan engines that exceed 26.7 kN rated thrust output at International Standard Atmosphere (ISA) sea-level static (SLS) conditions.

Emissions have been regulated in the USA since the use of gas turbine engines became widespread about three decades ago (NRC, 2002). Engine emissions standards are usually formulated in a way that ensures consistency with international standards. The first federal regulations to limit the effects of aircraft on air quality were issued in 1973 by the FAA and the EPA. These regulations required the use of low-smoke combustors and banned intentional fuel venting after normal shutdown. In accordance with the CAA, the EPA Office of Air Quality Planning and Standards (OAQPS) sets National Ambient Air Quality Standards for six "criteria" pollutants: PM2.5, PM10, CO, SO_x, NO_2 and lead. With the exception of lead, all the criteria pollutants are emitted in the course of commercial aviation-related activities. Most recently, in 1997, the EPA set standards for NO_x and CO emissions on newly manufactured engines (EPA, 2005). Currently almost all aviation-related emissions sources are regulated through equipment-specific regulations, standards and recommended practices, and operational guidelines established by a variety of organizations (FAA, 2005). For example, stationary sources of emissions at airports, such as power boilers and refrigeration chillers, are required to meet various state regulations. FAA certification is required for virtually all aviation equipment (e.g., engine design, instrumentation and control) and processes (e.g., landing procedures). Since 1997, airport construction plans requiring FAA approval have had to show that both direct and indirect emissions resulting from the project would be consistent with state implementation plans (SIPs) for meeting air quality standards.

Many airports in the USA have also instituted voluntary programs defined at the national or airport level to reduce emissions from ground support and other vehicles (FAA, 2005). Airports are experimenting with hybrid electric, compressed natural gas and alternative fuel vehicles. In particular, airport shuttle buses have been converted to natural gas at a number of airports. Recently, the FAA and EPA instituted an initiative to reduce emissions at airports in non-attainment areas (i.e., areas where CAA pollution limits are exceeded). The Voluntary Airport Low-Emission (VALE) program uses funding and emission credit incentives to encourage airports to convert to low-emission vehicles, modify airport infrastructure for alternative fuels, supply gate electricity and air for parked aircraft (so they do not have to use their auxiliary power units), and consider other emissions reductions options.

The EU has instituted similar regulations and programs to limit aviation-related emissions. Various Directives regulate several atmospheric pollutants including SO_2, NO_2, PM, lead, ozone, benzene, CO, poly-aromatic hydrocarbons, cadmium, arsenic, nickel and mercury. The first Directive limiting emissions of NO_x, SO_2, lead and PM10 in ambient air came into force in July 1999 and the second Directive limiting emissions of benzene and CO in ambient air came into force in December 2000. Prior to the creation of EU Directives, emissions standards were set and enforced at the national level.

Efforts to expand airports and increase capacity may be hampered if the projected increase in traffic would place the airport in conflict with local emissions standards. For example, the UK government turned down plans for a third runway at Heathrow in 2003 on the grounds that NO_x levels around the airport would exceed the limits set in an EU Directive that will come into force in 2010.[5] Currently further environmental assessments are under way. The Heathrow airport operator, BAA, is planning to submit a planning application for a third runway in 2008, and to open the runway in 2017.[6]

14.5.5 Emerging Issues

Two areas of increasing importance and high uncertainty relating to local air quality have emerged for aviation in the last decade. The first is fine PM, which has a negative impact on respiratory health (e.g., Lighty et al., 2000). There are also indications that PM could contribute to contrail and cirrus cloud formation and thus contribute to climate change (see Section 14.5).

While the EPA has introduced increasingly stringent national ambient air quality standards for PM, there are currently no uniformly accepted methods for measuring PM from aviation. It is therefore difficult to assess the magnitude of aviation's contribution to PM emissions, or to determine whether mitigation strategies are needed, and what those mitigation strategies should be. To help address these limitations in the USA, the FAA, NASA, EPA, industry and academic institutions have recently joined together to develop a National Roadmap for Aviation Particulate Matter Research (FAA, 2005) to outline the efforts required in this area.

The second emerging local air quality concern is the potential for aviation to contribute hazardous air pollutants (HAPS) to local environments. The EPA has identified 188 HAPS, which are defined as air pollutants that are known or suggested to cause adverse health or environmental effects, such as cancer, birth defects, miscarriages or damage to the nervous system. Recent airport environmental assessments have not found significant health impacts from HAPS associated with emissions from airports. However,

---

[5] "Row takes off over Heathrow runway claims," *Independent* (London), Michael Harrison, May 6, 2006.
[6] " 'Secret pact' over Heathrow's third runway," *The Times*, London, Ben Webster, June 6, 2007.

the estimates of HAPS emissions used in these reviews were developed using measurements from 35-year-old engine technology because no other data was available. There is currently no generally accepted way of characterizing HAPS emissions from aircraft engines. Ambient HAPS concentrations can be monitored, but it is difficult to determine what proportion of HAPS are due to aviation because of the lack of engine HAPS emissions models and because aircraft HAPS emissions have been difficult to distinguish based on a unique chemical signature (Tesseraux, 2004). It is therefore difficult to determine whether or not aviation-related HAPS are a problem and, if so, what should be done about HAPS emissions. Nonetheless, in recent airport environmental assessments, HAPS reviews have figured more prominently (see, e.g., FAA, 2003).

14.6   Impact of Aviation on Climate

The topic of greatest uncertainty and contention at this time concerns the impact of aviation on climate change. In Europe, where aviation's share of greenhouse gas emissions is 3% and growing, climate change is considered the single most important environmental impact from aviation (CFIT, 2007), while in the USA many still regard it as less important and less urgent than community noise and surface air quality.

Anthropogenically induced climate change is considered to be a problem because it is occurring at a much faster rate than during previous climate change episodes. Living organisms therefore do not have time to adapt and many species may go extinct. Climate change also affects the ability to cultivate crops – while some regions may become more suitable for cultivation, other regions may become less suitable. Global warming is expected to result in a rise in sea levels, placing some of the world's large cities in jeopardy. Climate change is also expected to affect human health in many, mostly negative, ways (McMichael *et al.*, 2006; Kovats and Haines, 2005). Higher temperatures result in increased thermal stress, more extreme weather events and allow some infectious diseases to spread more easily. Health risks arise from the social, demographic and economic disruptions of climate change. Any assessment of the impact of aviation on health and the environment must therefore take the impact of aviation on climate change into account.

14.6.1   Effects

Aircraft emit chemical species and produce physical effects, such as contrails, that affect climate. This section discusses the most important emissions from a climate viewpoint and the mechanisms by which these emissions affect climate.

CO_2 is a greenhouse gas that causes an increase in global temperatures. Because of the very long timescales for the global carbon cycle, CO_2 may remain in the atmosphere and affect the climate for 500 years after it is emitted. Figure 14.10 shows the worldwide distribution of aircraft CO_2 emissions in 2000, as modeled by the FAA's System for assessing Aviation's Global Emissions (SAGE), which calculates aircraft emissions on a flight-by-flight basis as a function of aircraft type and detailed flight profile information. As shown in Figure 14.10, there are strong regional differences in the emissions. These differences are important for some of the shorter-lived climate impacts discussed below.

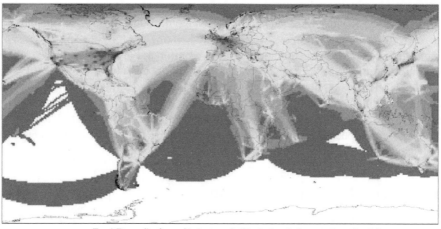

Fuel Burn (kg/year/1 degree latitude by 1 degree longitude)

0.00e+000 – 3.01e+005	5.81e+006 – 7.22e+006	2.49e+007 – 3.01e+007	9.49e+007 – 1.14e+008
3.02e+005 – 7.66e+005	7.23e+006 – 8.79e+006	3.02e+007 – 3.59e+007	1.15e+008 – 1.38e+008
7.67e+005 – 1.31e+006	8.80e+006 – 1.05e+007	3.60e+007 – 4.20e+007	1.39e+008 – 1.69e+008
1.32e+006 – 1.93e+006	1.06e+007 – 1.24e+007	4.21e+007 – 4.89e+007	1.70e+008 – 2.23e+008
1.94e+006 – 2.65e+006	1.25e+007 – 1.48e+007	4.90e+007 – 5.64e+007	2.24e+008 – 2.86e+008
2.66e+006 – 3.51e+006	1.49e+007 – 1.75e+007	5.65e+007 – 6.60e+007	2.87e+008 – 4.02e+008
3.52e+006 – 4.54e+006	1.76e+007 – 2.07e+007	6.61e+007 – 7.88e+007	4.03e+008 – 5.95e+008
4.55e+006 – 5.80e+006	2.08e+007 – 2.48e+007	7.89e+007 – 9.48e+007	5.96e+008 – 1.09e+008

Figure 14.10   Worldwide distribution of aircraft carbon dioxide emissions for 2000
Source: Reprinted from *Transportation Research Part D: Transport and the Environment*, 12:5,
B.Y. Kim *et al*, System for Assessing Aviation's Global Emissions (SAGE), Part 1: Model
Description and Inventory Results, 325–346, 2007, with permission from Elsevier

Other types of engine exhaust emissions produce impacts with shorter atmospheric lifetimes than CO_2 – on the order of a day (for contrails) to 10 years (for the removal of methane by NO_x) rather than centuries. However, they may still be important contributors to aviation's impact on climate:

- **Water vapor** affects the formation of contrails and also contributes directly to warming by acting as a greenhouse gas, especially in the stratosphere. Unlike CO_2, water vapor is naturally present in relatively high concentrations in the troposphere and aviation does little to perturb these concentrations.
- **NO_x emissions** from aircraft operating in the troposphere increase the net concentration of *ozone*. In the stratosphere, ozone directly affects the atmosphere's ability to filter harmful ultraviolet radiation, but it also acts as a potent greenhouse gas at tropospheric altitudes. Ozone breaks down relatively quickly (days to weeks). Thus this impact of NO_x produces a net warming effect which is regional, occurring predominantly in the northern hemisphere (where most of the air traffic is).
- **NO_x emissions** from aircraft operating in the troposphere also act to accelerate the removal of *methane* (CH_4) which is a strong greenhouse gas with many natural and anthropogenic sources. Methane is a globally well-mixed gas with a lifetime of tens of years (IPCC, 1999). Thus this impact of NO_x is global and produces a net cooling effect.

- **Condensation trails**, or **contrails**, are formed when the relatively warm and humid air in an aircraft's engine plume mixes with colder and less humid ambient air in the atmosphere, causing water vapor in the engine exhaust to condense and nucleate on particles – both those from the ambient atmosphere and those from exhaust soot and volatile aerosol particles, such as sulfates. The formation of contrails is affected by environmental conditions (pressure, temperature and relative humidity at a given flight level), fuel combustion properties and the overall efficiency of the engine. The persistence of contrails, or how long they last, is affected by the local atmospheric conditions (i.e., whether the thermodynamic conditions of the atmosphere are such that it is saturated with respect to ice).

  Contrails and other thin clouds tend to have a warming effect because they reduce the amount of heat that is radiated from the earth into space (also referred to as upward terrestrial radiative flux), while having only a small effect on the amount of heat that is received by the earth from the sun (downward solar flux) (IPCC, 1999). The long-term effect of contrails on climate change is limited by their relatively short lifetimes (on the order of hours) as compared to that of CO_2. Contrails may also contribute to the formation of cirrus clouds, which have a similar effect on heat reflection and retention. The extent and impacts of aviation-induced cirrus cloudiness are uncertain, but most estimates place the impacts at a level equal to or greater than those associated with the more readily observed linear contrails.

- **Soot emissions** may also contribute directly to climate change, in addition to their indirect impact through contrail formation. There is still large uncertainty on whether soot has a net cooling or warming effect and what the magnitude of the effect is.

- **Sulfate emissions**, like soot, also contribute directly to climate change, in addition to their indirect impact through contrail formation. Sulfate aerosol scatters incident radiation from the sun back to space, and therefore tends to have a cooling effect.

- **Stratospheric emissions of water vapor and NO_x** have different impacts. In particular, NO_x emissions in the stratosphere contribute to ozone depletion. The ozone in the stratosphere is an important shield against ultraviolet radiation and thus depletion of stratospheric ozone contributes to increased skin cancer incidence and ecological changes. Water emissions in the stratosphere have a strong warming impact as noted above. Roughly one-third of aviation emissions are injected into the lower stratosphere (IPCC, 1999).

To summarize, the impacts of a single flight are as follows. The most important impact over the first days is the warming contrail/aviation-induced cirrus effect, if it occurs. After that, the warming impact of ozone changes is most important. Both of these effects are regional in nature, occurring where the aircraft are flown. Then the longer-timescale impacts are felt globally. On the timescale of years, the reduced methane has a cooling effect. On a timescale of decades to centuries, a small warming effect due to the persistence of carbon dioxide remains as the sole impact. The net effect of all these impacts is still the subject of scientific study.

14.6.2 Measurement

Climate change can be measured in several ways, but the most common summary metric is the change in globally averaged surface temperatures resulting from perturbations in the radiative balance of the atmosphere. Changes in the radiative balance are indicated in terms of radiative forcing, measured in watts per unit of surface area (e.g., W/m^2). Positive radiative forcing is associated with a net warming tendency, as compared to temperatures in pre-industrial times. Different chemical species produce different levels of radiative forcing per unit change in concentration, and, again, different temperature change at the surface of the earth per unit change in radiative forcing (Hansen *et al.*, 2005). Radiative forcing and the resulting temperature change over time for each of the effects discussed above can be estimated using simplified models derived from combined atmospheric and carbon cycle models (e.g., Sausen and Schumann, 2000; Marais *et al.*, 2008).

In the past, the relative impacts of aviation effects on climate change have been estimated based on radiative forcing at a given time due to all prior and current industrial activity (e.g., effects of accumulated CO_2 emissions, plus present-day, short-lived impacts like contrails) (see, e.g., IPCC, 1999). However, these estimates of radiative forcing for aviation can be misleading when evaluating relative effects, since they do not account for the full impacts of the aviation activity (e.g., the future impacts of activity related to the long-lived effects of CO_2). When assessing policies, it is more appropriate to assess the full future marginal costs of a new unit of activity. Thus, the marginal change in climate response (e.g., in globally averaged surface temperature, or sea-level rise) as a function of time may be a more useful basis for judging the relative impacts of aviation (Marais *et al.*, 2008).

14.6.3   Greenhouse Emissions Controls

There are currently no national or international regulations for greenhouse gas emissions that apply to aircraft or other aviation-related sources (FAA, 2005). However, the aviation industry has made significant progress in reducing CO_2 and H_2O emissions, primarily as a side effect of improving fuel efficiency. The Kyoto Protocol does not currently include aviation emissions of greenhouse gases.

ICAO CAEP continues to evaluate options from technology and standards, operational measures and market-based measures such as voluntary agreements, open emissions trading and emissions-related levies, to limit or reduce greenhouse gases from aviation (ICAO, 2005b). Preliminary results indicate that emissions-related levies are not cost effective, but that voluntary arrangements and emissions trading may be a cost-effective way of limiting or reducing aviation greenhouse gas emissions (ICAO, 2005a). Currently ICAO is recommending that any emissions-related levies should be in the form of charges rather than taxes, and that funds thus collected should be applied to the mitigation of environmental impacts (ICAO, 2005a).

Instituting effective controls is difficult because there are significant uncertainties regarding the impacts of aviation on climate. However, it is not unreasonable to assume that aviation greenhouse gas emissions will continue to increase, as they have done on a global-average basis for the past decade (FAA, 2005). The uncertainties make it difficult to determine the most appropriate technological, operational and policy options for mitigation of aviation's impact on climate change. As a result, most mitigation options currently being pursued focus on reducing fuel burn. Further, although fuel use per passenger mile has been reduced by 60% in the last 35 years, most projections suggest a slower rate of improvement in the next 15 to 20 years of about 1% per year, which falls short of the expected growth in demand (IPCC, 1999; Lee *et al.*, 2001).

NASA has a five-year goal to deliver technologies that will enable the reduction of CO_2 emissions of new aircraft by 25%. However, significant challenges remain with respect to demonstrating technological feasibility and economic reasonableness, before these concepts can be employed in aircraft fleets. It may take an additional 5 to 15 years and significant industrial investment before these NASA technologies can be introduced into new aircraft.

The European Commission has instituted a Greenhouse Gas Emissions Trading Scheme (ETS) across several major industries involving about 11 000 industrial installations and is currently considering incorporating commercial aviation into this scheme. The ETS places an overall cap on greenhouse gas emissions, under which operators can buy and sell emissions allowances according to their need (Climate Policy, 2005). Each operator is allocated a limited number of emissions allowances. Operators emitting more than their allotted quota may choose to buy additional emissions allowances or invest in reducing emissions, while operators emitting less than their allowed quota may sell their emissions credits. Operators are therefore free to choose the most cost-effective way to control their emissions, and also have a permanent incentive to reduce emissions. Emissions trading schemes are a way of better distributing costs and have the potential to be more effective in environmental terms, as well as being more cost efficient than other market-based solutions such as airline ticket taxes, departure taxes or emissions charges. Currently, aviation is not included in this ETS, but in December 2006 the EC adopted a proposal for legislation to include aviation (European Commission, 2006b).

In the USA emissions trading schemes for SO_2 and NO pollution have also been instituted (Burtraw *et al.*, 2005), but have not been expanded to include aviation.

Other marketplace incentives have also been considered. For example, some major European airports have implemented landing fees that reward operators using ultralow-NO_x combustors and penalize operators using standard combustors (NRC, 2002). In this case the cost differential is generally not sufficient to change operator behavior dramatically because of the limited number of airports involved.

There are also several active research efforts under way to advance the use of alternative fuels for aviation. Some of these fuels are derived from renewable feedstocks such that the net life cycle CO_2 emissions are approximately zero (i.e. the plants from which the fuel is derived remove an approximately equal amount of CO_2 from the atmosphere during their growth to that which is released when the fuel is combusted). Nonetheless, there are significant hurdles to overcome, most especially those related to assessing broader life cycle impacts, competition for arable land with food crops, the costs of producing such fuels, competition with other uses for these environmentally attractive fuels, and the challenge of producing enough biofuel to significantly offset the use of fossil-fuel-derived energy.

14.7   Summary and Looking Forward

The impact of aviation on the environment is a growing concern because of the projected two- or three-fold increase in demand for air transportation over the next 20 years (JPDO, 2006). The attendant rate of growth is expected to outpace gains in aviation environmental performance. Despite significant progress in reducing aircraft noise over the past three or four decades, noise continues to be a critical limiting factor to aviation growth through the limitations that local communities place on airport operations and expansion. Emissions of CO_2 and H_2O per passenger kilometer traveled have decreased significantly over the past decades, primarily as a result of increases in fuel efficiency. Conversely, emissions of NO_x, which tend to increase with increasing fuel efficiency, have not changed as significantly. Reducing NO_x poses challenging technological problems.

There are several emerging concerns regarding noise and emissions. Low-frequency aviation-related noise is blamed for structural "rattling" and there is also some evidence that it could have adverse health effects. But the links are not clear and more research is needed to establish a clear relationship between low-frequency noise and possible adverse health effects. It has also been suggested that interdependencies between noise and emissions impacts on health may mean that the total adverse health impact of aviation is larger than currently assumed (Schwela et al., 2005). For example, both noise and emissions may have an impact on cardiovascular health.

Hazardous air pollutants (HAPS) are chemicals that have severe health and environmental effects. Which HAPS are included in aircraft exhaust and to what degree are the subjects of ongoing research. Aviation is now also known to have an impact on global climate change. However, how much aviation contributes to the total anthropogenic effect, and what the best approaches to limit aviation's impact would be, are not clear.

The reduction of noise and the reduction of emissions are goals that also pose interesting policy questions. Noise and emissions impacts result from an interdependent set of technologies and operations, so that action to address impacts in one domain can have negative impacts in other domains (Waitz et al., 2005). For example, both operational and technological measures to reduce noise can result in greater fuel burn, thus increasing aviation's impact on climate change and local air quality (SBAC, 2001). Interdependencies among emissions make it difficult to reduce environmental impacts by modifying engine design, because they force a trade-off among individual pollutants as well as between emissions and noise (FAA, 2005). To date, interdependencies among various policy, technological and operational options and the full economic consequences of these options have not been appropriately assessed; however, the FAA and other international organizations are actively pursuing the development of tools to advance the state of the art in this area.

References

ACARE – Advisory Council for Aeronautics Research in Europe (2001) *European Aeronautics: A Vision for 2020, Report of the Group of Personalities*, January. Available online at: http://www.acare4europe.org/html/background.shtml, accessed May 2007.

Aviation International News (2004) "Europe's View of Green Skies Could Have Global Influence," Thierry Dubois, AS3/GSE, May 18–20. 2004. Available online at: http://www.ainonline.com/publications/as3/as3_04/as3_04_europesviewp3.html, accessed May 2006.

BAA (2003) British Airports Authority Heathrow, Airport Water Quality Strategy 2003–2008, Beacon Press, Lickfield, East Sussex.

Babikian, R., Lukachko, S.P. and Waitz, I.A. (2002) "Historical Fuel Efficiency Characteristics of Regional Aircraft from Technological, Operational, and Cost Perspectives," *Journal of Air Transport Management*, Vol. 8, No. 6, pp. 389–400.

BenMAP, Abt Associates Inc. (2005) BenMAP: Environmental Benefits Mapping and Analysis Program. Prepared for US EPA, Office of Air Quality Planning and Standards, Research Triangle Park, NC.

Berglund, B., Hassmen, P. and Job, R.F.S. (1996) "Sources and Effects of Low-Frequency Noise," *Journal of the Acoustical Society of America*, Vol. 99, No. 5, pp. 2985–3002.

Betts, K.S. (1999) "Airport Pollution Prevention Takes Off," *Environmental Science and Technology*, Vol. 33, No. 9, pp. 210A–212A.

Boeing Commercial Aircraft (2006a) "Growth in Airport Noise Restrictions." Available online at: http://www.boeing.com/commercial/noise/airports2005.pdf, accessed May 2006.

Boeing Commercial Aircraft (2006b) "Noise Regulations Timeline." Available online at: http://www.boeing.com/commercial/noise/timeline.pdf, accessed June 2006.

Burtraw, D., Evans, D.A., Krupnick, A., Palmer, K. and Toth, R. (2005) "Economics of Pollution Trading for SO_2 and NO_x," *Annual Review of Environment and Resources*, Vol. 30, pp. 253–289.

Casalino, D., Diozzi, F., Sannino, R. and Paonessa, A. (2008) "Aircraft Noise Reduction Technologies: A Bibliographic Review," *Aerospace Science and Technology*, Vol. 12, No. 1, pp. 1–17.

CFIT – Commission For Integrated Transport (2007) "Transport and Climate Change: Advice to Government from The Commission," London, UK, CEMT, 07/220, 009911.

Climate Policy (Various Authors) (2005) Special Issue on EU Emissions Trading, *Climate Policy*, Vol. 5, No. 1.

Eagan, M.E. (2007) "Supplemental Metrics to Communicate Aircraft Noise Effects," *Transportation Research Record*, Vol. 2011, pp. 175–183.

EPA – US Environmental Protection Agency (1994) "Addendum to Municipal Wastewater Management Fact Sheets: Storm Water Best Management Practices," EPA-832-F-93-013, Office of Water, Washington, DC.

EPA – US Environmental Protection Agency (1995) "Emerging Technology Report: Preliminary Status of Airplane De-icing Fluid Recovery Systems," EPA 832-B-95-005, Office of Water, Washington, DC.

EPA – US Environmental Protection Agency (1999a) "The Benefits and Costs of the Clean Air Act, 1990 to 2010," EPA Report to Congress, November, EPA-410-R-99-001.

EPA – US Environmental Protection Agency (1999b) "Evaluation of Air Pollutant Emissions from Subsonic Commercial Jet Aircraft," EPA420-R-99-013, April. Available online at: http://www.epa.gov/otaq/regs/nonroad/aviation/r99013.pdf, accessed June 2006.

EPA – US Environmental Protection Agency (2002) "Source Water Protection Practices Bulletin: Managing Aircraft and Airfield Deicing Operations to Prevent Contamination to Drinking Water," EPA 816-F-02-018, August.

EPA – US Environmental Protection Agency (2005) "Control of Air Pollution from Aircraft and Aircraft Engines; Emission Standards and Test Procedures," 40 CFR Part 87, Federal Register, Vol. 70, No. 221, 17 November. Available online at: http://www.epa.gov/fedrgstr/EPA-AIR/2003/September/Day-30/a24412.htm, accessed June 2006.

EPA – US Environmental Protection Agency (2006) "EPA Green Book." Available online at: http://www.epa.gov/air/oaqps/greenbk/o3co.html, accessed May 2006.

European Commission (2005) ExternE, Externalities of Energy, Methodology 2005 Update. European Commission, DGXII, EUR 21951, Brussels.

European Commission (2006a) Environment: Policies: Air: Air Quality. Available online at: http://europa.eu.int/comm/environment/air/ambient.htm, accessed May 2006.

European Commission (2006b) Official Press Release – Climate change: Commission proposes bringing air transport into EU Emissions Trading Scheme. Available online at: http://europa.eu/rapid/pressReleasesAction.do?reference=IP/06/1862&format=HTML&aged=0&language=EN&guiLanguage=en, accessed September 2006.

FAA – Federal Aviation Administration (2001) "Noise Levels for US Certificated and Foreign Aircraft," FAA Advisory Circular AC36-1 H, Office of Energy and Environment, Washington, DC, November. Available online at: http://www.faa.gov/about/office_org/headquarters_offices/aep/noise_levels/, accessed June 2006.

FAA – Federal Aviation Administration (2003) "Select Resource Materials and Annotated Bibliography on the Topic of Hazardous Air Pollutants (HAPs) Associated with Aircraft, Airports, and Aviation," July 1.

FAA – Federal Aviation Administration (2005) "Aviation and Emissions, A Primer," Office of Environment and Energy, Washington, DC, January.

FICAN – Federal Interagency Committee on Aircraft Noise (1992) "Federal Agency Review of Selected Airport Noise Analysis Issues," Washington, DC, August.

FICAN – Federal Interagency Committee on Aircraft Noise (2002) FICAN on the Findings of the Minneapolis–St. Paul International Airport (MSP) Low-Frequency Noise (LFN) Expert Panel, Washington, DC, April.

Findeis, H. and Peters, E. (2004) "Disturbing Effects of Low Frequency Sound Immissions and Vibrations in Residential Buildings," *Noise Health*, Vol. 6, No. 23, pp. 29–35.

GAO – US General Accounting Office (2000) "Aviation and the Environment – Results from a Survey of the Nation's 50 Busiest Commercial Service Airports," GAO/RCED-00-222, Washington, DC, August.

Haines, M.M., Stansfeld, S.A., Job, R.F.S., Berglund, B. and Head, J. (2001) "Chronic Aircraft Noise Exposure, Stress Responses, Mental Health and Cognitive Performance in School Children," *Psychological Medicine*, Vol. 31, No. 2, pp. 265–277.

Hansen J., Sato M., Ruedy R. *et al.* (2005) "Efficacy of Climate Forcings," *Journal of Geophysical Research – Atmospheres*, Vol. 110 (D18), Art. No. D18104, September 28.

Hubbard, H.H. (1982) "Noise Induced House Vibrations and Human Perceptions," *Noise Control Engineering Journal*, Vol. 19, No. 2, pp. 49–55.

ICAO – International Civil Aviation Organization (2004) "Work on Environmental Protection Reflects Need for Balancing Many Factors," J. Hupe, ICAO Secretariat, *ICAO Journal*, Vol. 59, August, pp. 4–6, 23–24.

ICAO – International Civil Aviation Organization (2005a) "ICAO Exploring Development of a Trading Scheme for Emissions from Aviation," A. Karmali and M. Harris, *ICAO Journal*, Vol. 59, pp. 11–13, 25–26.

ICAO – International Civil Aviation Organization (2005b) Statement from the International Civil Aviation Organization (ICAO) to the Twenty-Second Session of the UNFCCC Subsidiary Body for Scientific and Technological Advice (SBSTA), Bonn, May 19–27. Available online at: www.icao.int/env/sbsta-22.pdf, accessed January 2009.

IPCC – Intergovernmental Panel on Climate Change (1999) *Aviation and the Global Atmosphere*, ed. J.E. Penner *et al.*, Cambridge University Press, Cambridge.

Ising, H. and Kruppa, B. (2004) "Health Effects Caused by Noise: Evidence in the Literature from the Past 25 Years," *Noise Health*, Vol. 6, No. 22, pp. 5–13.

Jarup, L. *et al.* (2008) "Hypertension and Exposure to Noise Near Airports: the HYENA Study," *Environmental Health Perspectives*, Vol. 116, No. 3, pp. 329–333.

JPDO – Joint Planning and Development Office (2006) Washington, DC, http://www.jpdo.aero, accessed May 2006.

Kovats, R.S. and Haines, A. (2005) "Global Climate Change and Health: Recent Findings and Future Steps," *Canadian Medical Association Journal*, Vol. 172, No. 4, pp. 501–502.

Lee, J., Lukachko, S., Waitz, I. and Schafer, A. (2001) "Historical and Future Trends in Aircraft Performance, Cost, and Emissions," *Annual Review of Energy and the Environment*, Vol. 26, pp. 167–200.

Lee, J., Lukachko, S. and Waitz, I. (2004) "Aircraft and Energy Use," Invited Chapter in the Encyclopedia of Energy, Academic Press/Elsevier Science, San Diego, CA.

Leventhall, H. (2004) "Low Frequency Noise and Annoyance," *Noise Health*, Vol. 6, No. 23, pp. 59–72.

Lighty, J.S., Veranth, J.M. and Sarofim, A.F. (2000) "Combustion Aerosols: Factors Governing Their Size and Composition and Implications to Human Health," *Journal of the Air & Waste Management Association*, Vol. 50, No. 9, pp. 1565–1618.

Lim, C., Kim, J., Hong, J. and Lee, S. (2008) "Effect of Background Noise Levels on Community Annoyance from Aircraft Noise," *Journal of the Acoustical Society of America*, Vol. 123, pp. 766–771.

Lynch, F.T. and Khodadoust, A. (2001) "Effects of Ice Accretions on Aircraft Aerodynamics," *Progress in Aerospace Sciences*, Vol. 37, No. 8, pp. 669–767. Available online at: http://www .sciencedirect.com/science/article/B6V3V-44M2CN3-1/2/257b55e1b893bad89729939ec7ba9813, accessed September 2006.

Marais, K., Lukachko, S.P., Jun, M., Mahashabde A. and Waitz, I. (2008) "Assessing the Impact of Aviation on Climate," *Meteorologische Zeitung*, Vol. 17, No. 2, pp. 157–172.

Matheson, M., Stansfeld, S. and Haines, M. (2003) "The Effects of Chronic Aircraft Noise Exposure on Children's Cognition and Health: 3 Field Studies," *Noise Health*, Vol. 5, No. 19, pp. 31–40.

Matsui, T., Stansfeld, S., Haines, M. and Head, J. (2004) "Children's Cognition and Aircraft Noise Exposure at Home – the West London Schools Study," *Noise Health*, Vol. 7, No. 5, pp. 49–57.

Maziul, M., Job, R.F. and Vogt, J. (2005) "Complaint Data as an Index of Annoyance – Theoretical and Methodological Issues," *Noise Health*, Vol. 7, No. 28, pp. 17–27.

McCarty, P.L. (1964) "Anaerobic Waste Treatment Fundamentals," *Public Works*, Vol. 95, No. 9, pp. 107–112.

McMichael A.J., Woodruff, R.E. and Hales, S. (2006) "Climate Change and Human Health: Present and Future Risks" *Lancet*, Vol. 367, No. 9513, pp. 859–869.

Miller, L.M. (1979) "Investigation of Selected Potential Environmental Contaminants: Ethylene Glycol, Propylene Glycols and Butylene Glycols," Final Report, EPA-68-01-3893, US EPA, Office of Toxic Substances, Washington, DC, May.

Morrison, S.A., Winston, C. and Watson, T. (1999) "Fundamental Flaws of Social Regulation: The Case of Airplane Noise," *Journal of Law & Economics*, Vol. 42, No. 2, pp. 723–743.

Nelson, P. (2004) "Meta-analysis of Airport Noise and Hedonic Property Values – Problems and Prospects," *Journal of Transport Economics and Policy*, Vol. 38, No. 1, pp. 1–27.

NRC – National Research Council (1977) "Guidelines for Preparing Environmental Impact Statements on Noise," Report of Working Group 69, Assembly of Behavioral and Social Science, Committee on Hearing, Bioacoustics, and Biomechanics, Washington, DC.

NRC – National Research Council (2002) "For Greener Skies, Reducing Environmental Impacts of Aviation," Committee on Aeronautics Research and Technology for Environmental Compatibility, Aeronautics and Space Engineering Board, Washington, DC. Available online at: http://books.nap.edu/catalog/10353.html, accessed May 2006.

NSTC – National Science and Technology Council (1995) "Goals For A National Partnership in Aeronautics Research and Technology," Office of Science and Technology Policy, Executive Office of the President, Washington, DC, September.

Persson, W.K. (2004) "Effects of Low Frequency Noise on Sleep," *Noise Health*, Vol. 6, No. 23, pp. 87–91.

Pillard, D.A. (1995) "Comparative Toxicity of Formulated Glycol Deicers and Pure Ethylene and Propylene Glycol to Ceriodaphnia Dubia and Pimephales Promelas," *Environmental Toxicology and Chemistry*, Vol. 14, No. 2, pp. 311–315.

Ramakrishna, D.M. and Viraraghavan, T. (2005) "Environmental Impact of Chemical Deicers – A Review," *Water, Air and Soil Pollution*, Vol. 166, Nos. 1–4, pp. 49–63.

Raschke, F. (2004) "Arousals and Aircraft Noise – Environmental Disorders of Sleep and Health in Terms of Sleep Medicine," *Noise Health*, Vol. 6, No. 22, pp. 15–26.

Rojo, J. (2007) "Future Trends in Local Air Quality Impacts of Aviation," Masters thesis, Massachusetts Institute of Technology.

Sausen, R. and Schumann, U. (2000) "Estimates of the Climate Response to Aircraft CO_2 and NO_x Emissions Scenarios," *Climatic Change*, Vol. 44, Nos 1–2, pp. 27–58, DOI 10.1023/A:1005579306109. Available online at: http://dx.doi.org/10.1023/A:1005579306109.

Schipper, Y., Nijkamp, P. and Rietveld, P. (2001) "Aircraft Noise Valuation Studies and Meta-analysis," *International Journal of Environmental Technology and Management*, Vol. 1, No. 3, pp. 317–320.

Schust, M. (2004) "Effects of Low Frequency Noise up to 100Hz," *Noise Health*, Vol. 6, No. 23, pp. 73–85.

Schwela, D., Kephalopoulos, S. and Prasher, D. (2005) "Confounding or Aggravating Factors in Noise-Induced Health Effects: Air Pollutants and Other Stressors," *Noise Health*, Vol. 7, No. 28, pp. 41–50.

Sequeira, C. (2008) "An Assessment of the Health Implications of Aviation Emissions Regulations," Master of Science Thesis, Massachusetts Institute of Technology.

Shield, B.M. and Dockrell, J.E. (2008) "The Effects of Environmental and Classroom Noise on the Academic Attainments of Primary School Children," *Journal of the Acoustical Society of America*, Vol. 123, No. 1, pp. 133–144.

Switzenbaum, M.S., Veltman, S., Mericas, D., Wagoner, B. and Schoenberg, T. (2001) "Best Management Practices for Airport Deicing Stormwater," *Chemosphere*, Vol. 43, No. 8, pp. 1051–1062.

Tesseraux, I. (2004) "Risk Factors of Jet Fuel Combustion Products," *Toxicology Letters*, Vol. 149, Nos. 1–3, pp. 295–300.

Transport Canada (2006) "Guidelines for Aircraft Ground-Icing Operations," TP 14052. Available online at: http://www.tc.gc.ca/civilaviation/publications/TP14052/menu.htm, accessed September 2006.

USDOT (2008) http://www.bts.gov/publications/national_transportation_statistics/html/table_04_05 .html, accessed January 2009.

Valarezo, W.O., Lynch, F.T. and McGhee, R.J. (1993) "Aerodynamic Performance Effects due to Small Leading-Edge Ice (Roughness) on Wings and Tails," *Journal of Aircraft*, Vol. 30, No. 6, pp. 807–812.

Waitz, I., Townsend, J., Cutcher-Gershenfeld, J., Greitzer, E. and Kerrebrock, J. (2004) "Aviation and the Environment: A National Vision Statement, Framework for Goals and Recommended Actions," Report to Congress, December.

Waitz, I.A., Lukachko, S.P. and Lee, J. (2005) "Military Aviation and the Environment: Historical Trends and Comparison to Civil Aviation," *AIAA Journal of Aircraft*, Vol. 42, No. 2, pp. 329–339.

15

Information Technology in Airline Operations, Distribution and Passenger Processing

Peter P. Belobaba, William Swelbar and Cynthia Barnhart

Airlines are leaders in the use of information technology, with the development of databases and decision support systems for a wide range of applications. This chapter first provides a review of the role of computerized systems in airline planning and operations, and then focuses on the evolution of information technology applications for airline distribution and passenger processing. The development of computer reservations systems is described, as a background for the dramatic changes in airline distribution that have occurred over the past decade – a shift away from traditional travel agencies to Internet websites for booking and ticketing. The differences between traditional and emerging airline distribution channels are explained, and the resulting cost reductions experienced by airlines are presented. In addition, electronic ticketing and recent innovations in passenger processing, including airport kiosks and web check-in, are discussed, with a focus on their implications for airline economics and passenger satisfaction.

15.1 Information Technology in Airline Planning and Operations

Several previous chapters have described in detail the steps in an airline's development of plans for operating a schedule of flight departures on a given future date. As shown in Figure 15.1, the airline planning process starts with longer-term decisions about fleet planning, followed by route evaluation and then schedule development, as described in Chapter 6. Once a schedule and operating plan have been developed, shorter-term commercial decisions about pricing and revenue management are required, as described in Chapter 4. On the operations side, the performance of the airline's schedule (including

The Global Airline Industry   P. Belobaba, C. Barnhart and A. Odoni
© 2009 John Wiley & Sons, Ltd

timetables, fleet assignments, crew schedules and aircraft maintenance rotations) is subject to the availability of airport resources as well as the many factors that can result in irregular operations that must be addressed by the airline's Systems Operations Control Center (SOCC), as described in Chapters 8 and 9.

Virtually all of the decision steps illustrated in Figure 15.1 are supported by computer databases, interactive decision tools and optimization models, depending on the size and sophistication of each airline. Even the smallest airlines have data reporting capabilities that can support analysis of most planning decisions. In this book, we have focused on the characteristics of advanced decision support models that are more typically employed by the largest of the world's airlines. Perhaps the greatest progress in the development of databases, decision support tools and advanced optimization models has been experienced in the areas of schedule planning, operations control and revenue management.

As described in Chapters 6 and 7, the goals of the schedule development and optimization process are to determine the origin, destination, schedule, and assigned aircraft and crews for each flight leg to be operated in the airline's network, for a series of departure dates in the future. Whether designing future schedules or operating current schedules, the challenge is to manage numerous heterogeneous resources, all interacting in complex ways over the flight network and subjected to numerous regulatory restrictions. This high degree of problem complexity contributed to the airline industry's early adoption and development of operations research techniques that rely heavily on the use of information technologies. Indeed it is difficult to think of any single sector, other perhaps than military operations, with which operations research has been linked more closely.

Initially, the optimization focus at airlines was on planning problems. This was due in part to the long solution times (many hours or even days) allowed to address planning

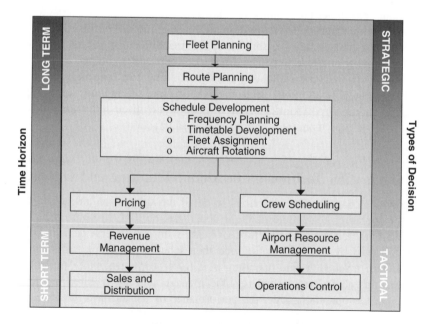

Figure 15.1   The airline planning process

problems, compared to those required (minutes or even seconds) for real-time operations problems. The initial focus on planning problems is also explained by the relative ease of compiling the necessary data that details the numbers of available aircraft and crew members, and the potential numbers of passengers, but does not provide up-to-the-minute information about their status, as required for operations problems.

More recently, especially in the last decade, there has been increased focus on the development and implementation of optimization-based decision support tools for operations control. The SOCC of an airline, as introduced in Chapter 8, is responsible for dealing with irregular operations caused by unexpected events that can include severe weather, non-scheduled maintenance, and delays of inbound aircraft, crews or even passengers. Airline SOCC functions are increasingly supported by IT and decision support systems to help with the complicated trade-offs involved in the rescheduling and rerouting of aircraft and crew resources, as well as passenger reaccommodation.

The reality is, however, that optimization approaches are useless to the airline without accurate data to populate the associated models. Aircraft and crew schedule planning and recovery optimization approaches require an enormous amount of input data including, for example:

- tens or hundreds of thousands of demand forecasts, one for each itinerary–fare class combination in the airline's network;
- estimated costs to operate each aircraft type on each flight leg;
- the flying range and seating capacity of each aircraft;
- the myriad of rules governing the assignment of crews to scheduled flights;
- locations and status (e.g., the numbers of hours worked and the amount of rest accrued) for each crew member, at all times;
- locations and status of each aircraft, at all times;
- locations of maintenance bases for each type of aircraft, and
- rules specifying maintenance requirements and the flying and elapsed time since the last maintenance for each aircraft.

Airlines have long been aware of the need for (and potential use of) this data, but limitations in computing capabilities and database sizes prevented most airlines from making full use of such detailed data. With the evolution of IT capabilities, especially since the 1970s, airline databases are increasingly able to capture and store these detailed data inputs, thereby enabling the use of optimization models for schedule planning and operations. Many of the largest airlines have undertaken large-scale projects to develop centralized data warehouses that can support all of the planning and operations control functions. Centralized data warehouses provide the same data, at the same time, to all decision makers at an airline, for long-term strategic planning, tactical planning and short-term operational decisions.

The development of computerized systems for revenue management provides another important example of the successful integration of IT and operations research in the airline industry. As introduced in Chapter 4, revenue management (RM) systems play a central role in the determination of the fares at which an airline's inventory of seats is sold to consumers and the ways in which this inventory is displayed in various distribution channels. The database requirements of advanced revenue management systems can exceed those of the scheduling optimization models described above. While a scheduling optimization problem typically involves demand and operational constraints for an extended schedule period (e.g., a future month), network revenue management systems must forecast demand by passenger itinerary and fare type for *each future departure date*. The database requirements can be staggering – a large network carrier with an O-D control RM system can be forecasting demand for hundreds of thousands of itinerary–fare combinations on thousands of scheduled flight legs for each day up to a year in advance of each departure date.

Fortunately, the network RM optimization problem can be solved separately for the airline's network (or sub-networks) for each departure date, under the assumption that the demand forecasts on different future dates are independent. Moreover, the network RM problem can be modeled as a "transportation problem" that is very tractable in mathematical terms. Relatively frequent reoptimization of RM booking limits and bid prices (see Chapter 4) over an entire network of flights for a future departure date has therefore become a reality with improvements in computational speeds. These developments have occurred only in the recent past, as even in 1999 a large US network airline reported that it could only perform a complete network optimization for RM purposes on the most advanced mainframe computer available at the time, once every seven days for any given future departure date (Saranathan *et al.*, 1999).

Another IT development critical to the implementation of advanced network RM systems by large network airlines was that of "seamless connectivity" or instantaneous communication of seat availability information between different airline reservations systems (as will be described below). This capability did not exist before the late 1990s, yet it was essential for O-D control RM systems – their effectiveness depends on the ability of an airline to respond to different requests for different itinerary and fare combinations, depending on the current demand forecasts, remaining capacity and optimized bid prices for the airline's network. Seamless connectivity among airline reservations systems has now become the norm, and has enabled airlines to improve their control over their seat inventories by responding to requests for availability based on the revenue value of each request.

15.2   Airline Distribution Systems

The airline distribution function is responsible for selling the airline's transportation service to its customers, by allowing them to obtain information about flight schedules, available seats and fares, and then enabling them to make reservations and purchase tickets for their travel itineraries. From the airline's perspective, the distribution function is critical to the management of its available flights and seat inventory, tracking of sales and revenues collected from passengers, and capturing market share by ensuring widespread dissemination of information about its services to consumers.

In this section, we provide a brief overview of the evolution of the airline distribution function, from its manual origins to today's sophisticated IT systems. We then describe the characteristics of current airline distribution channels and the development of worldwide global distribution systems (GDSs). The relationship of these channels to the recent rapid evolution of Internet-based distribution channels, in terms of both technology and costs, is discussed, as well as the use of "code-share" flights for distribution by airline alliances. Finally, the challenges that these recent developments in airline distribution have created for airlines, and their IT systems in particular, are discussed.

15.2.1 Evolution of Computer Reservations Systems

Historically, the primary role of the airline distribution function has been to provide passengers with the ability to make reservations and purchase tickets for future airline flights. At the same time, the airline needed to keep track of the number of bookings and remaining available seats on each future flight departure, along with information about each passenger's reservation (including the passenger's name, contact details, flight itinerary, special requests and ticket payment details). These functions were originally performed at each airline's central reservations office manually (with pen and paper), but as airlines grew in size it became increasingly difficult to manage this process. Telephone and teletype messages were the modes of communication between the reservations office and either consumers or, more commonly, travel agents.

The development of computerized reservations systems (CRSs) began in the 1950s, when American Airlines partnered with IBM for this purpose, and the first CRS was implemented by American in 1962 (Copeland et al., 1995). The "Semi-Automated Business Research Environment" (SABRE) allowed airline reservations agents to manage the distribution process, not only at the centralized reservations office, but at airports and city ticketing offices. SABRE was "the first real-time business application of computer technology, an automated system with complete passenger records available electronically to any agent connected to the SABRE system" (Smith et al., 2001).

The real-time communications capabilities of SABRE, followed by similar computer reservations systems developed by other large airlines (e.g., Apollo at United and PARS at TWA) led to the extension of these capabilities outside of the airline organization, to travel agents. By the 1970s, several airlines with CRSs made their systems available for use by travel agents (at a cost). Travel agency CRSs showed flight schedules and fare information not just for one airline, but for virtually all airlines, and allowed travel agents to book reservations and sell tickets directly to customers, on behalf of the airlines.

By making these capabilities available to travel agencies, the airline owners of CRSs were able to generate incremental revenues (and profits) in several ways:

- CRS rental and/or usage fees charged to travel agencies.
- Booking fees for each flight segment transaction charged to the airline accepting the CRS reservation.
- Revenues from bookings made due to CRS "display bias" in which the flights of the airline owner of the CRS were given preferential display, which influenced the way in which travel agents presented options to consumers.
- Revenues from bookings made due to a "halo effect": travel agents giving preference to the flights of the airline owner of the CRS.

Travel agencies, on the other hand, received commission payments from airlines for tickets issued, paid as a percentage of the total ticket value. In addition, various commission "override" payment schemes offered by the airlines provided travel agencies with even higher commission rates for achieving airline-specified booking and revenue targets. This travel agency commission structure raised questions about the extent to which travel agencies were in fact motivated to search for the lowest fare options for their customers in an unbiased manner.

With increasing competition among CRSs and continued advancement of both computer and communications technologies, CRSs began to merge and/or combine their communications networks to become what are now referred to as global distribution systems (GDSs). GDSs allow travel agencies around the world to have access to airline information and distribution capabilities. For example, the Apollo CRS originally developed by United for US travel agencies merged with several other US and European CRSs to become the Galileo GDS. Similarly, the Worldspan GDS is the result of several mergers of previous CRSs owned by TWA, Northwest and Delta. The SABRE GDS has grown from being the CRS of American Airlines in the 1970s to its current dominant global position primarily through internal growth, whereas the Amadeus system initially combined several European airline-owned CRSs, but is today an independent GDS provider.

The US Department of Transportation in 1984 implemented regulations governing airline CRSs, in an effort to eliminate display bias and preferential treatment for its airline owner. These rules required that each CRS make public its algorithm for showing the order of flight options (e.g., shortest elapsed times, flight times closest to time requested), and prevented CRSs from showing any airline's special fares, if those fares were not also available through other distribution channels. These CRS regulations remained in place until 2005, when they were eliminated, setting the stage for a new and deregulated competitive environment for airline distribution systems, both traditional ones and those newly developed as Internet distribution engines.

Throughout the 1970s and 1980s, and well into the 1990s, the distribution of airline flight and fare information, along with the ticketing of airline bookings, was dominated by these traditional systems, which were used primarily by travel agents. In the early 1990s, over three-quarters of all US airline tickets were purchased through travel agencies, which relied almost exclusively on CRSs and GDSs. The GDSs were in turn each owned by one or more airlines, thus allowing the airlines to exert considerable control over the distribution process. In some years, the airline owners made a greater profit from their GDS transaction fees than from their airline operations! Despite the economic deregulation of the US airline industry, the distribution of flight and fare information, and control of the booking and ticketing process, remained under the control of a few dominant providers.

Recent developments in the airline distribution world have alleviated some of these concerns about airline ownership of distribution channels. Several of the largest GDSs are no longer owned exclusively or even partially by airlines – for example, the SABRE GDS was spun off by AMR into a separate company in 2004, while both Amadeus and Worldspan have also reduced to minority status the proportions of their ownership stake held by airlines. Furthermore, new players in the airline distribution field are developing alternatives to the traditional GDSs, such as ITA and G2 Switchworks.

Still, the worldwide airline distribution market continues to be dominated by large GDS competitors – Amadeus, SABRE, Galileo, Worldspan and Abacus. Figure 15.2 shows the market shares of total GDS airline bookings made by each GDS in the first half of 2008. In 2006, Galileo and Worldspan announced a planned merger, which would further concentrate GDS booking shares. It is important to note that the GDS share of total bookings made via all distribution channels continues to decrease with the growth of airline and other Internet channels. By 2005, the GDS share of total bookings made in the USA had dropped to less than 60%, although GDS shares of bookings made throughout

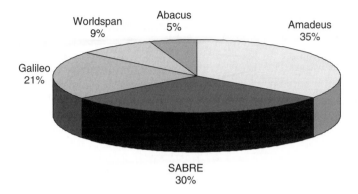

Figure 15.2   World GDS market shares, 2008
Data Source: MIDT Data, January–June 2008

the rest of the world remained well above 70%. This shift in distribution channels and its implications for airline distribution costs will be discussed further in Section 15.3.

15.2.2   Terminology and Capabilities of Reservation Systems

The evolution of airline distribution systems has fostered terminology that might seem confusing – what is the difference between an airline's own internal reservations system and the external CRS and GDS distribution systems? Before we continue the discussion of recent developments in airline distribution, we clarify these terms and describe the most important functionalities of the electronic airline distribution system. An example of a flight availability display from a traditional GDS will help illustrate these capabilities.

Virtually every airline makes use of an internal IT system that is typically known within the airline as its own "reservations system" or "RES." The airline's own reservations system contains all of its flight schedules, prices, seat inventories, real-time operational information ("FLIFO") and departure control systems (used for airport check-in). The RES is proprietary to each airline and typically occupies a (large) part of a mainframe IT system, either owned by the airline or "hosted" by an outside IT provider. The airline's own RES is, in many respects, the heart of the airline's operation, combining distribution and inventory functions with information required for operations control, as well as customer service information.

Figure 15.3 illustrates the relationships among major components of the airline's RES and various airline planning systems internal to the airline on the right side, as well as the links to external CRSs/GDSs on the left side. As shown, planning systems such as the schedule optimization, crew/aircraft rotation planning and RM systems described in previous chapters feed their solutions into the RES, which stores the airline's schedules, operational plans and inventory data. Other IT systems that provide pricing decision support, for example, also feed the fares database. The departure control, check-in and flight dispatch systems are examples of IT systems that also feed off the information available in the RES, and which are used by airline personnel at the airport on the day of each flight's departure.

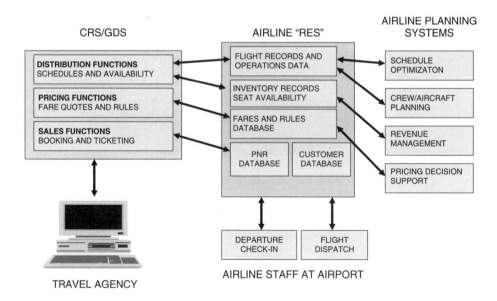

Figure 15.3 Reservations systems functions (PNR = Passenger Name Record)

On the left side of Figure 15.3, we see the linkages from the airline's own RES to the external CRS/GDS environment. "Computer Reservations System" or CRS is the term originally used to refer to the public (non-proprietary) versions of certain airline RESs, developed for travel agencies to use for distribution. As introduced above, travel agency CRSs (e.g., SABRE, Worldspan, Apollo) show schedules, prices and seat availability information for all participating airlines in an "unbiased" manner. The difference between the airline's own RES and a public CRS is that users of the CRS have only limited access to the complete functionality of the airline's internal system, whereas users within the airline itself can have full information about the airline's own seat inventories and booking totals. Public CRS users also do not have access to certain operational information contained in the airline's own RES, including details of crew and aircraft rotations, for example.

Thus, the development of airline RESs, followed by the emergence of CRSs and then their growth to become GDSs, served to define the characteristics of the traditional airline distribution system. Today, the term "GDS" is most commonly used to refer to the large airline distribution systems external to the airline's own RES. Perhaps the most important function of these systems historically has been the display of flight availability information to travel agents (and, in turn, consumers). In the following paragraphs, we provide an example of a GDS availability display to give the reader a flavor of GDS capabilities and their importance for airline distribution and competition.

Figure 15.4 shows an example of a flight availability display from an airline GDS, in this case a SABRE display of flights for travel from Boston to Fort Lauderdale on January 31, 2007. This display's date is January 26, 2007, i.e., five days before the flights' departure. The flight display shows six flight alternatives on the first page (or "screen"), three of which are operated by JetBlue Airways (B6) and the other three operated by

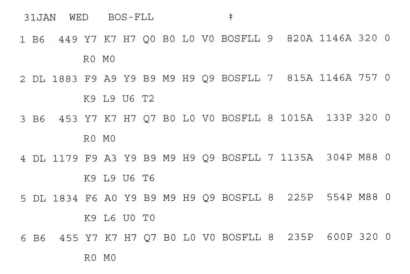

```
 31JAN   WED     BOS-FLL                  ‡

1 B6    449  Y7  K7  H7  Q0  B0  L0  V0  BOSFLL  9   820A 1146A 320 0
             R0  M0

2 DL  1883  F9  A9  Y9  B9  M9  H9  Q9  BOSFLL  7   815A 1146A 757 0
             K9  L9  U6  T2

3 B6    453  Y7  K7  H7  Q7  B0  L0  V0  BOSFLL  8  1015A  133P 320 0
             R0  M0

4 DL  1179  F9  A3  Y9  B9  M9  H9  Q9  BOSFLL  7  1135A  304P M88 0
             K9  L9  U6  T6

5 DL  1834  F6  A0  Y9  B9  M9  H9  Q9  BOSFLL  8   225P  554P M88 0
             K9  L6  U0  T0

6 B6    455  Y7  K7  H7  Q7  B0  L0  V0  BOSFLL  8   235P  600P 320 0
             R0  M0
```

Figure 15.4   GDS availability BOS–FLL, 31 January 2007
Source: Adapted from SABRE GDS

Delta Air Lines (DL). Each of the six lines provides specific and detailed information about the schedule and seat availability of a single flight departure. In this example, the first line shows information for JetBlue (B6) flight 449, which is scheduled to depart BOS at 08:20 and arrive FLL at 11:46, and is planned to operate with an Airbus 320 aircraft.

The seat availability for this flight is shown for nine different booking classes (Y, K, H, Q, B, L, V, R, M) corresponding to different fares associated with each booking class. At the moment this availability display was recorded, JetBlue flight 449 on January 31 had seats available in Y, K and H booking classes, but not in any of the other booking classes associated with lower-priced fares. This example illustrates the critical relationship between the control of seat inventory in the GDS and the airlines' use of differential pricing and RM concepts, as described in Chapter 4. By showing different availability for different flights and booking classes, even the low-fare carrier JetBlue is able to limit the number of seats it sells at the lowest prices on peak demand flights.

Each of the six different flight departures shown on the availability display example of Figure 15.4 has different booking classes "open" or "closed" to sale. While JetBlue's 08:20 departure does not have "Q-class" fares available, both its 10:15 and 14:35 departures are available in "Q-class." At the same time, the competing Delta flights appear to have seats available in many more fare classes, suggesting that Delta has lower fares available than does JetBlue at this moment. This would occur either because Delta is more willing to sell low-fare seats than JetBlue or because JetBlue's flights are more heavily booked, with relatively few seats remaining. In the latter case, JetBlue's RM system could be protecting the few remaining seats on the flight for potential purchase by higher-fare passengers.

A few words of caution about the example shown on Figure 15.4 – it is just the "tip of the iceberg" in terms of GDS capabilities. From this first availability screen, the agent (airline or travel) can issue dozens of different commands to find more information about fares, fare rules, seat assignments, on-time performance, ticketing requirements, etc. Note that the numbers shown beside each booking class represent the maximum number of seats the airline is making available for sale in each class. Different airlines use different default values – the DL flights show nine seats as the maximum default (unless fewer seats are actually available), whereas the B6 flights show seven seats in each booking class.

Note also that these seat availability numbers are not additive – DL1883 could have only nine total seats left in the economy cabin, all of which could be available in any booking class from Y through L (F and A are first-class cabin booking classes). As described in Chapter 4, airlines typically employ "nested booking limits" for seat inventory control. Thus, the GDS availability displays do not reveal to customers or competitors the actual number of seats remaining on the flight or the total number of seats available in each booking class. These details are proprietary to the airline's own RM system and RES.

15.2.3 Alternative Airline Distribution Channels

The evolution of airline distribution from centralized airline reservations offices to GDS networks with global coverage and, most recently, to Internet distribution channels has increased the range of alternative distribution channels employed by airlines and available to consumers today. In Figure 15.5, these distribution channels are illustrated in simplified form. The most traditional channels (and most costly to the airline) involve a consumer using a travel agency as an intermediary. When a consumer purchases a ticket from a travel agency that uses a GDS, the operating airline pays booking fees to the GDS and also pays commissions to the travel agency. If the travel agency makes the booking through the airline's reservations call center, GDS booking fees might be avoided, but the airline incurs the costs of its own telephone reservations agents.

Slightly less costly for the airline is the case in which the consumer calls the airline reservations office directly. The airline thus avoids commission payments and perhaps GDS booking fees (if it is using its own RES), but it nonetheless incurs the fixed and variable costs of operating its own telephone reservations operation. Not shown on Figure 15.5 is the most costly of traditional distribution channels – the case in which a customer calls

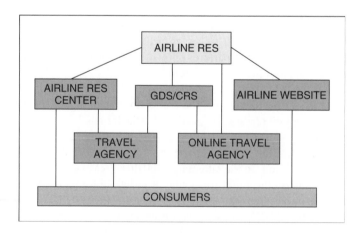

Figure 15.5 Airline distribution channels

the airline reservations office to make a booking, and then goes to a travel agent to have this booking ticketed. The travel agency then receives a commission from the airline despite not having made the booking, while the airline ends up paying for its own reservations office, travel agency commission and perhaps even GDS booking fees (if the travel agency "claims" the existing passenger name record from the airline and makes use of a GDS to make changes to the reservation).

To the extent that direct telephone reservations by customers reduced the costs of distribution for airlines, the more recent development of web-based channels for airline booking and ticketing was an extension of this drive to reduce distribution costs. Internet travel distribution channels replaced to a large extent the services of "brick and mortar" travel agencies with web-based intermediaries. Travelocity and Expedia are two US examples of such web-based intermediaries, both of which rely on GDSs for flight and fare information. Many of the largest airlines established "direct connect" links between online travel agencies and their own RES (shown in Figure 15.5) in an effort to reduce GDS booking fees. The cost savings to airlines of these online travel agencies, as compared to traditional travel agencies, depended on the extent to which the negotiated booking fees and/or commissions were lower than those paid to brick and mortar agencies.

Even if the savings in distribution costs for airlines were substantial, there still existed an intermediary between the airline and the consumer. It is the goal of many airlines, in their effort to further reduce distribution costs, to encourage their customers to book and purchase tickets directly on the airline's own website. If the airline's website does not require access to a GDS for information, booking and ticketing, then the airline avoids the costs of travel agency commissions, GDS booking fees, as well as its own telephone reservations office. Direct booking by consumers on airline websites represents the closest the airline can come to eliminating all intermediate players and charges for the distribution and sales of its product.

Despite efforts to shift bookings to their own websites, most legacy airlines still rely heavily on GDS bookings, and the fees charged to airlines for booking passengers have remained a cost concern. Given that GDS booking fees can range from $2 to $4 per transaction and that many reservations involve multiple transactions for booking, ticketing and changing itineraries, the total GDS fees per passenger can easily reach $15–20. This represented a growing proportion of the total ticket value, especially in domestic markets where low-cost carriers (LCCs) had pushed the lowest available fares to well below $100. The high costs of traditional GDSs thus generated a great deal of interest among airlines in lower-cost technologies under development by several companies that plan to build alternative systems to existing GDSs. The growing penetration of online booking portals also improved the bargaining position of network airlines in the negotiation of booking fees with the traditional GDS companies.

The complexity of developing any type of reservations environment has delayed the implementation of completely new next-generation airline inventory and distribution systems. At the same time, traditional GDS companies have responded by modifying their fee structures and providing airline customers with enhanced capabilities to make more effective use of various emerging online distribution functions. And GDSs continue to be an important source of bookings and ticket revenues for airlines, particularly in the corporate and business travel segments which tend to use travel agencies. Even some low-cost airlines, while still relying primarily on their own websites for ticket distribution, have recognized the potential value provided by the GDS. For example, JetBlue made the strategic decision in August 2006 to participate in three of the major GDSs, with the expectation of improved access to the corporate travel market with higher-yield business traffic.

15.2.4   Impacts of Alliances and Code Sharing on Distribution

One of the most common marketing practices of airline alliances is to sell or distribute a flight under a "code-share" arrangement. Under such an arrangement, one of the airlines in the alliance or partnership operates the flight with its own aircraft, while one or more non-operating partner airlines place their own airline code on an alliance flight for distribution purposes. The code-share partner can then market and sell its own tickets for the flight, which is actually operated by another alliance airline. Thus, a single flight is listed twice (or more) in airline schedules and GDSs.

An example of a GDS code-share flight availability display is provided in Figure 15.6. In this example, the Lufthansa-operated flight LH423 is also listed in a SABRE GDS display as United flight UA/LH8852 (i.e., UA flight operated by LH). Thus on the flight availability display for Boston (BOS) to Frankfurt (FRA) on January 31, 2007, there are two "different" non-stop flight departures shown, both leaving at 16:40. Of course, there was only one aircraft departing Boston for Frankfurt at 16:40 on that date, but the different GDS listings allow both Lufthansa and United to sell tickets for this Lufthansa-operated flight under their own name (and on their own websites). Such code sharing is common within global alliances. In this case, Lufthansa and United are partners in the Star Alliance.

Lines 3 and 4 of the GDS display example in Figure 15.6 show a connecting flight alternative for the same BOS–FRA request for January 31 availability. Air France flight

```
        31JAN  WED    BOS-FRA              ‡

  1 LH       423 F9 A9 C9 D9 Z9 BOSFRA   440P   530A‡1 747 M 0

                  Y9 B9 M9 H9 Q9 V9 W9 S9

  2 UA/LH 8852 F2 C4 D4 Y4 B4*BOSFRA     440P   530A‡1 744 M 0

                  M4 H4 Q4 V4 W4 S4 T0 K0 L0

  3 AF       337 J0 C0 D0 I0 Z0*BOSCDG   540P   620A‡1 744 MB 0

                  O0 Y9 B9 K9 H9 W9 T9 V9 X9

  4 AF      1418 C8 D0 Z0 O0 Y9*   FRA   735A   900A   318 B 0

                  S9 B9 U0 K9 H9 T9 V9 L9 X9

  5 DL/AF 8303 C0 D0 I0 Y9 B9 BOSCDG     540P   620A‡1 744 D 0

                  M9 H9 Q9 K9 L9 U9 T9

  6 AF      1418 C8 D8 Z8 F0 O7*   FRA   735A   900A   318 B 0

                  Y9 S9 B9 R0 U0 K9 M9 H9 T9
```

Figure 15.6   BOS–FRA GDS display, January 31, 2007
Source: Adapted from SABRE GDS

AF337 departs BOS to Paris Charles de Gaulle Airport (CDG) at 17:40, and connects to AF1418 from CDG to FRA the following morning at 07:35. This two-line display is an illustration of how connecting flights are shown on GDS screens – seats must be available in a desired booking class on both flight legs in a connecting itinerary for an agent to be able to book a seat and issue a ticket from Boston to Frankfurt. Lines 5 and 6 of this example provide another example of the impacts of code sharing on GDS displays. The connecting alternative shown on lines 5 and 6 is in fact the same AF-operated flight from BOS to CDG, connecting to the same AF flight from CDG to FRA. However, the BOS–CDG flight shown on line 5 carries a Delta code-share label, which Delta can market and sell as part of its alliance membership in SkyTeam, on behalf of its partner, Air France.

Supporters of the code-sharing practice claim that it increases the number of options available to consumers for travel in a city pair. If the partner airlines do not have antitrust immunity, then they must price their own code-share flights competitively, to the potential benefit of consumers. Opponents of this practice claim that it only serves to clutter GDS displays and confuse consumers. In the example shown, the entire first page of this GDS display is used up by what are effectively only two different flight options, thereby displacing other alternatives lower on the screen. For consumers, there is also the possibility of brand confusion – one can purchase a United ticket from Boston to Frankfurt on the United website, with little warning that the flight will actually be a Lufthansa-operated flight offering a different on-board product and requiring passengers to check in at a different terminal.

15.2.5 IT Challenges of Evolving Distribution Channels

The growth and evolution of the airline distribution systems have led to major IT challenges, the vast majority of which involve technical problems of communication and coordination between the different GDSs, airline websites and other distribution channels. These problems inevitably lead to consumer confusion, frustration and even a perception that the entire distribution process is becoming less user-friendly and transparent, contrary to technological expectations.

Perhaps the greatest IT challenge in the distribution area is the synchronization of flight, fare and passenger information among airline RESs and multiple GDSs. Changes in airline marketing practices (including pricing, RM and alliances) have handily outpaced the ability of most airline IT systems to adapt. Airline RESs were originally programmed in the 1960s in now-obsolete computer languages, making large-scale changes to the software both difficult and expensive. Airline alliances have found synchronization of their IT systems and reservations information to be a particular problem. These problems led the Star Alliance, for example, to undertake the development of a "common IT platform" for its airline members.

Synchronization problems are most commonly noticed by passengers when seat assignments or boarding passes issued by one airline are not recognized by the partner airline, or when frequent-flyer miles from a partner airline go missing. But consistency of information is a problem that is not limited to airline alliances. Even for a single airline, inconsistencies can occur in the fares being quoted for its own flights on different distribution channels. The development of "direct access" and "seamless connectivity" communications links between major airline RESs and GDSs, starting in the 1990s, has eliminated most of these problems for bookings made through the GDS networks. Seamless connectivity allows the users of a GDS to receive accurate real-time information about booking class availability and fare quotes for each flight request directly from the operating airline's own internal reservations inventory. Under seamless connectivity, each travel agency request is transmitted to each operating airline's own system, which must send a reply within approximately 300 milliseconds.

Seamless connectivity represented a major technological advance in airline distribution that allowed airlines much improved control over their inventories – airlines could now improve their RM systems to offer different availability responses (number of seats available in different booking classes) to different requests, even for the same flight departure. For example, the airline's RM system might choose to maximize revenues by responding with seat availability according to the expected revenue value of each GDS request – connecting vs. one-leg itinerary, point of sale (fare and currency differences). For the travel agent and consumer, seamless connectivity reduced dramatically the potential for inconsistent seat availability or fare information across the GDS networks.

Unfortunately, the rapid growth of Internet-based airline distribution channels placed the airline distribution systems once again in a position of trying to catch up with technology. Whereas seamless connectivity improved the accuracy of seat inventory and fare quote information for what, in retrospect, was a relatively small volume of travel agency availability requests, the volumes of requests originating from websites were unprecedented and enormous. The shift from thousands of travel agents making requests for information from an airline's RES each day, to millions of consumers doing the same thing, put unexpected loads on the airline systems. Even worse, the searches performed by large travel websites like Orbitz, Travelocity and Opodo over thousands of different airline flight options for the lowest possible fares, each time a consumer inputs a different set of search parameters, are well beyond the IT capabilities of the vast majority of airline RESs.

The makeshift solution to this unprecedented new demand on airline distribution systems was to provide travel websites with only periodic updates of seat availability and fare information for future flight departures. The implication for consumers using these websites is that the initial fare quote shown for a given flight departure could well change once that flight is actually selected for booking – only at this point does seamless connectivity ensure an accurate availability and fare quote. This might be perceived by consumers (unfairly) as "bait and switch" marketing by the offending airline and/or website. More broadly, these IT constraints have given rise to a distribution environment in which consumers can and do receive different fare quotes for the same flight options on different websites, as compared to airline websites, as compared to travel agency GDS information.

In the short term, many airlines responded to this potential for consumer confusion by guaranteeing that their own websites will show the lowest available fares for a flight request. Assuming that the airline website obtains its information from the airline's own RES, this guarantee should be relatively easy for an airline to honor. In the medium term, several of the largest travel websites are developing more sophisticated sampling and caching (i.e., storage of previous request and seat availability data) strategies that allow them to minimize the risk of inconsistencies between the information they receive from each airline and what they display to consumers. And, in the longer term, technological improvements to allow for real-time accurate communication between all websites and airline RESs will be required if total consistency and transparency of airline inventory and pricing information is the goal. On the other hand, with the recent deregulation of most distribution channels and the possibility of selective marketing and pricing practices, total transparency might not be in the profit-maximizing interests of either the airlines or the distribution systems.

15.3   Distribution Costs and E-commerce Developments

The global airline industry's downturn that began in 1991 with the Gulf War and subsequent recession led many carriers to look for cost reduction opportunities outside their traditional focus on labor and aircraft ownership expenses. A cost center that had remained largely untouched since the start of airline deregulation was that of sales and distribution. Fear of retribution for changing the historic travel agent–airline relationship was the principal reason for airlines leaving in place the long-held business practice of paying to travel agencies significant commissions based on a proportion of ticket price.

Historically, total airline sales and distribution costs exceeded 25% of total passenger revenues, with the biggest portion of these costs consisting of airline payments of ticket commissions to travel agencies. As shown in Figure 15.7, US airline commission payments reached their peak in 1993, when commission costs represented almost 13% of passenger revenues, as the US industry paid $7.6 billion in commissions on passenger revenues of nearly $60 billion. In an effort to reduce distribution costs, carriers began to reevaluate the entire process of product delivery – from accepting passenger reservations to ticketing and

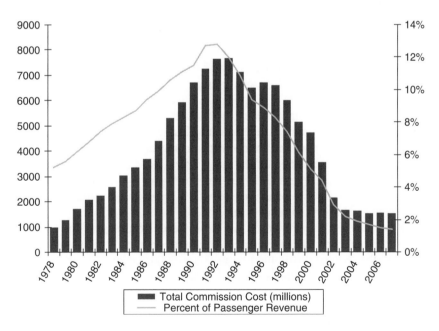

Figure 15.7   US airline commission costs, 1978–2007
Data Source: US DOT Form 41

passenger processing. At the same time, advances in IT were beginning to be introduced, with the promise of improved inventory management and the potential to ultimately remove the traditional travel agency "intermediary" from the distribution process. In 1994, Delta Air Lines became the first airline to introduce the concept of travel agency commission caps.

The initial travel agency caps introduced by Delta were quickly matched by virtually all of its legacy airline competitors, which were eager to reduce this component of distribution costs. In the years that followed, further reductions in commission rates and more restrictive caps on commission payments to travel agencies led to very substantial cost savings to the industry. By 2000, travel agency commissions had largely been eliminated for US domestic tickets, and total annual commission costs for US airlines dropped by 44%, despite a 37% increase in total revenues. The US industry had achieved nearly $3 billion in annual savings as commissions fell from almost 13% to about 5% as a percentage of revenue.

Another driver of significant change to distribution practices during this time was the onset of a second large wave of low-cost new-entrant airlines in the US domestic market. ValuJet was launched and was ultimately merged with AirTran Airways, Southwest began to grow east of the Mississippi River, and JetBlue entered the market by 2000. These low-cost airlines made the goal of keeping distribution costs low a central element of their business plans, and many used the strategy of limiting their ticket distribution to their own websites and call centers. The newer LCCs could do this given that they did not have a historical relationship with the travel agent community.

Forced to reduce their own distribution costs to remain competitive, legacy network airlines took advantage of the same technological advances as their low-cost competitors. Like their low-cost competitors, they focused increasingly on direct distribution of tickets through expanded call centers and the development of their own websites. To provide consumers an incentive to use their websites, the most cost effective of all distribution channels, some network legacy carriers offered bonus miles for online bookings and imposed an additional fee for all tickets booked through their call centers and ticket offices.

As the airline industry entered the new millennium, the growth of Internet usage and improvements in technology allowed even greater changes to traditional airline distribution practices. Major advances occurred in inventory management and "business to business" sales to corporate travel accounts. At the same time, online shopping was beginning to be a way of life for consumers, more a rule than an exception, whether the consumer was using an airline's own website or any number of other online distribution sites such as Orbitz, Priceline or Hotwire, to name a few.

The shift to Internet distribution channels provided airlines with an unprecedented opportunity to simultaneously take increased control of the distribution of their own product and further reduce distribution costs. Since 2000, major US airlines have saved an *additional* $4 billion in annual sales and distribution costs, representing a reduction of almost 40% between 2000 and 2007. With the elimination of most domestic travel agency commissions, total airline commission costs dropped to less than 1.5% by 2007 (see Figure 15.8). And, with the additional cost savings made possible by Internet distribution, total sales and distribution costs as a percentage of passenger revenues have plummeted in just 12 years, from 28% in 1995 to 19% in 2001 and just about 10% in 2007. The overall

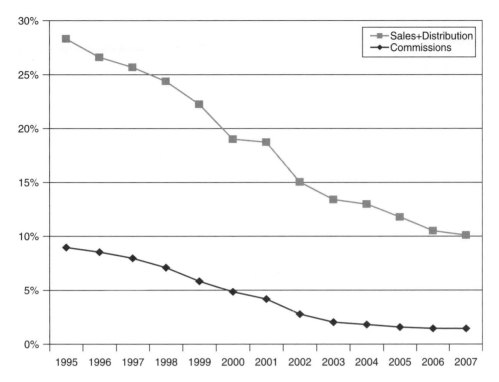

Figure 15.8 US airline sales and distribution costs as a percentage of passenger revenues, 1995–2007

Data Source: US DOT Form 41

reduction in annual distribution costs for US airlines since the first commission caps were imposed is astonishing – over $7 billion of cost savings per year from changes to the distribution process alone! The magnitude of this cost reduction is particularly impressive given that the US industry's largest-ever annual operating profit was $9 billion, in 1998.

While the US airlines have led the way in both the reduction in travel agency commission payments and the shift to Internet distribution, the same trends have begun to spread worldwide, as all airlines reexamine their sales and distribution practices in an effort to reduce costs. Although the categorization and availability of detailed operating cost data for world airlines makes a direct comparison with US carriers difficult, "ticketing, sales and promotion" costs accounted for just over 9% of world airlines' total operating costs in 2005 (ICAO, 2007). This figure had dropped from 16.4% in 1992 to 10.7% by 2002, due in large part to reductions in commission payments and GDS fees (see also Figure 5.5).

The progress of airlines outside the USA in reducing sales and distribution costs has been limited by several factors. First, they have not been able to reduce travel agency commission rates as aggressively as US carriers did in the US domestic market. While some of the largest European carriers successfully cut commission rates in their own domestic markets (over which they have substantial control), it was more difficult to do so for travel agents outside their home countries. Given the greater reliance on travel agency distribution in many countries outside Western Europe, both competitive pressures and greater travel agency resistance have slowed the pace of commission reductions. The second major factor limiting sales and distribution cost cuts in other parts of the world is in fact related to the continued reliance on travel agency channels – that is, the lower penetration and consumer acceptance of Internet distribution channels. As will be discussed below, other regions of the world lag behind the USA in the proportion of airline bookings made through web channels, which limits the ability of airlines to take full advantage of evolving Internet distribution technologies.

15.3.1 Internet Distribution Channels

With the growth of Internet sales portals for airline ticket sales, legacy airlines and LCCs alike began to focus on this new distribution channel for a variety of reasons. By promoting the use of their own websites (and in some cases, limiting the purchase of tickets to only their own websites) successful upstart LCCs were increasingly using Internet technology to maintain a significant cost advantage over their legacy competitors. This forced legacy airline companies to set even more aggressive goals for cost containment in the areas of sales and distribution expense. Whereas JetBlue, the most formidable LCC upstart since Southwest, sold nearly 75% of its tickets through its own site in 2004 and Southwest sold nearly 60% of its tickets through its own portal, the US legacy airlines were only selling 20% of their ticket revenue through their own respective websites.

The other major type of Internet distribution channel for airlines includes online travel agents that offer travel booking services that can include hotel accommodations and car rentals. Travelocity and Expedia were among the first and largest online travel retailers in the USA, growing quickly at the end of the 1990s and beginning of the new century. They were later joined by competitors launched by the airlines themselves such as Opodo, owned by nine European flag carriers and the European GDS Amadeus, and Orbitz, which was founded in 2001 by five US network carriers and later sold to US travel conglomerate Cendant in 2004.

In many respects these intermediary channels simply replace most of the existing travel agency functions. Initially, airlines saw relatively little in the way of incremental cost savings from passengers using these sites, as airlines still had to make payments to these third-party distributors for each ticket purchased. Still, third-party online websites had a large impact on the distribution of air tickets and price transparency in air travel markets. Some of these sites make use of fare search engines that allow air travelers to compare fares offered by a wide array of carriers, including combinations of tickets on several carriers for a single itinerary.

Other third-party websites have developed "web scraping" technologies that search over both airline and online third-party travel agent websites, and let travelers compare fares across travel sites. Examples of these sites include Sidestep, Mobissimo and Kayak, and they earn revenues through referral fees paid by the airlines or the online travel retailers for directing potential travelers to their website. More recently, a new travel website has gone a step further by providing prospective travelers shopping the Internet with guidelines on when to book airline tickets. Launched in 2006, Farecast has developed a large historical database of airlines fares and is using statistical techniques to predict whether an airline fare is likely to increase or drop in the future.

Yet another category of travel website involves the sale of "distressed inventory" for airlines; that is, flights with forecasted empty seats that are not likely to be sold at regular fares prior to departure. Priceline was the first and best-known example of such sites, and its business objective of distributing distressed inventory to price-sensitive travelers initially fulfilled a unique role for airlines. This "electronic wholesaler" of empty seats sold those seats to the public at net fares that it negotiated with airlines. For the airline, it offered yet another opportunity to fill otherwise empty seats, with the inventory made available to Priceline still being controlled by airline RM systems. As the airlines improved their own RM capabilities and expanded their website functionality to better promote their own special fares for "distressed inventory," the attractiveness of Priceline and other third-party web channels designed to offer cheap tickets on empty flights began to wane. With the evolution of the airlines' own online distribution and promotion tools, the importance of such third-party websites has diminished in recent years.

The proportion of airline tickets booked and purchased on the Internet has grown rapidly since 2000, when only 5% of all bookings were made online. As shown in Figure 15.9, the percentage of online bookings worldwide reached 35% in 2007, of which over three-quarters involved direct booking on the airline's own website. These data indicate that airline efforts to direct consumers to their own websites are succeeding. This tremendous growth in online sales has exhibited some differences, however, both between different types of airlines and among different regions of the world.

As a general rule, legacy airlines around the world remain substantially behind their low-cost competitors in terms of the proportion of direct and online bookings, as many of their customers, especially business travelers, still rely on traditional travel agents. Worldwide, network legacy carriers in 2007 booked 28% of their passengers through Internet channels (both their own websites and third-party online channels), while LCCs reached almost 75% of bookings made online (Jenner, 2008). Even more impressive is the fact that, for LCCs, virtually all of these online bookings are made through the airline's own website, further reducing costs. JetBlue reported that 80% of its ticket revenue in 2007 came from online sales, while European LCC Ryanair sold an amazing 99% of its tickets through its own website (Sobie, 2008).

There are also differences in online booking penetration in different countries. In North America, consumer acceptance of airline websites for purchasing air travel continues to soar – by 2007, more than 63% of all airline tickets in North America were sold via

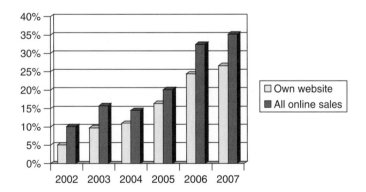

Figure 15.9 Percentage of tickets sold online – world airline survey
Data Source: *Airline Business*, July 2008

the Internet. In other regions of the world, online ticket sales remain substantially lower, due to lower home computer penetration levels, as well as lower levels of trust among consumers concerned about online security of credit card purchases. For example, in both Europe and South America, online ticket sales account for approximately 30% of total bookings. And the online booking rates are even lower in Asia–Pacific countries, at 14%, and in Africa and the Middle East, at 7% (Sobie, 2008).

For both legacy and low-cost airlines, the shift of much of the distribution function to the Internet thus provided opportunities for substantial further reductions in the portions of distribution costs that remained after travel agency commissions were reduced or eliminated. The use of online channels has also provided airlines with opportunities to increase revenue from their distressed (i.e., likely to remain unsold) seat inventories and to better manage passenger demand versus available capacity. Because bookings made on Internet travel sites are more likely to be made by price-sensitive leisure travelers, and given that most search engines identify flight options with the lowest available fares, airlines have seen a distinct reduction in the day-of-week variability of passenger travel in recent years (Brunger, 2008). As shown in Figure 15.10 for a legacy US airline, load factors on what were in the past considered to be "off-peak" days of the week increased between 2003 and 2007, while they remained relatively unchanged on "peak" days. At the same time, average yields (revenue per passenger kilometer) increased on peak days, as RM systems continued to protect the most valuable seats for passengers willing to pay higher fares for the most desirable departure days and times. Brunger found that these shifts were driven in large part by significantly greater growth in Internet bookings for flights on off-peak days.

The fact that online bookings are more likely to be made by leisure travelers – with greater price sensitivity and greater willingness to shift days of travel, airlines and even

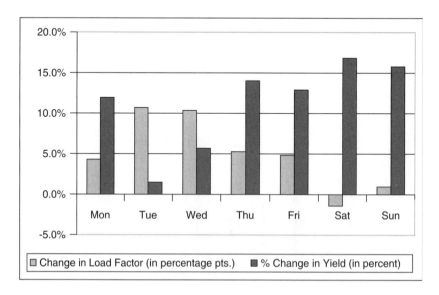

Figure 15.10   Change in load factor and yield by day of week, 2007 vs. 2003
Source: Brunger (2008)

destinations given information about lowest-fare options – has proven to be a mixed blessing for airline revenues. On the one hand, to the extent that passengers who book online are able to find the airline's "distressed inventory" and purchase travel that they would otherwise not have booked, the airline can generate incremental revenues at very low cost. On the other hand, if online bookings are being made for the lowest-fare options by passengers who previously would have chosen a higher-priced set of flight and fare options for the same trip, the increased access to more complete information on the Internet can result in revenue "dilution," meaning lower average fares for the airline.

Brunger (2008) performed an extensive analysis of airline booking and ticket data which revealed that bookings made online consistently had average fares more than 5–8% lower than bookings made through traditional travel agencies, even though the fares and seat availability offered by the airline were identical via each channel. Brunger's analysis isolated passenger itineraries that were "clearly leisure" (trips booked more than 14 days in advance and involving a stay at the destination over a weekend), and controlled for various other explanatory factors. This "Internet price effect" persisted across a large sample of US domestic markets and over multiple years of bookings examined from the recent past. There are several plausible reasons for this Internet price effect. The willingness to pay of consumers that use online booking channels could be different from those that book at traditional travel agencies (even for very similar leisure itineraries). Even if their willingness is not different, the information about low-fare options provided by travel agencies might not be as complete as that provided by an Internet search engine. In either case, it is the access to information about flight and fare alternatives provided by the Internet that contributes to the fare differentials.

Increasing use of airline websites and third-party online distribution channels has provided consumers with direct access to current information about airline flight schedules, fare rules and seat availability, giving them increased choice and control over their travel planning. These channels have in effect shifted much of the responsibility (and cost) of the passenger choice process away from traditional travel agents and airline reservations agents to the passengers themselves. Many passengers perceive this shift as positive – if they have the time to spend on searching the Internet, they can obtain much more information about flight and fare options than they were able to in the past and can make bookings at their convenience in the comfort of their own home. On the other hand, the complexity of airline fare structures, rules and restrictions can make the self-booking option more frustrating for some passengers, who can end up buying air travel services with conditions and/or levels of service quality that could prove to be very different from their expectations.

15.3.2 Electronic Ticketing

The shift by airlines worldwide away from traditional paper tickets to electronic tickets ("e-tickets") was driven by two major trends discussed in the preceding sections – the desire to reduce distribution costs and the rapid growth of Internet distribution channels. The average cost of issuing a paper ticket is estimated to be over $10 compared to about $1 for an electronic ticket, a cost saving of $3 billion per year for the world's airlines (Grossman, 2008). At the same time, Internet distribution could only reach its full potential in terms of both consumer acceptance and cost efficiency if the delivery of a physical ticket to the consumer was not required.

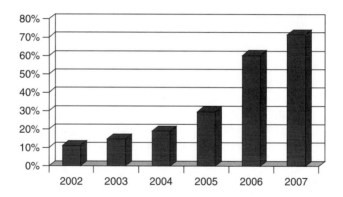

Figure 15.11   Percentage of e-tickets issued – world airline survey
Data Source: *Airline Business*, July 2008

Figure 15.11 shows the recent growth of electronic ticketing among world airlines: it increased from about 10% of all tickets issued in 2002 to almost 72% in 2007. The use of electronic ticketing was initially aimed at shorter-haul and leisure travelers that typically had simpler itineraries and were less likely to require changes to their tickets. Business travelers using more flexible fares were more reluctant to accept e-tickets, due to the difficulties of using an e-ticket issued by one airline for travel on another airline. These difficulties were caused by IT obstacles to communication and exchange of e-ticket details among competing airlines. These obstacles have now been overcome through various e-ticket data exchange agreements. The penetration of electronic tickets continues to expand as various challenges related to IT complexity, storage and exchange of e-ticket information, as well as trust and security, are overcome.

In 2007, over 90% of US domestic tickets were issued electronically, meaning no paper tickets were printed or delivered to passengers before their flights. The progress toward complete paperless ticketing has not been as rapid outside the USA, but IATA still hopes to eliminate all paper tickets with 100% electronic ticketing by all of its world airline members by 2010. By the deadline of June 1, 2008 originally set by IATA, the global e-ticketing rate had reached 96.5%, as smaller and less sophisticated airlines struggled to deal with concerns about IT automation and exchange of ticket information for inter-airline itineraries (Ezard, 2008). Paper tickets are no longer being produced and provided to these airlines, but it could take a year or more for existing ticket stock to be expended.

E-ticketing has had major implications for both the growth of online booking channels and reduced distribution costs over the past decade, as mentioned. With over 95% e-ticket penetration worldwide, the airline industry can all but eliminate the infrastructure and personnel required to collect, process and record paper ticket transactions. And, as will be discussed in the next section, e-ticketing has also contributed to the rapid growth of various "self-service" approaches to passenger processing enabled by recent IT innovations.

15.3.3   Implications for Airlines and Consumers

Cost efficiency will remain at the forefront of future developments of airline distribution, whether through intermediaries or airline direct web channels. As introduced in the previous sections, changes to distribution technology can also have revenue and passenger service implications. For example, by choosing to allocate seat inventory to an intermediary, an airline might be diverting that inventory away from its own direct channel. At the same time, providing an intermediary with access to inventory that it will sell at extremely low fares could divert the airline's own loyal customers to the intermediary, leading to both revenue dilution and increased distribution costs for the airline.

On the other hand, failure to remain competitive in terms of fares and availability on one or more intermediary websites can result in loss of market share, mainly among passengers who are not inclined to shop directly from airline websites. Whereas airlines historically competed for travel agency bookings by offering higher commission rates and "override" payments, the development of multiple travel booking websites, distressed inventory intermediaries and direct airline web channels has changed this competitive dynamic. Airlines must now ensure that they offer competitive fares and seat availability on these numerous distribution channels, trading off the differential costs of each channel and the risks of revenue dilution against the potential for market share loss.

From the consumer's point of view, this shift away from travel agents to multiple online channels has made purchasing air travel simultaneously simpler and more complicated. The largest travel booking sites do indeed search over millions of possible flights and fares to find the lowest available fare for a passenger's itinerary at a given moment in time, bringing the consumer closer to "one-stop shopping" for air travel. In this respect, airline prices have become more transparent to the marketplace, forcing airlines to compete on price more than ever. At the same time, however, the regulation of distribution channels is quickly disappearing, and historical agreements ensuring that all (or at least most) travel websites would have access to all of the fares offered by the largest airlines will be expiring soon.

It is already the case that some large airlines offer "web-only" fares available only on their own websites, not on third-party intermediaries. Many low-cost airlines have built their distribution model on this concept – for example, tickets for easyJet flights in Europe can only be purchased on the airline's own website. What this means for consumers is in fact less, not more, price transparency. Finding the best flights with the lowest fares already requires consumers to develop their own "search protocol" of visiting several travel websites before making a booking. Some consumers today already perceive that different websites offer different fare quotes for the same flights. This perception will only increase as the deregulated "wild west" of airline distribution continues to evolve.

15.4   Innovations in Passenger Processing

Much of consumer dissatisfaction with airline service can be traced to problems with airport processing of passengers and baggage. Most notorious for frequent travelers are check-in delays and seat assignment problems, lost baggage or slow delivery at destination, along with inadequate information flow and poor treatment during unusual events (irregular operations, misconnects). At the same time, many passengers have come to believe that in-flight service is a commodity, equally good (or equally bad) among most airlines. As a result, airline innovations in passenger and baggage processing, together with integration of IT used for distribution, represent a potential area for gaining competitive advantage.

Recent innovations by airlines in pre-flight passenger processing have transformed the check-in process, while reducing airline staffing requirements at airports and, in turn, labor costs. Before the passenger even arrives at the airport, automatic upgrade notifications and flight status information can be delivered to passengers via the Internet or telephone. Internet check-in at home, office or hotel allows passengers to check in for their flights, select seats, provide international passport information and print boarding passes. By 2009, 30% of all airline passengers worldwide are expected to use Internet check-in prior to arriving at the airport (Jenner, 2008).

As airlines enhance the functionality of their websites, they can take advantage of the potential to generate incremental revenues by offering new or previously available but "unbundled" services to passengers. For example, airport lounge access can be sold during the check-in process to non-member passengers at different fees based on their fare levels and perhaps even on the basis of forecasts of excess lounge capacity at the time of the passenger's planned departure. New services such as access to priority rebooking and reaccommodation in the event of irregular operations provide another example. And the introduction of fees for checked baggage by many US airlines in 2008 provides an example of "unbundling" of the air transportation product and an opportunity to generate additional revenues for the airline.

For passengers unable to check in via the Internet, self check-in machines at the airport perform the same tasks while allowing for faster passenger processing and reduced costs for the airline. In 2007, over 50% of US domestic passengers used self check-in capabilities, either on the Internet or at airport machines. Worldwide, this 50% threshold is expected to be surpassed by 2009, with 26% of all passengers using self check-in machines at airports (Jenner, 2008). Because both Internet check-in and airport check-in machines perform the same functions for passengers, airlines have had to decide where to focus their investments in new check-in technology. In fact, the rapid growth and acceptance by passengers of Internet check-in before airport arrival has prompted some US airlines to reduce their plans to further expand the number of self check-in machines at airports.

Further complicating the assessment of future technology requirements is the recent emergence of mobile or cell phones as yet another alternative for passengers to use for check-in. Web-enabled mobile phones not only give passengers access to the same check-in functionality as personal computers, but can also generate bar-coded displays of boarding pass information and eliminate the need for paper boarding passes.

The automation of passenger processing (and the corresponding reduction in the need for airline staffing at airports) has also affected the passenger boarding process at the airport. Electronic boarding pass readers at airport gates, for example, improve the speed and accuracy of the boarding process, and further reduce the manual labor required to board passengers. In fact, some European airlines allow passengers with machine-readable boarding passes to pass through boarding turnstiles with no interaction with a human gate agent. This level of automation is possible on domestic flights, but might have implications for passenger bag match and security requirements on international flights.

Other recent passenger processing innovations made possible by significant IT developments include airline tracking of passengers throughout the airport pre-departure process.

At least one European airline has tested the use of radio (RFID) chips embedded in the frequent-flyer program membership cards of selected elite-status passengers. This technology allows the airline to automatically issue boarding passes and baggage tags as soon as the passenger passes through the terminal door. The passenger then proceeds directly to a special check-in desk for more rapid processing. The technology can also track the whereabouts of passengers in the terminal, which can be useful for airlines making decisions about waiting for connecting passengers and for airport operators interested in the shopping patterns of passengers. Of course, the use of such technology also raises some concerns about privacy on the part of the tracked passengers.

IT developments have also begun to improve the passenger experience in flight, on board the aircraft. Several airlines have experimented with on-board Internet and e-mail access and, despite some teething problems for the first systems, several airlines have more recently announced plans to equip their aircraft with second-generation systems that will provide Wi-Fi access to passengers in flight. Along similar lines, the installation of on-board live satellite TV systems is expanding to more airlines. These enhancements not only provide passengers with a better on-board experience, but also give the airlines the potential for generating incremental revenues by charging for these services.

Technological advances have made these innovations in passenger processing (and entertainment) possible, at an ever-increasing pace particularly during the past decade. For airlines considering such innovations, the primary questions they face are: (1) will investment in these new technologies reduce costs, improve productivity and/or increase revenues; and (2) can implementation lead to greater passenger satisfaction and even loyalty?

If the answer to both questions is "yes," then these and other IT innovations in passenger processing represent a win–win situation for airlines and passengers. Airlines are becoming increasingly interested in how to speed up the penetration and acceptance of such innovations, without compromising passenger privacy and similar concerns. The proliferation of these innovations has already had noticeable impacts on airline and airport staffing, as well as on terminal check-in and passenger processing configurations. Looking ahead, it may have a potentially even greater impact on the design of airport terminals and passenger flows through the airport in the future.

References

Brunger, W.G. (2008) "The Impact of the Internet on Airline Fares Part II: Understanding the 'Internet Price Effect'," Doctoral dissertation, Case Western Reserve University.

Copeland, D.G., Mason, R.O., and McKenney, J.L. (1995) "Sabre: The Development of Information-based Competence and Execution of Information-based Competition," *IEEE Annals of the History of Computing*, Vol. 17, No. 3, pp. 30–55.

Ezard, K. (2008) "Papering Over," *Airline Business*, July, p. 52.

Grossman, D. (2008) "The End of the Paper Airline Ticket," *USA Today*, May 30.

ICAO – International Civil Aviation Organization (2007) *Outlook for Air Transport to the Year 2025*, September, Montreal, p. 31.

Jenner, G. (2008) "Mobilised Change," *Airline Business*, July, pp. 48–51.

Saranathan, K., Peters, K., and Towns, M. (1999) "Revenue Management at United Airlines," Presentation to AGIFORS Reservations and Yield Management Study Group, April 28, London.

Smith, B.C., Gunther, D.P., Venkateshwara Roa, B., and Ratliff, R.M. (2001) "E-Commerce and Operations Research in Airline Planning, Marketing and Distribution," *Interfaces*, Vol. 31, No. 2, pp. 37–55.

Sobie, B. (2008) "Weaving the Web," *Airline Business*, March, pp. 46–49.

16

Critical Issues and Prospects for the Global Airline Industry

William Swelbar and Peter P. Belobaba

The global airline industry has undergone numerous changes since airline markets were first deregulated in the USA 30 years ago. The liberalization of airline markets has become a global phenomenon, and growing competition has forced airlines around the world to search for ways to operate more efficiently. Throughout this book, which has covered many elements, processes and issues related to the global airline industry, it has been made clear that a multitude of stakeholders have been affected by the industry's transformation. Yet its transformation is far from complete.

Among the industry's many stakeholder groups, the one clear winner has been the air travel consumer, as the real price of air travel has, on average, fallen dramatically. While consumers have benefited from increased competition, airlines have not been able to retain the financial benefits from the many cost and productivity efficiencies they developed and implemented. As a result, it can be argued that the clearest losers in this transition to a more competitive global airline industry have been the shareholders of airline companies.

At the same time, there is evidence that the strong linkages between economic growth and airline industry prosperity might be breaking down. The traditional assumption has been that airline market liberalization leads to new and better air services, which lead to traffic growth, which leads to economic growth, which in turn creates the need for new and better air services. It is no longer clear that these linkages are sufficient to sustain an airline industry that continues to falter financially with every new economic cycle and world event. The historical economic fluctuations that have driven change in the airline industry might ultimately pale in comparison to the volatility being experienced in commodity, credit and currency markets as this book goes to press.

We begin this concluding chapter with a look at the major lessons provided by the evolution of the US airline market and its implications for the global industry. The US market was the first to deregulate and remains the largest in the world. While it may

The Global Airline Industry   P. Belobaba, C. Barnhart and A. Odoni
© 2009 John Wiley & Sons, Ltd

no longer be the strongest market, it nonetheless offers a most important case study as the airline industry continues its transformation from a regionally centered to a globally focused industry. As mentioned in several previous chapters, many of the trends observed in the US airline market have already spread to Europe and Asia.

In the second section of this concluding chapter, we look ahead and discuss some of the critical challenges facing the global airline industry. This industry has fought its way through a variety of macroeconomic challenges in the past. With each new challenge, airlines have adapted their operating strategies, reduced costs and responded to new competitive forces in order to survive. Upward pressures on labor costs, combined with volatile fuel prices, unstable financial markets and an economic slowdown, will force the industry to change once again, and could well be the catalyst for completing the transformation envisioned by the deregulators and liberalizers of the industry decades ago.

16.1   Evolution of US and Global Airline Markets

A case can be made that US airlines have lost their leadership position relative to the new global elite carriers based in several other regions of the world. Still, since the US airline industry was deregulated in 1978, the US market has repeatedly proven to be a template for the evolution of airline services and operating strategies around the world. The US industry has been an incubator for many of the most important airline developments that have now spread to the global industry. The pursuit of industry consolidation, the growth of low-cost carriers (LCCs), the deterioration of the linkage between economic growth and industry revenues, and the restructuring of costs and productivity are important trends that have defined the evolution of the US airline industry since deregulation. Because each of these trends has spread (or will spread) to other world regions, it is important to put in perspective the influence of the US model in shaping the global airline industry.

After the US domestic market was deregulated in 1978, several new entrants and incumbents were forced to file for bankruptcy due to the new competitive pressures on fares and operating costs that forced changes to network structure and operating strategies. By the mid-1980s, the first wave of airline mergers and consolidation was well under way, driven by the need to build regional dominance within US borders. Many of these transactions involved direct competitors merging with one another. This type of merger activity has historically been unique to the US market, although other regions of the world are now beginning to see proposed and actual consolidation. For example, in Europe Lufthansa purchased Swiss, and Air France and KLM have effectively merged their networks. Several secondary flag carriers such as SAS, Iberia, Austrian and Alitalia could well see their ownership structures change to involve one of Europe's three largest carriers: Air France/KLM; Lufthansa/Swiss; or British Airways.

The LCC phenomenon started in the USA, but has spread to virtually every liberalized airline market in the world. When the US industry was deregulated, only two airlines had the characteristics of what would come to be known as the "low-cost carrier model." Southwest Airlines and Pacific Southwest Airlines (PSA) had already been operating within the borders of Texas and California, respectively, exempt from the regulations that applied only to interstate airlines. Each had developed innovative operating strategies that came to be viewed as models of what could be expected as the barriers to entry were removed for all US domestic markets.

Although many new-entrant airlines emerged after deregulation, most failed during the 1980s and 1990s. After 2000, the LCC sector of the US industry grew rapidly and captured significant market share as the larger, less efficient and less nimble legacy carriers cut capacity. The rapid market penetration of the LCCs all but assured that if the legacy airlines were going to survive, they would have to compete on price. Due in part to the rapid growth of LCCs and the inability of legacy carriers to maintain their historical fare premium, the US industry experienced a structural revenue issue that was unprecedented. The relationship between US airline industry revenue and gross domestic product began to diverge in 2000, as shown in Figure 16.1. Historically, annual airline industry revenues had equaled approximately 0.95% of GDP, but this relationship deteriorated after 2000. By 2003, US airline passenger revenues dropped to 0.70% of GDP, translating into a loss of nearly $25 billion in revenue for the US industry as a whole.

Between 2002 and 2007, bankruptcy and related restructuring by the US carriers resulted in nearly $20 billion in operating expenses being eliminated, representing more than one-quarter of their non-fuel expenses, with the majority of the cuts coming in labor wages and benefits. But another macroeconomic phenomenon began to emerge in 2005. Just as the revenue environment had fundamentally changed early in the decade, the global demand for crude oil and its derivatives caused jet fuel prices to triple in less than three years. The economics of the industry thus changed dramatically after 2000, on both the revenue side and the expense side.

Although the US airline industry was deregulated more than 30 years ago, it remains in many respects a regulated industry. The fact that the industry is so visible and important to the economy provides motivation for legislative and regulatory bodies to continue to be

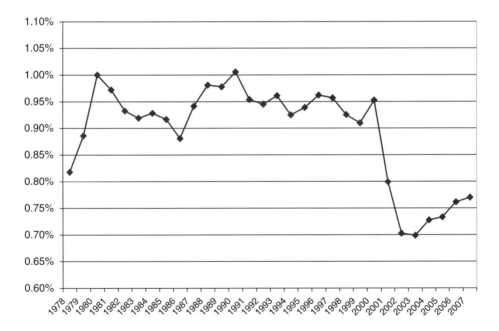

Figure 16.1   US airline passenger revenues as a percentage of US GDP
Data Source: ATA (2008)

involved in commercial decisions, generally in the name of protecting competition and the consumer. Airline labor relations are governed by the Railway Labor Act, the major provisions of which have remained essentially unchanged since 1934. Infrastructure constraints remain an impediment to operational efficiency, in many cases due to regulatory requirements and the inability of major stakeholders to resolve policy differences. Strategies for improving aviation infrastructure and air traffic management systems require agreement between the branches of government as well as labor organizations, local governments and organizations representing environmental concerns. Despite all of the efforts of US carriers to restructure themselves in recent years, the remnants of 60 years of regulation continue to play a major role in the evolution of the US airline industry.

16.1.1 Implications for the Global Airline Industry

The US industry model might be only one of many, but its characteristics have been emulated at least in part by virtually all carriers of prominence around the world. Fundamental to the evolution and lessons of the US airline industry is the sheer size of its domestic market. The evolution of the US domestic market has driven most of the operating strategies that have been employed by US legacy airlines and LCCs alike. Despite the reductions in domestic capacity by the legacy carriers and the outsourcing of domestic flying to smaller regional carriers, US carriers still have networks that are nearly 60% dependent on the hypercompetitive US domestic market.

Airlines around the world have not experienced the extreme boom and bust cycles in financial performance of the US industry due in large part to the diversity of international flying that the non-US carriers perform. This is not to say that there have not been sustained periods of troubled economics for carriers operating in each of the world's regions. Rather, the non-US carriers have operated within a structure largely governed by international regulations and bilateral agreements that have arguably protected many carriers from the pressures of unfettered competition and the tendency to overexpand during the macroeconomic growth cycles. Non-US carrier growth has been more measured, as capacity expansion was largely dictated by new route opportunities that emanated from relaxed bilateral restrictions between countries. This is in contrast to the US airline industry after its deregulation, which has emulated other capital-intensive commodity industries by overexpanding during the up cycles and then not removing the uneconomic capacity in the down cycles.

While the US industry stagnated as it focused on the restructuring of its costs and productivity in the 2002–2007 period, airlines in other regions of the world continued to grow. Carriers in Latin America and the Middle East grew at aggressive rates while most carriers in Europe and Asia grew at steady and sustainable rates, reflecting the relative maturity of their respective regional markets. Throughout the restructuring period, the industries in Asia and Europe remained profitable with only a slight loss for the European airlines in 2002 largely attributable to their dependence on international service to the US market.

During this period, the European carriers expanded their networks into many non-North American markets. China has become a growth opportunity for carriers in all regions and secondary markets in Asia have developed to the point where there is sufficient demand to justify service. Africa is presenting new growth opportunities particularly for the European airlines that can expand based on their historical presence in that region.

Similar to the US, Europe saw rapid growth in new intra-Europe city pairs. Just as the LCC sector expanded aggressively at the expense of the legacy carriers in the USA, a significant low cost sector grew even more rapidly in Europe. Certain LCCs entered the more mature markets in Europe while Ryanair expanded to secondary markets surrounding the largest population centers. Europe has many geographic areas left in which LCCs can expand further, such as Eastern Europe and the Mediterranean, whereas the US domestic market is mature and offers few remaining growth opportunities for the further stimulation of air travel demand.

LCCs have been started in nearly every world region, including Australia, Asia, South America, Latin America and, most recently, even in the Middle East. The coexistence of legacy and low-cost operating models can be expected to continue for some time. To be sure, some LCCs will entertain thoughts of starting transoceanic services that will again challenge legacy carrier incumbency on these routes. At the same time, however, LCCs alone cannot and will not replace the extensive services provided by large network legacy carriers, as recent developments have made clear that even LCCs are not immune from financial problems.

16.1.2 Government Policy and International Liberalization

The US government has been a force in negotiating liberalized ("Open Skies") agreements since 1979 in its efforts to facilitate the free (or a freer) flow of capacity between it and countries around the world. The first "Open Skies" Agreement of significance was negotiated between the USA and The Netherlands in 1992. In 2007, the first mega-multilateral agreement between the USA and the European Union was negotiated and ultimately implemented in March of 2008. Phase I of the US–EU Open Skies Agreement was significant in that it provided freedoms for airlines on both sides of the Atlantic to operate from various points within the other's territory and it opened up previously closed London Heathrow Airport to all US carriers.

What Phase I did not address were some difficult issues that include: (1) cabotage rights or the ability for a US carrier to operate between points within the EU or for EU carriers to operate within the US domestic market; (2) the limits on foreign ownership of a US carrier by an EU carrier; (3) certain tax-free provisions for US and EU aviation; (4) ending the Fly America program which forces US government employees to fly on a US carrier unless no other alternatives are available; and (5) developing an adequate framework for antitrust concerns that might arise from future mergers and consolidation, should foreign ownership restrictions be relaxed. The negotiations over Phase II of the US–EU Agreement promise to be contentious and highly charged politically. Within Phase II, however, are the remaining elements that will permit the industry to continue to evolve from a regionally focused structure to a globally oriented one.

16.1.3 Evolution of Global Airline Markets

The US market was the first to open itself to both internal and external competition, and the US industry experienced substantial consolidation as airlines sought to achieve economies of scale in the new competitive environment. In Europe, liberalized competition was slower to develop and the European industry still has not gone through the consolidation process that occurred in the US industry. Whether through bankruptcies or mergers, European consolidation appears to be inevitable. Airlines in Europe will likely have to consolidate and protect regional traffic flows in order to compete on a global scale. Like the US airlines, the largest European carriers have built their hubs/gateways into formidable traffic distribution networks to all world regions. Large-scale mergers, like Air France with KLM and Lufthansa with Swiss, have already taken place, and other European carriers (e.g., British Airways and Iberia) are seeking approval for similar transactions.

It is not as if Europe has not had casualties – the former Belgian flag carrier Sabena has ceased to exist. Swiss is but a shell of the former Swissair, which had to cease operations in 2001 after declaring bankruptcy. Olympic Airways in Greece and Alitalia, the flag carrier of Italy, continue to sit on the brink of financial calamity. Questions remain about the sustainability of three London Heathrow-based carriers given the liberalization agreement between the USA and the EU. And questions, such as whether secondary carriers like SAS and Austrian can survive as stand-alone entities, will certainly be asked within the European marketplace.

As this book goes to press, there is a new aviation power emerging in the Middle East comprising three carriers of significance: Emirates, Etihad and Qatar. These carriers offer all of the comforts that the world elite carriers provide and operate to dozens of important world destinations from their hubs. The US airlines will be impacted somewhat as these carriers expand into more US points. However, it will be the hubs and gateways in Europe that will most feel the competitive effects of growing airlines from this region.

Each of these three carriers has very ambitious expansion plans over the next 10 years and strong positions for delivery of new aircraft. Strategically, their growth promises to introduce new competitive pressures on carriers based in Europe, Oceania and Asia. The aggressive capacity growth of developing hubs at Dubai, Abu Dhabi and Doha with their geographically desirable global locations will inevitably shift traffic flows connecting at traditional hubs such as Frankfurt, Paris and London. These new hubs will provide new connecting opportunities for traffic flows between North and South America on the one hand and India, Southeast Asia and Australia on the other. The load-consolidating power of hub networks will enable these Middle East airlines to provide connecting service not only to large markets such as New York–Delhi, but to smaller secondary markets, e.g., Sao Paolo–Osaka. Furthermore, the geographic location of their growing hubs could facilitate the development of an African commercial aviation market that has until now been generally devoid of meaningful carriers of its own and which remains an extension of European carrier networks.

With few exceptions, unstable economies in South America have led to many airline casualties. Iconic names such as Varig and Aerolineas Argentinas have been crushed under financial pressures and significant inefficiencies in their operations. That being said, two airlines have emerged with different attributes that hold promise for the future in this region: LAN Airlines is the result of the consolidation of the former LAN Peru, LAN Argentina and LAN Ecuador under the umbrella of what was previously LAN Chile; and GOL, the Brazilian start-up LCC. The LAN model of building its network by gaining access to the bilateral rights of failing airlines on the same continent has been quite successful. This is an operating model with promise as other regions suffer similar fates or in countries where sufficient scale and scope is not present to operate an airline under a single flag. GOL remains the continent's strongest LCC and has set its visions on developing its brand in the weaker Latin regions of Central America and Mexico.

The Asian region will be among the most interesting to watch as it, too, will evolve in the face of global industry transformation. Although the economic importance of this region to the world is unquestioned, airline access to the region remains highly regulated today. With the exceptions of India, Korea, Taiwan and Singapore, along with a number of smaller countries in the region, Asia remains largely a limited-entry aviation space. China, of course, remains the market with arguably the greatest potential as the earning power of its populace, and thus the inherent demand for air travel, increases.

Japan has historically provided an important gateway, as both US and Japanese carriers have developed international connecting hubs for access to Asian destinations. But can Japan support two airlines in an increasingly competitive environment? The development of new and meaningful competition in Korea already has had some effect on the Japanese carriers. As hubs and gateways continue to grow in Korea, Taiwan and China, traffic that has been transiting Japan for access to Asia will begin to overfly Japan. These shifts in international traffic flows in Asia will be similar to those expected as the Middle East carriers expand their networks – both developments will redefine the competitive traffic flows that have been in place for several decades.

In Oceania, the New Zealand government's policy has been more liberal than that of its Australian counterpart. However, that is changing as the Australian industry has been strong and is home to one of the world's elite carriers, Qantas. The Australians have demonstrated some willingness to entertain new and different ownership regimes in recent history. It is the Australians that could prove to be the international leaders in terms of liberalized aviation policy, which could in turn have a great impact on global airline competition.

Finally, the continent of Africa has been home to few successful carriers as the global industry has grown. But like other regions of the world, this undeveloped region will hold promise for some, as South African Airways is doing well today and Kenya is often mentioned as the next promising African market. It will take more developed economies to sustain an airline that can compete effectively with larger global airlines. They will emerge, but until they do, this region will continue to be dependent on its connections to the air transportation system via Europe and the Middle East.

16.2   Looking Ahead: Critical Challenges for the Global Airline Industry

In Chapter 1, we identified some current challenges for the global airline industry – sustaining airline profitability and developing adequate air transportation infrastructure. As we conclude this book, we return to these challenges, and add another increasingly critical issue to the list – the environmental impacts of aviation. Looking ahead, we provide some final thoughts as to the stakeholders, forces and processes that will shape the global airline industry in the years and decades ahead.

16.2.1   *Strategies for Sustained Profitability*

This book has made clear that the airline business is both capital intensive and labor intensive, and is subject to a tremendous cyclicality driven primarily by macroeconomic forces. Repeated cycles of record profitability followed by huge losses have left many of the world's airlines in a weakened financial situation. Although many elements of this cyclicality will not change for the airline managers of tomorrow, perhaps the greatest challenge to the airline industry as a whole is to find a model for planning, operations and competition that can provide sustained profitability and greater stability. In this section, we explore some of the specific challenges airlines will have to face in this quest.

At the time of this book's publication, the airline industry is about to enter yet another phase in its continuing transformation, with the development of global networks served by new airlines from emerging regions and by new alliances of existing carriers. Mature (North America) and maturing (Europe and parts of Asia) markets will test the extent to which new capacity can continue to be added. Energy costs will play a large role in the development of markets and service, as inevitable increases in ticket prices will begin to test the elasticity of air travel demand as never before in many regions of the world.

As the industry evolves, the airline planning process will not undergo dramatic change, aircraft fleet decisions will remain long term, route planning decisions will remain medium term, and short-term decisions will continue to be driven by unpredictable events. What will change is how the global airline market develops, as many new route opportunities will emerge. Global alliances, and possibly even global mergers, will present different decision-making challenges for airline management.

These global networks will of course be shaped by competition, both among airlines and between the principal aircraft manufacturers. The ability of Boeing and Airbus, along with smaller manufacturers such as Embraer and Bombardier, to deliver new products with different range and payload capabilities will affect the network strategies of airlines. Airlines facing infrastructure constraints will focus on consolidating traffic through larger hubs with larger aircraft or using secondary airports. Carriers in emerging regions with fewer concerns about airport capacity are more likely to pursue new point-to-point route opportunities with smaller long-range aircraft. Certainly the manufacturers will have a strong say in how demand will be satisfied and how airline networks evolve.

Critical to the success of any airline strategic plan is ensuring that the invested capital can be utilized to the maximum extent possible. Flexibility for airlines in accepting delivery of aircraft types with different capacities will continue to be important in reducing the risks of uncertain future demand. Increasing this flexibility by adopting more sophisticated fleet assignment optimization models to match the right-sized aircraft to the stochastic and dynamic booking patterns of different routes will become even more critical in the future. Improved capacity management strategies to optimize revenue and load factors, maximize aircraft utilization, and make the most productive use on of an airline's largest capital investment will remain paramount.

As networks and competition have evolved, pricing and revenue management techniques have had to evolve as well. The more traditional approaches embraced the concept of filling excess capacity with incremental revenues at almost any fare, and the growth of LCCs has made price a primary differentiator of airline services. However, with increases in energy costs and the expectation that environmental costs will play a significant role in the future, airline pricing and revenue management will have to refocus on maximizing overall profitability, not simply on the generation of incremental revenues.

Differential pricing techniques will have to respond to the need for maximizing revenues to cover higher energy costs, particularly for airlines with very little room to make further cuts to other operating cost components. Historically high and extremely volatile energy prices have already caused the industry to begin reinstituting the more differentiated (some would say more complex) fare structures of the past in order to maximize revenue according to basic economic theory. High and volatile energy prices are arguably challenging the LCCs in a more profound way than the legacy carriers, forcing even this sector to reevaluate its "simple" approach to pricing and revenue management.

The attitude and choice behavior of the air travel consumer has largely been shaped by price, and this will continue. However, continued real decreases in the price of air travel will be difficult to achieve in a higher energy cost environment as the global industry faces the need for billions of dollars in reinvestment. The use of marginal economic analysis for route evaluation and even revenue management has facilitated the rapid growth of networks around the world, but significant changes will be required if the industry is to achieve anywhere near an acceptable return on invested capital. Assessment of longer-term profitability of flights, routes and hubs will have to be the standard, replacing the previous emphasis on incremental traffic and revenue.

Anything that airlines can do to improve the air travel experience for their customers will give them a competitive advantage. As airlines unbundle the services they provide, the very nature of an airline ticket and what it provides to consumers is evolving. Successful airlines must redefine the "value proposition" to consumers in a way that provides services and options desired by different demand segments, at price levels that can generate revenues to cover costs and ensure profitability. The opportunities for innovation in service delivery and revenue generation need not be limited to aircraft and in-flight amenities. They will also include the airport experience for passengers, including elements of security processing and border control.

In the area of ticket distribution, the industry has made enormous progress in cost reduction due in large part to the emergence of the Internet as a significant distribution channel. For the US airlines, a focus on distribution costs resulted in nearly $6 billion savings per year. While there appears to be little room left for further distribution cost reductions of similar magnitude for US airlines, any incremental efficiency in this area will clearly be driven by technological developments. On the other hand, carriers in virtually every other region of the world have yet to realize much of the distribution cost reduction made possible by the Internet and emerging information technologies.

With little fat left in airline cost structures there will be an increasing need to continue to find new technologies and new processes that can improve on operating practices that have been in place for decades. In the coming years, global alliances will evolve and possibly global mergers will take place. Today's industry and business practices recognize borders. Tomorrow's industry may not and that will present significant new opportunities to improve efficiency through innovations in information technology and operations research.

Labor relations will be perhaps the most critical issue in the quest for sustained profitability by airlines. The labor negotiations that follow the restructuring period of the early 2000s may provide the most important milestone in the US airline industry since deregulation. The labor concessions – many of them made under Chapter 11 bankruptcy – led to significant changes in labor costs and productivity for legacy airlines. With those changes, airline employees helped contribute to a short-term recovery of the industry. Finding a new model for compensation that is durable and works to address the cyclicality of the industry will be critical.

Even with increased fuel price volatility, labor and related expenses for nearly all the world's airlines remain the single largest controllable cost component. Prior to governments either deregulating or liberalizing their airline industries, labor cost increases could largely be passed on to the consumer through increases in fares. As a result, distortions to competitive market labor rates and inefficient work rules became entrenched. With increased liberalization, competition and new entrants, market forces have established a very different airline labor market, in terms of both wage rates and productivity standards. Over time, a global labor market will establish itself. But until that occurs, there will continue to be differences in the prices paid for labor both within and among regions of the world. These differences contribute to the competitive advantage of emerging airlines in regions with lower labor rates, and will remain a concern for organized labor in the more established regions of North America and Europe.

This need to manage labor costs in order to compete in the global airline market has placed increasing stress on the labor–management relationship. No longer can increases in labor costs be passed on to the consumer as new-entrant carriers with lower unit labor costs use their cost advantages to offer substantially lower fares to passengers and penetrate incumbent networks. The disparity in rates of pay and productivity levels has caused airlines around the world to outsource certain types of flying. In the USA, regional airlines now fly a significant portion of the domestic schedules of the largest legacy airlines, reflecting a major shift from previous practices. In Europe, several legacy airlines have created multiple subsidiaries under their corporate umbrella to perform intra-European flying in an effort to circumvent restrictions contained in labor contracts.

Airline labor–management relations promise to become even more complicated as the industry continues its transformation into a global market. As described in Chapter 10, labor laws are different around the world, and labor practices can differ even within regions. With little or no commonality of labor laws, labor–management tensions will persist well into the twenty-first century as global alliances, global mergers and a freer flow of capital between and among airlines around the world become more common. Resolving these tensions is perhaps the most critical challenge for airlines searching for sustained profitability, given that many of the efficiencies associated with innovative planning and operational practices cannot be achieved without the participation of labor. The way in which labor is compensated will be a determining factor in the profitability and, in some cases, survival of many of the world's major airlines.

16.2.2 Infrastructure and the Environment

Airports are at the epicenter of the air transportation infrastructure. Yet the augmentation of airport capacity is constrained by environmental concerns, funding issues for large capital projects, and the availability of land and airspace to name but a few. In some cases, traditional approaches to how airlines compensate airports are sure to be challenged. As the airport is both the beginning and end for each passenger journey, continued improvements in the airport experience will be required to reduce the wait times and inconvenience so often associated with the ground portion of air travel. Baggage-related services are in constant need of improvement and are often cited by passengers as one of the most troubled areas of the airport experience, and runways will need to be built in order to accommodate future demand, particularly in the largest metropolitan centers around the globe.

The lack of adequate infrastructure capacity – airports and airspace – and the rapidly growing costs of maintaining and expanding this infrastructure are two of the most critical problems for the future of air transportation, nationally and internationally. The prospects for substantial relief on the capacity front are not good – at least in the medium term (next 10 years). While the FAA and other air navigation service providers around the world have been working, with some success, toward increasing the capacity of the en route airspace, the real bottlenecks of the air transportation system are the runway systems of the major commercial airports in North America, Europe and Asia and the terminal airspace around them. The only clear way to increase the capacity of runway systems at these airports substantially, i.e., at rates similar to those at which demand is growing, is through the construction of new runways at existing airports or of additional airports in the same metropolitan areas. But obtaining approval for and eventually opening additional runways and new airports is an extremely difficult and time-consuming proposition in most developed countries. Barring these, airports and national civil aviation authorities may have to resort to increasingly stringent "demand management" measures, such as slot restrictions, congestion pricing and even the auctioning of access to major airports.

On the cost side, the enormous investments required in order to expand and maintain the capacity of existing airports or to build new ones have been one of the main reasons for the airport privatization trend that has been in evidence in much of the world (but, for statutory reasons, not in the USA) since the late 1980s. A growing tendency to tax airline passengers and cargo directly is another consequence of the rapidly increasing costs of aviation infrastructure (airports and air traffic control). Various taxes and fees for infrastructure support and security currently increase the cost of the average domestic airline ticket in the USA by about 16%. The situation in the EU is roughly the same.

Given that the immediate potential for further development of new aircraft technology is limited, the management of airspace can be viewed as a significant impediment for an industry struggling to find new ways to improve efficiencies. In the USA and around the globe, air traffic management is simply not keeping pace with the demand for air travel. The lack of adequate air traffic management systems and infrastructure will contribute to growing delays and passenger inconvenience. Without major improvements, it will be extremely difficult to accommodate the expected growth in air travel demand, a possibility that has important implications for nearly every stakeholder group in the global airline industry.

Similarly, the growing concerns about the environmental impacts of aviation will need to be addressed. Most forecasts suggest that future demand for air travel is likely to grow faster than the pace of development of new technologies required to reduce the impacts of aviation noise and emissions. Noise compliance has imposed a constraint on airline operations for several decades, and emerging emissions standards and regulations will almost certainly create more constraints in the future. Adoption of more stringent environmental standards will encourage the development of new aircraft technologies, and force airlines to adopt these new technologies more quickly. However, without a return to sustained profitability for the world's airlines, it will be difficult for many carriers to fund the investment required to renew their fleets.

In the end, tomorrow's airline managers will struggle with many of the same issues that confront this exciting, dynamic and perplexing industry today. Fundamental to the success and survival of virtually every airline will be the identification of an operating and financial model that can be sustainably profitable and which allows the airline to reinvest in itself. Managers will still face an industry that is both capital and labor intensive and subject to the cyclicality of macroeconomic forces. And many commercial and strategic decisions will continue to face political scrutiny in an industry that is perhaps the most regulated deregulated industry in the world.

Perhaps only one thing is certain – after 30 years of airline regulatory liberalization and the resultant industry evolution, there remains much more transformation ahead. Over the past few decades, strong economic times and continued government protection have helped to mask the possibility that the airlines have not done enough to restructure or rethink a business model that was developed under regulated conditions. Looking ahead, airline markets around the world should not fear individual carrier failures or consolidation. Indeed, this industry has demonstrated time and time again that where competition is vulnerable, a new entrant will exploit that vulnerability. Where there are market opportunities, there will be a carrier to leverage that opportunity. What will be different for tomorrow's managers is that decisions will have to be made with a global focus, rather than a regional focus.

References

ATA – Air Transport Association (2008) "US Passenger Airline Cost Index," www.airlines.org.

Index